SPACE
NUCLEAR
SAFETY

Albert C. Marshall
With
F. Eric Haskin, U.S. coeditor
Veniamin A. Usov, Russian coeditor

KRIEGER PUBLISHING COMPANY
Malabar, Florida
2008

Original Edition 2008

Printed and Published by
KRIEGER PUBLISHING COMPANY
KRIEGER DRIVE
MALABAR, FLORIDA 32950

Library of Congress Cataloging-in-Publication Data

Marshall, Albert C., 1943–
    Space nuclear safety / Albert C. Marshall with F. Eric Haskin, U.S. coeditor, Veniamin A. Usov, Russian coeditor. — Original ed.
        p. cm.
Includes bibliographical references.
ISBN-13: 978-0-89464-061-2 (alk. paper)
ISBN-10: 0-89464-061-5 (alk. paper)
1. Space vehicles—Nuclear power plants—Safety measures. 2. Astronautics—Accidents—Prevention.
3. Astronautics—Accidents—Risk assessment. 4. Radioactive substances—Safety measures.
5. Radioactive pollution—Prevention. 6. Space vehicles—Nuclear power plants—Environmental aspects.
I. Haskin, F. E. II. Usov, Veniamin A. III. Title.
    TL1102.N8M37 2008
    363.12'4—dc22
                                                                                    2007030618

10 9 8 7 6 5 4 3 2

For
Suzanne, Melanie, and Jason
and my parents,
Albert and Elizabeth

Contents

Contents

Chapter 8—**Reactor Criticality Safety**
Albert C. Marshall

Chapter 9—**Reactor Transient Analysis**
Albert C. Marshall and Edward T. Dugan

Chapter 10—**Risk Analysis**
F. Eric Haskin

Contributing Russian authors:
        Edward I. Grinberg, Vladimir I. Kuznetsov, Vadim S. Nikolaev, Nikolai A. Sokolov,
        Evgeny M. Strakhov, Veniamin A. Usov
Russian-English translator:
        Vladaslov Malakhov

Preface

In 1992 Albert C. Marshall and F. Eric Haskin taught a graduate course on space nuclear safety at the University of New Mexico. Although our combined experience seemed well suited to the task, the absence of an applicable text presented a challenge. The lack of an appropriate text is in part due to the esoteric nature of space nuclear safety. Space missions requiring nuclear power are infrequent and, compared to the terrestrial nuclear power industry, are modest enterprises. As a consequence, the specialized knowledge and expertise required for this area is confined to a small community of engineers and scientists. Although texts on terrestrial nuclear safety are valuable supplements, they do not address some of the most important space nuclear safety considerations; and some practices applicable to terrestrial nuclear plants are inappropriate for space nuclear systems. Engineers new to the field typically go through a protracted learning phase, gleaning information from fragmented documentation and from discussions with experienced engineers. Given these considerations, we chose to produce a textbook on space nuclear safety as an introduction for students and as a compendium of vital knowledge. This vital knowledge includes not only safety analysis and testing methods; but also the philosophies, practices, review procedures, and programmatic approaches that are necessary to provide safety assurance without creating useless operational and design restrictions.

The breadth of disciplines required to address space nuclear safety and the diverse backgrounds of potential readers led to several key decisions on the writing of this book. First, we felt that the exposition of diverse topics could be improved by enlisting specialists to contribute chapters in their areas of expertise. The contributing authors are recognized experts from the United States and Russia who have conducted safety assessments for major space nuclear programs. The United States and Russia are, at present, the only two countries with extensive space nuclear experience. The second key decision was to omit detailed discussions of nuclear engineering basics in order to keep the book readable and concise. Brief discussions of nuclear engineering principles and concepts are included for readers without nuclear engineering backgrounds. The third key decision was to emphasize simplified, approximate analysis approaches to provide a more intuitive grasp of basic principles.

A background discussion is provided in the first few chapters of this book; i.e., the introduction, and Chapters 1, 2 and 3. Chapter 1 reviews basic nuclear concepts needed to understand nuclear power systems and nuclear safety. Students and professionals familiar with basic nuclear concepts may choose to skip this chapter. The second chapter presents an overview of space nuclear power systems, and Chapter 3 provides an overview of space nuclear safety issues.

Chapters 4 through 11 address space nuclear safety issues, safety approaches, and safety analysis methods. Chapters 4 through 9 cover deterministic safety analysis methods, and Chapters 10 and 11 discuss probabilistic safety assessments and consequence analysis. These chapters can be covered in any order; however, the reader unfamiliar with radiation protection concepts and terminology should begin with the discussion of these topics in Chapter 4. Although we have focused on simple safety analysis methods, each chapter provides a brief discussion of computer analysis methods used in space nuclear safety programs. Chapters 4 through 9 also include a discussion of safety issues and scenarios, protection and mitigation methods, and safety testing. Student exercises that can be solved using a hand-held calculator are given at the end of each chapter.

The last chapter, Chapter 12, gives a review of the process used to assure safety for space nuclear missions. The organization of space nuclear safety programs and the launch safety review process are described. Chapter 12 also provides an assessment of the current safety process and a discussion of trends in space nuclear safety.

This textbook should be useful to professors, university students, and professionals in the fields of nuclear engineering and aerospace safety. Although the book is oriented toward nuclear engineers, the material covered should be accessible to engineers and scientists without nuclear engineering backgrounds. This book is the first written on the topic of space nuclear safety, and the need for future revisions and additions is anticipated. The editors and authors welcome any suggestions or comments for future improvements and supplements to this book.

ALBERT C. MARSHALL
F. ERIC HASKIN
VENIAMIN A. USOV

Acknowledgements

I am sincerely grateful for the assistance provided by Professor Mohamed El-Genk at the University of New Mexico for his help establishing the nuclear engineering course that led to the writing of this book. I am also grateful for the guidance provided by Mary Roberts at Krieger Publishing Company; her assistance and good-natured patience kept me on track through the slow and difficult process of producing the finished product. I am especially thankful for the contributions, assistance, and friendship of Dr. Eric Haskin. I also wish to thank Dr. Veniamin Usov and all contributing authors; their contributions were essential to the successful completion of this book. I am very grateful to Professor Ed Dugan at the University of Florida for his careful and professional review of this text. I also wish to express my appreciation to Vladaslov Malakov for his competent translation of the Russian contributions into English, and to my son, Dr. Jason Marshall, for his final review of the text.

Figure I Saturn photographed on October 30, 1980 by NASA's Voyager I
(powered by radioisotope power sources).
Courtesy of NASA Jet Propulsion Laboratory.

Introduction

Albert C. Marshall

Space nuclear power systems are generally used when solar or chemical power sources cannot meet the high power or energy requirements of a mission. Nuclear power sources also offer the advantage of rugged construction, high reliability, and the ability to operate in harsh or sunless environments. Both the United States and Russia have used nuclear power in space since the United States launched the first space-based nuclear power source in June 1961. Over the next 45 years, the United States deployed one reactor into space and a total of 45 *radioisotope thermoelectric generators* (RTGs) on 26 separate missions. During this period, the former Soviet Union placed 35 reactors into orbit and deployed several radioisotope sources into space. Nuclear power sources have been used for a variety of civilian and military missions. Most U.S. nuclear power sources were used for scientific studies of the moon and planets. For example, RTGs aboard the Voyager spacecraft provided the power needed to acquire and transmit spectacular images of Saturn (Figure I), Mars, Jupiter, and other planets. Russian space reactor systems were used primarily for ocean reconnaissance.

For space nuclear programs, the fundamental safety goal is the protection of humans and Earth's biosphere from nuclear radiation and potential radiological contamination by space nuclear power systems. The importance of safety was recognized since the first use of nuclear power sources in space; consequently, a meticulous safety review and approval process was established. Safety processes and practices have evolved, but the basic approach has remained unchanged since the launch of the first nuclear power source. Mission plans are guided by the need to minimize risk from postulated accidents, and systems are designed to remain in a safe configuration during postulated severe accident scenarios. All space nuclear systems must be rigorously analyzed and tested to demonstrate that the mission will be safe from launch through operation and disposal. For U.S. space programs using on-board nuclear power systems, program developers must prepare comprehensive safety analysis reports. These reports are presented to an independent safety review panel who, in turn, prepares a safety evaluation report. Using these reports, the President or the President's science advisor must decide that benefits clearly outweigh any residual safety or environmental risks before a nuclear system can be launched. Similar safety review and approval processes are used in Russia (formerly the Soviet Union), including the issuance of safety reports, review by an independent safety committee, and high-level government approval.

The careful safety approach and the outstanding accomplishments made possible by space nuclear power over the past several decades have placed humankind on a path toward yet greater achievements. The future of space nuclear power will depend, in part, on the power and energy requirements of future missions. Possible future missions requiring high power levels include space-based surveillance, air traffic monitoring, massive data processing, advanced science missions, and direct broadcasting from space. Systems capable of generating many kilowatts of electrical power will be needed for human life support and for other functions for a lunar or Mars outpost. If human exploration of space is to advance, both nuclear electric power and nuclear propulsion will be needed. Human exploration missions to Mars will benefit from high-velocity, low-mass nuclear propulsion systems that reduce astronaut exposure to cosmic radiation and substantially reduce launch costs. Nuclear propulsion may also play an important role in reducing the cost associated with changing satellite orbits. For some missions, nuclear-powered systems may be the only feasible approach.

Public acceptance is equally important to the future of space nuclear power. Public concern over nuclear safety has intensified since the Three Mile Island and Chernobyl accidents. When this common fear of nuclear energy is coupled with the image of the *Challenger* and *Columbia* space shuttle accidents, winning public trust in the safety of space nuclear missions presents a daunting challenge. A credible and workable space nuclear safety program must be maintained and open communication with the public is needed to assure safety and to garner public acceptance.

Engineers and scientists new to the field of space nuclear safety often focus on issues and approaches associated with terrestrial nuclear plants. Although terrestrial nuclear safety approaches have some relevance to space nuclear programs, the very different environments for space-based and Earth-based nuclear systems result in a different set of safety considerations. This difference in focus has produced design and operational strategies that are unique to space nuclear enterprises. For example, launch and deployment failures are relatively common occurrences for all spacecraft; consequently, incidents with nuclear systems on board must be anticipated. A failure of the launch vehicle prior to orbital insertion could subject the on-board nuclear system to severe environments such as reentry aerodynamic forces, propellant explosions and fires, or high-speed impact. The safety concerns associated with launch, reentry, and the space environment have no counterparts in the terrestrial nuclear field. On the other hand, many of the most important postulated accident considerations for terrestrial nuclear power plants have no safety consequences for some space reactor missions. Some safety practices regarded as sacred for Earth-bound nuclear systems can be counterproductive if applied to nuclear systems deployed in space.

Safety assessment methods and safety strategies developed and refined over several decades must be available to designers, safety analysts, mission planners, and others. The goal of this book is to provide a summary of the current knowledge and to explain the basic principles underlying safety analysis methods and safety strategies.

Bibliography

1. Angelo, Jr., J. A. and D. Buden, *Space Nuclear Power*. Malabar, FL: Orbit Book Co., 1985.
2. El-Genk, M. S. (Ed.), *A Critical Review of Space Nuclear Power and Propulsion, 1984–1993*. New York: AIP Press, American Institute of Physics, 1994.
3. Bennett, G. L., "A Look at the Soviet Space Nuclear Power Program." 24th Intersociety Energy Conversion Engineering Conference IECEC-89, Aug. 6–11, 1989, Washington, DC, New York: IEEE, 1989, vol. 2, p. 1187–1194.
4. Marshall, A. C., "Projected Needs for Space Nuclear Power and Propulsion." Space Technology & Applications International Forum (STAIF 96), Proceedings of the Conference. Albuquerque, NM, Jan. 1996. Part 3: 13th Symposium on Space Nuclear Power and Propulsion. (n.p.).
5. U. S. Dept of Energy. *Atomic Power in Space: A History*. DOE/NE/32117-H1, Planning and Human Systems, Inc., Washington, DC, Mar. 1987.

Contributors

Gary L. Bennett is a consultant in aerospace power and propulsion systems with over 40 years of experience in space and terrestrial power and propulsion systems. He has held management and technical positions at NASA, the U.S. Department of Energy (DOE), the U.S. Nuclear Regulatory Commission (NRC), the U.S. Atomic Energy Commission (AEC), and what is now the Idaho National Laboratory (INL). At NASA, he managed programs in advanced space power and propulsion. While at the DOE, he directed the safety and nuclear operations for the successful radioisotope power programs for the Galileo and Ulysses missions. At the NRC he led the reactor operational safety research program. While at the AEC, he was the flight manager for the radioisotope power sources currently in use on the Voyager 1 and 2 spacecraft. Bennett also worked on the nuclear rocket program at what is now NASA's Glenn Research Center and did fundamental reactor safety research at INL. For 9 years he was a member of or adviser to U.S. delegations to the United Nations dealing with the use of nuclear power sources in outer space. He has received numerous awards from NASA, DOE, NRC, and professional societies. He has authored one book and contributed chapters to three other books. He has authored or coauthored over 150 technical papers, reports, and articles on power, propulsion, and space missions.

Neil W. Brown is a nuclear power specialist at Lawrence Livermore National Laboratory where he has worked since 1994. His work has focused on development of small secure, transportable reactors for use in developing countries. He has also lead an effort to convert Department of Energy seismic design standards into national consensus standards for the design of nuclear facilities. From 1993 to 1994 he worked in the Space Power Programs Division of Lockheed Martin. This division continued development of the SP-100 space power reactor whose development had been initiated by General Electric, where he worked from 1960 until the SP-100 project was acquired by Lockheed Martin. His General Electric career involved numerous advanced reactor development projects, including the Clinch River Breeder Reactor where he was responsible for management of a team of safety analyst working on reliability of safety systems and severe accident analysis. During this period he developed an approach to design certification for advanced reactors with little operational history. He also, participated in the safety and licensing of the SEFOR reactor project that demonstrated the safety of sodium cooled fast reactors.

James R. Coleman died on 9 March 2004 after a long battle with cancer. His professional career spanned 45 years, and included: service as a commissioned officer in the U.S. Public Health Service; Manager, Special Projects, Environmental Safeguards Division, NUS Corporation; Assistant Environmental Health Director, Minnesota Department of Health; President and Chief Executive Officer, Sigma Associates, Inc. (environmental and safety assessment, risk analysis, and health effects); and owner and Principal Consultant, James R. Coleman Consulting. Coleman participated in the Presidential launch safety review/evaluation of every NASA nuclear-powered space mission launched from 1969-2003, i.e., Apollo 12, 13, 14, 15, 16 and 17, Pioneer 10 and 11, Viking 1 and 2 missions, Voyager 1 and 2, Galileo,

Ulysses, Mars Pathfinder, Cassini, and the Mars Exploration Rover A and B missions. He also conducted pioneering environmental and radiological safety studies for the U.S. NERVA nuclear rocket test program, and developed a novel Energy Interaction Model for quickly predicting the mechanical response of U.S. GPHS-RTG hardware to potential pre-launch, launch, and ascent accidents. Coleman, a chemical and nuclear engineer and certified health physicist by training, was a nationally recognized expert and author in the areas of radionuclide (especially plutonium-238) atmospheric and soil transport; radiological uptake, dose delivery and health effects; and space nuclear system accident analysis and safety testing. His probing independent analyses, insightful perspectives, gentlemanly calm demeanor, and genuine interest in understanding nature more fully will be sorely missed by the U.S. space nuclear safety community.

Leonard W. Connell is a Distinguished Member of the Technical Staff in the National Security Studies Department at Sandia National Laboratories. He is a technical advisor to Senior Sandia Management on nuclear weapons, nuclear terrorism, and unconventional nuclear warfare. He has a BS with High Honor in Mechanical Engineering from Michigan Technological University (1976) and a PhD in Nuclear Engineering from the University of New Mexico (1996). Connell has written several scientific journal and conference papers on space nuclear power and propulsion, the space radiation environment, and reentry aerothermodynamics along with several recent classified reports on nuclear terrorism and unconventional nuclear warfare. Connell performed the analysis of a postulated reentry accident for the U.S./Russian TOPAZ II space reactor program. He is a contributing author/editor on two textbooks dealing with space nuclear power.

Sandra M. Dawson worked for 20 years on launch approval, environmental, and risk communication issues at Jet Propulsion Laboratory (JPL). She worked on the Galileo, Ulysses, Cassini, Mars Rover, and New Horizons missions, as well as on the Outrigger Telescope Project and Project Prometheus. She was supervisor of the launch approval group at JPL for 6 years. In recognition of her expertise in launch approval and risk communication, she was made a Principal at JPL. Dawson moved to Caltech in 2006 and now works for the Thirty Meter Telescope project, which is in the design stage of building and operating the world's largest ground telescope. She is responsible for all nontechnical issues (environmental, legal, and public) surrounding the siting of the telescope. She holds a BS in Political Science from West Virginia University and a Masters of International Study from Claremont University.

Edward T. Dugan has been a full-time faculty member in Nuclear and Radiological Engineering at the University of Florida since 1977. His education includes a BS in Mechanical Engineering and an MS and PhD in Nuclear Engineering. Areas of expertise include reactor analysis and nuclear power plant dynamics and control; space nuclear power and propulsion; radiation transport and Monte Carlo analysis; and radiographic imaging techniques applied to non-destructive examination. Over the past 20 years he has written over 90 refereed technical papers and has been Principal Investigator or Co-Principal Investigator on contracts and grants in these areas totaling over $10 million. He has also done consulting work with the nuclear industry, including nuclear power training and the development and teaching of nuclear power plant technical courses. Other consulting involved work on nuclear reactor power supplies, including thermal nuclear propulsion. Most recent work, starting in May 2003, has included the development of a specialized Back Scatter X-ray (BSX) imaging systems for the Lockheed Martin Space Systems Company in New Orleans. These BSX systems were developed following the *Columbia* accident for the purpose of performing nondestructive examination (NDE) on the foam thermal insulation on the Space Shuttle external tank. Production BSX units were delivered to Lockheed Martin and BSX scanners are used to perform NDE on the external tanks at the Machoud Assembly Facility in New Orleans.

Mehdi Eliassi is a mechanical engineer and Principal Member of Technical Staff at Sandia National Laboratories, where he has been employed since 1996. His current work involves computational modeling of shock physics and high strain-rate phenomena applied to hypervelocity impact on space and reentry objects. From 1996 through 2003 he worked on a variety of projects developing theoretical models for the subsurface flow and transport of contaminants originating from unexploded ordinance, long-term nuclear waste storage for the Yucca Mountain Project, and the Japan Nuclear Cycle Development Institute. From 1991 to 1995, as a part of the TOPAZ II space reactor safety team, he examined the response of the TOPAZ II reactor to prelaunch accidents related to hard surface and water impacts, fires, and explosions. During this period, he was also involved in developing computational fracture and fragmentation models to simulate the hypervelocity impact of space debris on satellites. Eliassi worked as a research engineer at New Mexico Engineering Research Institute from 1981 through 1987. He taught laboratory fluid mechanics, as graduate assistant, at the University of Maryland (1987–1988). He was a staff scientist at S-cubed Division of Maxwell Laboratories from 1988 to 1991, where he performed thermal analysis of a pellet-bed space thermal propulsion reactor concept for the U.S. Strategic Defense Initiative. He received his BE, MS, and PhD, all in Nuclear Engineering, from University of New Mexico.

Edward I. Grinberg, Head of Laboratory, FSUE (Krasnaya Zvezda) Rosatom, participated in the safety analysis (accidental reentry, fire at the launch pad, and radiation doses and risks) for a number of space nuclear programs. These programs included the Orion space radioisotope nuclear power sources, radioisotope heating units for Lunokhod moon rover, the source for the Mars-96 spacecraft, and gamma sources for the Salyut-1 manned orbital station. The radioisotope sources were successfully operated in space. Grinberg was also participated in analytical verification of the operability of a backup radiation safety system (BRSS) provided in the reactor system used onboard the Cosmos-176 spacecraft. In total, 16 spacecraft with this BRSS were launched. The BRSS efficacy was confirmed during the accidental reentry of the Cosmos-1402 spacecraft in 1983.

F. Eric Haskin is a registered professional engineer and a self-employed consultant to both private companies and national laboratories. He has been involved in modeling and analyzing the safety of both terrestrial and space nuclear power sources since 1970. His work has primarily dealt with accident progression and consequence analysis, design and development of accident mitigation systems and strategies, full-scope risk analysis, qualitative and quantitative characterization of uncertainties in both deterministic and probabilistic model predictions, and development of risk-informed regulations. He developed and teaches a course entitled "Perspectives on Reactor Safety" for the U.S Nuclear Regulatory Commission. Before becoming a consultant, he was a Research Professor of Nuclear and Chemical Engineering at the University of New Mexico (1990-1998), a Manager at Sandia National Laboratory (1980–1990), a Mechanical/Nuclear Engineering Supervisor at Bechtel (1973–1980), a Visiting Assistant Professor of Nuclear Engineering at the University of Arizona (1971–1973), and an Instructor of Nuclear Engineering at Kansas State University (1970–1971). He has managed a number of major research projects including an assessment of public risks associated with a proposed orbital mission of the Topaz II space reactor, safety analysis and testing of a space nuclear reactor for the Strategic Defense Initiative, initial development of the NRC's MELCOR and MACCS computer codes, and development of methods and computer codes used to characterize uncertainties in the NRC's NUREG-1150 risk assessments.

Vladimir I. Kuznetsov graduated from the Theoretical and Experimental Physics Department of the Moscow Engineering and Physics Institute (University) with specialization in experimental nuclear physics. Beginning in 1960, he has taken part in ground testing of space nuclear power systems. Since 1970, he has been directly involved in the assembly of reactors and pre-flight preparations of nuclear power systems. He has worked in the field of development and nuclear safety of new-generation space nuclear power systems since 1992.

David L. Y. Louie completed a master's thesis on modeling fission gas venting for the SP-100 space reactor system. Louie has worked as a consultant after graduating from the University of New Mexico with a PhD in nuclear engineering. Much of his consulting work was for the U.S. Department of Energy. His focus is in phenomenological modeling and software development. He has expertise in the areas of explosions, reactor criticality, radiation shielding, and thermal-hydraulics. Since 2000, he has modeled and analyzed many postulated explosion accidents for facility safety at Sandia National Laboratories, Los Alamos National Laboratories, Lawrence Livermore National Laboratories, and Nevada Test Site. He also served as a nuclear criticality safety engineer at various DOE sites and has evaluated space reactor designs for criticality safety. Louie worked on many large thermal hydraulic severe accident reactor computer codes (such as CONTAIN, MELCOR and TRAC) for the U.S. Nuclear Regulatory Commission.

Vladaslov Malakhov received his BS in Mechanical Engineering from the University of Missouri with a minor in metallurgy. After working for Western Electric he enrolled at the University of California at Berkeley, where he received an MS in Nuclear Engineering. His master's thesis was an investigation of the effect of radiation on propulsion rings of ionic rockets. He then worked for General Atomics as a high temperature gas cooled reactor (HTGR) core designer. He participated in the analysis of the Peach Bottom reactor and the design, fabrication, and operation of the Fort St. Vrain core. Malakhov was also Project Manager in charge of refueling. At the same time, he was Nuclear Safety Manager for the oversight of all fuel activities at General Atomics. He also performed a safety evaluation for nuclear systems in space. Being fluent in Russian, Malakhov participated in nuclear and chemical disarmament projects with Russia. Malakhov continues to consult for General Atomics and Russia on the design of an HTGR at Tomsk-7.

Albert C. Marshall was employed as a Nuclear Engineer and Distinguished Member of Technical Staff at Sandia National Laboratories between 1976 and 2005. During this period he worked in the areas of reactor physics, space nuclear power, space nuclear safety, reactor testing, terrestrial reactor safety, depleted uranium dispersal and health effects, assessment of terrorist threats, thermionic physics, and several other areas. From 1977 through 1984 he was project leader for in-reactor simulation experiments for Three Mile Island-type accidents. He was the Sandia project leader for space reactor systems for the Strategic Defense Initiative between 1985 and 1989. He created a suite of system models (RSMASS) for predicting masses of a broad variety of space nuclear systems. Marshall was the Space Reactor Safety Technical Advisor to the U.S. Department of Energy from 1989 through 1984 and was chairman of the U.S. Interagency Nuclear Safety Policy Working Group in 1991. In 1992, he taught a graduate course in space nuclear safety at the University of New Mexico. From 1991 through 1994, he served as project leader for reactor safety and neutronics for the TOPAZ II space reactor program. After his retirement from Sandia Laboratories in 2005, he served as a consultant for Sandia Laboratories and the Department of Energy on space nuclear safety and U.N. nuclear safety principle guidance. He was a representative for the U.S. Department of Energy at a U.N. meeting on space nuclear safety in Vienna in 2006. Before working at Sandia Laboratories, Marshall worked as a reactor physicist at Bettis Atomic Power Laboratory (1967–1969) and Gulf General Atomics (1969–1976). He received his BS in Physics and MS in Nuclear Engineering from Pennsylvania State University.

Vadim S. Nikolaev, Lead Scientist at FSUE (Krasnaya Zvezda) Rosatom, is a safety expert for space reactor and radioisotope power systems. His expertise includes physics and thermal analysis of aerodynamic disruption of nuclear power systems during accidental reentry and assessments of associated radiation doses and risks. He was a key expert and leader of an analytical and experimental verification effort on the efficacy of safety systems relating to aerodynamic, explosive, and chemical dispersion of a space reactor system. He formulated guidelines and requirements on safety assurance during operation, removal from service and accidental reentry of nuclear power and propulsion systems in accordance with provisions of international agreements, conventions, treaties and guidelines adopted by the United Nations and IAEA with due regard for ICRP recommendations and current trends in development of safety criteria. He developed draft regulations (notification rules, sanitary rules, safe use concepts, technical requirements) on safety assurance of space reactor and radioisotope systems. He participated in the development of working documents from Russia and papers on the application and safe use of space nuclear power systems. These documents were submitted to the U.N. Committee on Peaceful Uses of Space and its Science and Technology and Legal Subcommittees between 1980 and 2006.

Joseph A. Sholtis spent 23 years in the U.S. Air Force (LtCol, retired) as a Nuclear Research Officer, Development and Acquisition Officer, and Program Manager for the development and oversight of a variety of advanced nuclear energy systems and technologies intended for missile, space and unique terrestrial applications. Upon retirement from the USAF in 1993, he formed Sholtis Engineering & Safety Consulting. The company provides nuclear, aerospace, systems engineering, and safety management and technical support services. During his professional career, he has participated in the INSRP safety reviews of every U.S. nuclear-powered space mission launched during the period 1974–2005. In addition, he has served as: Director of the Reactor, Linear Accelerator, Cobalt-60, and X-ray facilities at the Armed Forces Radiobiology Research Institute (1980–1984); DOE Program Manager of the SP-100 Space Reactor Power System Development Program (1984–1987); a member of the U.S. delegation and technical expert to the U.N. Working Group on Nuclear Power Sources in Outer Space (1984–1987); and Director of the USAF Nuclear Regulatory Authority (1991–1993), which licenses/regulates all USAF nuclear facilities and radioactive sources worldwide. He is a former NRC-licensed Senior Reactor Operator with more than 2000 hours of total console time, including 100+ pulse operations. Currently, he is providing technical safety support to the New Mexico Office for Space Commercialization for the New Mexico Spaceport, and Sandia National Laboratories for the nuclear-powered NASA Mars Science Laboratory (MSL) mission safety/risk analysis effort.

Nikolai A. Sokolov graduated from the Moscow Power Engineering Institute. Since 1973 he has been involved in analytical and experimental verification of radiation safety of space nuclear systems. At present he is a Senior Scientist at FSUE (Krasnaya Zvezda). He is an expert in ballistics and dynamics. He participated in the safety analysis for the space nuclear power systems designed for Cosmos series spacecraft.

Evgeny M. Strakhov, Deputy Director of Department FSUE (Krasnaya Zvezda) Rosatom, is an expert in space nuclear power systems. He leads the development of nuclear power systems and analytical verification of safety systems and safety-related structural members. From 1985 through 1990 he was in charge of preparation and performance of full-scale ground tests of a system for explosive dispersion of space reactor systems, processing and analysis of test results, and documentation of findings in a report. Since 2001 he has been a member of the Working Group on Space Nuclear Power Sources of the Science and Technology Subcommittee of the U.N. Committee on Peaceful Uses of Space.

Veniamin A. Usov is the Deputy Director for Space Nuclear Power, High-temperature Power Systems Department, Institute of Nuclear Reactors, RRC (Kurchatov Institute). He was a member of the research team that built the world's first Romashka high-temperature nuclear reactor with thermoelectric energy conversion using silicon-germanium semiconductor elements. He participated in the development and nuclear ground testing of the Enisey thermionic reactor system designed for use in space communications and television. He took part in the U.S./Russia/U.K./France International Program on the evaluation of the TOPAZ II space system experimental units in an electrically heated facility in Albuquerque, New Mexico, U.S.A. From 1991 through 1997 he was Director General of Intertek U.S./Russian Joint Venture for managing the TOPAZ II tests, development of higher power space system designs (Space-R) using TOPAZ II technology, and development of a design of flight tests of the upgraded TOPAZ II system for the proposed NEPSTP mission.

Chapter 1

Basic Nuclear Concepts

Albert C. Marshall

The objective of this chapter is to provide the reader with an understanding of fundamental nuclear concepts important to space nuclear safety. Topics include the atomic nucleus, relevant nuclear processes, important types of nuclear radiation, common sources of radiation, radiation interaction with matter, and energy from nuclear processes.

1.0 The Atomic Nucleus

An understanding of relevant nuclear processes is essential to the study of space nuclear safety. We begin with an elementary description of the atomic nucleus, primarily for the benefit of readers unfamiliar with nuclear concepts, terminology, and units.

1.1 Atomic and Nuclear Structure

An atom can be described simply in terms of its constituents. As illustrated schematically in Figure 1.1, an atom consists of a positively charged nucleus surrounded by negatively charged electrons e. The nucleus of an atom is made up of positively charged protons p and uncharged neutrons n; both protons and neutrons are referred to as nucleons. Hydrogen is the only element that possesses a stable nucleus containing no neutrons. The masses of protons, neutrons, and electrons are 1.00866, 1.00727, and 0.00055 atomic mass units (u), respectively [1]. An atomic mass unit is equal to 1.6605×10^{-24} g. Protons and electrons carry a unit electrical charge of equal magnitude, but of opposite sign; consequently, a neutral atom contains an equal number of electrons and protons. The number of protons in the nucleus is called the atomic number Z, and the total number of nucleons (protons plus neutrons) is called the mass number A. All atoms possessing the same atomic number are identified as the same chemical element and have identical chemical characteristics, regardless of their mass number.

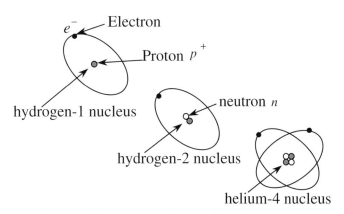

Figure 1.1 Schematic illustration of neutral hydrogen and helium atoms.

Nuclear characteristics of an atom depend on the number of protons and neutrons in the nucleus. Species of atoms having the same atomic number, but different mass numbers, are referred to as isotopes. For example, the nucleus of the element uranium contains 92 protons ($Z = 92$), and the nuclei of naturally occurring uranium isotopes are found to contain either 142, 143, or 146 neutrons. Thus, the naturally occurring isotopes of uranium have mass numbers of 234, 235, and 238, respectively. In this text an isotope will be identified by writing the mass number after the element name or by writing the mass number as an anterior superscript to the symbol for the chemical element. Thus, uranium-234, uranium-235, and uranium-238 identify the natural uranium isotopes; or we can use ^{234}U, ^{235}U, and ^{238}U. When it is useful to include the atomic number as well as the mass number, the atomic number is given as an anterior subscript; e.g., $^{234}_{92}U$, $^{235}_{92}U$, and $^{238}_{92}U$.

1.2 Binding Energy

The masses of atomic nuclei, determined experimentally, are found to be less than the masses computed as the sum of the constituent neutron and proton masses. For example, the relative atomic mass A_r of uranium-235 is 235.04394 u [1], while the mass of the sum of the constituent nucleons is

$$
\begin{aligned}
\text{Neutrons:} \quad & 143 \times 1.00866 = 144.2384 \\
\text{Protons:} \quad & 92 \times 1.00727 = 92.6688 \\
\hline
\text{Total:} \quad & 236.9072 \text{ u,}
\end{aligned}
$$

and
$$\Delta m = 236.9072 - 235.04394 = 1.8633 \text{ u.}$$

This mass difference m, known as the mass defect, is associated with the nucleon binding energy; i.e., the energy binding the nucleons together in the nucleus. The relationship of the mass defect to binding energy is given by Einstein's equation relating mass and energy,

$$E = \Delta m c^2, \tag{1.1}$$

where E is the binding energy, and c is the speed of light. For nuclear processes, the common unit of energy is the *electron volt* (eV), equal to the energy attained by an electron accelerated through a potential of one volt. One eV is equal to 1.603×10^{-19} joules. From Eq. 1.1, a mass of one atomic mass unit is found to be equivalent to to 931.494 million electron volts (MeV). For uranium–235, the binding energy per nucleon is

$$\frac{931.494 \text{ MeV/u}}{235 \text{ nucleons}} (1.8633 \text{ u}) = 7.386 \text{ MeV/nucleon.}$$

1.3 Nuclear Stability

Data points are shown in Figure 1.2 for the nuclear constituents of naturally occurring nuclides (nuclear species); the ordinate indicates the number of protons, and the abscissa gives the number of neutrons. Most naturally occurring nuclides are stable, although a few (such as uranium-235) undergo very slow nuclear decay. The data points in Figure 1.2 indicate a general stability trend; i.e., almost all stable nuclides contain more neutrons than protons. Unstable nuclides, produced in nuclear reactors or by particle accelerators, will emit particles to form nuclides with a stable neutron to proton ratio indicated by the data points in Figure 1.2.

The reason stable nuclei typically possess a greater number of neutrons than protons is understood by considering the attractive strong nuclear force between all nucleons and the repulsive Coulomb force

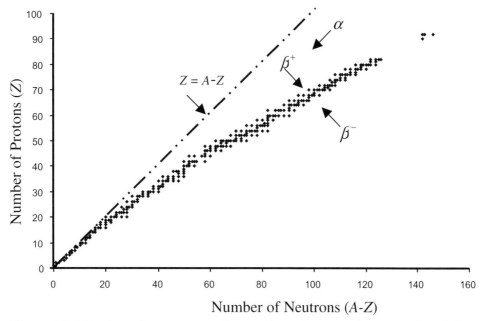

Figure 1.2 Number of neutrons vs. the number of protons for natural nuclides.

between protons. The short-range ($\sim 10^{-13}$ cm) strong attractive nuclear forces between neutrons and neutrons, neutrons and protons, and protons and protons must counterbalance the repulsive Coulomb force between protons. For low atomic numbers, the balance is obtained when the number of protons and neutrons is about equal. The Coulomb force, however, increases in proportion to Z^2, resulting in very large repulsive forces for nuclides with high atomic numbers. For high-Z nuclides, the number of neutrons relative to protons in the nucleus must increase to counterbalance the greater repulsive force between protons. Too few or too many neutrons, for a particular atomic number, results in an unstable nucleus and in the emission of particles to achieve stability. Elements with atomic numbers greater than or equal to 84, such as $_{92}U$, are also unstable [2].

2.0 Radioactive Decay

The process of radiation emission by unstable nuclides is called radioactive decay, and radioactive isotopes are called radioisotopes. In a radioisotope power source, particles emitted during nuclear decay interact with atoms in the fuel material to produce thermal energy for power production. Following shutdown of an operating reactor, radioactive decay heating of fuel and containment materials can be an important safety consideration. Essentially all radiation from radioisotope power sources is produced by radioactive decay or by subsequent interactions of decay radiations with materials. Radioactive decay is also an important contributor to radiation produced by an operating reactor, and radiation from a shutdown reactor is almost entirely the result of radioactive decay.

2.1 Nuclear Decay Processes

Nuclear decay typically results in the emission of alpha or beta particles and gamma rays (α, β, and γ, respectively). An alpha particle is actually a highly stable cluster of two neutrons and two protons, identical to a helium nucleus. Alpha emission is often exhibited by nuclides with high atomic numbers, such as $_{92}^{234}U$. The decay of uranium-234 into thorium-230, by alpha emission, is illustrated schematically in Figure 1.3. A beta particle is a negatively or positively charged electron emitted from the nucleus. Emission of a negatively charged beta particle β^- from the nucleus results from the spontaneous

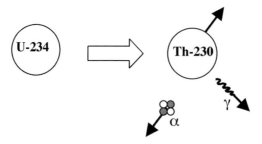

Figure 1.3 Schematic illustration of alpha particle emission from uranium-234 decay.

transformation of a neutron into a proton, an electron, and a particle called an antineutrino $\bar{\nu}_e$ (the anti particle of the electron neutrino ν_e).

$$n \rightarrow p^+ + \beta^- + \bar{\nu}_e$$

Negative beta particle emission is typically observed when the neutron-to-proton ratio is too high for nuclear stability. If the neutron-to-proton ratio is too low for nuclear stability, nuclei typically decay by transforming a proton into a neutron and emitting a positively charged beta particle β^+ and a neutrino ν_e. A positively charged beta particle is an antielectron (commonly called a positron), identical to an electron except that the charge is positive rather than negative. Although neutrinos and antineutrinos carry off energy, they interact so little with all forms of matter that they are of no consequence to radiological health and safety.

Quantum theory identifies discrete energy states for nucleons within the nucleus and for orbital electrons in an atom. The lowest allowed energy state is referred to as the ground state and higher energy levels are called excited states. Energy transitions for nucleons and atomic electrons occur in discrete energy increments. Gamma rays are emitted when the nucleus is left in an excited state following alpha or beta emission (as shown in Figure 1.3). The emitted gamma rays carry away the energy difference between excited states and the ground state of the nucleus. Gamma radiation is high-energy electromagnetic radiation produced by nuclear transitions. According to quantum theory, gamma radiation can exhibit either wave or particle characteristics. When regarded as individual particles, gamma rays are called photons. Figure 1.4 provides an energy diagram for the emission of gamma rays following beta decay of cobalt-60. Note that two gamma photons are emitted when cobalt-60 decays, corresponding to a drop in energy from two excited states in succession.

Several other types of nuclear decay are possible, but less common. For example, some unstable nuclei produced by nuclear fission will decay by emitting a neutron. Nuclei having an excess of protons, but insufficient energy to emit a positron, can capture an orbital atomic electron and transform a proton into a neutron while releasing a neutrino (see Figure 1.5). An X-ray is emitted when an electron from a higher energy shell drops into the lower energy vacancy left by the captured electron. Note that the daughter product nucleus produced by decay processes can contain a different number of protons than the parent nucleus; consequently, nuclear decay commonly results in a transmutation of one element into another element. If the daughter product nuclide produced by decay is unstable, it will subsequently decay into another nuclide. The nucleus will continue to decay, typically producing a different chemical element at each stage, until a stable nuclide is obtained.

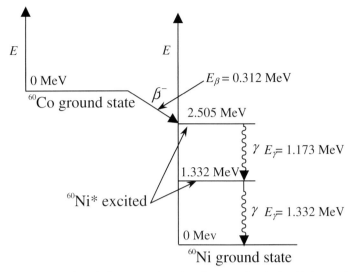

Figure 1.4 Energy level diagram for the decay of cobalt-60 into nickel-60.

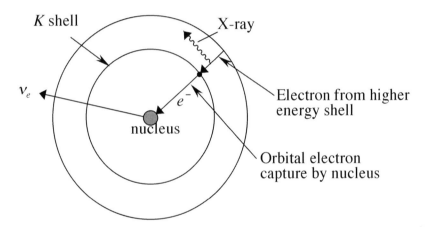

Figure 1.5 Schematic illustration of electron capture.

2.2 Activity

Radioactive decay is governed by the statistical law that the rate of decay for a particular species of radionuclide is directly proportional to the number of atoms of the radionuclide. Thus, the rate of decay in disintegrations per second (dis/s), is

$$\frac{d\mathcal{N}(t)}{dt} = -\lambda \mathcal{N}(t), \tag{1.2}$$

where $\mathcal{N}$ is the total number of atoms of a radioisotope, t is time in seconds (s), and λ is the proportionality constant, referred to as the *decay constant* (s^{-1}). The decay constant is a fundamental property for each species of radionuclide. This differential equation has the solution

$$\mathcal{N}(t) = \mathcal{N}_0\, e^{-\lambda t}, \tag{1.3}$$

where $\mathcal{N}_0$ is the initial number of atoms.

The number of disintegrations per second of an isotope is called its *activity* $\mathcal{A}$. If only one nuclide in a substance is radioactive and its daughter product is stable, then its activity is given by

$$\mathcal{A}(t) = \lambda \mathcal{N}(t). \tag{1.4}$$

Activity is often given in units of *curies* (Ci), equal to 3.7×10^{10} disintegrations/second (dis/s). Another unit of radioactivity, the *becquerel* (Bq), is equal to 1 dis/s. We can also express activity in terms of an initial activity $\mathcal{A}_0$ by

$$\mathcal{A}(t) = \mathcal{A}_0\, e^{-\lambda t}. \tag{1.5}$$

Equation 1.2 can be written in terms of an atom density N in atoms/cm^3, rather than total atoms $\mathcal{N}$. With this substitution we obtain the solution

$$N(t) = N_0\, e^{-\lambda t}. \tag{1.6}$$

Here, N_0 is the initial atom density of the radionuclide given by

$$N_0 = \frac{\varsigma N_A}{A_r}. \tag{1.7}$$

The parameter N_A is the Avogadro constant, equal to 6.022×10^{23} atoms/mole, A_r is the relative atomic mass in atomic mass units, and ς is the density of the isotope in g/cm^3. In order to avoid confusion with the almost universal use of "ρ" for reactivity discussed in Chapter 9, the symbol "ς" is used throughout this text for density, rather than the more common symbol for density "ρ".

A specific activity $\tilde{A}$, in Bq/g, can be defined in terms of the atom density N as,

$$\tilde{A} \equiv \lambda\, \frac{N}{\varsigma}. \tag{1.8}$$

If we take ς as the density of the radioisotope at any time t, then N at any time is equal to the expression given by the right-hand side of Eq. 1.7. We can combine Eqs. 1.7 and 1.8 and divide by 3.7×10^{10} to obtain

$$\tilde{A} \equiv \lambda\, \frac{N_A}{A_r(3.7 \times 10^{10})} = \frac{\lambda(1.63 \times 10^{13})}{A_r}, \tag{1.9}$$

with units of Ci/g. The specific activity is time-independent because it is given per gram of the radioisotope at the time under consideration. The specific activities are computed using this equation and presented in Table 1.1 for several plutonium isotopes and for the natural uranium isotopes. Equation 1.9 must be modified if more than one decay mode or more than one radioisotope is present.

2.3 Half-life

Radionuclides are commonly characterized by a half-life $t_{1/2}$. The half-life is the time in which half the number of atoms of a particular isotope decays to form another nuclide or to decay to a lower energy state. It follows from Eq. 1.5 that the half-life is inversely proportional to the decay constant,

$$t_{1/2} = \frac{\ln(2)}{\lambda} = \frac{0.693}{\lambda}. \tag{1.10}$$

Table 1.1   Radioactivity Characteristics of Uranium and Plutonium Isotopes

Species	Abundance (%)	Emission	Half-life (y)	$\tilde{A}$ (Ci/g)
^{234}U	0.0057	α and γ	2.50×10^5	6.1×10^{-3}
^{235}U	0.714	α and γ	7.10×10^8	2.1×10^{-6}
^{238}U	99.28	α and γ	4.51×10^9	3.3×10^{-7}
^{238}Pu	0	α	8.77×10^1	1.7×10^1
^{239}Pu	0	α	2.43×10^4	0.062
^{240}Pu	0	α	6.60×10^3	0.226

The half-lives for the naturally occurring uranium isotopes are also provided in Table 1.1, along with their natural abundance. The majority of natural uranium consists of the U-238 radioisotope, which decays with a half-life of 4.51×10^9 years. Uranium-234 and uranium-235 are far less abundant, and plutonium does not occur naturally. The isotopes of plutonium are produced in reactors by neutron capture in uranium or by using accelerators to bombard heavy nuclei with nuclear particles.

Example 1.1

The half-life and relative atomic mass of thorium-232 are 1.45×10^{10} y and 232.11 u, respectively. Assume 30 g of pure ^{232}Th metal with a density of 11.3 g/cm^3. Compute the specific activity, the initial activity, and the mass of thorium after 100 million years.

Solution:

From Eq. 1.10, $\lambda = \dfrac{0.693}{1.45 \times 10^{10} \text{ y}} = 4.78 \times 10^{-11} \text{ y}^{-1} = 1.52 \times 10^{-18} \text{s}^{-1}.$

Using Eq. 1.9, $\tilde{A} \equiv \dfrac{(1.52 \times 10^{-18}\text{s}^{-1})(1.63 \times 10^{13} \text{ Ci·s})}{232.11 \text{ g}} = 1.06 \times 10^{-7} \text{ Ci/g},$

and $\mathcal{A}_0 = (30 \text{ g})(1.06 \times 10^{-7}\text{Ci/g}) = 3.2 \times 10^{-6} \text{ Ci}.$

The mass of Th-232 left after 10^8 years is

$$m = (30 \text{ g}) \, e^{-(4.78 \times 10^{-11} \text{ y}^{-1})(10^8 \text{ y})} = 29.86 \text{ g}.$$

Thus, the mass of thorium changes by only 0.14 g ($\sim$0.5%) in 100 million years.

3.0 Directly Ionizing Radiation

All nuclear safety issues ultimately relate to the possibility of exposure to ionizing radiation and to potential health effects from exposure. Ionizing radiation includes all types of radiation capable of displacing electrons from atoms to form ions (an ion is an atom possessing a net electrical charge). For space nuclear systems, important types of ionizing radiation include neutrons, gamma radiation, alpha particles, beta particles, and X-rays. Alpha and beta particles are directly ionizing radiations, and their interactions with matter are discussed here. Uncharged particles, such as neutrons and gamma rays, interact with the atomic nucleus or atomic electrons, creating charged particles that subsequently ionize atoms in the substance. Because of this two-step process, neutrons, gamma rays, and X-rays are called indirectly ionizing radiation. Neutron and gamma ray interactions with matter are discussed in Sections 4 and 5, respectively.

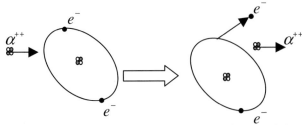

Figure 1.6 Schematic illustration of ionization by an alpha particle.

3.1 Ionization

Charged particles incident upon matter interact with atomic electrons of the constituent atoms of the substance. If an incident charged particle, such as a beta particle, passes close to an atom it can transfer a fraction of its kinetic energy sufficient to remove an orbital electron from the atom. This process will produce an ion-pair (the ionized atom and liberated electron) and result in a loss in energy for the incident radiation. Ionization is illustrated schematically in Figure 1.6. The degree of ionization produced by radiation will depend on the type of radiation, the kinetic energy of the incident particles, and the material irradiated. Specific ionization, given in ion-pairs/cm (ip/cm), is commonly used as a measure of the ability to produce ionization. Specific ionization increases with increasing mass and charge of the incident radiation. For example, the mass of an alpha particle is more than 7000 times greater than the mass of a beta particle. An alpha particle produces on the order of 50,000 ip/cm in air, compared to on the order of 100 ip/cm produced by beta particles. In general, specific ionization increases with increasing density and increasing atomic number of the irradiated material [3].

3.2 Other Energy Transfer Mechanisms

Energy loss by charged particle interactions can result from processes other than ionization. These processes include excitation, bremsstrahlung, and positron annihilation. If energy transmitted by incident charged particle radiation is insufficient to ionize the atom, radiation energy loss can result from excitation of an orbital electron to a higher energy state. The electron then falls back to its ground state giving off an X-ray with energy equal to the difference between the two atomic energy levels. At energies greater than 1 MeV, beta particles deflected by an atomic nucleus give up energy in the form of X-rays, by a process known as bremsstrahlung. Bremsstrahlung (braking or slowing down radiation) results when charges are accelerated or decelerated in an electric field. Positrons (positive beta particles) ultimately combine with negatively charged electrons, resulting in matter-antimatter annihilation. By Einstein's equation (Eq. 1.1), the entire mass of the electron and positron are converted into energy, and the annihilation energy is carried off by two 0.511 MeV gamma rays.

3.3 Alpha Particle Range

Charged particles directly interact with orbital electrons; as a consequence, they do not penetrate deeply into matter. Generally simple methods can be used to estimate the maximum ranges of alpha particles [3, 4]. The specific ionization caused by alpha particles can be plotted as a function of penetration distance into a material. The resulting plot, illustrated in Figure 1.7, is called a Bragg curve. Ionization initially increases as penetration increases, then drops rapidly to zero. The rapid increase in ionization occurs because the alpha particle loses energy and speed as it traverses the material. At slower velocities, the probability for interacting with atomic electrons increases, resulting in increased ionization. When the alpha particle's energy becomes sufficiently small, the alpha particle will capture two electrons to form a helium atom in thermal equilibrium with the medium. Alpha particles emitted during nuclear decay have energies typically in the range of 4 to 9 MeV. Although the kinetic energy for alpha particles produced by decay is appreciable, the massive alpha particle moves relatively slowly compared to

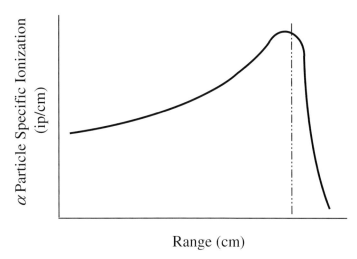

Figure 1.7 Qualitative Bragg curve. Specific ionization vs. range for alpha particles.

other, less massive, particles of the same energy. As a consequence, an alpha particle loses its energy by ionization far more rapidly than other types of radiation, resulting in a very small range of penetration. Alpha particles produced by nuclear decay do not have sufficient energy to penetrate human skin. The range for alpha particles in air can be estimated using

$$R_\alpha^{air} = 0.56\,E \qquad \text{(for } E < 4 \text{ Mev)}$$

$$R_\alpha^{air} = 0.318\,E^{3/2} \qquad \text{(for } 4 < E < 7 \text{ Mev).}$$

(1.11)

Here, R_α^{air} is the range of an alpha particle in air (cm) and E is the emission energy (in MeV) of the alpha particle. For other materials the range R_α^M can be estimated using the Bragg-Kleeman rule [4],

$$R_\alpha^M = \frac{3.2 \times 10^{-4}\,\sqrt{\bar{A}_r}}{\varsigma_M}\,R_\alpha^{air}.$$

(1.12)

Here, ς_M and $\bar{A}_r$ are the density (g/cm³) and relative atomic mass averaged over the constituents of material M.

3.4 Beta Particle Range

Unlike alpha particles, beta particles are not emitted at some characteristic energy for each decay of a particular type of radioisotope. Instead, beta particles exhibit a spectrum of energies, as shown in Figure 1.8. The maximum energy for emission equals the energy increment associated with the mass difference between the parent nuclide and the daughter product. The spectrum of energies is due to the division of the available energy between the beta particle and an emitted neutrino. In contrast with alpha particles, the low-mass beta particles undergo frequent changes in direction due to electrostatic interactions with electrons and nuclei. Because beta particles lack a specific emission energy and undergo random scattering collisions, a narrowly defined penetration range cannot be assigned to beta particles. Nonetheless, a simple formula can be used to estimate the maximum range of a beta particle in any material. The ability of a material to stop beta particles is directly proportional to the density of the material; consequently, the range of beta particles is often expressed as a density thickness $R_\beta \times \varsigma_M$ (in g/cm²), given by the approximate relationship

$$R_\beta \times \varsigma_M = 0.412\,E_{\max}^{[1.265 - (0.0954)\ln E_{\max}]}.$$

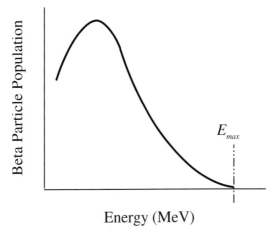

Figure 1.8 Typical beta particle energy spectrum.

Here, R_β is the maximum range for a beta particle (cm) and E_{max} is the maximum energy for beta emission in MeV. Equation 1.13 is valid for beta particle emission with energy 0.01 MeV $< E_{max} <$ 2.5 MeV. The maximum range is estimated for any material by dividing the range density by the material density. A typical maximum range for a beta particle in tissue is about 5 mm, compared to 0.04 mm for an alpha particle [3].

Example 1.2

Estimate the maximum range of a 2 MeV beta particle in water.

Solution:

From Eq. 1.13, $R_\beta \times \varsigma_M = 0.412(2.0)^{[1.265-(0.0954)\ln(2)]} = 0.9465$ g/cm^2.

Thus, the range in water is $R_\beta = \dfrac{0.9465 \text{ g/cm}^2}{1 \text{ g/cm}^3} = 0.95$ cm.

4.0 Neutron Interactions

Neutrons are uncharged particles that interact only with the nuclei of atoms, rather than directly interacting with orbital electrons. As a consequence, the range for neutrons in matter can be much greater than the range for charged particles. Nuclear fission in reactors is the most common and copious source of neutrons relating to space nuclear systems. Neutron production in reactors, however, requires a neutron source to initiate a chain reaction. Neutron emission can be induced by bombarding the nucleus of some atomic species (beryllium, boron, lithium, and oxygen-18) with an alpha particle. A few isotopes, such as beryllium-9 and deuterium (^{2}H), will emit neutrons when their nuclei are struck by a gamma ray [3]. Neutron interactions are generally more complex than charged particle interactions, particularly in an operating reactor. Interactions include a variety of absorption reactions and scattering events with nuclei.

4.1 Neutron Absorption

Once neutrons are released from a nucleus, they are unstable and decay (with a half-life of about 12 minutes) into a proton, a beta particle, and an antineutrino [4]. In most situations of interest, neutrons will be absorbed by the nucleus of another atom before the neutron decays. The complete process of a neutron absorption reaction is a two-step process. When a neutron strikes the nucleus of an atom, the neu-

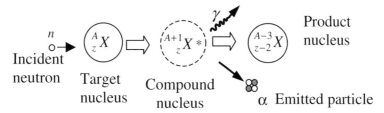

Figure 1.9 Neutron absorption followed by charged particle emission (n, α).

tron can be absorbed to form a compound nucleus. The compound nucleus is left in an excited (high-energy) state, commonly identified by using an asterisk as the symbol for the compound nucleus; e.g., $^{236}U^*$. The excitation energy of the compound nucleus is produced from the binding energy and kinetic energy associated with the absorbed neutron. The compound nucleus does not remain in an excited state and very quickly releases energy by emitting a particle, a gamma ray, or by fissioning (breaking into two parts).

Neutron Induced Particle Emission

Charged particles may be emitted from a compound nucleus following neutron absorption. The reactions for alpha particle and proton emission following neutron absorption are identified by the notations (n,α) and (n,p), respectively. Neutron induced alpha particle emission is illustrated in Figure 1.9. Although neutron-induced charged particle emission is not common, important n,α reactions occur for the isotopes boron-10 and lithium-6 [3]. The isotope boron-10 makes up approximately 20% of naturally occurring boron. Because boron-10 readily absorbs neutrons, it is often used as a neutron absorber for the control of nuclear reactors. Boron control materials in reactors will build up helium gas as a result of neutron absorption and the subsequent emission of alpha particles. The buildup of helium gas in boron control materials must be considered in the design of reactor systems. For space reactors, LiH is a common choice for a neutron radiation shield material, and some designs use lithium as a reactor coolant. Reactor system designs using LiH for shielding, or incorporating a lithium coolant, must accommodate the accumulation of helium gas resulting from the n,α reaction with Li-6. Lithium-6 constitutes about 7% of naturally occurring lithium. Using depleted lithium (increased Li-7/Li-6 ratio) can significantly reduce helium gas buildup.

Neutron absorption can also result in subsequent emission of another neutron (n,n) with less energy than the incident neutron. For some nuclides, the compound nucleus will emit two neutrons $(n,2n)$. Neutron emission following neutron absorption (n,n) is treated as an inelastic scattering event and is discussed in Section 4.2.

Radiative Capture and Activation

In contrast to charged particle emission, neutron absorption followed by gamma-ray emission (n,γ), is quite common. An (n,γ) reaction is called radiative capture, and the emitted gamma photon is called a capture gamma ray. Gamma rays carry off the excess energy of the excited compound nucleus. Several photons are typically emitted with energies corresponding to increments between energy levels of the excited nucleus. With the exception of hydrogen, the total gamma ray energy emitted during radiative capture is typically in the range of 6 to 8 MeV. The capture gamma ray energy for hydrogen is 2.2 MeV [3].

The product nucleus from radiative capture is often unstable and will decay with the characteristic half-life of the radioisotope. The process of making materials radioactive by radiative capture is called *neutron activation*. Decay of activated materials typically results in the emission of a negative beta particle although alpha particle decay is exhibited by some nuclides. The processes of radiative capture forming a stable nucleus and radiative capture followed by radioactive decay are illustrated in Figure 1.10

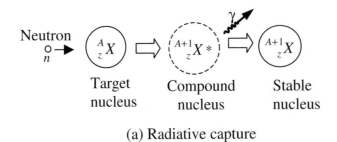

(a) Radiative capture

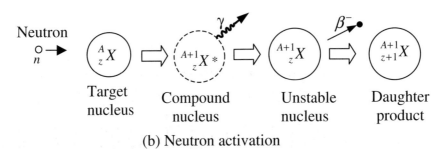

(b) Neutron activation

Figure 1.10 Radiative capture (n,γ) (a) forming a stable nucleus and (b) forming an unstable nucleus followed by decay (activation).

(a) and (b), respectively. Materials activated by neutron absorption must be considered in the safety analysis of postulated accidents. Even for space reactors that use highly enriched uranium (^{235}U content in uranium increased to ~93%), the fuel will contain some uranium-238. During reactor operation the radioisotope plutonium-239 will be produced by radiative capture in uranium-238 and two subsequent β^- decays. Plutonium-239 may subsequently absorb a neutron resulting in nuclear fission; however, plutonium-239 decays by alpha particle emission with a half-life of 2.44×10^4 y. As a consequence, any unfissioned plutonium-239 will be an alpha emission source following reactor operation.

Nuclear Fission

For some heavy nuclides, such as uranium-235 or plutonium-239, neutron absorption can cause nuclear fission (n,f), forming two smaller nuclei called fission products (see Figure 1.11). Two or three neutrons are released during fission, along with beta particles, gamma rays, and neutrinos. The division of

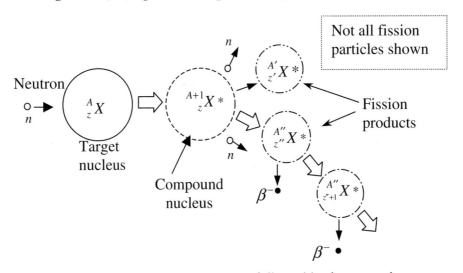

Figure 1.11 Neutron absorption followed by fission (n,f).

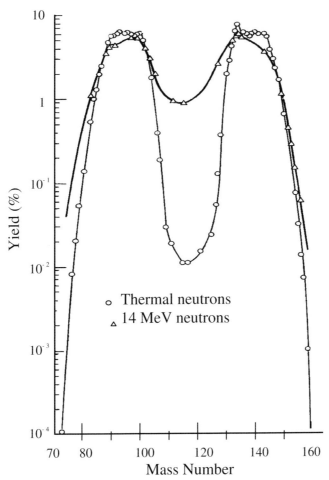

Figure 1.12 Fission yield as a function of mass number. *Courtesy of Argonne National Laboratories* [5].

the nucleus into two pieces can occur in more than 40 different ways. Fission yields from uranium-235 are plotted for the various fission products as a function of mass number in Figure 1.12. The yields presented in the figure are for fission following absorption of thermal neutrons and for fission following absorptions of high-energy neutrons (14 MeV). *Thermal neutrons* are low energy neutrons in near-thermal equilibrium with their environment. Note that fission products fall into two groups. One group of fission products covers a range of mass numbers between 80 and 110, the other group covers the range between 125 and 155. Similar curves have been obtained for other fissionable nuclei. Most fission products are radioactive and usually decay by the emission of a negative beta particle. The nuclei of nuclides produced by the decay of fission products are usually unstable and will decay as well. Fission products typically undergo several decay stages before forming a stable nucleus.

The release of neutrons during fission is an essential feature of nuclear reactor operation. For uranium–235, fission induced by absorption of thermal neutrons yields an average of 2.43 neutrons per fission. The average yield for plutonium-239 is 2.9 neutrons per fission (for thermal neutron absorption). When a nucleus fissions, each of the two smaller nuclei are in an excited state and contain too many neutrons. Most fission products quickly release excess neutrons ($< 10^{-14}$s) as well as accompanying gamma radiation. The energy spectrum of these *prompt neutrons* and *prompt gamma rays* are presented in Figures 1.13 and 1.14, respectively. Prompt neutrons constitute more than 99% of all the neutrons produced by fission. A small fraction of fission products decay by emitting neutrons instead of beta particles. These *delayed neutrons* are emitted over an extended period of time (a mean time

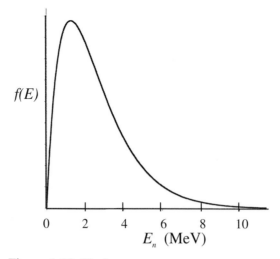

Figure 1.13 Fission neutron energy spectrum.

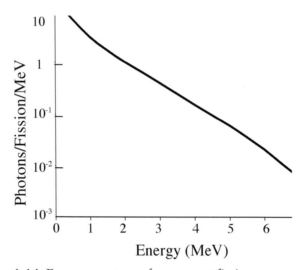

Figure 1.14 Energy spectrum for prompt fission gamma rays.

after fission of about 10 to 13 s) by a number of species of fission product radioisotopes. For uranium-235, only 0.65% of all the neutrons produced by fission are delayed neutrons. Although the fraction of fission-produced neutrons that are delayed is small, delayed neutrons play an extremely important role in reactor operation.

4.2 Neutron Scattering

Most neutrons released during fission are *fast neutrons* with energies well above the thermal energy range. Fast neutrons can give up some of their kinetic energy during elastic or inelastic collisions with the nuclei of atoms. The two scattering processes are shown schematically in Figure 1.15 (a) and (b), respectively.

Inelastic Scattering

During inelastic scattering collisions, the incident neutron is absorbed by the target nucleus and momentarily forms a compound nucleus. The nucleus then emits a neutron with an energy E_{ns} less than the incident energy E_n. Some of the kinetic energy E_γ of the incident neutron is carried off by a gamma ray released from the nucleus. Inelastic scattering occurs most commonly in elements with high mass

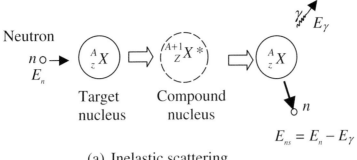

(a) Inelastic scattering

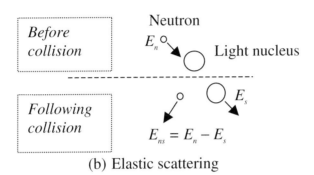

(b) Elastic scattering

Figure 1.15 Neutron scattering: (a) inelastic scattering of fast neutrons, and (b) elastic scattering of fast and epithermal neutrons.

numbers and only for high neutron energies. For high mass number isotopes, the threshold for inelastic scattering is in the range of 0.1 to 1 MeV [2].

Elastic Scattering

In elastic neutron scattering, neutrons can give up some of their kinetic energy to the target nucleus during impact, but the total energy of the neutron and target nucleus is conserved (no gamma ray emission occurs). Elastic scattering can result from the process described for inelastic scattering, or from potential scattering. In potential scattering, a neutron-nucleus collision does not result in a compound nucleus; this type of elastic scattering can be likened to the collision of two billiard balls. By conservation of momentum and energy it can be shown that the maximum neutron energy loss during an elastic collision ΔE_{max} is approximately

$$\Delta E_{\mathrm{max}} \approx (1 - \alpha_s)E. \tag{1.14}$$

The parameter α_s is defined by

$$\alpha_s \equiv \frac{(A + 1)^2}{(A - 1)^2}, \tag{1.15}$$

and A is the mass number of the nuclide. For hydrogen $A = 1$, $\alpha_s = 0$, and $\Delta E_{max} = E$. In other words, a neutron can lose all of its kinetic energy in a single elastic collision with the nucleus of a hydrogen atom. The maximum possible energy loss per collision decreases with increasing atomic mass [3]. Note that Eq. 1.14 is approximately correct only for energies above the thermal range, where the target nucleus can be considered to be at rest relative to the neutron.

Materials called *moderators* are often used in reactors to reduce the average neutron energy in order to increase the probability of absorption by fissile materials. Reactor moderators include hydrogenous materials and other low-mass-number materials. Water and graphite are common moderators for terrestrial reactors. For space reactors, common moderator choices include zirconium hydride, beryllium, beryllium oxide, and graphite.

For highly moderated reactor systems, neutron scattering results in a significant population of low-energy thermal neutrons. Thermal neutrons take on an approximate Maxwell-Boltzmann energy distribution. The Maxwell-Boltzmann distribution is given by

$$f(E) = \frac{2\pi\sqrt{E}}{(\pi kT)^{3/2}} \, e^{-E/kT}, \tag{1.16}$$

where $f(E)$ is the fraction of neutrons per unit of energy at energy E, k is Boltzmann's constant, and T is the absolute temperature. When Eq. 1.16 is applied to neutrons, the temperature represents the *neutron temperature* rather than the temperature of the medium. The neutron temperature (somewhat higher than the temperature of the medium) approximately accounts for hardening (higher energy) of the neutron spectrum due to generally increased neutron absorption at lower energies. A Maxwell-Boltzmann energy distribution for a temperature of 2000 K is presented in Figure 1.16.

4.3 Reaction Rates and Cross Sections

Neutron interactions in materials containing fissile isotopes are generally complex and are governed by a dynamic process of neutron production by fission and neutron loss by absorption or leakage from the system. The concepts of neutron flux, cross sections, and reaction rates are required to describe the complex interactions of neutrons with matter.

Microscopic Cross Sections

A microscopic cross section $\sigma_{ij}(\mathbf{r},E)$ for a particular reaction j can be defined as the effective cross sectional area of a nucleus of isotope i at location $\mathbf{r}$ as it appears to a neutron with kinetic energy E. The microscopic cross section is not the true cross sectional area associated with the size of the nucleus. Nuclear interactions must be understood as complex quantum mechanical interactions that are conveniently characterized by an effective microscopic cross section. The magnitude of the microscopic cross section for a particular isotope can vary by orders of magnitude, depending on the energy of the incident neutron and the type of neutron-nucleus reaction. Microscopic absorption cross sections are defined for each

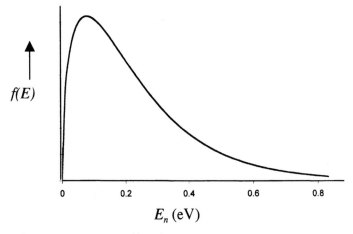

Figure 1.16 Maxwell-Boltzmann energy distribution.

type of interaction. The symbols σ_γ, σ_α, σ_f, and σ_{2n} refer to microscopic absorption cross sections for (n,γ), (n,α), (n,f), and $(n,2n)$ reactions, respectively. Microscopic cross sections are also defined for all neutron scattering events σ_s. Some scattering cross sections are used to characterize neutron scattering from one specific energy range to another specific energy range (discussed in Chapter 8). Microscopic absorption cross sections σ_a are defined as the sum of the capture and fission cross sections; i.e.,

$$\sigma_a = \sigma_c + \sigma_f. \tag{1.17}$$

We can define a macroscopic cross section as

$$\Sigma_{ij}(\mathbf{r},E) \equiv N_i \sigma_{ij}(\mathbf{r},E). \tag{1.18}$$

The macroscopic cross section is the total effective cross section area for all isotope i nuclei per unit volume of a medium. When all dimensional units are given in centimeters, the microscopic cross sections for nuclei are typically very small and atom densities are usually very large. Reactor physicists have defined a more convenient unit called a barn (b), equal to 10^{-24} cm^2. Microscopic cross sections are commonly expresses in terms of barns, and atom densities are often expressed in the somewhat odd units of atoms/b·cm. The units for macroscopic cross sections are cm^{-1}.

Flux and Reaction Rates

The neutron flux ϕ is a useful and pervasive concept in the analysis of neutron interactions. Neutron flux is defined as the product of the neutron density n (neutrons/cm^3) and the neutron speed v (cm/s); i.e.,

$$\phi(\mathbf{r},E) \equiv n(\mathbf{r},E) \mathrm{v}(\mathbf{r},E). \tag{1.19}$$

The flux has units of neutrons per cm^2 per second per eV (n/cm^2·s·eV); and the variables $\mathbf{r}$ and E are the position vector and the neutron kinetic energy, respectively. From a comparison of Figure 1.13 (MeV units) with Figure 1.16 (eV units) it is clear that the neutron energy spectrum is strongly influenced by the composition of the medium. Thus, the energy distribution of the flux can vary significantly among different reactor types. Furthermore, the flux can change appreciably from one region to another within a particular reactor.

The rate of neutron interaction $\mathcal{R}$(reactions/cm^3·s) at position $\mathbf{r}$ can be computed using the neutron flux and the macroscopic cross section $\Sigma(\mathbf{r},E)$; i.e., for a particular type of reaction j and isotope i in a medium

$$\mathcal{R}_{ij}(\mathbf{r}) = \int_0^\infty \Sigma_{ij}(\mathbf{r},E) \, \phi(\mathbf{r},E) dE. \tag{1.20}$$

We can compute the total reaction rate for type j reactions as the sum of the reaction rates for each isotope in the medium

$$\mathcal{R}_j(\mathbf{r}) = \sum_{all\ i} \mathcal{R}_{ij}(\mathbf{r}). \tag{1.21}$$

Cross Section Energy Dependence

Microscopic absorption cross sections are strongly dependent on the energy of incident neutrons and are frequently characterized by three energy ranges. These ranges include a thermal, an epithermal, and a fast energy range. The boundaries of these energy ranges are not clearly defined because the cross section characteristics typifying each energy range begin and end at different energies for different isotopes. The thermal energy range covers all energies below an upper bound of about 1 to 2 eV. For many materials, absorption cross sections for the thermal energy range are inversely proportional to the square root of the neutron kinetic energy; i.e., the energy dependence is inversely proportional to the neutron speed v,

$$\sigma_a \sim \frac{1}{E^{1/2}} \sim \frac{1}{v}. \tag{1.22}$$

Thermal cross sections for absorption and for scattering are presented in Table 1.2 for a variety of elements. The cross sections in Table 1.2 are for neutrons with a speed of 2200 m/s. A neutron speed of 2200 m/s corresponds to the most probable energy of 0.0252 eV at 293 K, assuming a Maxwell-Boltzmann energy distribution. The cross sections in Table 1.2 are based on measurements that include all naturally occurring isotopes of each element. Cross sections of different isotopes of the same element can differ drastically from each other. Thermal absorption cross sections for uranium isotopes and for plutonium-239 are presented in Table 1.3. The fission and capture cross section constituents of the absorption cross sections are also provided.

The epithermal range spans energies from the upper end of the thermal energy range to about 10 thousand electron volts (10 keV). Many isotopes display a number of sharp spikes in the absorption cross section in the epithermal range. These peaks, called resonances, are associated with the quantum states in the compound nucleus. At certain neutron energies the probability of forming a compound nucleus is much greater than for other energies to either side of the resonance peak. The resonance peaks for the fission cross section of uranium-235 are clearly observed in Figure 1.17. The epithermal range is also called the resonance range. In the fast energy range, directly above the epithermal range, resonance structure is absent and cross sections are usually small. The three approximate cross section energy ranges for uranium-235 are identified in Figure 1.17.

Table 1.2   Thermal (2200 m/s) Cross Sections for Some Naturally Occurring Elements

Element	σ_a (b)	σ_s (b)	Element	σ_a (b)	σ_s (b)
H	0.332	38.0	Fe	2.62	11.0
He	0.007	0.8	Cu	3.77	7.2
Li	71.0	1.4	Zr	0.180	8.0
Be	0.010	7.0	Cd	2.45×10^3	7.0
B	7.6×10^2	4.0	Hf	1.05×10^2	8.0
C	0.0034	4.8	Ta	21.0	5.0
O	2.19×10^{-4}	4.2	W	19.2	5.0
Na	0.53	4.0	Hg	3.8×10^2	20.0
Al	0.230	1.4	Th	7.56	12.6
Cl	33.8	16.0	U	7.68	8.3

Table 1.3 Thermal (2200 m/s) Cross Sections for Uranium and Plutonium Isotopes

Species	σ_f (b)	σ_γ (b)	σ_a (b)
^{233}U	527.0	54.0	581.0
^{235}U	577.0	106.0	683.0
^{238}U	0.0	2.71	2.71
^{239}Pu	742.0	287.0	1029.0

Example 1.3

Assuming a thermal neutron flux of 1.8×10^{11} n/cm²s, compute the reaction rate in full density liquid sodium. Use a sodium density of 0.82 g/cm³ and a relative atomic mass of 22.99 u. Assume that the reaction rate is entirely due to thermal absorption of neutrons, characterized by its 2200 m/s absorption cross section.

Solution:

From Eq. 1.9 $\qquad N = \dfrac{(0.82 \text{ g/cm}^3)(6.022 \times 10^{23} \text{ atoms/mole})}{22.99 \text{ g/mole } (10^{24} \text{ b/cm}^2)} = 2.15 \times 10^{-2}$ atom/b·cm,

and from Table 1.2, $\sigma_{\text{Na}} = 0.53$ b. Using Eqs. 1.18 and 1.20, we obtain

$$\mathcal{R} = (0.0215 \text{ atoms/b·cm})(0.53 \text{ b})(1.8 \times 10^{11} \text{ n/cm}^2\text{·s}) = 2.05 \times 10^9 \text{ captures/s}$$

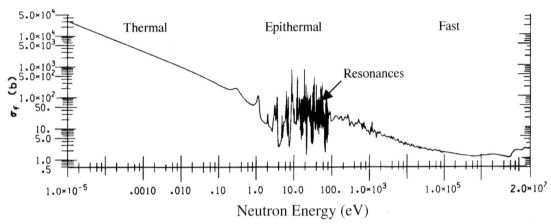

Figure 1.17 Uranium-235 fission cross sections as a function of neutron energy. *Courtesy of Brookhaven National Laboratory* [6].

5.0 Gamma Ray Interactions

Gamma rays are quanta of electromagnetic radiation emitted from the nucleus of an atom. Gamma radiation is emitted during nuclear decay, neutron capture, and nuclear fission. Although the energy of gamma radiation is typically higher than the energy of X-rays, gamma rays are only distinguished from X-rays by their origin; i.e., X-rays are not emitted from the nucleus. Instead, they result from atomic electrons changing energy states. Matter-antimatter annihilation radiation is often referred to as gamma radiation; however, annihilation radiation does not originate from a nucleus and, strictly speaking, is not gamma radiation. The historical usage of this terminology for electromagnetic radiation has no bearing on the physics of photon interactions with matter. This discussion on the physics of gamma ray interactions applies to all electromagnetic radiation with energy greater than 10 keV, regardless of its origin.

5.1 Interaction Processes

Gamma rays, like neutrons, are uncharged and are far more penetrating than alpha and beta particles. The three principal processes for gamma ray interaction with matter are the photoelectric effect, Compton scattering, and pair production. These three processes, illustrated schematically in Figure 1.18, result in the ejection, scattering, or creation of a high-speed electron. The electron behaves like a beta particle (i.e., like an electron ejected from the nucleus), transferring its kinetic energy to the medium by ionization, and other processes.

Photoelectric Effect

At relatively low energies (< 1 MeV) the photoelectric effect is important. The energy of a photon is equal to the product of Plank's constant h (6.6255 × 10^{-34} joules·s) and the frequency ν. For the photoelectric effect, the energy of an incident gamma photon is completely transferred to an orbital electron of an atom in the medium. The electron is ejected from the atom with a kinetic energy equal to the

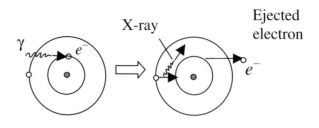

(a) Photoelectric effect

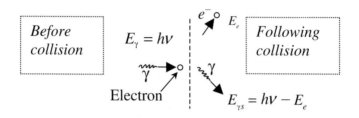

(b) Compton scattering

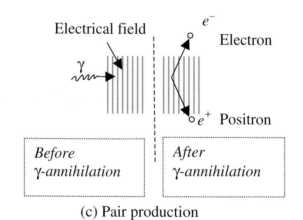

(c) Pair production

Figure 1.18 Gamma ray interactions with matter.

energy of the incident gamma photon minus the binding energy of the electron. The probability of a photoelectric interaction is directly proportional to the fourth power of the atomic number of the atoms in the medium and inversely proportional to the cube of the energy of the gamma photon; i.e., $\sim Z^4/E_\gamma^3$.

Compton Scattering

Compton scattering of electrons results when a gamma photon makes an elastic collision with an electron. Energy and momentum are conserved; consequently, some of the kinetic energy is transferred to the electron, and the photon is scattered in another direction. The energy of the scattered photon ($E_{\gamma s} = h\nu_{\gamma s}$) is equal to the energy of the incident photon ($E_\gamma = h\nu_\gamma$) minus the kinetic energy transferred to the electron E_e. The probability of a Compton scattering event is directly proportional to the atomic number of the atoms and inversely proportional to the photon energy (Z/E_γ).

Pair Production

A gamma photon entering the strong electric field near the nucleus of an atom can create an electron-positron pair. For this process to occur, the energy of the gamma photon must be at least 1.02 MeV, equal to the combined rest mass of the electron and positron. The photon is completely absorbed during pair production, and most of the excess energy is carried away in the form of kinetic energy of the electron and positron. A small fraction of the excess energy is transferred to the nucleus associated with the electric field. Pair production electrons and positrons interact with the medium by ionization and other mechanisms. Positrons ultimately combine with an atomic electron in the medium and annihilate, producing two photons with an energy of 0.511 MeV each. The probability of pair production for gamma rays interacting with the electrical field of nuclei in a medium is proportional to $Z^2(E_\gamma - 1.02)$. A small fraction of pair production electrons and positrons are also produced in the weaker electric field of the orbital electrons [3].

5.2 Gamma Ray Attenuation

For a narrow beam of gamma rays incident upon a medium, the differential fractional change in the gamma ray intensity dI_γ/I_γ is directly proportional to the infinitesimal distance traveled in the medium dx,

$$\frac{dI_\gamma}{I_\gamma} = -\mu_\gamma dx. \tag{1.23}$$

The proportionality constant μ_γ for the medium is called the linear attenuation coefficient. The units for intensity I_γ are photons/cm^2 or MeV/cm^2. Typically the thickness is expressed in cm, and the linear attenuation coefficient is given in units of cm^{-1}. Equation 1.23 can be integrated to obtain

$$I_\gamma(x) = I_{\gamma 0} e^{-\mu_\gamma x}, \tag{1.24}$$

where $I_{\gamma 0}$ is the incident gamma ray intensity. This simple relationship expressed by Eq. 1.24 is not strictly valid for broad beams of gamma rays [3]. A more general equation for gamma ray attenuation is discussed in Chapter 4.

The linear attenuation coefficient includes the attenuation effect of all gamma absorption processes; consequently, the energy dependence of the coefficient includes the energy dependence of the photoelectric effect, Compton scattering, and pair production. The energy dependence of μ_γ is shown for lead in Figure 1.19. The linear attenuation coefficients for several materials are given as a function of

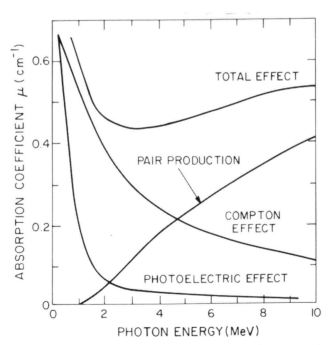

Figure 1.19 Linear absorption coefficient for gamma rays in lead as a function of energy for principal interaction processes [4].

Table 1.4   Gamma Ray Total Linear Attenuation Coefficients for Various Materials Adapted from [5]

$\mu_\gamma(cm^{-1})$

Material	Gamma Ray Energy (MeV)								
	0.1	0.5	1.0	1.5	2.0	3.0	5.0	8.0	10.0
Carbon	0.335	0.196	0.143	0.117	0.100	0.080	0.061	0.048	0.044
Aluminum	0.435	0.227	0.166	0.135	0.116	0.095	0.076	0.065	0.062
Iron	2.704	0.651	0.468	0.381	0.333	0.284	0.246	0.232	0.231
Tungsten	81.25	2.413	1.235	0.950	0.843	0.782	0.789	0.845	0.898
Lead	59.99	1.644	0.776	0.581	0.518	0.477	0.483	0.521	0.555
Uranium	19.82	3.291	1.416	1.025	0.905	0.832	0.834	0.896	0.956
Water	0.167	0.097	0.071	0.058	0.049	0.040	0.030	0.024	0.022
Concrete	0.397	0.205	0.149	0.122	0.105	0.085	0.067	0.057	0.054

Table 1.5   Gamma Ray Total Mass Attenuation Coefficients for Various Materials Adapted from Reference [5]

μ_m/ς_m (cm^2/g)

Material	Gamma Ray Energy (MeV)								
	0.1	0.5	1.0	1.5	2.0	3.0	5.0	8.0	10.0
Hydrogen	0.295	0.173	0.126	0.103	0.088	0.069	0.050	0.037	0.032
Carbon	0.149	0.087	0.064	0.052	0.044	0.036	0.027	0.021	0.019
Aluminum	0.161	0.084	0.061	0.050	0.043	0.035	0.028	0.024	0.023
Iron	0.344	0.083	0.060	0.049	0.042	0.036	0.031	0.030	0.029
Tungsten	4.21	0.125	0.064	0.049	0.044	0.041	0.041	0.044	0.047
Lead	5.29	0.145	0.068	0.051	0.046	0.042	0.043	0.046	0.049
Uranium	1.06	0.176	0.076	0.055	0.048	0.045	0.045	0.048	0.051
Air	0.151	0.087	0.066	0.052	0.045	0.036	0.027	0.022	0.020
Water	0.167	0.097	0.071	0.058	0.049	0.040	0.030	0.024	0.022
Concrete	0.169	0.087	0.064	0.052	0.044	0.036	0.029	0.024	0.023
Tissue	0.163	0.094	0.168	0.056	0.048	0.038	0.029	0.023	0.021

energy in Table 1.4. In the energy range from 0.1 to 10 MeV, the linear attenuation coefficient divided by the density of the material μ_m/ς_m is relatively insensitive to the type of material for light elements [3]. The quantity μ_m/ς_m, called the mass attenuation coefficient, is often used in the characterization and analysis of gamma ray interactions with matter. Values for μ_m/ς_m for several materials (in cm^2/g) are presented in Table 1.5.

6.0 Nuclear Energy

The energy released in nuclear processes is enormous compared to chemical processes. The net energy obtained per molecule in an exothermic chemical reaction is measured in electron volts; for nuclear reactions it is measured in millions of electron volts per atom. Nuclear energy is released during the spontaneous radioactive decay of unstable nuclei and during the fission of heavy nuclei or the fusion of two light nuclei. Power systems using nuclear fusion are still in the speculative stage and will not be discussed in this text. Only the nuclear disintegration processes, radioactive decay and nuclear fission have been used for space nuclear power. The energy released during nuclear processes is associated with the mass difference between nuclear reactants and nuclear products, according to Einstein's equation (Eq. 1.1). For nuclear decay of a radioisotope, the reactant is the original unstable nucleus, and the products are the nucleus following decay and the particles released during decay. For nuclear fission, the reactants are the incident neutron and the target nucleus; the products are the fission products and all of the particles released as a result of nuclear fission.

6.1 Radioisotope Power Sources

Radioisotope power sources produce nuclear energy by the spontaneous radioactive decay of fuel nuclei. In the United States, plutonium-238 has been used as the heat source for all radioisotope power sources launched into space. About 5.5 MeV is released during radioactive decay of a plutonium-238 nucleus. The nuclear energy from radioactive decay is manifest primarily as kinetic energy of the emitted alpha particle. The emitted alpha particle gives up its kinetic energy by interaction with fuel atoms (e.g., ionization), resulting in heating of the nuclear fuel material. Heat from the radioisotope fuel can then be used with an energy conversion device to produce electric power.

Since each radioactive decay deposits energy in the fuel, the power density in a radioisotope fuel source is directly proportional to the decay rate of all radioactive species i,

$$q'''(t) = \sum_{all\ i} \kappa_i^D \lambda_i N_i(t). \tag{1.25}$$

Here, q''' is the power density (W/cm^3) and κ_i^D is energy yield per nuclear disintegration (W·s/dis) for nuclide i. If it is assumed that essentially all of the radioactive decay in a radioisotope source is produced by one nuclide and the decay product nuclide is stable, then by combining Eqs. 1.6 and 1.25 the power density is

$$q'''(t) = \kappa^D \lambda N_0 e^{-\lambda t}. \tag{1.26}$$

Although radioisotope sources can involve more than one unstable isotope and unstable daughter products can contribute to the total energy production, Eq. 1.26 is a good approximation for many radioisotope fuel choices. Table 1.1 shows that the half-life for plutonim-238 is 87.7 y. Using Eq. 1.10 we can compute the decay constant as

$$\lambda = \frac{0.693}{87.7y(3.1536 \times 10^7 s/y)} = 2.5 \times 10^{-10} s^{-1}.$$

Assuming a nominal density for PuO$_2$ of 10 g/cm^3, the atom density of plutonium-238 is

$$N_{Pu} = \frac{\varsigma_{PuO_2} N_A}{M_r} = \frac{10.0\ g/cm^3 (6.022 \times 10^{23}\ atoms/mole)}{[238.05 + 2(15.994)]\ g/mole} = 2.23 \times 10^{22}\ atoms/cm^3.$$

Note that the relative molecular mass M_r is used to compute atom density, rather than the relative atomic mass A_r. Equation 1.26 gives the power density

$$q'''(t) = (5.5\ MeV/atom)(2.5 \times 10^{-10} s^{-1})(2.23 \times 10^{22} atom/cm^3)\ e^{-(2.5 \times 10^{-10} s^{-1})t}$$

$$= 3.07 \times 10^{13}\ MeV/cm^3 \cdot s(1.603 \times 10^{-13} W \cdot s/MeV)\ e^{-(2.5 \times 10^{-10} s^{-1})t}$$

$$= 4.9\ W/cm^3\ e^{-(2.5 \times 10^{-10} s^{-1})t}.$$

Hence, power density is approximately 4.9 W/cm^3 at beginning-of-life and decreases exponentially with time.

6.2 Reactor Power Systems

Space reactor systems produce energy by controlled nuclear fission. Approximately 200 MeV are released per fission. This value is more than 35 times greater than the energy release per disintegration by nuclear decay of plutonium-238 fuel in a radioisotope source. As shown in Table 1.6, most of the en-

Table 1.6 Uranium-235 Fission Energy Distribution [4]

Energy Form	Energy Released (MeV)	Recoverable Energy (MeV)
Fission fragments kinetic energy	168	168
Decay of fission products		
—Beta radiation	8	8
—Gamma radiation	7	< 7
—Neutrinos	12	0
Prompt (fission) gamma radiation	7	< 7
Fission neutron kinetic energy	5	< 5
Capture gamma radiation	—	< 12
TOTAL	207	~ 200

ergy released during fission is in the form of kinetic energy of the fission products. For uranium-235, the total kinetic energy for fission products is an average of 168 MeV per fission. Fission product nuclei are unstable and most commonly decay by emitting beta particles, gamma radiation, and neutrinos, accounting for another 27 MeV per fission. The total kinetic energy of all neutrons released per fission averages about 5 MeV, with an average kinetic energy per neutron of about 2 MeV. Neutron activation of reactor materials produces an additional 3 to 12 MeV per fission event [4]. Because a fraction of the gamma rays and neutrons resulting from fission can escape the core, much of the gamma ray energy and some of the neutron energy cannot be recovered. Neutrinos have essentially no interaction with reactor materials, and none of their energy is recoverable. The additional energy released during decay of activated materials is usually assumed to roughly compensate for energy lost through escaping radiation. Thus, the net recoverable energy resulting from fission of a nucleus is approximately 200 MeV.

Fission products and nuclear particles released during fission collide with atoms within the fuel material resulting in fuel heating. Heat from the fuel can then be used with a power conversion device to produce electrical power. Alternatively, reactor heat can be used to heat a fluid for propulsive thrust. Power densities possible for reactors are enormous. The power density q''' in the reactor fuel is directly proportional to the reaction rate for fission. Using Eq. 1.20, the power density (in W/cm^3) can be determined from

$$q''' = \int_0^\infty \kappa \Sigma_f(E) \phi(E) dE. \tag{1.27}$$

The parameter κ is the recoverable energy released per fission. For a thermal reactor, assuming that most of the fissions are caused by thermal neutrons, Eq. 1.27 can be written

$$q''' \approx \kappa \bar{\Sigma}_f \phi, \tag{1.28}$$

where $\bar{\Sigma}_f$ and ϕ are the average macroscopic fission cross section for the thermal energy range and the thermal neutron flux, respectively. Assuming uranium dioxide (UO$_2$) fuel with a density of 10 g/cm^3, 93% enriched in uranium-235, the fuel atom density is

$$N_{235} = \frac{(0.93)(10 \text{ g/cm}^3)(6.022 \times 10^{23} \text{ atoms/mole})}{[235.044 + 2(15.994) \text{ g/mole}](10^{24} \text{ b/cm}^2)} = 0.021 \text{ atom/b·cm}.$$

For an average microscopic fission cross section of about 350 b we obtain

$$\bar{\Sigma}_f = (350 \text{ b/atom})(0.021 \text{ atom/b·cm}) = 7.35 \text{ cm}^{-1}.$$

Assuming a thermal neutron flux of about 2×10^{12} n/cm²s, Eq. 1.28 predicts a fuel power density of

$$q''' = (200 \text{ MeV})(7.35 \text{ cm}^{-1})(2 \times 10^{12} \text{n/cm}^2 \cdot s)(1.603 \times 10^{-13} \text{W} \cdot \text{s/MeV}) = 471 \text{ W/cm}^3.$$

For this example the power density produced in the reactor fuel is about 100 times greater than a typical radioisotope power density. By increasing the flux level, much higher reactor fuel power densities can be achieved. The flux level and power density attained will be restricted by the operational constraints of the reactor system, such as material temperature limitations.

7.0 Summary

The nucleus of an atom consists of positively charged protons and uncharged neutrons. Most stable nuclei contain more neutrons than protons. An unstable nucleus will decay to form a more stable nucleus by emitting nuclear particles. The rate of nuclear decay of an atomic species is directly proportional to the number of atoms of the radioisotope, and the decay constant λ is the proportionality constant. The number of atoms of an isolated radioisotope decreases exponentially with time with a half-life characteristic of the particular isotope. Nuclear radiations released during decay include negative and positive beta particles, alpha particles, and gamma rays. Beta particles are negatively or positively charged electrons emitted from a nucleus during the conversion of a neutron to a proton or a proton to a neutron, respectively. Positive beta particles are also called positrons. An alpha particle is a stable cluster of two protons and two neutrons; and gamma rays are electromagnetic radiation. According to quantum theory gamma rays may be regarded as individual particles called photons. Neutron radiation is produced primarily during nuclear fission.

All nuclear safety issues ultimately relate to the possibility of exposure to ionizing radiation. Alpha and beta particles are directly ionizing radiation; i.e., they are charged particles that interact with matter by ionizing orbital electrons in a medium. As a consequence, the depth of penetration of alpha and beta particles in dense matter is very small. Uncharged neutrons and gamma rays interact with the atomic nucleus or atomic electrons creating charged particles that subsequently ionize atoms in the substance. Because of this two-step process, neutrons and gamma rays are called indirectly ionizing radiation. These indirectly ionizing radiations are highly penetrating.

The three principal processes for gamma ray interaction with matter are the photoelectric effect, Compton scattering, and pair production. For the photoelectric effect, the energy of an incident gamma photon is completely transferred to an orbital electron ejecting it from an atom. Compton scattering of electrons results when a gamma photon makes an elastic collision with an electron. For pair production, annihilation of a gamma photon results in the creation of an electron-positron pair. For a narrow beam of gamma rays incident upon a medium, the fractional decrease in the gamma ray intensity decreases as an exponential function of the thickness of the medium. The proportionality constant μ_γ is called the linear attenuation coefficient.

Neutron flux and cross sections are used to describe the interactions of neutrons with the nuclei of atoms. Neutron flux ϕ is the product of the neutron density and the neutron speed. The microscopic cross section σ is the effective cross sectional area exhibited by a nuclide for a particular neutron-nucleus interaction. A macroscopic cross section Σ is the product of the microscopic cross section and the atom density of a nuclide. Neutrons may be absorbed or scattered by nuclei of atoms in a medium. When a neutron is absorbed a compound nucleus is formed in an excited (high-energy) state. The compound nucleus releases its excess energy by either emitting a particle, emitting a gamma ray, or by fissioning. Gamma ray emission following neutron absorption (called radiative capture) is exhibited by virtually all elements. Radiative capture can result in activation of materials; i.e., a stable isotope can be transformed into a radionuclide after absorbing a neutron. Neutron absorption in some heavy nuclides can result in fission of the compound nucleus, forming two smaller nuclei called fission products. The fission products are unstable and usually decay by emitting negative beta particles. Two or three neutrons, and other nuclear particles, are released during fission. Neutrons may lose energy by inelastic or elastic scattering interactions with atomic nuclei. During inelastic scattering, a neutron is absorbed by a nucleus; and the compound nucleus emits a neutron at a lower energy, along with a gamma ray. Elastic scattering can be likened to the collision of two billiard balls. For elastic scattering, the total kinetic energy of the neutron and the nucleus is conserved, but neutrons can give up some of their kinetic energy to the target nucleus during impact.

Nuclear energy is released during the decay of radioisotopes and during nuclear fission. The energy released during nuclear processes is enormous compared to chemical processes. The power density in the nuclear fuel from nuclear fission is typically orders of magnitude greater than the power density in radioisotope fuel.

Symbols

A	mass number	M_r	relative molecular mass
A_r	relative atomic mass	N	atom density
$\mathcal{A}$	activity	N_A	Avogadro's number
$\tilde{A}$	specific activity	n	neutron
b	barn	n	neutron density
c	speed of light	$\mathcal{N}$	number of atoms
Ci	curie	q'''	power density
e^-	electron	$\mathbf{r}$	position vector
e^+	positron	R	range of particles
eV	electron volt	$\mathcal{R}$	reaction rate
E	energy	$t_{1/2}$	half-life
$f(E)$	neutron spectrum	t	time
h	Plank's constant	u	atomic mass unit
I_γ	gamma ray intensity	v	neutron speed
k	Boltzmann's constant	Z	atomic number
m	mass		

α	alpha particle	μ_γ	γ-ray attenuation coefficient
α_s	$(A+1)^2/(A-1)^2$	μ_m	γ-ray mass attenuation coef.
β	beta particle (electron)	ν	frequency for γ photon
Δm	mass defect	$\nu_\varepsilon \bar{\nu}_e$	neutrino, antineutrino
ϕ	neutron flux	σ	microscopic cross section
γ	gamma ray	σ_γ	σ for capture
κ	energy yield per fission	Σ	macroscopic cross section
κ^D	energy yield/nuclear decay	ς	density
λ	decay constant		

Special Subscripts/Superscripts

a	neutron absorption	M	material type
f	nuclear fission	0	initial value
i	isotope identifier	*	excited state
j	reaction type		

References

1. Walker, F. W., J. R. Parrington, and F. Feiner, *Chart of the Nuclides: With Physical Constants, Conversion Factors and Table of Equivalents*. 14th rev. Ed., San Jose, CA: General Electric Co., Nuclear Energy Marketing Dept. 1988.
2. El-Wakil, M. M., *Nuclear Power Engineering*. New York: McGraw-Hill, 1962.

3. Glasstone, S. and A. Sesonske, *Nuclear Reactor Engineering*. Princeton, NJ: Van Nostrand, 1967.
4. Angelo, Jr., J. A. and D. Buden, *Space Nuclear Power*, Malabar, FL: Orbit Book Co., 1985.
5. Reactor Physics Constants Center (U.S.), *Reactor Physics Constants*. 2nd ed. ANL-5800, U.S. Atomic Energy Commission, Division of Technical Information, 1963.
6. Magurno, B. A., R. R. Kinsey, and F. M. Scheffel, *Guidebook for the ENDF/B-V Nuclear Data Files*. EPRI-NP-2510; BNL-NCS-31451; ENDF—328. Upton, NY: National Nuclear Data Center, 1982.

Student Exercises

1. A space reactor operates for 10 years at 100 kW_e, with a net efficiency of 17%. The critical mass of the reactor equals 100 kg of uranium-235. (a) Compute the amount of U-235 consumed (kg fissioned) after 10 years. (b) Compute the mass of U-235 converted to energy. (c) Compare the mass consumed and the mass converted to energy with the critical mass.

2. Polonium-210 was once considered as a potential radioisotope power source for brief-duration missions. The half-life of Po-210 is 138.4 d, and the initial power density is 1320 W/cm^3. (a) Compute the specific activity of Po-210 and compare to Pu-238. (b) Assuming a 6 month mission and 200 g of pure Po-210 metal at beginning-of-mission (BOM), compute the activity at BOM and at end-of mission (EOM). (c) Compute the power density at EOM.

3. Estimate the range of 5 MeV alpha particles and the maximum range of 1.25 MeV beta particles in air, aluminum, water, iron, and uranium metal. Use the following values for relative atomic masses and densities.

	Aluminum	Water	Iron	Uranium
$\bar{A}_r$ (u)	26.98	18.016	55.85	238.03
ς (g/m)	2.699	1	7.8	18.9

4. Compute the total reaction rate and the reaction rate for neutron capture in the constituents of a device located near a lunar-based reactor. Assume that the reaction rates are dominated by thermal neutron capture. Also assume that the device is sufficiently thin, such that the flux within the device is approximately equal to the flux incident on the surface, equal to 3.2×10^9 n/cm²·s. Use the thermal cross sections in Table 1.2, and the compositions, relative atomic masses, and densities from the following table.

Constituent	Al	Fe	C	W	Cu	BeO	Void
Volume %	21	28	12	7	5	17	10
$\bar{A}_r$	26.98	55.85	12.011	182.8	663.54	9	0
ζ^* (g/cm³)	2.7	7.86	1.6	19.3	8.94	3.0	0

* Unsmeared density

5. Assume that a flux of 5 MeV gamma rays are incident on a shield consisting of a 1.5 cm thick W slab and a 2.2 cm thick Fe slab. What fraction of the flux penetrates the W slab, and what fraction will penetrate the entire shield. Repeat the calculations assuming a 0.05 MeV gamma ray flux.

Chapter 2

Space Nuclear Systems

Albert C. Marshall

The objective of this chapter is to provide the reader with sufficient background in space nuclear systems to facilitate understanding of safety discussions. This background includes a brief historical review of space nuclear programs and a discussion of the types of nuclear systems used or considered for use in space. Three general types of space nuclear systems are described: (1) radioisotope sources, (2) space power reactor systems, and (3) nuclear thermal propulsion systems. Basic design considerations, principles, and some of the more important design options are also presented.

1.0 Historical Review

In 1955 the U.S. Atomic Energy Commission began what became known as the Systems for Nuclear Auxiliary Power (SNAP) program. The SNAP program developed both radioisotope sources and reactor systems to produce electrical power for spacecraft. On June 29, 1961, the United States launched a Navy Transit 4A navigational satellite, the world's first nuclear powered system in space. A SNAP-3B7 radioisotope source (illustrated in Figure 2.1) provided the 2.7 W of electrical power required by the Transit satellite. Four years later, the United States launched the SNAP-10A, the first space-deployed reactor system, and the only U.S. reactor system launched to date. The 3.5-m-long SNAP-10A system, illustrated in Figure 2.2, produced 500 W of electrical power. The mission objective for SNAP-10A was technology evaluation and demonstration. Between 1961 and 1973 the United States carried out 16 missions with radioisotope powered systems on board. Milestones included three Apollo missions with radioisotope power sources placed on the moon and the Pioneer 10 and 11 missions to explore the outer solar system [1].

The United States has not launched another reactor system since SNAP-10A; however, the SNAP program developed and ground-tested five space power reactors during the 1960s. One of the reactors,

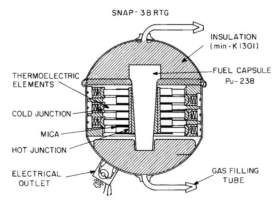

Figure 2.1 SNAP-3B7 RTG. *Courtesy of Mound Laboratory, Monsanto Research Corporation.*

28

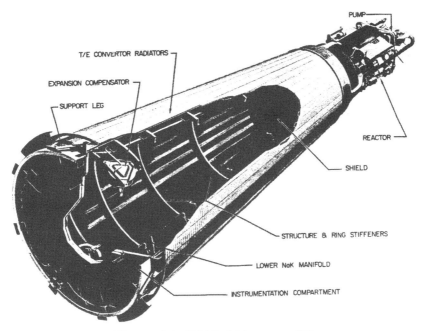

Figure. 2.2 SNAP-10A system [2].

SNAP-8, was designed to produce 30 kW of electrical power. Other space power reactor concepts were explored between the late 1950s and the early 1970s. Alternative concepts included a wide variety of core configurations, fuel types, and power conversion approaches. Detailed system design, extensive fuel testing, and non-nuclear component testing were carried out, but full-core testing of these alternative designs was never initiated. Planned electrical power levels for the alternative designs ranged from a few kW up to 10 MW. The United States also began a program for nuclear rockets in 1955. The U.S. Rover nuclear engine for rocket vehicle applications (NERVA) program produced and ground-tested 20 reactors. One of the reactors in this series produced over 4000 MW of power. Though considered a success, NERVA was never used on a space mission [1].

The entire U.S. space reactor program was shut down in 1973 because of budgetary considerations and a change in government priorities. Only small research programs remained for space reactors until the early 1980s. Missions with radioisotope sources continued after the shutdown of the space reactor program. Eight missions powered by radioisotope sources were carried out between 1975 and 1990. Milestone accomplishments included the use of radioisotope sources aboard the two Viking spacecraft placed on Mars, and aboard the two Voyager spacecraft placed on an interplanetary trajectory to explore the outer solar system. After a decade hiatus, the U.S. Strategic Defense Initiative revitalized the U.S. space reactor program and spawned a number of other programs and studies involving space reactors. The most significant reactor programs included the SP-100 space power reactor program [3], the Space Nuclear Thermal Propulsion program [4], the U.S. Air Force bimodal program [5], and a program to launch a Soviet reactor aboard a U.S. spacecraft [6]. All of these programs ended prematurely, again due to funding constraints and the lack of a high priority mission.

The Soviet space nuclear program began much like the U.S. program but soon diverged. The Soviets launched their first radioisotope power source on September 3, 1965, aboard the Cosmos 84 spacecraft. Unlike the United States, where the focus was on radioisotope power sources, the USSR deployed only a few radioisotope sources in space. On the other hand, the Soviets have orbited 33 space reactors to provide electrical power to spacecraft. Thirty-one 3-kW$_e$ Buk reactor systems were deployed in space to provide electrical power for ocean reconnaissance satellites. An illustration of the Buk reactor system is presented in Figure 2.3. Two 5-kW$_e$ TOPAZ reactors, shown in Figure 2.4, were launched into

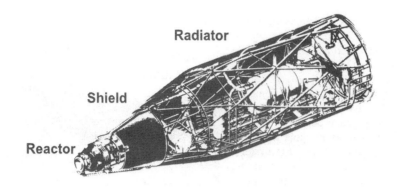

Electric power	3 kW
Thermal power	100 kW
Uranium-235 loading	30 kg
Mass	930 kg

Figure. 2.3 Buk reactor power system.

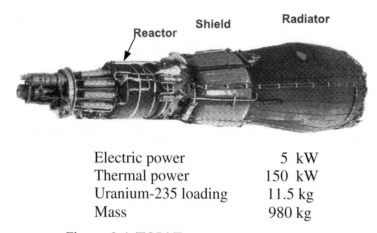

Electric power	5 kW
Thermal power	150 kW
Uranium-235 loading	11.5 kg
Mass	980 kg

Figure. 2.4 TOPAZ reactor power system.

space by the Soviets in 1987 and 1988. The TOPAZ reactors were of a fundamentally different design from the Buk reactors [7]. The USSR, like the United States, carried out extensive development and ground-testing of nuclear thermal propulsion systems without ever using nuclear thermal propulsion in space.

Tables 2.1 and 2.2 present a list of all U.S. and Soviet launches of spacecraft with nuclear power systems on board. At present, the U.S. space nuclear system research programs remain modest efforts. Space nuclear system design and analysis has continued at U.S. national laboratories and universities. In this post-Soviet era, space reactor research is still underway at some of the major Russian scientific institutes. France, Japan, and China are also conducting studies on the use of nuclear power in space. Future space missions using nuclear power may include international cooperation.

2.0 Radioisotope Power Sources

We begin our discussion of space nuclear systems with the relatively simple design of a typical radioisotope power source. The basic components of a radioisotope power source for electrical power production are illustrated schematically in Figure 2.5. The components include the radioisotope fuel,

Table 2.1   Summary of Space Nuclear Power Systems Launched by the United States (1961-1997) [1]

Spacecraft Designation	Mission Type	Launch Date	Power Source (# Sources/ Power)	Status
Transit 4A	Navigation	29 Jun 61	SNAP-3B7 (1/2.7We)	Successfully achieved orbit
Transit 4B	Navigation	15 Nov 61	SNAP-3B8 (1/2.7We)	Successfully achieved orbit
Transit 5BN-1	Navigation	28 Sep 63	SNAP-9A (1/25We)	Successfully achieved orbit
Transit 5BN-2	Navigation	5 Dec 63	SNAP-9A (1/25We)	Successfully achieved orbit
Transit 5BN-3	Navigation	21 Apr 64	SNAP-9A (1/25We)	Failed to achieve orbit; RTG burned up on reentry as designed
SNAPSHOT	Experimental	3 Apr 65	SNAP-10A (1/500We)	Successful orbit; spacecraft voltage regulator malfunction after 43 days resulted in reactor shutdown as designed
Nimbus B-1	Meteorological	18 May 68	SNAP-19B2 (2/40We)	Vehicle destroyed during launch; RTGs retrieved intact; fuel used on later mission
Nimbus III	Meteorological	14 Apr 69	SNAP-19B3 (2/40We ea.)	Successfully achieved orbit
Apollo 12	Lunar Exploration	14 Nov 69	SNAP-27 (1/70We)	Successfully placed on Moon
Apollo 13	Lunar Exploration	11 Apr 70	SNAP-27 (1/70We)	Mission aborted en route to Moon, RTG survived reentry and sank in deep ocean
Apollo 14	Lunar Exploration	31 Jan 71	SNAP-27 (1/70We)	Successfully placed on Moon
Apollo 15	Lunar Exploration	26 Jul 71	SNAP-27 (1/70We)	Successfully placed on Moon
Pioneer 10	Outer Solar System Exploration	2 Mar 72	SNAP-19 (4/40We ea.)	Successfully placed on interplanetary trajectory
Apollo 16	Lunar Exploration	16 Mar 72	SNAP-27 (1/70We)	Successfully placed on Moon
Transit	Navigation	2 Sep 72	TRANSIT-RTG (1/30We)	Successfully achieved orbit
Apollo 17	Lunar Exploration	7 Dec 72	SNAP-27 (1/70We)	Successfully placed on Moon
Pioneer 11	Outer Solar System Exploration	5 Apr 73	SNAP-19 (4/40We ea.)	Successfully placed on interplanetary trajectory
Viking 1	Mars Exploration	20 Aug 75	SNAP-19 (2/40We ea.)	Successfully placed on Mars
Viking 2	Mars Exploration	9 Sep 75	SNAP-19 (2/40We ea.)	Successfully placed on Mars
LES 8	Communications	14 Mar 76	MHW (2/150We ea.)	Successfully achieved orbit
LES 9	Communications	14 Mar 76	MHW (2/150We ea.)	Successfully achieved orbit
Voyager 2	Outer Solar System Exploration	20 Aug 77	MHW (3/150We ea.)	Successfully completed mission
Voyager 1	Outer Solar System Exploration	5 Sep 77	MHW (3/150We ea.)	Successfully completed mission
Galileo	Jovian Exploration	18 Oct 89	GPHS-RTG (2/275We ea.)	Successfully completed mission
Ulysses	Solar Polar Exploration	6 Oct 90	GPHS-RTG (1/275We)	Successfully completing mission 4/11/07
Cassini	Saturn Exploration	Oct 97	GPHS-RTG(3/275We ea.)	Successfully completing mission 4/11/07
New Horizons	Pluto Exploration	Jan 06	GPHS-RTG (1/240We)	Successfully completing mission 4/11/07

RTG—Radioisotope Thermoelectric Generator   LES—Lincoln Experimental Satellite
SNAP—System for Nuclear Auxiliary Power   GPHS-RTG—General Purpose Heat Source RTG
MHW—Multi-Hundred Watt RTG

Table 2.2   Summary of Soviet Space Nuclear Launches (1965–1996) [8]

Spacecraft Designation	Launch Date	Lifetime	System Type
Cosmos 84	3 Sep 65	—	Radioisotope Thermoelectric Generator
Cosmos 90	18 Sep 65	—	Radioisotope Thermoelectric Generator
Cosmos 367	3 Oct 70	< 3 hours	Buk-Reactor
Luna 17	10 Nov 70	10 months	Radioisotope Heater Unit
Cosmos 402	1 Apr 71	< 3 hours	Buk-Reactor
Cosmos 469	25 Dec 71	9 days	Buk-Reactor
Cosmos 516	21 Aug 72	32 days	Buk-Reactor
Luna 21	8 Jan 73	—	Radioisotope Heater Unit
Cosmos 626	27 Dec 73	45 days	Buk-Reactor
Cosmos 651	15 May 74	71 days	Buk-Reactor
Cosmos 654	17 May 74	74 days	Buk-Reactor
Cosmos 723	2 Apr 75	43 days	Buk-Reactor
Cosmos 724	7 Apr 75	65 days	Buk-Reactor
Cosmos 785	12 Dec 75	< 3 hours	Buk-Reactor
Cosmos 860	17 Oct 76	24 days	Buk-Reactor
Cosmos 861	21 Oct 76	60 days	Buk-Reactor
Cosmos 952	16 Sep 77	21 days	Buk-Reactor
Cosmos 954	18 Sep 77	43 days	Buk-Reactor
Cosmos 1176	29 Apr 80	134 days	Buk-Reactor
Cosmos 1249	5 Mar 81	105 days	Buk-Reactor
Cosmos 1266	21 Apr 81	8 days	Buk-Reactor
Cosmos 1299	24 Aug 81	12 days	Buk-Reactor
Cosmos 1365	14 May 82	135 days	Buk-Reactor
Cosmos 1372	1 Jun 82	70 days	Buk-Reactor
Cosmos 1402	30 Aug 82	120 days	Buk-Reactor
Cosmos 1412	2 Oct 82	39 days	Buk-Reactor
Cosmos 1579	29 Jun 84	90 days	Buk-Reactor
Cosmos 1607	31 Oct 84	93 days	Buk-Reactor
Cosmos 1670	1 Aug 85	83 days	Buk-Reactor
Cosmos 1677	23 Aug 85	60 days	Buk-Reactor
Cosmos 1736	21 Mar 86	92 days	Buk-Reactor
Cosmos 1771	20 Aug 86	56 days	Buk-Reactor
Cosmos 1818	1 Feb 87	~6 months	TOPAZ-Reactor
Cosmos 1860	18 Jun 87	40 days	Buk-Reactor
Cosmos 1867	10 Jul 87	~1 year	TOPAZ-Reactor
Cosmos 1900	12 Dec 87	124 days	Buk-Reactor
Cosmos 1932	14 Mar 88	66 days	Buk-Reactor

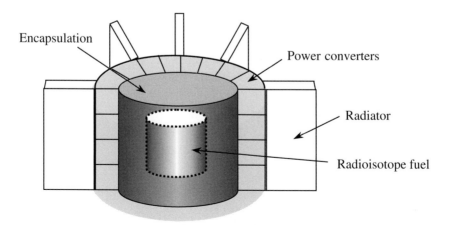

Figure 2.5. Schematic cutaway of a radioisotope power source.

encapsulating materials, a power conversion device, and a heat rejection radiator. Radioisotope sources are used for other purposes, such as heat sources for maintaining components at desired temperatures; for these applications power conversion devices and heat rejection radiators are not required.

2.1 U.S. Radioisotope Sources

By the end of the 20th century, the United States had deployed 45 radioisotope power sources in space; consequently, system developers and safety reviewers accumulated considerable safety experience relating to space-based radioisotope sources. Although radioisotope systems are fairly simple, great care is needed in the design, analysis, and testing of these devices to assure safety under all credible accident conditions. In this section, the topics of fuels, encapsulation, power conversion, system assemblies, and radioisotope heaters are discussed using U.S. radioisotope system designs as examples.

Fuels and Encapsulation

Radioisotopes that emit alpha particles can exhibit high power densities with very little radiation escaping the source. Because of these advantages, radioisotope power sources are commonly designed to use an alpha-emitting fuel. The United States developed plutonium-238, cerium-144, polonium-210, curium-242, strontium-90, and curium-244 radioisotope fuels [9]; however, all U.S. radioisotope systems launched into space have used plutonium-238 as the nuclear heat source. Plutonium-238 was chosen for U.S. missions because of its desirable decay rate and other attributes. Radioactive decay data for plutonium-238 are given in Table 2.3. The half-life of ^{238}Pu, 87.74 years, is long enough to maintain a relatively constant power level for several years, but not so long that the decay rate would be too low to provide an adequate power density. Another advantage of ^{238}Pu is that its radiation emissions are sufficiently low that radiation shielding is generally not required. Penetrating radiations emitted from a ^{238}PuO$_2$ source include relatively low level neutron and gamma emission from spontaneous fission and from α,n reactions with oxygen in the fuel compound.

The fuel for the first U.S. radioisotope source was designed to burn up and disperse in the upper atmosphere in the event of an inadvertent reentry. Plutonium was used in its metallic form to satisfy this requirement. For later designs higher operating temperatures were required, and the reentry safety practice changed to intact reentry. The fuel form was then changed to PuO$_2$ and PuO$_2$-molybdenum cermet.

Table 2.3 Radioactive Decay Data for Plutonium-238 [1]

Half Life: 87.75 years
Decay Constant: $7.899 \times 10^{-3} y^{-1}$
Typical Heat Source Isotope Composition:

Pu Isotope	Weight (%)
238	90.0
239	9.1
240	0.6
241	0.03
242	<0.01

Major Radiations:

Type	Energy (MeV)	Yield (%)
α_1	5.36	0.1
α_2	5.46	28.3
α_3	5.50	71.6

Alpha Activity: 3.9×10^7 α/min·mg
Spontaneous Fission Rate (Metal): 3420 n/s·g
PuO$_2$ Neutron Production (α,n Reaction in Oxygen): 1.4×10^4 n/s·g

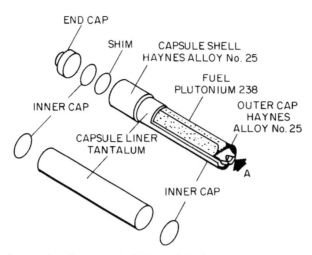

Figure 2.6 SNAP-9A fuel capsule. *Courtesy of Mound Laboratory, Monsanto Research Corporation.*

(Cermets consist of a metal matrix, such as tungsten or molybdenum, containing a dispersion of a ceramic fuel particles.)

Radioisotope fuel shapes are typically cylindrical. The SNAP-9A heat source, for example, consisted of six cylindrical fuel capsules. Each fuel capsule, illustrated in Figure 2.6, is 14.6 cm long, 2.5 cm in diameter, and contains about one-half kg of plutonium-238 metal encapsulated in a tantalum-lined Haynes 25 metal alloy. Some spherical fuel forms have also been used. For SNAP-27A the fuel is in the form of PuO_2 microspheres, 50 to 250 microns in diameter. The fuel microspheres are encapsulated in an annular super alloy structure 6.3 cm in diameter and 41.9 cm long, as shown in Figure 2.7. Spherical fuel was also used for the Multihundred Watt Generator (MHW) used for the Voyager missions. The MHW shown in Figure 2.8 used large fuel spheres, 3.7 cm in diameter. Each sphere consists of a central PuO_2 sphere surrounded by an iridium shell and a graphite impact shell. Twenty-four fuel balls are contained in a cylindrical container 38 cm in diameter and 61 cm long.

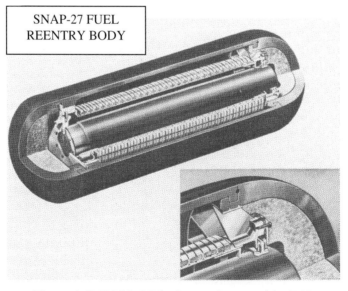

Figure 2.7 SNAP-27 fuel capsule assembly [10].

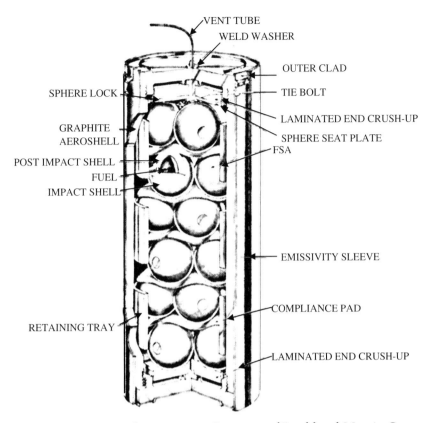

Figure 2.8 MHW heat source. *Courtesy of Lockheed Martin Co.*

The encapsulating materials surrounding the radioisotope fuel provide containment of the radioactive materials. The constituents of the encapsulation include: a liner, a strength member, cladding, and helium accommodation features. Figure 2.9 illustrates the encapsulation used for the Viking/SNAP-19 heat source. The liner is a thin metallic capsule serving as the innermost barrier for the fuel. The liner must be chemically compatible with the fuel and decay products at operating temperatures. The strength member is designed to withstand high velocity impact, fragment impact, explosion overpressure, and fires. Since 1968 U.S. radioisotope assembly designs have used refractory strength members with an oxygen resistant cladding for reentry protection. Helium accommodation is required because helium buildup, due to alpha emission, can create high pressures within the fuel capsule. Helium accommodation can be accomplished by vents that allow helium escape without releasing fuel particles. For contemporary designs, a reentry aeroshell heat shield is also required. The reentry aeroshell is used to assure intact reentry through the atmosphere. The aeroshell is made from an ablative material, such as graphite, that can withstand aerodynamic forces and high thermal stresses that occur during reentry. The shell must permit heat conduction from the fuel during normal operation, yet it must prevent melting of the metallic encapsulation during a postulated reentry accident. Additional insulators can be used with the aeroshell to meet these requirements [9].

Power Conversion

Although a variety of power conversion devices could be used with radioisotope power sources, only thermoelectric power conversion devices have been used on U.S. missions. Thermoelectric semiconductor materials produce electrical power when placed in a temperature gradient. Both n-type and p-type semiconductors are now used in thermoelectric devices to increase the voltage output. As shown in Figure 2.10, a common hot shoe is bonded to one end of the semiconductor materials; and separate cold shoes are bonded to the other end of the semiconductors. Heat is transferred from the nuclear heat

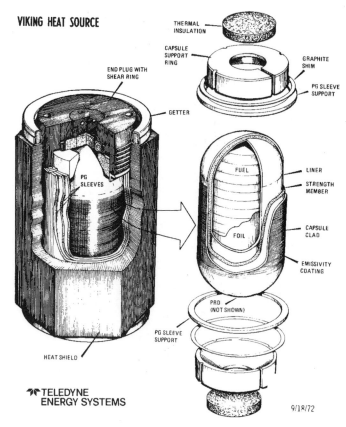

Figure 2.9 SNAP-19 heat source. *Courtesy of Teledyne Energy Systems, A Teledyne Technologies Company.*

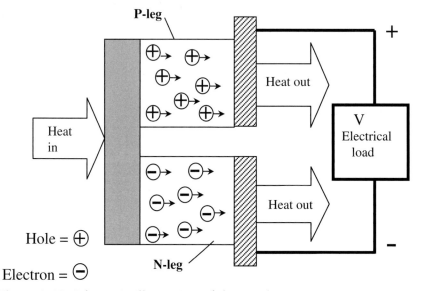

Figure 2.10 Schematic illustration of thermoelectric power conversion.

source to the hot shoe by thermal radiation, and waste heat is removed from the cold shoe by conduction to a heat rejection radiator. Electrons in the n-type semiconductor and holes in the p-type semiconductor migrate from the hot end toward the cold end creating a voltage across the cold shoe plates. Power is extracted by connecting an electrical load between the two cold shoes. A thermoelectric *unicouple* design is shown in Figure 2.11.

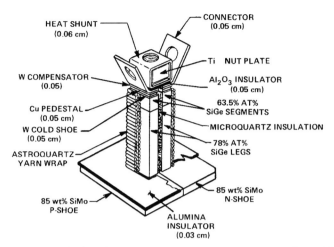

Figure 2.11 SiGe unicouple for the MHW generator. *Courtesy of Lockheed Martin Co.*

Popular thermoelectric materials include: lead telluride and tellurides of antimony, germanium and silver, as well as lead-tin telluride and silicon germanium. Telluride materials are limited to a maximum temperature of 825 K; whereas, silicon germanium thermocouples can operate up to 1300–1400 K. All of these materials have been used with radioisotope sources deployed in space [9]. Radioisotope systems using thermoelectric power conversion are commonly called Radioisotope Thermoelectric Generators, or RTGs.

Dynamic conversion devices first convert heat to mechanical energy which is then converted to electrical energy by a generator. Although dynamic power conversion for radioisotope power sources has not been used in space, the United States developed a Dynamic Radioisotope Power System (DIPS) for space applications. DIPS was 132 cm long, 24 cm in diameter, and weighed about 215 kg. The DIPS system was designed to produce 1 to 2 kilowatts-electrical (kW_e) using a Rankine power conversion device with an efficiency of 18.1% [1].

An entropy-temperature (*ST*) diagram for a Rankine cycle is presented in Figure 2.12. The sub-cooled working fluid enters the nuclear heat source where it gains sensible heat (1-1'). Between point 1' and 2 continued heating produces a phase change. The vapor expands isentropically in the turbine (2-3), and the pressure from the expanding vapor causes a turbo-alternator to spin, generating electri-

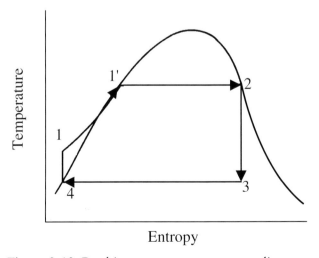

Figure 2.12 Rankine entropy-temperature diagram.

cal power. The vapor-liquid working fluid exiting the turbine rejects heat isothermally in a condenser-radiator (3-4) and the cooled fluid is pumped back to the heat source (4-1). The principal advantage of dynamic systems is that the high conversion efficiencies reduce fuel mass requirements, resulting in a cost savings and reduced activity. However, dynamic systems do not provide system mass advantages at low power levels. Furthermore, operational issues for dynamic systems must be addressed in order to develop flight-qualified dynamic power conversion devices.

Example System Assemblies

Figure 2.13 presents an illustration of an assembly of RTG components for the SNAP-19 radioisotope power system used on the Viking mission. The complete SNAP-19 assembly is 59 cm in diameter, including radiator fins, and 28 cm long. SNAP-19 produced a little over 40 watts of electrical power at

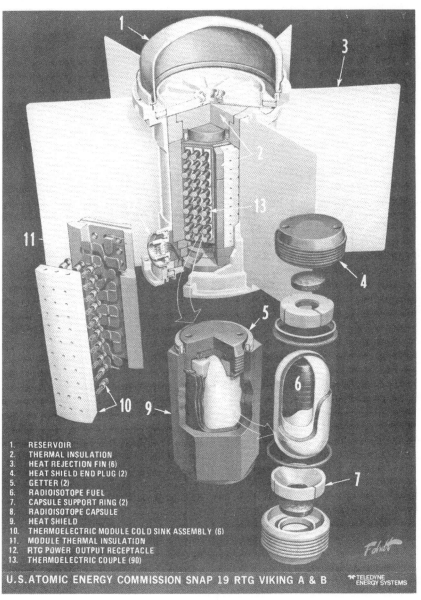

1. RESERVOIR
2. THERMAL INSULATION
3. HEAT REJECTION FIN (6)
4. HEAT SHIELD END PLUG (2)
5. GETTER (2)
6. RADIOISOTOPE FUEL
7. CAPSULE SUPPORT RING (2)
8. RADIOISOTOPE CAPSULE
9. HEAT SHIELD
10. THERMOELECTRIC MODULE COLD SINK ASSEMBLY (6)
11. MODULE THERMAL INSULATION
12. RTG POWER OUTPUT RECEPTACLE
13. THERMOELECTRIC COUPLE (90)

U.S. ATOMIC ENERGY COMMISSION SNAP 19 RTG VIKING A & B

Figure 2.13 Viking SNAP-19 radioisotope power system. *Courtesy of Teledyne Energy Systems, A Teledyne Technologies Company.*

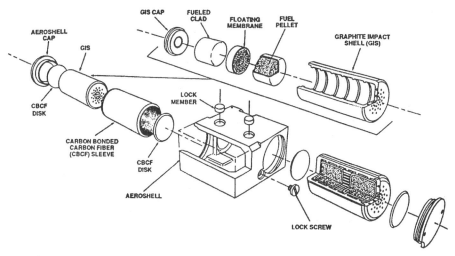

Figure 2.14 General Purpose Heat Source (GPHS) module. *Courtesy of U.S. Department of Energy* [11].

beginning-of-mission; it weighs 15.2 kgs, and contains approximately 20,000 curies of ^{238}Pu in the form of plutonium dioxide-molybdenum cermet disks [1].

Until recently, radioisotope sources were custom designed for missions. Whenever power requirements changed significantly from previous missions, new radioisotope power sources had to be developed to meet the new power needs. The last four U.S. missions with onboard radioisotope power sources used a modular design called the General Purpose Heat Source-Radioisotope Thermoelectric Generator (GPHS-RTG). The basic modular approach allows the power requirements of new missions to be met without the costly development of an entirely new power source. The GPHS-RTG incorporates an assembly of self-contained GPHS modules. Each module produces 250 watts of thermal power. As illustrated in Figure 2.14, each GPHS module includes an aeroshell containing two impact shells. Plutonium-238 dioxide fuel pellets are contained within the impact shells [1].

The GPHS-RTG power system illustrated in Figure 2.15 contains 18 modules stacked together to produce 4500 watts of thermal power. Thermoelectric converters and heat rejection radiator fins sur-

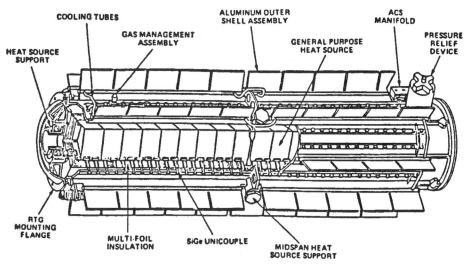

Figure 2.15 General Purpose Heat Source-Radioisotope Thermoelectric Generator (GPHS-RTG). *Courtesy of U.S. Department of Energy.*

Table 2.4 Radioisotope Generator Characteristics [1]

	SNAP–3B	SNAP–9A	SNAP–19	SNAP–27	Transit-RTG	MHW	GPHS-RT
Mission	Transit	Transit	Nimbus Pioneer Viking	Apollo	Transit	LES 8/9	Galileo
Fuel Form	Pu metal	Pu metal	PuO_2-Mo cermet	PuO_2 microspheres	PuO_2-Mo cermet	Pressed PuO_2	Pressed PuO_2
Thermoelectric Material	PbTe	PbTe	PbTe-TAGS	PbSnTe	PbTe	SiGe	SiGe
BOL Output Power (W_e)	2.7	26.8	28-43	63.5	36.8	150	290
System mass[1] (kg)	2.1	12.2	13.6	30.8[2]	13.5	38.5	54.1[4]
Specific Power (W_e/kg)	1.3	2.2	2.1-3.0	3.2[3]	2.6	4.2	5.2
Conversion Efficiency (%)	5.1	5.1	4.5-6.2	5.0	4.2	6.6	6.6
Radiological Inventory (Ci)	1800	17,000	34,400–80,000	44,500	25,500	77,000	130,000[4]

1. BOL=Beginning of Life   2. Without cask   3. Includes 11.1 kg cask   4. Per RTG

round the column of GPHS modules. The system is 114 cm long, 42.2 cm in diameter, and weighs 54.1 kgs. The thermoelectric converter produces 290 watts of electrical power at the beginning of mission, and is designed to assure a minimum of 250 watts after 40,000 hours of mission operation. The thermoelectric efficiency is 6.8% with hot side and cold side temperatures of 1275 K and 575 K, respectively. A comparison of the GPHS-RTG with other U.S. RTG designs is presented in Table 2.4.

Heater Units

Discussions of radioisotope power sources in this text focus on electrical power sources; however, small radioisotope sources are also used to maintain spacecraft components within acceptable temperature ranges. The Cassini spacecraft, for example, contains 157 Light-Weight Radioisotope Heater Units (LWRHU). Each LWRHU produces 1 watt of power and consists of a platinum-rhodium clad $^{238}PuO_2$ fuel pellet within an aeroshell/impact body. The entire unit is a cylinder with a diameter of 2.6 cm and a length of 3.2 cm [11].

2.2 Russian Radioisotope Sources

Although space nuclear programs in Russia (formerly the Soviet Union) were focused on space reactors, the Russians developed and space-deployed several radioisotope power sources. Figure 2.16 illustrates the design of a 2.5 kg radioisotope power source for electrical power generation (20 W_e) launched on-

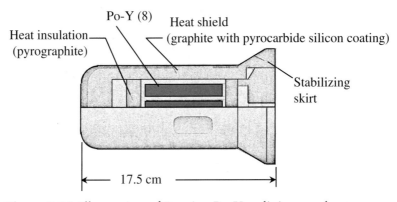

Figure 2.16 Illustration of Russian Po-Y radioisotope heat source.

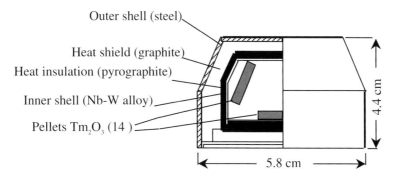

Outer shell (steel)

Heat shield (graphite)

Heat insulation (pyrographite)

Inner shell (Nb-W alloy)

Pellets Tm_2O_3 (14)

4.4 cm

5.8 cm

Figure 2.17 Illustration of Russian Tm_2O_3 gamma radiation source.

board the Cosmos-84 and Cosmos-90 spacecraft. The radioisotope source consists of eight clad [210]Po-Y capsules enclosed within a graphite heat shield coated with silicon carbide. The design assures integrity and hermetic sealing of the fuel in the event of an explosion, rocket fuel fire, reentry, and impact on a concrete surface at 100 m/s. Another Russian [210]Po-Y radioisotope source was used on Mooncar-1 and Mooncar-2, as a 1 kW heat source. The [210]Po-Y capsules were also enclosed within a graphite heat shield. A 20,000 Ci, 300 g thulium-170 gamma source, illustrated in Figure 2.17, was used aboard the Salyut-1 to enable an automatic docking procedure. Electrical power for the Russian Mars-96 spacecraft was provided by plutonium-238 radioisotope power sources. These sources incorporated a liquid metal coolant and thermoelectric energy conversion devices to produce 0.2 W of electrical power. The fuel was in the form of clad PuO_2. As for other Russian radioisotope sources, a graphite heat shield and insulation were used for reentry protection.

3.0 Reactor Power Systems

Space power reactor systems are used to provide electrical power for space for missions requiring kilowatts or megawatts of electrical power. Unlike radioisotope sources that produce power by passive nuclear decay, reactor power production requires dynamic control of the neutron production and loss rate. The subsystems of space power reactor systems are illustrated schematically in Figure 2.18. The major subsystems include the reactor, instrumentation and control, radiation shield, heat transport system, power conversion system, heat rejection system, and power conditioning.

In the following discussion of reactor systems we will refer to a number of reactor system designs developed in United States and Russia. As noted in Section 1, not all of these systems were launched into

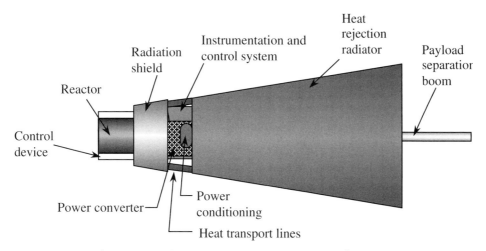

Heat rejection radiator

Instrumentation and control system

Radiation shield

Payload separation boom

Reactor

Control device

Power converter

Power conditioning

Heat transport lines

Figure 2.18 Space power reactor system schematic.

space. In the United States, only the SNAP-10A was space deployed and operated. The Former Soviet Union space deployed and operated a number of Buk and TOPAZ reactors. The Russian Enisey (also called TOPAZ II) and Romashka reactor power systems were extensively ground tested, but never launched.

3.1 Reactor Subsystem

The principal components of a typical reactor subsystem, illustrated schematically in Figure 2.19, include the reactor core, neutron reflectors, reactor control elements, and reactor vessel. The reactor core contains nuclear fuel and coolant passages. Some designs also include a neutron moderator.

In order to explain the function of reactor system components, we begin with a brief introduction to the concept of criticality. The subject of criticality is discussed in greater detail in Chapter 8.

Criticality

The fission process is usually initiated in a terrestrial nuclear reactor using a discrete neutron source (e.g., a PoBe source) to introduce neutrons into the reactor core, whereas space reactor designs often use natural sources, such as spontaneous fission neutrons, rather than an installed neutron source. Neutrons from the source are absorbed by the nuclei of fuel atoms (e.g., ^{235}U), resulting in nuclear fission. Each nuclear fission releases energy and two to three additional neutrons. Some of the neutrons released during fission are absorbed by the nuclei of other fuel atoms causing their nuclei to fission and release more neutrons. Hence, a fission chain reaction can be sustained, resulting in continuous release of neutrons and nuclear energy. A reactor is said to be *critical* when the neutron production rate from fission equals the neutron loss rate. Neutron loss is due to absorptions by nuclei within the reactor, as well as neutron leakage from the surface of the reactor. For a critical reactor, the neutron flux will be sustained at a constant level without the need for an additional source of neutrons. When neutron losses exceed neutron production, the reactor is *subcritical*, and is unable to sustain a neutron flux without an addi-

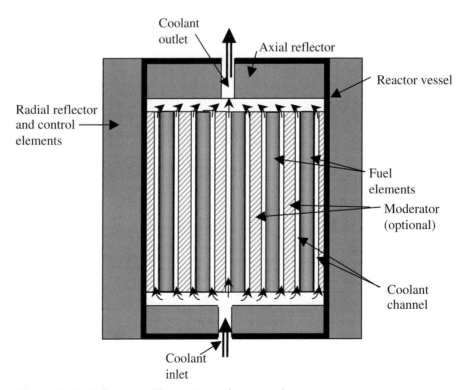

Figure 2.19 Schematic illustration of a typical space power reactor system.

tional neutron source. A reactor is *supercritical* whenever the production rate exceeds the loss rate. For a supercritical reactor, the flux level will increase with time.

For any particular core geometry and composition, the fraction of core neutrons lost by leakage will decrease as core size increases. Thus, for any reactor core composition where the neutron production rate exceeds the loss rate by absorption, a core size exists for which the reactor core is exactly critical; i.e., the neutron loss rate by leakage plus absorption will equal the production rate. The mass of fuel contained within the dimensions of the exactly critical core is called the critical mass. Launch costs, as well as launch vehicle size and mass limitations, provide an incentive to develop designs with low critical masses and volumes.

All power and propulsion reactors are designed with the capability to achieve supercriticality; i.e., they are designed and fabricated to contain more fuel than required for criticality. The excess of fuel is needed to compensate for the burnup of the nuclear fuel, the buildup of fission products that possess parasitic (radiative) capture cross sections, and other effects that can decrease the ratio of neutron production rate to loss rate. Excess fuel is also needed to achieve supercriticality in order to increase the reactor power to significant levels. A subsequent section (Reflectors and Reactor Control) discusses methods used to achieve and maintain criticality, to shut down the reactor, or to increase power level.

Fuel Options

The reactor fuel serves as the heat source for the power system. Although plutonium-239, uranium-233, and other isotopes have been discussed as potential space reactor fuels, the uranium-235 isotope is the standard fuel choice for space reactors. This choice is based on the relatively low curie content (activity) of U^{235} fuels, fuel availability, and other considerations. A number of fuel compounds have been proposed, developed, and used. In the United States, uranium dioxide (UO_2), uranium nitride (UN), uranium carbides (UC and UC_2), and uranium-zirconium hydride (U-ZrH) have been studied and developed for space reactor applications. The former Soviet Union used U-Mo fuel in their Buk space reactors and explored a number of advanced fuel compounds, such as uranium-tantalum-carbide ($U_{0.8}Ta_{0.2}C$), as well as the more common fuel compounds. The choice of fuel composition depends on operating temperature requirements, lifetime needs, chemical compatibility concerns, criticality requirements, development time and cost, and many other considerations.

A variety of fuel geometries have been developed for space reactors. Fuel elements for space reactors are often in the form of cylindrical rods, consisting of ceramic fuel pellets contained inside a cylindrical metal cladding. The SP-100 fuel element illustrated in Figure 2.20, for example, consists of UN fuel pellets within a cylindrical rhenium-lined, niobium-alloy cladding. The fuel elements are 0.78 cm in diameter, separated by a wire wrap spacer, and stacked together, as shown in Figure 2.21, into hexagonal fuel subassemblies of the core assembly. The maximum fuel surface temperature during operation is 1450 K. The U.S. SNAP-10A space reactor and the proposed Medium Power Reactor Experiment (MPRE) space reactor also used cylindrical fuel rods. The SNAP-10A fuel rods consist of a solid U-ZrH fuel rod within a cylindrical Hastelloy cladding tube. The fuel element proposed for the medium power reactor experiment, illustrated in Figure 2.22, is a 1.27 cm diameter stainless steel clad fuel rod containing a 29.5 cm long stack of UO_2 fuel pellets. The maximum fuel surface temperature for the MPRE design was 1133 K [1].

Prismatic shaped fuel elements have been developed for space reactors. These fuel elements typically consist of a fuel compound dispersed in a matrix material and formed into prismatic blocks with internal coolant channels. Cermets are commonly proposed for prismatic fuels. Cermet materials combine the high strength and high thermal conductivity of the metal matrix with the high temperature capability of ceramic fuels. The U.S. 710 program developed prismatic cermet fuel elements for a gas cooled space reactor. The 710 fuel elements (Figure 2.23) consisted of 40 volume percent UO_2 in a tungsten matrix

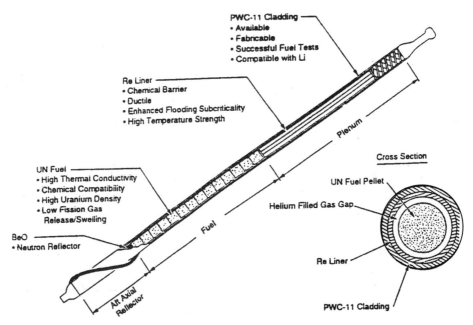

Figure 2.20 SP-100 fuel rod [3].

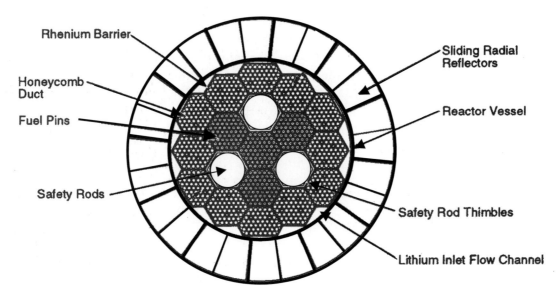

Figure 2.21 SP-100 reactor core and radial reflector. *Courtesy of Lockheed Martin Co.*

formed in the shape of a long hexagonal rod with internal coolant channels. The width across flats of the hexagon was 2.3 cm. A bonded tungsten alloy was used to clad the fuel element surface to enhance fuel and fission product retention. Fuel tests were conducted up to 1920 K [1].

Moderator Choices

The large population of thermal neutrons for a moderated reactor implies that the energy-averaged microscopic fuel cross section will be large (see Figure 1.17) relative to the energy-averaged cross sections for unmoderated reactors. An increase in the average fuel cross section usually reduces the required critical mass because the probability of absorption by the fuel increases relative to the probability of leakage from the core. Reactors that produce fission energy primarily due to thermal neutron absorp-

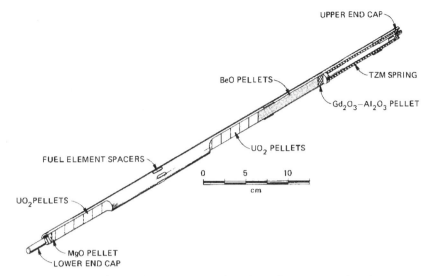

Figure 2.22 MPRE fuel rod. *Courtesy of Oak Ridge National Laboratory* [12].

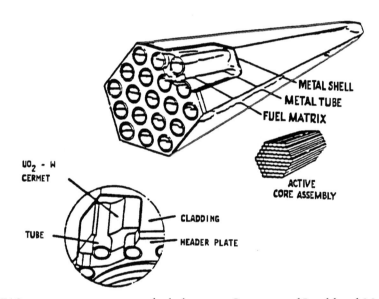

Figure 2.23 710 space reactor cermet fuel element. *Courtesy of Lockheed Martin Co.* [13].

tion in the fuel are called thermal reactors. Reactors that produce fission energy primarily due to neutron absorption in the fast and epithermal energy ranges are called fast and epithermal reactors, respectively. All three reactor types (thermal, epithermal, and fast reactors) can be used for space power and propulsion.

Using a moderator to reduce critical mass does not always reduce system mass. For some mission characteristics and design choices, a moderator can increase system size and mass. The designer must determine the system impact of reactor design choices before fixing the reactor design (this precaution applies to all components and subsystems). When a moderator is clearly beneficial, a number of different moderator materials may be used. Hydrogenous materials are excellent neutron moderators, but typically are limited to moderate temperatures. The zirconium hydride moderator for the SNAP-10A reactor is integral to the fuel (U-ZrH). The fuel operating temperature for SNAP-10A was limited to about 975 K to prevent significant hydrogen loss due to thermal dissociation. Also, the interior surface of the Hastelloy cladding was coated with a ceramic to reduce hydrogen loss. Two Russian space reactors, TOPAZ and Enisey, used a ZrH moderator is in the form of stacked cylindrical blocks. For Enisey, thermionic fuel elements are contained within cylindrical holes in the moderator, as shown in Figure 2.24. The hot UO_2 fuel, with peak temperatures over 2000 K, is thermally isolated from the moderator. The ZrH moderator blocks are coated and contained in a steel can to enhance hydrogen retention. Although the low atomic weight of hydrogen is an attractive feature for a moderator, other characteristics, such as high temperature stability and very small parasitic capture cross sections, can favor other moderator materials. Common non-hydrogenous moderator choices include beryllium, beryllium oxide, and graphite.

Reflectors and Reactor Control

In order to minimize critical mass requirements reactor cores are surrounded by materials called reflectors that reflect a fraction of escaping neutrons back into the core. Reflector materials are chosen that have large scattering cross sections and small capture cross sections. The arrangement of axial and radial neutron reflectors around the core is illustrated schematically in Figure 2.19, and the position of the radial reflectors for the Enisey reactor is shown in Figure 2.24. Beryllium and beryllium oxide are common reflector material choices.

Reactor startup, power increase, maintenance of steady power level, power level changes, and shutdown are accomplished by devices that control the neutronic state of the reactor. Reactor control can be accomplished by any process that influences the ratio of neutron production to neutron loss. For most terrestrial reactors, the ratio is controlled by adjusting the position of *neutron poisons* in the reactor core. A neutron poison is a material with a large parasitic capture cross section, such as boron. Neutron poisons are often in the form of cylindrical rods, called control rods. Control rods can be motor-driven into channels within the core to increase neutron absorptions, or withdrawn to decrease absorptions. When control rods are fully inserted, neutron absorptions are sufficiently high to result in a subcritical condition; and the reactor is said to be *shut down*. Control rods can be partially withdrawn to a position such that the neutron loss rate from absorption and leakage exactly equals neutron production rate; i.e., a critical reactor. The neutron flux level and the core power level can be increased by further withdrawing the control rods so that the neutron production rate exceeds the loss rate.

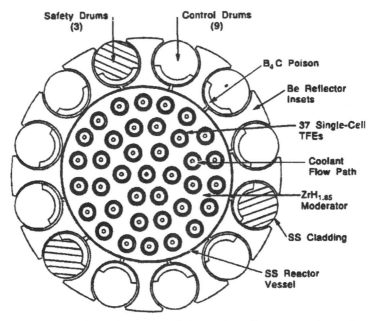

Figure 2.24 Enisey reactor top view. *Courtesy of the U.S. Ballistic Missile Defense Organization* [6].

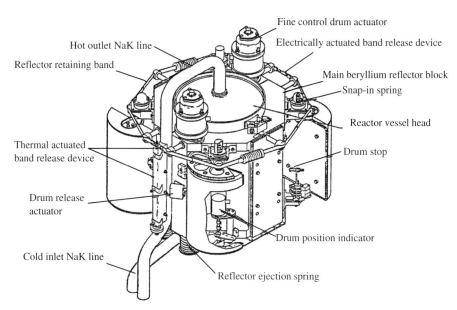

Figure 2.25 SNAP-10A reactor. *Courtesy of U.S. Air Force Weapons Laboratory* [14].

The use of in-core control rods in space reactors increases core size and presents additional design constraints when core temperatures are very high. Space reactors more commonly use reflector movement to control reactor operation. One common approach uses reflectors that incorporate rotatable reflector drums with a strip of neutron poison attached to a segment of the periphery. A typical reflector control drum arrangement is illustrated in Figure 2.24 for the Enisey space reactor. When the drums are rotated so that the poison strip is adjacent to the core, neutrons escaping from the core are absorbed rather than reflected. In this configuration neutron loss rate by core leakage plus absorption is greater than the neutron production rate; i.e., the reactor is subcritical. By rotating the reflector's poison strip away from the core, a position can be found where the neutron loss rate exactly equals the production rate. Thus, reflector movement can be used to achieve all control functions.

Other types of reflector control have been used or proposed for space reactors. The U.S. SNAP-10A space reactor, shown in Figure 2.25, used reflector drum segments that, when rotated, opened neutron leakage paths to space to increase neutron loss by leakage [1]. Regulating neutron leakage by sliding reflectors and hinged reflectors has also been proposed for reactor control for the U.S. SP-100 reactor concept [3] and other reactors. The fraction of neutrons from the core reaching the reflector depends on the size and composition of the core. For some large space reactors reflector control will be ineffective and in-core control rods will be required.

3.2 Radiation Protection and Shielding

Methods must be provided to protect equipment and astronauts from the effects of ionizing radiation emitted from the reactor core. This function is carried out by the use of shielding materials and other strategies. Sources of radiation, radiation shielding, and protection strategies are summarized here. A more detailed discussion of these topics is provided in Chapter 4.

Reactor Radiation and Protection Strategies

For a reactor that has never been operated, radioactivity is limited to the low level natural radioactivity from the decay of uranium fuel materials. For a space reactor, the activity level of unirradiated fuel is typically in the range of 1 to 50 curies. During reactor operation, however, the fission process produces intense neutron and gamma radiation fields. Radiation due to fission is terminated after reactor shutdown, but the reactor contains radioactive fission products produced during operation. Radiative capture of neutrons by core materials, such as structural materials, can also create radioisotopes called *activation products*. Once a reactor has been operated at high power for an extended period of time, the reactor will contain a large inventory of fission products and activation products. Fission product activity decays fairly rapidly, but decay down to the activity level of the unirradiated fuel may take several hundred years for some missions. In order to minimize radiological risk during launch, common practice is to prohibit high power operation prior to space deployment.

Although radiation shielding is typically designed as an integral component of the reactor system, reactors deployed on the surface of a planet or moon may use an existing crater for shielding. Another possible shielding approach is to fabricate a shield from the native regolith (surface rock), or to excavate a subsurface reactor chamber. Large separation distances from radiation sensitive equipment and human habitats have also been suggested for radiation protection from surface deployed reactors.

Shield Design

Typical radiation shields utilize a combination of high-density materials and low-relative atomic mass (low-A_r) materials. Relatively thin layers of high-density materials, such as tungsten, are capable of significant attenuation of gamma radiation. These high-density shields also serve to slow down high-energy neutrons by inelastic scattering. The function of low-A_r shield materials is to reduce neutron energy to the thermal energy range where they can be readily absorbed by neutron poisons. Low-A_r materials, such as lithium hydride, are especially efficient moderators. Even when an optimum material selection is made, radiation shields are relatively thick, heavy volumes of materials. In order to minimize shield mass, most designs shield only the portion of the reactor facing the payload or crew. This type of shield is referred to as a *shadow shield*. Another method used to reduce shield mass is to separate the payload from the reactor by a long boom to take advantage of the spatial attenuation of reactor radiation. In the vacuum of space, radiation intensity decreases approximately as the inverse square of the distance from the reactor. The mass of the boom and power transmission lines can be significant. System designs typically incorporate a combination of radiation shielding and a separation boom with dimensions optimized to minimize the total system mass.

3.3 Heat Transport Systems

A combination of several heat transport methods is commonly employed in an operating reactor. Heat produced by nuclear fission within the fuel is transported to the surface of the fuel element by thermal conduction. For most reactor concepts, heat is then transferred from the fuel element surface to a flowing coolant. The interstitial spaces between cylindrical fuel rods are often used as core coolant passages for rod-geometry cores. For prismatic fuel elements, coolant passages are typically provided as channels within the fuel element volume (see Figure 2.23). After exiting the reactor core, the coolant flows through coolant pipes to a power conversion device to convert heat to electricity. Waste heat is rejected to space and the cooled fluid is pumped back into the core to continue the cycle. For space reactors, high temperature operation is usually employed to achieve an acceptable energy conversion efficiency and to reduce radiator size and mass (discussed in Sections 3.4 and 3.5). The emphasis on high temperatures has favored the use of liquid metal coolants for space reactors. Liquid metals have good heat transfer characteristics and relatively low vapor pressures at high temperatures. High pressures would require heavy walls for pressure vessels, coolant lines, and other components. Choices for liquid metal coolants include potassium, sodium, sodium potassium eutectic (NaK), and lithium. NaK was used for the SNAP-10A reactor coolant, and lithium was the coolant choice for the SP-100 reactor. Noble gases, such as helium and xenon, have also been considered as reactor coolants for space power reactors. Noble gases are chemically inert and operating pressures are manageable for systems employing noble gases.

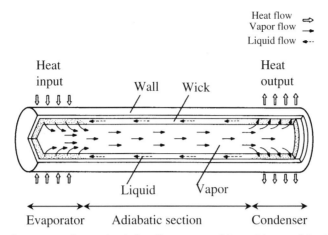

Figure 2.26 Heat pipe operating principle. *Courtesy of Los Alamos National Laboratories.*

Other heat transport approaches have been developed. For example, the proposed NEBA-1 reactor concept was designed to utilize heat pipes for convective heat transport [5]. As shown in Figure 2.26, a heat pipe transports heat by evaporating a liquid from the heated zone and condensing the vapor in a heat rejection zone. Capillary action is used to re-circulate the liquid back to the heated zone. Natural heat transport methods were used for the Russian Romashka reactor shown in Figure 2.27. For Romashka, heat is conducted through the solid reactor core, radiated across a gap to the inner surface of the radial reflector zone, and conducted through the reflector to thermoelectric devices attached to the outer surface of the reflector. For reactors employing thermionic energy conversion devices, cooling by thermionic emission of electrons plays an important role in heat transport. Thermionic energy conversion is discussed in the following section.

Designers must also account for continued reactor heating following reactor shutdown due to the energy released during decay of fission products and activation products. Post-shutdown heating by fission products is often referred to as *decay heating*. For some designs, active features are required to cool the fuel following shutdown until the decay heating has dropped to a level such that passive cooling processes are adequate to prevent overheating of the fuel.

Figure 2.27 Romashka reactor and fuel.

3.4 Power Conversion Options

Reactor heat can be converted to electrical energy using a variety of static and dynamic power conversion approaches. Dynamic approaches require the conversion of heat to mechanical energy and subsequent conversion of mechanical energy into electrical energy (e.g., a turbo-alternator). Static power conversion does not require the intermediate step; heat is converted directly to electrical energy without the need for a moving mechanical device. All devices used to convert heat to electrical energy, whether dynamic or static, are heat engines. The efficiency of any heat engine cannot exceed the efficiency of an ideal heat engine, η_c. The ideal efficiency is referred to as the Carnot efficiency and is given by

$$\eta_c = 1 - \frac{T_C}{T_H}. \tag{2.1}$$

Here T_H is the hot (heat supply) temperature and T_C is the cold (heat rejection) temperature. The efficiency of an actual power conversion device will be less than that of an ideal heat engine. The limiting efficiency described by Eq. 2.1 illustrates a fundamental characteristic of all heat engines; increasing the heat source temperature can increase energy conversion efficiency. Higher efficiencies reduce reactor power and heat rejection requirements, potentially resulting in a significant reduction in reactor power system mass.

Although the same type of thermoelectric power conversion devices used for radioisotope sources (unicouples) have also been used for space reactor systems, the power densities and efficiencies for these devices are quite low. In order to improve power densities, designers have explored advanced thermoelectrics, and a variety of alternative power conversion approaches have been explored to improve power conversion efficiency.

Static Power Conversion

The two most developed static energy conversion approaches for space nuclear systems are thermoelectric and thermionic conversion. Thermoelectric power conversion was used for the U.S. SNAP-10A reactor and the Soviet Buk reactor systems [7]. In order to improve power densities, the SP-100 space reactor program began development of advanced thermoelectric devices. Instead of the single-unit-radiatively-coupled thermoelectric devices (unicouple), used for radioisotope sources, the advanced devices consisted of conductively-coupled-multiple-unit thermoelectric devices (multicouple). Multicouples, shown in Figure 2.28, were designed to produce nearly 30 times the power of the unicouple thermoelectric devices used in the Galileo RTGs. The multicouple assembly included compliant pads to conductively couple the device to the heat transport systems. The compliant pads accommodated the thermal expansion in the power conversion assembly which otherwise would have damaged the brittle thermoelectric materials.

Thermionic energy conversion has also been developed for space reactor power systems. Thermionic energy conversion produces electricity from heat by evaporating electrons off a hot emitter surface and collecting them on a cold collector surface, as illustrated in Figure 2.29. If a vacuum is used in the interelectrode gap between the emitter and the collector, an electrical charge (space charge) in the gap results from electrons in transit between the emitter to the collector. The space charge repels low-energy electrons leaving the emitter back to the emitter surface, reducing the net current relative to the emitted current. The space charge effect is usually neutralized by using a positively charged cesium vapor in the interelectrode gap.

The basic equation for thermionic emission is

$$J = AT^2 \exp(-\phi/kT), \tag{2.2}$$

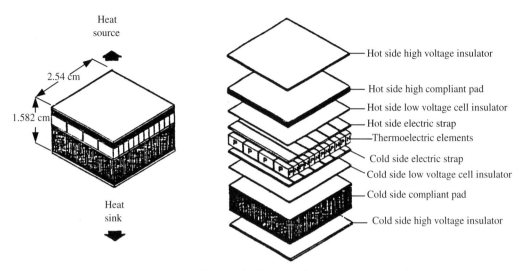

Figure 2.28 SP-100 multicouple thermoelectric converter design [3].

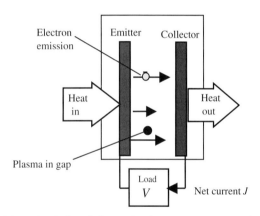

Figure 2.29 Principle of thermionic energy conversion device.

where J is the current density in amps/cm^2 (A/cm^2), T is the emitter temperature (K), A is the Richardson constant = 120.4 A/cm^{2}K^2, ϕ is the work function of the emitter, and k is Boltzmann's constant = 8.62×10^{-5} eV/K. The work function is a measure of the energy required for an electron to escape the surface of a material. As for any heat engine, efficiency of a thermionic device can be improved by increasing the heat supply temperature. Equation 2.2 also suggests that current density and efficiency can be enhanced by choosing electrode materials with a low work function, such as cesium-coated tungsten.

In-core thermionic reactor designs commonly employ thermionic fuel elements (TFEs). When TFEs are used, power conversion occurs in-core within the fuel element rather than transporting heat to a thermionic power conversion device external to the core. A TFE design is illustrated in Figure 2.30. Cylindrical fuel pellets are contained within a emitter tube surrounded by a cesium vapor-filled interelectrode gap. The outer wall of the gap is formed by a cylindrical collector tube coated on the outer surface with an electrical insulator, which in turn is surrounded by a metal sheath. Heat produced by nuclear fission in the fuel during reactor operation is conducted to the surrounding emitter tube. The emitter transmits energy to the collector in the form of electron kinetic energy and radiant heat. Waste heat is conducted from the collector through the electrical insulator to the metal sheath. The sheath is cooled by a flowing liquid metal coolant, and the heated coolant is pumped to a heat rejection radia-

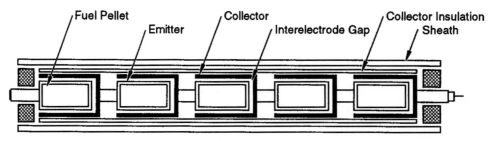

Figure 2.30 Schematic illustration of a TOPAZ thermionic fuel element (TFE).

tor. The two Soviet TOPAZ reactor systems deployed in space used in-core thermionic power conversion. Reactor concepts have also been proposed that transport reactor heat to ex-core (external to the core) thermionic diodes.

In addition to thermoelectric and thermionic power conversion, several other static power conversion approaches have been proposed for space nuclear power systems. Examples include Alkali Metal Thermoelectric Converters (AMTEC) [15], Hydrogen Thermal-To-Electric Converters (HYTEC) [16], thermo-photovoltaics [17], and Magneto-Hydro-Dynamics (MHD) [18]. A discussion of these options is provided in the referenced documents.

Dynamic Power Conversion

Dynamic power conversion devices include Rankine, Brayton, and Stirling engines. For a direct Rankine cycle (discussed in Section 2.1 for a radioisotope heat source), the coolant is vaporized directly as it passes through the nuclear heat source. For space reactors, in-core two-phase flow in a micro-gravity environment presents potential issues; consequently, a secondary-side Rankine cycle is often proposed. For this approach the coolant on the primary (reactor) side remains in the liquid state, and heat is transferred to a working fluid on the secondary side by means of a heat exchanger. Vapor is formed in the secondary side of the heat exchanger, and the balance of the cycle is identical to the Rankine cycle described in Section 2.1. Generally, liquid metal coolants and working fluids have been proposed for Rankine cycle power conversion for space reactors. The SNAP-2 and SNAP-8 space reactor programs carried out ground demonstration tests of a mercury Rankine cycle.

Stirling heat engines have also been proposed for space nuclear power conversion [19]. Stirling engines can operate at the highest efficiencies of all existing heat engines because an ideal Stirling engine most closely approaches a Carnot cycle. The Stirling engine is typically designed as a reciprocating free piston engine with a regenerator. The regenerator allows heat to be stored while reducing temperature and then recovered when the working fluid passes through the regenerator (see the schematic illustration in Figure 2.31). Stirling engines are often proposed for use with linear alternators to produce electrical power. These highly efficient engines are mass competitive, relative to other options, in the low to medium power ranges (10-100s kWe).

Brayton cycles use a single phase gas working fluid with a gas-turbine coupled to an alternator. An illustration of a Brayton turbo-alternator is presented in Figure 2.32. The gas working fluid, typically a mixture of helium and xenon, is heated as it flows through a heat source. The gas may be heated directly in-core by a gas cooled reactor, or a liquid metal cooled reactor may be used with a heat exchanger to transfer heat from the liquid metal coolant to a gas working fluid. The heated gas working fluid expands in the turbine causing the turbo-alternator to spin and generate electrical power. The gas exiting the turbine is cooled by a heat rejection radiator, and a compressor is used to recycle the gas back to the reactor or heat exchanger. Some Brayton system designs incorporate a regenerator; i.e., the gas exiting the turbine is used to preheat the inlet-gas before it enters the reactor. Heat addition and heat

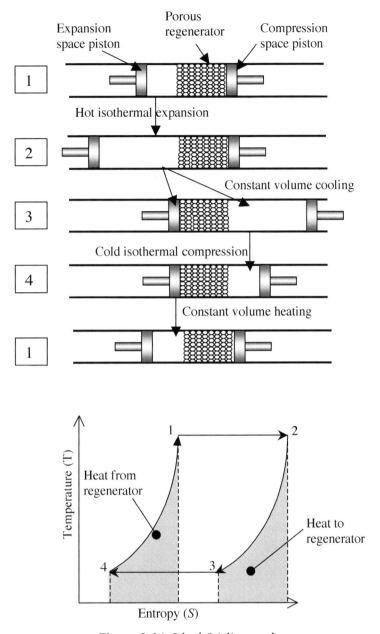

Figure 2.31 Ideal Stirling cycle.

rejection are at constant pressure; and, for an ideal system, expansion and compression are isentropic. Brayton cycles were studied for the 710 gas cooled reactor program [13].

3.5 Heat Rejection System

Because all thermal-to-electric power conversion devices are heat engines, all space power reactor systems must reject heat to produce electrical power. For space reactor systems operating on the surface of a planet supporting an atmosphere (e.g., Mars), convective cooling may be possible. If power is required for only a few minutes (burst mode applications), the release of the coolant into space may be an option for heat rejection. However, for reactor systems operating for an extended period of time in the vacuum of space, the only possible method of heat rejection is by thermal radiation to space. The

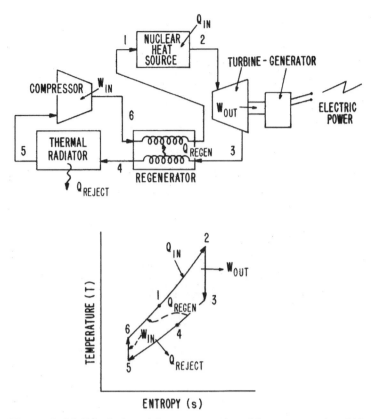

Figure 2.32 Ideal closed Brayton cycle with regeneration [1].

size and the mass of heat rejection radiators can be significant and measures must be taken to minimize radiator mass and volume. The heat rejection from a radiator to space is given by the expression,

$$q_R = \sigma \varepsilon A_s (T_R^4 - T_S^4). \tag{2.3}$$

Here, q_R is the waste thermal power that must be rejected, σ is the Stefan-Boltzmann constant (5.669×10^{-8} W/m^{2}K^4), ε is the emissivity of the radiator surface, A_s is the radiating surface area, T_R is the heat rejection temperature (K), and T_S is the effective temperature of space (K).

Returning to Eq. 2.1 we observe that another way to improve efficiency is to reduce the heat rejection temperature. However, if the heat rejection temperature is decreased, Eq. 2.3 indicates that the surface area must increase in order to reject heat at the same rate. The inverse fourth power relationship of temperature to surface area implies that small reductions in heat rejection temperature will require a substantial increase in radiator surface area to maintain the same heat rejection rate. Hence, reducing heat rejection temperature to increase power conversion efficiency can significantly increase radiator surface area, possibly resulting in a net increase in system mass. The optimal choice for heat rejection temperature will depend on the trade-off between the effect on efficiency and radiator heat flux, as well as other factors.

Radiator design is also constrained by radiation shielding requirements because nuclear radiation can scatter off the radiator structure. In order to protect the payload from excessive scattered nuclear radiation, the radiator is normally configured to fit within the cone of space protected by the shadow shield. As a consequence of this consideration, space reactor radiators commonly have the shape of a

Figure 2.33 Enisey reactor system.

truncated cone. The Enisey reactor system radiator shown in Figure 2.33 is an example of conical radiator. For very high power systems, exotic radiator designs have been proposed. Some of these exotic designs are described by Juhasz and Peterson [20].

3.6 Systems Options

The previous sections described a number of the design options for space reactor components and subsystems. Possible space power reactor systems, using combinations of components and subsystems, are presented in Table 2.5 for four basic reactor types. The basic types include liquid cooled reactors, heat pipe cooled reactors, gas cooled reactors, and naturally cooled reactors. The principal system options are given for the reactor core, the cooling subsystem material, and power conversion devices. Although most space power reactor development has focused on electrical power systems, very similar designs have been suggested to provide process heat for space applications. For process heat reactor systems, the reactor heat is used directly to drive some process, such as the extraction of oxygen from regolith. Process heat is included in the table as a system option. Cooling subsystem materials in Table 2.5 refer to coolant lines, reactor vessel, and other components that contain heat transport fluids. Steels and nickel-based super alloys are often chosen for system materials. Refractory metals, such as niobium or tungsten, are favored when very high temperature fluids are used. Design parameters for the SNAP-10A, Buk, TOPAZ, and SP-100 are summarized in Table 2.6. The SP-100 reactor-shield system is illustrated in Figure 2.34.

Table 2.5   Principal Design Options for Space Power Reactor Systems

ReactorType	Pumped Liquid Cooled	Heat Pipe Cooled	Gas Cooled	Combustion/ Radiation Cooled
Reactor Options				
Geometry: Pin	X	X	X	
Prismatic or Solid	X	X	X	X
Other (pebble bed, etc.)	X		X	
Fuel: UO_2	X	X	X	
UN	X	X	X	X
UC or UC_2	X	X	X	X
U-ZrH	X	X		
Other	X	X	X	X
Clad or Matrix:				
Stainless Steel	X	X		
Super Alloy	X	X	X	
Refractory Metal	X	X	X	X
Graphite or Ceramic		X	X	X
Moderator: None	X	X	X	X
ZrH				
Be, BeO	X	X	X	
Other	X	X	X	X
Cooling System Material				
Stainless Steel	X	X		
Super Alloy	X	X	X	
Refractory Metal	X	X	X	
Power Conversion				
Thermoelectric	X	X		X
In-core Thermionic	X	X		
Ex-core Thermionic	X	X		X
AMTEC	X			
Stirling	X	X		
Rankine	X*	X		
Brayton	X		X	
Other	X	X	X	X
None (Process heat)	X	X	X	X

* With heat exchanger   X = Possible applications

4.0 Nuclear Propulsion Systems

Nuclear reactors have been developed for propulsion in space. Two basic types of nuclear propulsion systems have been studied: Nuclear Electric Propulsion (NEP) and Nuclear Thermal Propulsion (NTP) systems. Both types of systems can provide performance advantages over conventional rockets. For example, nuclear rockets have been proposed to reduce trip times for piloted missions to Mars. The crew and the nuclear propulsion system are first placed into Earth orbit by conventional launch vehicles. The high velocity nuclear propulsion system is then used to transport the crew to Mars. Shorter trip times, made possible by nuclear propulsion, reduce crew exposure to cosmic radiation and reduce life support system requirements.

Table 2.6 Space Power Reactor System Parameters for
the SNAP-10A, Buk, TOPAZ and SP-100 Systems [1, 3, 7, 8]

SYSTEM	SNAP-10A	Buk	TOPAZ	SP-100
Power Level (kW$_e$)	0.5	3	5	100
Operational Life (y)	1	~ 0.5	3	7
Power Conversion	Thermoelectric (unicouple)	Thermoelectric	Thermionic (multicell)	Thermoelectric (multicouple)
System Efficiency (%)	1.6	3	3.3	4
Reactor Power (kW$_{th}$)	30	150	150	2500
First Launch	1965	1967	1987	project terminated
Fuel Type	U-ZrH	U-Mo alloy	UO$_2$	UN
Fuel Geometry	Pin	Pin	TFE (pin)	Pin
Cladding or Emitter	Hastelloy N		Mo/W Emitter	Re-lined Nb Alloy
Clad Diameter (cm)	3.17	~ 2	1.0	0.775
Clad Temperature (K)	833		1733	1450
Moderator	ZrH (Fuel Matrix)	None	ZrH	None
Coolant	NaK	NaK	NaK	Li
Max Coolant Temp. (K)	810	973	873	1375
Core Length (m)	0.41	~ 0.2	0.30	0.393
Core Diameter (m)	0.226	~ 0.6	0.26	0.325
System Length (m)	3.5		3.9	12 (stowed=6 m)
System Mass (kg)	436	930	980	4600
Reactor Mass (kg)	125		290	650
Shield Mass (kg)	98		390	890
System (W/kg)	1.1	3.2	5.1	21.7

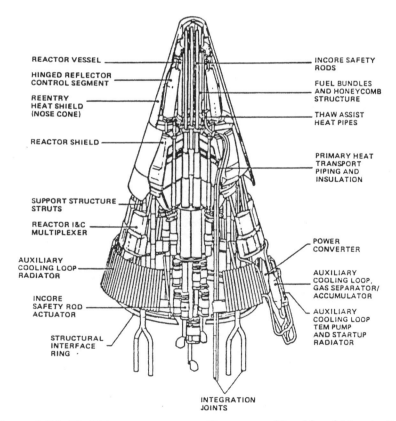

Figure 2.34 SP-100 reactor system. *Courtesy of Lockheed Martin Co.*

4.1 Basic Principles

Specific impulse I_{sp} is an important performance parameter for rockets; it is defined as

$$I_{sp} = \frac{F}{\dot{m}}. \tag{2.4}$$

F is the thrust (in Newtons), $\dot{m}$ is the propellant mass flow rate (in kg/s), and I_{sp} has the units of m/s. When English units are used and I_{sp} is measured at sea level, the more traditional engineering units of *seconds* (s) may be used (pounds force divided by pounds mass per second). A higher specific impulse implies that higher velocities and lower propellant mass can be obtained. The propellant mass can be the dominant contributor to propulsion system mass.

Nuclear Electric Propulsion

For NEP the reactor is used with a power conversion device to produce electricity. Electrical power is then used to create electric fields or arcs to accelerate charged particle or ionized gas propellants to very high velocities. Reactor power systems used to provide electric power for NEP systems have been described in Section 3; consequently, only nuclear thermal propulsion systems are discussed in this section.

Nuclear Thermal Propulsion

A schematic illustration of a nuclear thermal propulsion system is presented in Figure 2.35. Nuclear thermal propulsion rockets provide thrust by heating a propellant to high temperatures in the reactor core

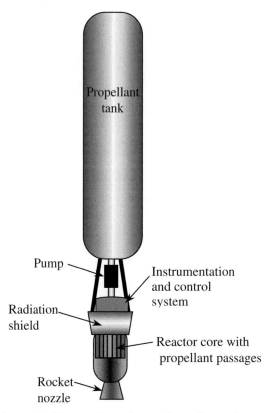

Figure 2.35 Schematic illustration of a nuclear thermal propulsion system.

and expelling the heated propellant through a nozzle. The specific impulse for conventional and NTP rockets can also be expressed as

$$I_{sp} = A_p C_F \sqrt{T_{ch}/M_r}$$

(2.5)

where, A_P is a propellant performance factor, C_F is a coefficient that is a function of nozzle design, T_{ch} is the chamber temperature (K), and M_r is the relative molecular mass of the propellant (u) [1]. Because nuclear reactors are capable of very high temperatures and use hydrogen as a propellant (lowest M_r), high specific impulses can be achieved. The hydrogen propellant for NTP systems is contained in liquid form in a tank maintained at cryogenic temperatures.

4.2 Propulsion Systems

A number of thermal propulsion reactor concepts have been explored. The U.S. Rover/NERVA program developed and ground-tested 20 nuclear thermal propulsion reactors. The XE′ nuclear engine developed in the Rover/NERVA program is shown in Figure 2.36. The NERVA reactor consisted of prismatic fuel elements approximately 130 cm long. The cross section of the fuel elements was hexagonal, approximately 1.91 cm across flats, with a number of axial coolant channels through each fuel element (Figure 2.37). During ground-testing, liquid hydrogen was pumped into the core inlet. The hydrogen propellant entered the coolant channels where it was heated to high temperatures. The high temperature hydrogen was then exhausted from a nozzle. Although a variety of fuel forms were explored, all of the NERVA reactor tests used graphite-coated UC micro-particles (50–150 m in diameter) imbed-

Figure 2.36 XE′ engine in its test stand. *Courtesy of Los Alamos National Laboratory* [21].

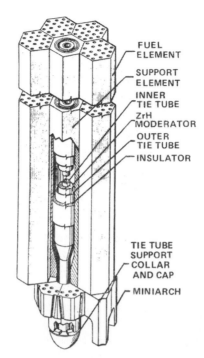

Figure 2.37 Fuel module for the small engine NERVA reactor. *Courtesy of Los Alamos National Laboratory.*

ded in a graphite matrix. The graphite fuel elements were coated with NbC to protect against corrosion from the hydrogen propellant. Fuel elements consisting of UC-ZrC dispersed in a graphite matrix and all-carbide fuels (UC-ZrC) were also developed to permit higher operating temperatures without extensive corrosion.

During NTP operation, the H_2 supply lines are heated by conduction from the hot reactor; consequently, the H_2 propellant is heated before it reaches the reactor core and enters the reactor in the gas phase. The low density of the hydrogen gas propellant places an emphasis on cores with large heat transfer areas in order to achieve the high propellant temperatures desired for nuclear propulsion reactors. A broad range of unusual fuel and core geometries have been proposed to maximize the fuel heat transfer surface area. For example, during the 1980s the United States explored a particle bed propulsion reactor concept [4]. This concept, shown in Figure 2.38, utilized ZrC-coated UC_2 micro-fuel particles, approximately 250 μm in diameter. The heat transfer surface area provided by micro-fuel particles is enormous. In order to utilize the micro-particles, fuel elements are formed by two concentric porous cylinders called *frits* with the fuel particles contained in the annular space between the frits. These fuel elements are inserted in cylindrical holes within a ZrH moderator block. During operation the hydrogen propellant-coolant first flows through coolant channels in the ZrH moderator (to cool the moderator) and then through an annular space between the fuel elements and the moderator. The propellant enters the fuel elements through the porous outer frit, is heated to high temperature by the fuel particles, and exits through the inner frit into a central orifice. The high temperature propellant is exhausted through a nozzle to provide thrust.

The Soviet Union developed a number of NTP fuels. Fuel elements were tested in loops within experimental reactors. Operational testing of an NTP fuel element in the Soviet IGV-1 test reactor is shown in Figure 2.39. For this test, a hydrogen temperature of about 3000 K was achieved. The Soviets developed *twisted strip* fuel elements shaped like a 2 mm diameter drill bit to optimize heat transfer to the hydrogen propellant [7].

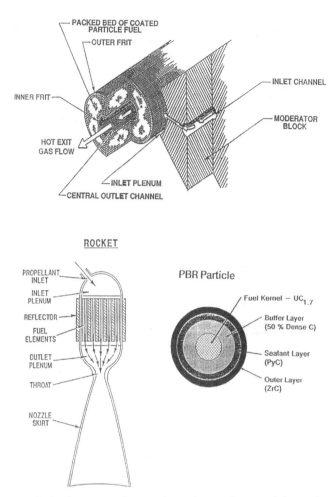

Figure 2.38 Illustration of particle bed reactor for nuclear thermal propulsion. *Courtesy of the U.S. National Aeronautics and Space Administration* [22].

Gascore propulsion reactors were studied in the United States and extensively explored in the Soviet Union. For this concept a gaseous fuel is contained by a confining buffer-gas-propellant, as illustrated in Figure 2.40. The advantage of this approach is that very high temperatures can be achieved, resulting in high specific impulse and thrust. Uranium hexafloride is typically proposed for temperatures in the range of a few thousand degrees Kelvin. For temperatures above about 5000 K, vaporized uranium metal is proposed as the nuclear fuel. Although the very high temperature capability of a gas core reactor is attractive for space nuclear thermal propulsion, the technical challenges associated with confinement of the extremely hot nuclear fuel have relegated the gas core reactor to the status of a speculative concept.

5.0 Summary

By the year 2006, the United States launched a total of 45 radioisotope power sources and one nuclear reactor system on 26 different missions. The Soviet Union space-deployed 31 3-kW$_e$ Buk reactors and two TOPAZ reactor systems, as well as several radioisotope sources. The United States and the Soviet Union developed a number of other reactor systems that were never launched.

For radioisotope power sources, nuclear energy is released by the decay of unstable nuclei. The basic components of radioisotope power sources include the radioisotope fuel, encapsulating materials, a power conversion device, and a heat rejection radiator. Alpha particle-emitting radioisotopes are

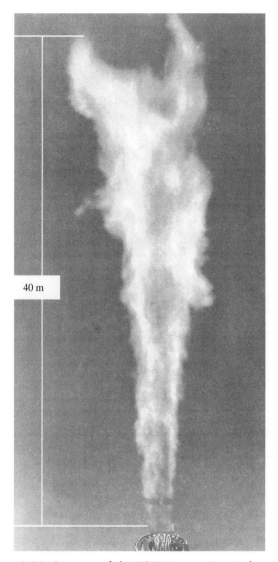

40 m

Figure 2.39 Startup of the IGV-1 experimental reactor.

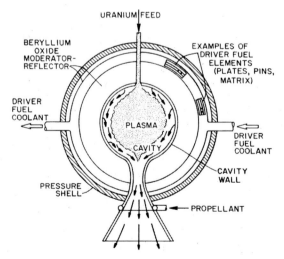

Figure 2.40 Illustration of a gas core reactor propulsion system [23].

attractive for radioisotope power sources because the power density can be fairly high, and emission of penetrating radiation is relatively low. All of the radioisotope power sources deployed in space by the United States used ^{238}Pu fuel. The first radioisotope sources were designed to burn up and disperse in the atmosphere in the event of an inadvertent reentry. Later, radioisotope power sources were designed to assure intact reentry.

All U.S. radioisotope electrical power systems to date employ thermoelectric power conversion devices. Thermoelectric energy conversion devices convert heat directly into electrical energy without an intervening mechanical device. Dynamic energy conversion devices have been developed for radioisotope power sources, but were never deployed in space. For the last four U.S. space missions with onboard radioisotope sources, General Purpose Heat Source-Thermoelectric Generators (GPHS-RTGs) were used. The modular design of the GPHS-RTG was developed to permit the use of prequalified radioisotope modules rather than custom designing and qualifying new radioisotope sources for each mission. Small radioisotope sources have been used as heat sources to maintain space system components within a desired temperature range.

The principal components of a reactor power system consist of the reactor, radiation shield, heat transport system, power conversion devices, instrumentation and control system, heat rejection subsystem, and power conditioning unit. The reactor is composed of the core, neutron reflectors, reactor vessel, and control elements. The core contains the nuclear fuel elements and coolant channels; some designs include a neutron moderator. Uranium-235 is the standard fuel choice for space reactors. Fuel materials developed for space reactors include UO_2, UN, UC, UC_2, and U-ZrH. Several advanced fuel materials have been developed by the former Soviet Union. Fuel forms include metal clad fuel rods containing ceramic fuel pellets, prismatic fuel elements, and spherical fuel balls. Zirconium hydride, Be, and BeO are common moderator choices.

For reactor systems, nuclear energy and two or three neutrons are released when the nucleus of a fuel atom fissions as a result of neutron absorption by the nucleus. The emitted neutrons can be absorbed by other fuel nuclei, inducing additional fission events and producing a neutron chain reaction. When the rate of neutron production by fission equals the loss rate by absorption and leakage, the reactor is said to be *critical*. If the neutron loss rate is greater or less than the production rate, the reactor is *subcritical* or *supercritical*, respectively. Neutron reflectors surround the reactor core to reduce the fuel mass required for criticality (critical mass). Some reactor designs incorporate a neutron moderator to reduce critical mass requirements; however, use of a moderator does not necessarily reduce system mass and size relative to unmoderated reactor systems.

In order to compensate for fuel burnup and other considerations affecting neutron loss and production, reactors generally contain more fuel than required to achieve criticality. For space reactors, a *just-critical* core is typically maintained by adjusting reflector control elements (for example, rotating drums with neutron poison segments). Control element maneuvers are used to place or maintain the reactor in a critical, subcritical, or supercritical configuration. For a critical reactor, the neutron flux and reactor power are maintained at a constant level. For a supercritical reactor, the neutron flux level increases with time; and a subcritical reactor cannot maintain a neutron flux without an external source.

Reactor power system designs usually include a radiation shield as an integral feature of the system to protect payloads and astronauts from radiation escaping the reactor. Typically a shadow shield and separation boom are used to reduce shielding mass. Heat rejection radiator designs for orbiting reactor systems are commonly shaped to fit within the cone of space protected by a shadow shield in order to reduce the potential dose from scattered radiation.

High operating temperatures are generally an advantage for space reactors because high temperatures can reduce radiator mass and improve reactor efficiency. Liquid metal coolants and inert gases are well-suited coolants for high temperature operation. Typically, reactor coolants are used to carry heat from the reactor core to a power conversion system where thermal energy is converted to electricity. Common choices for static power conversion devices include thermoelectrics and thermionic energy converters. Possible dynamic power conversion approaches include Rankine, Stirling, or Brayton cycles.

Although never used in space, a number of nuclear thermal propulsion reactors were developed and tested in the United States and the Soviet Union. For nuclear thermal propulsion, a hydrogen propellant is heated in the reactor core. The very hot propellant escapes the reactor system at high velocity providing propulsive thrust. The specific impulse, a rocket performance parameter, is directly proportional to the square root of the chamber temperature divided by the square root of the relative molecular mass. High temperatures possible in reactor systems and the ability to use the lowest relative molecular mass propellant (hydrogen) enable nuclear rockets to achieve very high specific impulses. Nuclear electric propulsion uses a reactor to produce electrical power; electricity is then used to create electric fields or arcs to accelerate charged particle or ionized gas propellants to very high velocities.

For a more detailed discussion of space nuclear systems see Angelo and Buden's *Space Nuclear Power* [1] and El-Genk's *A Critical Review of Space Nuclear Power and Propulsion 1984–1993* [24].

Symbols

A_s	surface area	k	Boltzmann's constant
A	Richardson constant	$\dot{m}$	mass flow rate
A_r	relative atomic mass	M_r	relative molecular mass
A_P	propellant performance factor	n	neutron
C_F	nozzle coefficient	q_R	radiated thermal power
F	force	T	temperature
I_{sp}	specific impulse	T_{CH}	chamber temperature
J	current density		
α	alpha particle	η_C	Carnot efficiency
ε	emissivity	σ	Stefan-Boltzmann constant
ϕ	work function		

Special Subscripts/Superscripts

C	cold	R	radiator
H	hot	S	space

References

1. Angelo, Jr., J. A. and D. Buden, *Space Nuclear Power*. Malabar, FL: Orbit Book Co., 1985.
2. Dieckcamp, H. M, *Nuclear Space Power Systems*. Canoga Park, CA: Atomics International, 1967.
3. Truscello, V. C. and L. L. Rutger, "The SP-100 Power Systems," Space Nuclear Power Systems; Proceedings of the 9th Symposium, Albuquerque, NM, Jan. 12–16, 1992. Pt. 1. New York: AIP Press, American Institute of Physics, 1992, p. 1–23.
4. Ludewig, H., "Particle Bed Reactor Nuclear Rocket Concept." Nuclear Thermal Propulsion: A Joint NASA/DOE/DOD Workshop, Cleveland, OH, Jul. 10–12, 1990, p .151–164.
5. Polansky, G. F., R. F. Rochow, N. A. Gunther, and C. H. Bixler. "A Bimodal Spacecraft Bus Based on a Cermet Fueled Heat Pipe Reactor." AIAA/SAE/ASME Joint Propulsion Conference and Exhibit. USDOE, San Diego, CA: Jul. 10–12, 1995, p. 9. Report Number SAND-95-1510C.
6. NEP Space Test Program Preliminary Nuclear Safety Assessment, SDI Report, Nov. 1992.
7. Ponomarev-Stepanoi, N. N., V. M. Talyzin and V. A. Usov, "Russian Space Nuclear Power and Nuclear Thermal Propulsion Systems," Nuclear News, 43:13, Dec. 2000, p. 13.
8. Bennett, G. L., "A Look at the Soviet Space Nuclear Power Program," Proceedings of the 24th Intersociety Energy Conversion Engineering Conference, IECEC-89, IEEE. Aug. 6–11, 1989, Washington, DC, vol. 2, p. 1187–94.

9. Lang, R. G. and E. F. Mastal, "A Tutorial Review of Radioisotope Power Systems," *A Critical Review of Space Nuclear Power and Propulsion, 1984–1993*. M. S. El-Genk, (Ed.), New York: AIP Press, American Institute of Physics, 1994, p. 1–20.

10. Pitrolo, A. A., B. J. Rock, W. Remini, and J. A. Leonard, "SNAP- 27 Program Review." Proceedings of 4th Intersociety Energy Conversion Engineering Conference, Washington, DC, Sept. 22–26, 1969 paper 699023, p. 153–170.

11. Halliburton NUS, Nuclear Safety Analyses for Cassini Mission Environmental Impact Statement Process. HNUS-97-0010. Gaithersburg, MD: Halliburton NUS Corp., Apr. 1997.

12. Frass, A. P., Summary of the MPRE Design and Development Program, ORNL-4048, Oak Ridge, TN: Oak Ridge National Laboratory, June 1967.

13. General Electric Co., 710 High-Temperature Gas Reactor Program Summary Report. Volume IV. Critical Experiment and Reactor Physics Development. GEMP600 (Vol. 4), Cincinnati, OH: General Electric Co., Nuclear Materials and Propulsion Operation, 1968.

14. Voss, S. S., SNAP (Space Nuclear Auxiliary Power) Reactor Overview. AFWL-TN-84-14. Kirtland AFB, Air Force Weapons Laboratory, 1984.

15. Svedberg, R. C., J. E. Pantolin, R. K. Sievers, and T. K. Hunt, "Enhancement of AMTEC Electrodes and Current Collectors." 12th Symposium on Space Nuclear Power and Propulsion Conference, Ballistic Missile Defense Organization. NASA and U.S. Dept. of Energy et al. AIP Conference Proceedings No. 324, Pt. 2, Albuquerque, NM, Jan. 8–12, 1995, p. 685–91.

16. Salamah, S. A., Sodium-Lithium HYTEC Application Studies, General Electric Report GEFR-0093 (IR), July 1991.

17. Francis, R. W., W. A. Somerville, and D. J. Flood, "Issues and Opportunities in Space Photovoltaics," Record of the Twentieth IEEE Photovoltaic Specialists Conference – 1988, Las Vegas, NV, Sept. 26–30, 1988, vol.1, p. 8–20.

18. Alemany, A., R. Laborde, P. Marty, J. Thibault and F. Werkoff, "Studies for the Definition of a Faraday Converter for Space Nuclear Systems." Transactions of the Fourth Symposium on Space Nuclear Power Systems. Albuquerque, NM: Institute for Space Nuclear Power Studies, Chemical and Nuclear Engineering Department, University of New Mexico, American Institute of Chemical Engineers and the American Nuclear Society. Jan. 12–16, 1987, p. 367–369.

19. Dudenhofer, J. E., D. L. Alger and J. S. Rauch, "Dynamic Power Conversion Systems for Space Nuclear Power," *A Critical Review of Space Nuclear Power and Propulsion, 1984–1993*. M. S. El-Genk, (Ed.), New York: AIP Press, American Institute of Physics, 1994, p. 305–369.

20. Juhasz, A. J. and G. P. Peterson, "A Review of Advanced Radiator Technologies for Spacecraft Power Systems and Space Thermal Control," *A Critical Review of Space Nuclear Power and Propulsion, 1984–1993*. M. S. El-Genk, (Ed.), New York: AIP Press, American Institute of Physics, 1994, p. 407–442.

21. Koenig, D. R., Experience Gained from the Space Nuclear Rocket Program (ROVER). Report LA-10062-H, Los Alamos, NM: Los Alamos National Laboratory, May 1986.

22. Clark, J. S., P. McDaniel, S. Howe, I. Helms and M. Stanley, Nuclear Thermal Propulsion Technology: Results of an Interagency Panel in FY1991. NASA Technical Memorandum 105711, Apr. 1993.

23. Chow, S., "Mini-cavity Plasma Core Reactors for Dual-Mode Space Nuclear Power/Propulsion Systems." Partially Ionized Plasmas including the Third Symposium on Uranium Plasmas. Princeton, NJ: National Aeronautics and Space Administration, June 10, 1976, p. 217–223.

24. El-Genk, M. S. (Ed.), A Critical Review off Space Nuclear Power and Propulsion 1984–1993, New York: AIP Press American Institute of Physics, 1994.

25. Marshall, A. C., "RS/MASS-D: An Improved Method for Estimating Reactor and Shield Masses for Space Reactor Applications," SAND 91-2876, Albuquerque, NM: Sandia National Laboratories, Oct. 1997.

Student Exercises

1. (a) Compute the Beginning-of–Mission (BOM) thermal power level required for two radioisotope power sources. One source must produce 0.25 kW_e and the other must produce 5 kW_e after 6 years of operation. Assume that ^{238}Pu makes up 90% of the plutonium isotopic composition, and 5.5 MeV is released by the decay of ^{238}Pu. Ignore the energy contribution from other radioisotopes. Power conversion is provided by SiGe thermoelectric devices, with a conversion efficiency of 6.7%. (b) Compute the mass of PuO_2 fuel required at BOM and

their BOM activity. Estimate the total power source mass, assuming that the ratio of total power source mass to PuO_2 mass is 5:1. (c) Repeat calculations (a) and (b) assuming a dynamic power conversion device is used with an efficiency of 18.1% and a ratio of system mass to PuO_2 mass of 12:1. Compare and comment on the differences between the systems using thermoelectric power conversion and systems using dynamic power conversion. Discuss why the conjectured 100-kWe radioisotope power system is impractical.

2. The following formula can be used to estimate the critical mass for a liquid metal-cooled fast reactor [25],

$$M_c = \frac{C_g(150)}{e^2}\left(\frac{5.43}{\zeta_F VF}\right)^{3/2}.$$

Here, M_C is the critical uranium mass in kg and e is the fractional enrichment of ^{235}U in the fuel. The parameters ζ_F and VF are the full density of the fuel (in g/cm^3) and the volume fraction of fuel in the core, respectively. The parameter C_g is a critical mass correction factor to account for geometric effects. The correction factor is given by

$$C_g = \frac{1}{3}\left(2.34\, a_r^{2/3} + a_r^{-4/3}\right).$$

where a_r is the aspect ratio equal to the core length-to-diameter ratio. (a) Estimate the critical mass of a core possessing the following characteristics: $e = 0.93$, UN fuel with $\zeta_F = 13.73$ g/cm^3, $VF = 0.4$, and $a_r = 1.0$. (b) Estimate the effect on the critical mass of using fuel with an enrichment of 20%. (c) Estimate the effect on critical mass, relative to (a), of using UO_2 fuel with a density of 10.0 g/cm^3. (d) Estimate the effect on critical mass, relative to (a), of incorporating empty safety rod channels (fuel volume fraction change). Assume that the empty channels increase the core volume by 20% . (See Figure 2.21, for example.) (e) Estimate the effect on critical mass of using core aspect ratios of 1.5:1, 2:1, and 2.5:1.

3. In Exercise 2, only the critical mass was considered for the fuel mass calculation. Temperature limits on fuel materials can also affect fuel mass requirements. For the reactor type described in Exercise 2, calculate fuel masses based on thermal limit considerations for required power levels of 5 kW$_e$ and 100 kW$_e$. Perform the calculations for both thermoelectric power conversion (with an efficiency of 4.4%) and Rankine power conversion (with an efficiency of 22%). Assume that $\hat{P}$, the local maximum-to-average core power, equals 1.5. For a pin type geometry, the specific power (power/mass) $\tilde{P}$ of the fuel is

$$\tilde{P} \approx \frac{4k_F\left(T_{\max} - T_c\right)}{\zeta_F r_F^2}$$

Here, k_F = thermal conductivity of the fuel = 0.26 W/cmK, r_F = fuel pin radius = 1.2 cm, ζ_F = fuel density = 13.73 g/cm^3, T_{max} = maximum allowed fuel temperature = 1650 K, and T_c = coolant temperature = 1350 K. Compare the required fuel masses to the critical masses calculated in Exercise 2 (a). What measures can be taken to reduce fuel mass based on thermal limit considerations?

4. Fuel mass calculations must account for fuel burnup and fuel damage limitations, as well as critical mass (Exercise 2) and thermal limitations (Exercise 3). (a) Assuming 200 MeV per fission and 12 years of operation, compute the mass of fuel burned up for the power levels and power conversion approaches described in Exercise 3. (b) Compute the fuel mass required at BOM if fuel damage considerations limit the quantity of fuel burnup to 7%. Include the effect of the local power peak-to-average. Compare calculated masses to fuel mass estimates obtained in Exercises 2 and 3.

5. For Exercises 2, 3, and 4, only the fuel mass was considered. The masses of other reactor components can have a significant impact on reactor mass. (a) Use the approach in Exercises 2, 3, and 4 to estimate the required (most limiting) fuel mass for the following reactor characteristics: $P = 150$ kW$_e$, t = lifetime = 15 years, η = net conversion efficiency = 11%, $e = 0.93$, $\zeta_F = 13.73$ g/cm^3, $VF = 0.52$, $a_r = 1.0$, maximum fuel burnup = 10%, $k_F = 0.26$ W/cm·K, $r_F = 1.1$ cm, $\hat{P} = 1.4$, $T_{max} = 1500$ K, and $T_c = 1350$ K. (b) For a fast reactor, the major mass components are the fuel, reflector, pressure vessel, and structure. Estimate the total reactor mass and dimensions assuming a cylindrical fast reactor (no moderator) . Assume that the sum of the pressure vessel and structure mass is approximately equal to the fuel mass. Also assume that the reactor is surrounded on all sides by a 10 cm thick BeO reflector, with a density of 3 g/cm^3.

6. A 200 kW$_e$ space reactor uses advanced thermionic devices. These devices are essentially equivalent to ideal thermionic devices, where the emitter, collector, and net current densities are given by

$$J_E = AT_E^2 \exp[-(\phi+eV)/kT_E], J_C = AT_C^2 \exp[-\phi/kT_C], \text{ and } J = J_E - J_C, \text{ respectively.}$$

Here V is the diode output voltage = 0.85 V. Use an emitter temperature of 1790 K and assume that the work function for both the emitter and collector equals 2.0 eV. For each of the following exercises, carry out calculations for a collector temperature of 1000 K and for a collector temperature of 400 K. (a) Compute the Carnot efficiency for the device. (b) Compute the emitter and collector current densities, net current densities, and power densities. (c) Assume an effective emissivity of 0.2 and compute the radiant heat flux between the emitter and collector surfaces using Eq. 2.3. Use the emitter and collector temperatures for T_R and T_S, respectively. (d) Estimate the device efficiency assuming that the only heat loss is due to radiant heat transfer between the emitter and collector. Compare to the Carnot efficiency.

7. For the cases described in Exercise 6: (a) For both collector temperatures, calculate the radiant heat flux from the radiator (assume that the collector temperature approximately equals heat rejection temperature). Use a radiator emissivity of 0.85 and an effective average temperature for space of 300 K. (b) Compute the required radiator surface area, and estimate the required radiator masses using an area density of 6.8 kg/m^3 for radiator temperatures below 700 K, and 8.2 kg/m^3 for temperatures above 700 K.

8. Estimate the power system mass of a reactor system producing 100 kW of electrical power, with an efficiency of 19%. Use the following approximations: (a) Calculate the required fuel mass assuming that the reactor has the characteristics given in Exercise 2 (a). Also assume that the fuel mass is only determined by critical mass considerations. (b) Use the approach given in Exercise 5 to estimate reactor mass. (c) Assuming an 800 K heat rejection temperature, use the approach and other parameters from Exercise 7 to estimate radiator mass. (d) For this reactor type and operating range, the radiation shield mass is approximately equal to the reactor mass. (e) Assume that the power conversion system mass plus miscellaneous system masses is approximately 25% of the total system mass. Estimate the total reactor system mass as the sum of the constituent masses and compare to radioisotope power system masses in Exercise 1. (f) Compute the fresh-fuel activity for the reactor systems and compare to radioisotope activities in Exercise 1. Use the specific activity of 7×10^{-5} Ci/g U for 93% enriched uranium.

Chapter 3

Space Nuclear Safety Perspectives

Albert C. Marshall

The objective of this chapter is to introduce space nuclear safety concepts and to provide an overview of space nuclear safety considerations. The scope and unique nature of space nuclear safety are discussed. The safety record of space nuclear missions is briefly reviewed, and common space nuclear safety issues and approaches are identified for each mission phase.

1.0 Definition and Scope

Before discussing space nuclear safety, we must have a clear understanding of what is meant when a condition or activity is said to be "safe," and we must clearly delimit the scope of space nuclear safety activities.

1.1 Definition of Safety

In this text, the following definition will be used:

> *Any activity or condition is safe if it is judged to entail a sufficiently low health risk.*

In this context, health risk includes illness, injury, death, and environmental contamination. This simple definition of *safe* may seem too vague to be of any value and begs the question, "Who makes the judgment?" Most people prefer the dictionary definition: *free from damage, danger, or injury* [1]. However, no activity or condition is completely free from damage, danger, or injury; and we consciously or unconsciously make risk judgments in all activities. For example, most of us judge the risk from food contamination to be sufficiently low that we do not concern ourselves when we purchase groceries or eat in restaurants. We make this judgment despite the fact that people die from contaminated food. If we are concerned, we can take extra precautions in selecting our food sources and preparing our food. The choice is up to the individual. However, we do not have a direct decision making capability for all activities that may affect us. For example, we may decide that airplanes are unsafe and decide not to fly. Our personal choice does not prevent the use of commercial airliners by others, even though a commercial airliner could crash into our house, killing us while we sleep. Although such accidents occur, commercial flight is permitted because the societal benefits are judged to outweigh the very small risk. As for all societal considerations, the judgment is made by a government agency acting in the public interest. The judgment often includes a comparison of potential hazards with risks from other activities. Public participation can be incorporated in the safety review process by establishing a public forum to air individual concerns and to communicate findings from safety reviews. Space missions incorporating nuclear systems fall into the category of societal considerations; consequently, such missions are required to undergo exhaustive safety analysis, testing, and government review and approval.

1.2 Scope of Activities

The scope of space nuclear safety covers all mission phases beginning with the pre-launch phase through the disposal phase. The scope includes protection of the public, workers, astronauts, and the environment from radiation produced by space nuclear systems. Much of the scope of space nuclear safety activities assumed here is consistent with the safety considerations reviewed by the U.S. Interagency Nuclear Safety Review Panel (INSRP). Specific responsibilities are summarized in the following:

Summary of Space Nuclear Safety Responsibilities

- Providing radiological protection for the public, workers, and astronauts during space nuclear missions from the prelaunch phase through the disposal phase.
- Providing measures to reduce the likelihood or mitigate the radiological consequences of postulated accidents involving space nuclear missions.
- Preventing radiological contamination of Earth's biosphere from radiation produced by space nuclear systems.
- Providing radiological protection during normal operations for space nuclear missions.

Considerations Addressed by Other Mission Activities

Exposure of mission crew to natural sources of radiation and radiological contamination of the space environment are important considerations, but these topics are not addressed in this book. Furthermore, non-radiological issues, even when they are associated with nuclear systems, are outside the scope of space nuclear safety responsibilities. Safeguarding special nuclear materials from diversion or unauthorized use is not classified as a safety issue. Risk of financial loss and risks to mission success are also beyond the scope of space nuclear safety. Radiological risks associated with manufacturing, transportation, and ground testing of nuclear systems are not within the purview of INSRP; however, in this text some discussion of ground phase activities is provided for issues unique to space nuclear systems. Issues and considerations outside the scope of space nuclear safety activities are summarized in the following:

- Exposure of astronauts to natural sources of radiation.
- Non-radiological safety issues.
- Non-radiological environmental issues.
- Nuclear safeguards (diversion of special nuclear materials).
- Radiological and non-radiological contamination of non-terrestrial environments that does not pose a risk to astronauts.
- Threats to mission success.
- Financial loss for other space assets and terrestrial assets.
- Ground phase activities
  - Manufacturing
  - Transportation
  - Prototype testing
  - Terrestrial criticality testing

Excluding these risk considerations from space nuclear safety assessments should not be interpreted as an indication that they are unimportant. All of these issues must be addressed by other mission activities. The limitation in scope is emphasized because requirements, approaches, and procedures appropriate for space nuclear safety issues may be inappropriate for other considerations.

2.0 Unique Aspects of Space Nuclear Safety

The bulk of the literature on nuclear safety was developed for terrestrial nuclear power plants. The reactors, systems, structures, operating requirements, and environments associated with terrestrial facilities are substantially different from those for space nuclear systems. Although much of the safety approach developed for terrestrial nuclear facilities is directly applicable to space reactor systems, space reactor missions pose unique safety considerations. Furthermore, terrestrial nuclear safety experience is rarely directly applicable to radioisotope sources used in space. These unique considerations have resulted in safety approaches somewhat different from those used for terrestrial nuclear power plants.

2.1 Space Reactors vs. Terrestrial Reactors

The volume of the reactor core of a terrestrial nuclear electric power plant is orders of magnitude larger than a typical space reactor. A large commercial water-cooled reactor core has an active height of about 3.7 m (12 feet) and an effective diameter of about 3.4 m. It is housed in a massive steel pressure vessel about 4.6 m in diameter and a wall thickness of 12–25 cm. The nuclear steam supply systems for such plants are usually surrounded by massive structures designed to contain any radionuclides that might be released from the reactor pressure vessel as a result of a postulated severe accident. The electrical output of modern nuclear power plants often exceeds 1000 MW_e. The power requirements of space nuclear systems are modest by comparison, ranging from kilowatts to a few megawatts. Commercial power reactors use uranium fuel slightly enriched (a few percent) in uranium-235; whereas space reactors, in order to minimize critical mass requirements, are typically designed to use fuel containing ~ 93% ^{235}U. Shipment of highly enriched fuel elements (e.g., from the fuel fabrication plant to the core loading facility) presents more significant safety and safeguard concerns than for typical slightly enriched commercial reactor fuel. On the other hand, operational buildup of long-lived actinide radioisotopes is much less for space reactors using highly enriched fuel than for commercial reactors using low enrichment fuel.

Terrestrial reactor systems must be protected against natural phenomena such as seismic events, floods, hurricanes, and tornadoes. These environments pose no risk to orbiting space systems. On the other hand, space reactor systems are, by necessity, in close proximity to propellants just before and during launch and ascent. A fire, explosion, or impact may have the potential to cause an inadvertent criticality or to cause reactor disruption and dispersal of the nuclear fuel. Furthermore, terrestrial nuclear power facilities are stationary, while space nuclear systems are often in motion through all mission phases. Space reactor movement includes ground transportation to the launch site, mating with the launch vehicle, launch, ascent, orbital operation, or following a trajectory through space. Reentry (return-entry) into Earth's atmosphere can be postulated for a variety of accident scenarios. Reentry into Earth's atmosphere can subject space systems to intense heating and high-speed impact.

The most important difference between terrestrial and space reactors is that space reactors typically are operated at significant power levels only after they are deployed in space. Reactors that have not been operated at high power possess relatively small radioactive inventories (typically less than 50 curies). Long-term operation of terrestrial commercial reactors can produce billions of curies of radioactive material. Terrestrial reactors are operated within Earth's biosphere and are much closer to human populations than space-based systems. An operational accident in space does not generally pose a significant threat to Earth's environment and populace. Rapid dispersal of radioactive materials into Earth's atmosphere is highly improbable for a system operating in high Earth orbit and is impossible for reactors based on another planet or moon. Many of the important safety concerns associated with terrestrial reactor operation, such as loss-of-coolant accidents, may not apply to a number of space reactor missions. Thus, safety features required for terrestrial reactors, such as auxiliary cooling systems, may be unnecessary for some classes of space reactor missions.

For some missions, space reactor operational accidents can be postulated that have the potential for safety consequences. For example, a reactor operational accident in Low-Earth-Orbit (LEO) may prevent proper disposal and could generate radioactive fragments. A hot reactor or fragments left in LEO may reenter Earth's biosphere before radioactive constituents have decayed to a low level. Operational accident analyses must include environmental factors that are not applicable to terrestrial safety analyses. For orbiting systems, these factors include operation in the vacuum of space under microgravity conditions and the possibility of meteoroid or space debris impact. The vacuum of space does not provide convective-path heat sinks, and the absence of gravitational effects precludes natural convection heat transfer within coolant loops. The effects of non-terrestrial environments must be considered for reactor systems based on other planets or moons.

At terrestrial reactor facilities, all accessible areas are protected from reactor radiation by massive radiation shielding. Radiation shielding is also required to protect astronauts and electronics during missions employing space reactors. In order to minimize space system mass, radiation shielding is usually limited to protect only those locations occupied by mission crew or sensitive electronics. Procedures may be used to assure that astronaut movement is limited to those areas protected by radiation shielding. A reactor based on the surface of a planet or moon may use surface materials for shielding. For these conditions, the possibility of neutron activation of surface materials must be evaluated. Another consideration is that an operational accident could present a radiological hazard for astronauts in the vicinity of the system; however, the astronauts' environment is controlled, thereby precluding the possibility of inhalation or ingestion of radioactive materials. The principle radiological risk to astronauts is direct radiation from external radiation sources.

2.2 Radioisotope Sources

Radioisotope systems are generally much simpler than reactor systems and present different safety considerations from both terrestrial and space reactor systems. Radiation from radioisotope sources results primarily from the natural decay of the radioisotope constituents of the nuclear fuel. The radioisotope inventory is greatest at the time of manufacture and slowly decays away. Prior to deployment and during launch, radioisotope sources often possess much larger radiological inventories than do space reactors. On the other hand accidental criticality is a virtual impossibility for typical radioisotope power sources. The principal issue for radioisotope space power systems is the possibility of disruption of containment barriers and dispersal of the radioisotope fuel due to propellant fires, propellant explosions, and high-speed impact. Thus, radioisotope systems are designed and tested to prevent or minimize dispersal of the radioisotope fuel for postulated launch and reentry accidents.

Safe disposal of both radioisotope and reactor systems is entirely different from disposal of terrestrial nuclear waste. Return of space nuclear systems to Earth for disposal is generally considered unnecessary and usually unwise. The most common disposal approach is to leave the system in place in high orbit or on the surface of a planet or moon. If the system operates in LEO, boosting to a high disposal orbit is the usual approach. When this option is used, failure to boost to high orbit is a safety issue. For interplanetary missions, disposal in a planetary or solar orbit may be used.

3.0 Safety Considerations by Mission Phase

For space nuclear systems, safety assessments are typically discussed in the context of mission phase. Mission phases can be categorized in a number of ways. For the purpose of this overview, the following broad mission phase categories will be used:

- Ground phase
- Prelaunch phase

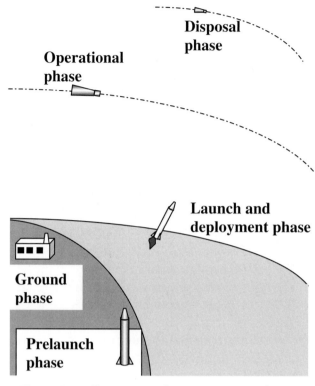

Figure 3.1 Illustration of major mission phases.

- Launch and deployment phase
- Operational phase
- Disposal phase

These broad categories are illustrated in Figure 3.1. Common safety considerations and safety approaches are introduced in the following discussion, and are summarized in Tables 3.1 through 3.9.

3.1 Ground Phase

The ground phase includes all activities prior to mating the nuclear system to the launch vehicle. Activities include fabrication, transportation, fuel loading, assembly, and testing. Most of these activities involve the same safety considerations and approaches that apply to conventional terrestrial nuclear systems. Although ground phase activities are beyond the scope of INSRP considerations, they are included in this overview to identify unique ground phase safety issues.

Radioisotope Sources

Standard safety practices are employed for radioisotope sources during the ground phase [Table 3.1 (a)]. The principal concern regarding radioisotope sources is the possibility of inhalation or ingestion of radioactive materials. Established methods for worker protection are used during manufacturing, such as glove boxes, to prevent intake of radioactive materials by workers and to prevent excessive radiation exposure to penetrating radiations.

Standard practices are also used to prevent radiological accidents, and established response procedures (e.g., evacuation in response to an air monitor alarm) can mitigate the consequences of postulated accidents [Table 3.1 (b)]. Radioactive materials must be transported in shipping containers that are designed and rigorously tested to assure containment of radioactive materials in the event of a trans-

Table 3.1   Ground Phase –Radioisotope Source Safety Considerations

Consideration	Typical Safety Approaches
(a) Normal Environment	
1. Direct exposure to neutrons and gamma radiation	• Use alpha emitters • Light shielding, exclusion areas • Other standard practices
2. Radioactive material inhalation or ingestion	• Glove boxes • Other standard practices
(b) Postulated Accident Environments	
3. Fabrication accident	• Standard practices for prevention and mitigation
4. Transportation accident leading to dispersal of radioisotope fuel – impact – fire – explosion – submersion – failure to maintain cooling	• Container or source integrity maintained for all credible accident conditions • Solid oxide fuel form • Provide highly reliable cooling system prior to launch

portation accident. The encapsulated radioisotope source itself, designed to withstand an inadvertent reentry accident, may provide adequate protection for some postulated transportation accidents.

Reactor Systems

Typical safety approaches for reactor systems in normal environments during the ground phase are summarized in Table 3.2 (a). Although the specific activity of highly enriched uranium is significantly greater than the activity of unirradiated commercial reactor fuel, little penetrating radiation is emitted from fresh enriched fuel. Critical assemblies and ground nuclear testing of reactor prototypes may be required for some space reactor programs. Well-established radiological safety practices are used for ground activities involving fuel manufacturing, fuel loading, critical assemblies, and operating prototype reactors. By prohibiting high power operation prior to space-deployment, only small fission product inventories are produced during criticality testing. Shielding, exclusion areas, and other practices limit potential exposure of ground personnel. A post-test *cooling period* may be required to allow for the decay of fission products before workers can approach the reactor system.

Typical safety approaches for postulated accidents are shown in Table 3.2 (b). Specially designed shipping containers are required to assure safety when shipping significant quantities of highly enriched fuel. Other precautions may be employed, such as the use of additional neutron poisons to preclude the possibility of an inadvertent criticality. Mostly standard practices are used to prevent or mitigate accident consequences during manufacture and prototype testing. Fuel loading and criticality testing of the flight reactor system at the launch site may require facility modifications and special procedures. Furthermore, testing nuclear thermal propulsion systems may require unique facilities to contain or confine potentially contaminated effluents.

3.2 Prelaunch Phase

The prelaunch phase includes all activities beginning with mating the nuclear system to the launch vehicle and ending just prior to launch. Normal environment considerations are similar to some of the ground phase considerations.

Table 3.2 Ground Phase–Reactor Safety Considerations

Consideration	Typical Safety Approaches
(a) Normal Environment	
1. Pre-test direct exposure to neutrons and gamma radiation	• Low dose for unirradiated fuel
2. Inhalation or ingestion of radioactive materials	• Standard practices and procedures
3. Prototype ground tests	• Standard practices and procedures
4. Exposure during fuel loading and criticality tests for flight reactor system	• Low power operation (just-critical) • Shielding and standard safety approaches and procedures
5. Post-test exposure from flight reactor	• Low activity from brief critical test • Shielding and exclusion zone until decay to low level emission • Sealed fuel elements
(b) Postulated Accident Environments	
6. Fabrication accident leading to – fuel dispersal, inhalation or ingestion – accidental criticality, radiation exposure	• Standard practices and procedures
7. Transportation accident leading to fuel dispersal, inhalation or ingestion – impacts – fires – explosions – submersion	• Container designed for all credible accident conditions • Fuel quantity limit
8. Transportation accident – impact – fires – explosions – submersion direct radiation exposure or release of leading to accidental criticality resulting in radioactive material	• Reactor or container designed for all credible accident configurations – disruption or reconfiguration – compaction – flooding moderation – submersion enhanced reflection – shutdown device movement • Fuel quantity limit • Neutron poisons during transportation • Fuel configuration selection
9. Prototype testing accidents, leading to direct radiation exposure or release of radioactive materials	• Standard practices and procedures • For nuclear thermal propulsion testing, scrubbers and containment for effluent
10. Accidents during fuel loading and criticality testing of flight reactor leading to direct radiation exposure or release of radioactive materials	• Standard safety approaches and procedures • Special considerations may be required for launch site loading and testing

Radioisotope Sources

Safety approaches for radioisotope sources during the prelaunch phase [see Table 3.3 (a)] are similar to the approaches for the ground phase. The principal concern for radioisotope sources is the possibility of accidental disruption and dispersion due to propellant fires, explosions, or dropping the source. The possibility of disruption due to shrapnel impact from propellant explosions must be addressed as well. Special containers are generally impractical for all phases following the ground phase; consequently, the

Table 3.3 Prelaunch Phase—Radioisotope Source Safety Considerations

Consideration	Typical Safety Approaches
(a) Normal Environment	
1. Direct exposure to neutrons and gamma radiation	• Use alpha emitters
	• Light shielding, exclusion space
2. Radioactive material inhalation or ingestion	• Encapsulated source, solid form
(b) Postulated Accident Environments	

3. Disruption and dispersion due to
  – propellant fires
  – propellant explosions
  – dropped source
resulting in inhalation or ingestion
	• Source containment resistant to disruption for most postulated accident conditions
	• Fuel material selected to mitigate consequences of radioisotope fuel release

system encapsulating materials and the form of the nuclear fuel are relied upon to mitigate prelaunch phase postulated accidents [see Table 3.3 (b)]. The encapsulating materials described in Chapter 2 are resistant to disruption during propellant fires, explosions, shrapnel impact, and dropping on a hard surface. In addition, fuel materials (e.g., PuO_2) are selected to mitigate the consequences of release for accidents postulated to release fuel particulate. This approach is discussed in Chapter 5.

Reactor Systems

Pre-launch safety approaches for reactors are shown in Table 3. 4 (a). By allowing a cooling period following criticality testing, the gamma and neutron dose from the reactor will be sufficiently low to permit handling without the need of significant additional shielding.

Table 3.4 Prelaunch Phase—Reactor Safety Considerations

Consideration	Typical Safety Approaches
(a) Normal Environment	
1. Direct exposure to neutrons and gamma radiation	• Allow cooling period • Added shielding • Special handling
(b) Postulated Accident Environments	

2. Inhalation or ingestion following disruption and dispersion of fuel due to
  – propellant fires
  – propellant explosions
  – dropping core
	• Low radiological hazard of essentially fresh fuel
	• Design core to resist disruption accidents
	• Safety procedures

3. Accidental criticality leading to direct radiation exposure or release of radioactive material, due to
  – propellant fires
  – propellant explosions
  – submersion
  – dropping core
	• Design core to prevent criticality for all credible accident configurations
	  – disruption or reconfiguration
	  – compaction
	  – flooding moderation
	  – submersion enhanced reflection
	  – shutdown device movement
	• Removable neutron poisons

4. Criticality due to accidental startup, direct exposure of workers or leading to release of radioactive materials	• Safety procedures, interlocks

Typical safety approaches for potential accidents are summarized in Table 3.4 (b). The principal concern is an inadvertent criticality. Potential initiators include dropping the reactor during mating with the launch vehicle, propellant fires, propellant explosions, and reactor fall to the launch pad following a launch vehicle propellant explosion. These initiators can be postulated to cause inadvertent criticality by core compaction, reconfiguration of fuel, or movement of shutdown devices. If the reactor can be subsequently submerged in a moderating fluid (e.g., water or propellant), the safety analysis must address the possibility of criticality due to enhanced core moderation or enhanced neutron reflection. Safety analyses must also consider the possibility of accidental startup due to spurious signals. Safety procedures are enforced to prevent accidents, and the reactor system is designed to remain subcritical under all credible accident conditions. Additional neutron poisons may be employed to prevent an inadvertent criticality. These additional poisons are removed after the reactor is safely deployed in space.

3.3 Launch and Deployment Phase

The launch and deployment phase has no terrestrial nuclear safety counterpart. The launch and deployment phase begins with rocket engine ignition and ends when the payload is placed in a stable orbit, put on an extraterrestrial trajectory, or deployed on the surface of a planet or moon.

Radioisotope Sources

Direct exposure to emitted radiation from a radioisotope source is only a safety consideration if both astronauts and the source are aboard the same launch vehicle. For normal environments, light shielding may be used if necessary [see Table 3.5 (a)]. Typical safety approaches for accidents during this phase are given in Table 3.5 (b). The issue of source disruption and dispersion of radioactive material due to propellant fires or explosions applies to the launch and deployment phase as well as the prelaunch phase. For the launch and deployment phase, safety analysis must also include the possibility of launch vehicle tipover and launch abort resulting in high-speed impact. Typically, launch trajectories follow a low-population flight path, which helps minimize risk to the general populace. Deviations from the intended flight path may require launch abort by intentional launch vehicle destruction.

Reactor Systems

Reactor safety approaches for normal environments during the launch and deployment phase are presented in Table 3.6 (a). Exposure of astronauts to neutrons and gamma radiation is generally not a consideration during the launch phase. Typically, direct neutron and gamma emission from the shutdown

Table 3.5   Launch and Deployment Phase—Radioisotope Source Safety Considerations

Consideration	Typical Safety Approaches
(a) Normal Environment	
1. Direct exposure to neutrons and gamma radiation	• Use alpha emitters • Light shielding for astronauts if necessary
2. Inhalation or ingestion following release of radioactive material	• Encapsulated source, solid form
(b) Postulated Accident Environments	
3. Inhalation or ingestion of radioactive material following disruption and dispersion due to – high speed impact – propellant fires – propellant explosions – launch abort	• Source containment resistant to disruption for most postulated accident conditions • Fuel material selected to mitigate consequences of radioisotope fuel release • Low-population flight path

Table 3.6 Launch and Deployment Phase—Reactor Safety Considerations

Consideration	Typical Safety Approaches
(a) Normal Environment	
1. Direct exposure to neutrons and gamma radiation	• Post-critical-test cooling period before contact • Light shielding, exclusion space
(b) Postulated Accident Environments	

2. Inhalation or ingestion of radioactive material following disruption and dispersion due to
   – fires
   – explosions
   – high-speed impact
	• Low radiological hazard of essentially fresh fuel
	• Design core to resist disruption accidents
	• Select low-population flight path

3. Accidental criticality leading to direct radiation exposure or release of radioactive material, due to
   – fires
   – explosions
   – high-speed impact
   – submersion and flooding
	• Design core to prevent criticality for all postulated credible accidents
	   – disruption or reconfiguration
	   – compaction
	   – flooding moderation
	   – submersion enhanced reflection
	   – shutdown device movement
	• Removable neutron poisons
	• Low-population flight path

reactor will be slight. If a reactor system is launched aboard a piloted mission, light shielding and an exclusion space can assure very little astronaut exposure to emitted radiation.

Typical reactor safety approaches for launch and deployment phase postulated accidents are given in Table 3.6 (b). Impact, propellant fires, and propellant explosions must be considered for postulated accident environments. Impact can result from launch vehicle tip-over, high-speed reentry, or explosively propelled fragments. These environments can lead to disruption and dispersion of reactor fuel. The *essentially fresh* reactor fuel does not present a major radiological hazard, and the reactor system can be designed to resist disruption for postulated accident environments. Although an inadvertent criticality resulting from these postulated environments is a typically improbable scenario, the radiological consequences of an inadvertent criticality can be significant. Impact, fires, and explosions can be postulated that may result in core reconfiguration or compaction, leading to accidental criticality. Disruption followed by submersion and flooding with ocean water, wet sand, or fresh water could result in accidental criticality by enhanced neutron moderation and reflection. Inadvertent criticality can be prevented for these postulated environments by proper reactor system design and validation testing. Removable neutron poisons can be used to provide shutdown assurance during all credible accident scenarios. Risk is also minimized by use of a typical low-population flight path.

3.4 Operational Phase

Nuclear systems provide electrical power, process heat, or propulsion during the operational phase. For reactor systems, this phase includes reactor startup, operation, and shutdown. On some missions, however, reactors may start up and shut down several times; and radioisotope sources are always undergoing radioactive decay. Furthermore, for some missions, nuclear systems may change orbits or trajectories during the operational phase. For these reasons, the operational phase is simply defined here as the period between space deployment and the time just prior to placement of the nuclear system in its disposal disposition.

Table 3.7   Operational Phase—Radioisotope Source Safety Considerations

Consideration	Typical Safety Approaches
(a) Normal Environment	
1. Direct exposure to neutrons and gamma radiation	• Light shielding • Exclusion space • Special handling
2. Inhalation or ingestion of onboard radioactive material	• Encapsulated source, solid form
(b) Postulated Accident Environments	
3. Debris or meteor-impact leading to possible premature reentry of debris, contamination of biosphere	• Select orbits with long orbital lifetime for debris
4. Earth swing-by trajectory error leading to hypervelocity impact, resulting in disruption and dispersion of radioactive material and contamination of biosphere	• Small trajectory adjustments during flight, reentry low-probability event

Radioisotope Sources

Astronauts can safely handle typical radioisotope sources during the operational phase using special handling features and light shielding [see Table 3.7 (a)]. Astronaut Gordon Bean is shown in Figure 3.2 deploying the SNAP-27 radioisotope heat source on the surface of the Moon.

Typical radioisotope safety approaches for operational phase postulated accidents are given in Table 3.7 (b). The principal issue for radioisotope sources is the possibility of system reentry and dispersal of the radioisotope fuel into Earth's biosphere. For example, if an Earth swing-by is used for gravitational assist during the operational phase, hypervelocity impact accidents must be considered. If an Earth-orbiting mission is planned, a meteor or space debris could collide with the radioisotope source. Direct reentry of the radioisotope system due to meteoroid or debris collision is not credible; however, the col-

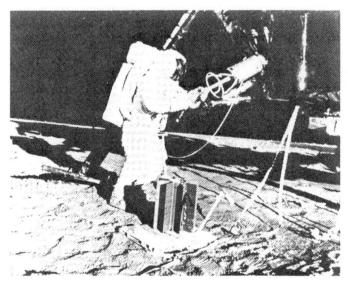

Figure 3.2 Astronaut Gordon Bean removing the SNAP-27 heat source from the Lunar Excursion Module during the Apollo 12 mission to the Moon, November 1969. *Courtesy of NASA.*

lision may generate fragments with shorter orbital lifetimes than the intact nuclear system (see Chapter 6 and 7). Orbit selection and trajectory planning can provide long-duration orbits for debris in the event of on-orbit impact. Radioisotope sources typically operate on the surface of a planet or moon, on an interplanetary trajectory, or in high orbits.

Reactor Systems

Common reactor safety approaches for normal environments during the operational phase are given in Table 3.8 (a). For reactor systems operating in the vicinity of astronauts, adequate radiation protection is required. Reactor shielding is commonly provided as an integral component of the reactor system. Some nuclear thermal propulsion concepts take advantage of the liquid hydrogen propellant as an additional radiation shield. For most designs of orbiting reactor systems, a boom is used to separate the crew and radiation-sensitive electronics from the operating reactor. Radiation shielding on the surface of a planet or moon can be provided using the native regolith. Astronaut movement may be restricted from exclusion areas that are not adequately protected by radiation shielding (e.g., regions not protected by a shadow shield). For surface based reactors, the possibility of neutron activation of surface materials must be addressed.

Table 3.8 Operational Phase—Reactor Safety Considerations

Consideration	Typical Safety Approaches
(a) Normal Environment	
1. Astronaut direct exposure to neutrons and gamma radiation	• Radiation shielding (integral shield, regolith, etc.) • Separation boom • Exclusion areas
2. Regolith activation for surface based systems	• Shielding if necessary
(b) Postulated Accident Environments	
3. Internally initiated accidents • Initiators: – transient overpower – loss of coolant or flow • Consequences: reactor disruption, and – debris reentry, biosphere contamination – crew exposure – inability to boost from LEO, resulting in failure to dispose	• Prohibit LEO operation • For LEO operation or operations near astronauts, use standard safety approaches and procedures for preventing and mitigating internally generated accidents (including auxiliary safety systems)
4. Externally initiated accidents • Initiators – meteoroid impact – debris impact • Consequences: reactor disruption, and – debris reentry, biosphere contamination – crew exposure – inability to boost from LEO, resulting in failure to dispose	• Meteoroid bumpers • Auxiliary safety systems • Prohibit LEO operation
5. Nuclear propulsion trajectory error resulting in reentry, disruption and dispersion, biosphere contamination	• Small trajectory adjustments during flight, reentry low-probability event

LEO = Low Earth Orbit

Typical reactor safety approaches for operational phase postulated accidents are summarized in Table 3.8 (b). Operational accidents may result from meteoroid or space debris impact or from internal failures of the system. As stated in Section 2.1, operational accidents do not always constitute a threat to Earth's biosphere. The principal issues are (1) the generation of fragments that may prematurely reenter Earth's biosphere, (2) the loss of ability to boost a system from a low Earth orbit (LEO) operational orbit to a high disposal orbit, and (3) radiological risk to mission crew. Issues (1) and (2) are typically postulated to initiate in the operational phase, but the safety consequences of these issues are more commonly a concern for the disposal phase. For LEO operations or operations near astronauts, reactivity induced accidents and coolant failure accidents must be considered. The possibility of reentry due to a trajectory or orbital maneuvering error must be addressed for nuclear propulsion missions that incorporate a return to LEO. Standard safety approaches can be used to prevent and mitigate possible internally generated reactor accidents. Protection for both internally and externally initiated accidents may include auxiliary safety systems, such as meteoroid bumpers and auxiliary cooling systems. The simplest and perhaps best operational safety approach is to refrain from LEO operation.

3.5 Disposal Phase

The disposal phase begins after mission termination. A number of interesting disposal options are possible (see Table 3.9). The best disposal options must be determined in the context of the application. Return of spent nuclear systems to Earth for disposal is not generally accepted as a viable option. Nuclear systems operated in a sufficiently high orbit (SHO) may be left in orbit, provided that their presence does not jeopardize other missions. A SHO is an orbit in which the activity of the system will have decayed to a low level before reentry can occur. The choice of a SHO will depend on the system design, the radiological inventory at the time of disposal, and the possibility of fragmentation of the disposed system by meteor or debris impact. Reactor activity is primarily due to relatively short-lived fission and activation products. Fission and activation product activity typically decays to very low levels within a few hundred years. However, for some reactor designs the decay of activation products to low levels may require thousands of years. Radioisotope sources typically require thousands of years to decay to low

Table 3.9   Disposal Phase—Safety Considerations for Radioisotope Sources and Reactors

Consideration	Typical Safety Approaches
(a) Normal Environment	
1. Astronaut direct exposure to radiation from disposed space reactor systems	• Prohibit human space operations near disposed reactor systems
(b) Postulated Accident Environments	
2. Premature reentry of nuclear system due to operational accidents in LEO leading to disruption and dispersion, contamination of biosphere	• Prohibit operation in LEO • Operational phase safety measures
3. Failure of boost system preventing boost from LEO to SHO, resulting in premature reentry of nuclear system, disruption and dispersion, contamination of biosphere	• Prohibit operation in LEO • Use highly reliable boost system
4. Debris or meteor-impact with disposed system leading to premature reentry of debris, disruption and dispersion, contamination of biosphere	• Use higher disposal orbit or leave on planet or moon surface

LEO = Low Earth Orbit   SHO = Sufficiently High Orbit

levels. As mentioned in Section 3.4, operational accidents in LEO may prevent boosting the system, resulting in premature reentry. Failure of the boost system could also strand the nuclear system in LEO with subsequent premature reentry. Nuclear systems can be safely disposed by placing them in an extraterrestrial planetary orbit or solar orbit.

4.0 The Safety Process

Here we summarize the basic features of the safety process for space nuclear systems; this topic is discussed in more detail in Chapter 12. The safety process differs from the terrestrial nuclear power approach in several important areas. First, the safety criteria established for terrestrial nuclear power do not apply to space nuclear systems. Secondly, much of the safety analysis required for space nuclear systems differs from the safety analysis required for terrestrial nuclear power. Lastly, the established approach for safety review and approval of space nuclear missions differs from the established approach used for terrestrial nuclear power.

4.1 Review and Approval

The unique nature of space programs using nuclear materials was recognized from the beginning. In 1960 the U.S. Atomic Energy Commission (AEC) established an Aerospace Nuclear Safety Board *to analyze and project the possible effects of nuclear space devices upon the health of the peoples of the world . . . and recommend standards of safe practice for the employment of nuclear powered space devices proposed by the U.S.* [2]. Systematic and comprehensive review and approval procedures followed, and by the mid-1960s an *ad hoc* safety review panel was formed. The panel was eventually named the Interagency Nuclear Safety Review Panel. The Interagency Nuclear Safety Review Panel (INSRP) now consists of representatives from the Department of Energy (DOE), the Department of Defense (DoD), the National Aeronautics and Space Administration (NASA), and the Environmental Protection Agency (EPA). The INSRP is supported by teams of scientists and engineers from several subpanels. Originally the DOE position was represented by the AEC, and the EPA was only recently established as a panel member. The Nuclear Regulatory Commission (NRC), responsible for licensing commercial nuclear power plants, serves as an INSRP technical advisor, but does not have regulatory authority for space nuclear missions.

The INSRP reviews Safety Analysis Reports (SARs) prepared by the DOE for the mission sponsor. Based on the SAR review and other sources of information, INSRP prepares a Safety Evaluation Report (SER) for approval by the Office of the President. In 1971, a requirement for an Environmental Impact Statement (EIS) was established under Public Law 91-190. An EIS is required for those activities that may have an adverse effect on the environment, and NASA executes a mission-specific EIS for each space nuclear application. Other safety review processes include approvals for transportation of nuclear materials to the launch site and launch site safety approval.

4.2 Safety Guidance

The Nuclear Regulatory Commission (NRC) uses a number of prescriptive safety criteria for the regulation of commercial nuclear power plants. The unique nature of space nuclear power, however, does not lend itself to prescriptive safety requirements. A very wide range of missions, environments, and systems are possible for space nuclear missions. Establishing a set of general requirements for all space nuclear missions would result in needless safety requirements in some cases and insufficient safety regulation in other situations. In lieu of codified safety criteria for space nuclear power, the safety process relies on established safety practices and in-depth safety analysis, and testing. Furthermore, both the DOE in conjunction with the mission sponsor and INSRP review and assess the safety of each mission to assure that the risks are extremely small.

For radioisotope power sources launched by the United States, the standard safety practice is to design systems to prevent or minimize release of radioactive materials for postulated accident conditions. Reactor systems offer a much broader range of design options and applications, compared to radioisotope sources. Furthermore, the only U.S. space-deployed reactor system was launched in 1965. These factors have resulted in the absence of an accepted simple guiding principle for space reactor safety. None of the U.S. space reactor programs following the launch of the SNAP-10A resulted in the launch of a reactor system; nonetheless, subsequent space reactor programs generally established their own safety approaches. Safety approaches typically incorporated guidance previously established for SNAP-10A, radioisotope power sources, and commercial reactors. In 1990, the U.S. DOE chartered an Interagency Nuclear Safety Policy Working Group (NSPWG) to recommend safety policy, requirements, and guidance for the nuclear propulsion program for the Space Exploration Initiative. The NSPWG recommendations [3] are non-prescriptive and are fairly general in their applicability. In order to clarify the intent of the recommended requirements, the NSPWG wording was revised in application to subsequent space reactor programs [4, 5]. A revised version of the NSPWG recommendations is presented in Table 3.10. The NSPWG was chartered to address some topics beyond the scope of space nuclear safety; these other topics are not included in Table 3.10.

Several important points must be made regarding the NSPWG guidance. First, the NSPWG safety requirements are recommendations that may or may not apply to other missions; they are not established policy. Furthermore, the NSPWG specified that the guidance could be modified, with the concurrence of safety reviewers. A hierarchical safety approach was recommended such that more specific safety design specifications would be established as mission plans and system design became more firmly established. The United Nations (UN) has also instituted several principles applicable to space nuclear safety. These somewhat controversial, non-binding principles are discussed in Chapter 12.

Table 3.10 Suggested Safety Guidance for Space Reactors (Based on NSPWG recommendations)

1. The reactor shall not be operated prior to space deployment, except for low-power testing on the ground, for which negligible radioactivity is produced.
2. The reactor system shall be designed to remain shut down prior to the system achieving its planned orbit.
3. Sustained inadvertent criticality shall be precluded for both normal and credible accident conditions.
4. For radiation produced by on-board radiological sources, radiation worker dose limits will apply to astronauts.
5. Radiological release from the spacecraft during normal operation shall have an insignificant effect on Earth.
6. The probability shall be extremely low for radiological accidents that affect the health of the crew.
7. For postulated accidents involving radiological release for which the crew is expected to survive, the radiological release shall not render the spacecraft unusable.
8. The consequence on Earth of a radiological release from an accident in space shall be insignificant.
9. Safe disposal of spent nuclear systems shall be explicitly included in mission planning.
10. Adequate protection for the reactor system shall be provided to prevent disruption or degradation for any credible accident conditions that could preclude safe disposal.
11. Planned Earth reentry shall be precluded from mission profiles.
12. Both the probability and the consequences of an inadvertent reentry shall be made as low as reasonably achievable.
13. For inadvertent reentry of a radiologically hot reactor, the reactor shall be essentially intact or reentry shall result in essentially full dispersal of radioactive materials at high altitude.
14. The reactor shall remain subcritical throughout an inadvertent reentry and Earth impact.
15. For an Earth impact, radioactivity shall be confined to a local area to limit radiological consequences.

4.3 Safety Analysis

Although some space nuclear safety analysis requirements are similar to the analyses required for commercial nuclear power, other space nuclear safety analysis categories have no terrestrial counterpart. Chapters 4 through 9 of this book each address a general space nuclear safety analysis category. These chapters address radiation safety, fires and explosions, orbital mechanics and reentry, impact, inadvertent criticality, and reactor dynamics, respectively. Chapters 10 and 11 discuss probabilistic safety assessments and consequence analysis, respectively. An alternative *chronological approach* could have been used; i.e., each chapter could have been dedicated to the analysis of a particular mission phase. The difficulty with the chronological approach is that some types of accidents can occur in more than one phase. For example, criticality accidents can be postulated for the ground phase, prelaunch phase, and the launch phase. Furthermore, these accident scenarios can include several types of analysis. A prelaunch accident scenario analysis may include separate analysis of propellant explosions, ground impact, water flooding criticality, reactor kinetics, and fuel dispersion. Recognizing these difficulties, the division of chapters into analysis categories was used rather than a chronological division.

5.0 The Safety Record

Space nuclear power programs select designs and mission plans to assure that safety risks are extremely small. This process is guided by safety assessments that identify all significant safety issues, as well as the most effective approaches to resolve these issues. Although no specific risk goals have ever been established for space nuclear missions, probabilistic safety assessments have consistently predicted risks from space nuclear missions that are far less than risks posed by other human activities. The efficacy of the U.S. and Soviet space nuclear safety programs has been tested over the last 35 years. The following brief summary reviews the safety record of U.S. and Russian space nuclear missions.

5.1 The U.S. Record

Launch or deployment failures have occurred during 3 of the 26 U.S. space missions with nuclear systems on board. In 1964 the Transit 5BN-3 failed to achieve orbit due to a guidance malfunction. The satellite's radioisotope source was designed to burn up and disperse its nuclear fuel at high altitude in the event of a deployment failure; the system responded as designed. Although fuel dispersal did not pose a threat to the biosphere, the safety practice was changed to a system design that assured intact reentry following a launch or deployment failure.

In May 1968, a launch failure occurred during the launch of a Nimbus B-1 meteorological satellite from the Vandenberg Air Force Base launch site. The launch vehicle's erratic behavior required an intentional destruction by the range safety officer. The launch vehicle and satellite were completely destroyed at an altitude of 30 km downrange from the launch site. The satellite was located, using tracking data, in the Santa Barbara Channel off the coast of California. The SNAP-19B2 radioisotope source onboard the spacecraft survived intact and was recovered 5 months later (see Figure 3.3). The capsule was designed for intact reentry and for the possibility of submersion in ocean water. Examination of the recovered device showed no deleterious effects; consequently, the fuel was reused on a subsequent mission.

The Apollo 13 mission was the most recent U.S. space mission to experience a failure with a nuclear system onboard. While on a trajectory to the Moon, an explosion in the spacecraft's service module required the unplanned return of the lunar module into Earth's atmosphere in order to save the lives of the crew. The lunar module was jettisoned during reentry with a SNAP-27 radioisotope source still on board. The source returned to Earth above the Pacific Ocean, and subsequent atmospheric monitoring revealed that no fuel was released. The radioisotope capsule is presumed to have reentered intact and still remains at a depth in excess of 2 kilometers at the bottom of the Tonga trench in the South Pacific. No adverse environmental effects have been observed [6, 7].

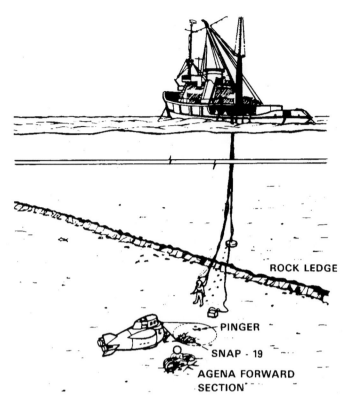

ROCK LEDGE

PINGER

SNAP - 19

AGENA FORWARD
SECTION

Figure 3.3 Illustration of recovery of SNAP-19 RTG. *Courtesy of NASA.*

5.2 The Soviet (Russian) Record

Russia, and its predecessor the Soviet Union, reported 4 failures out of about 40 space nuclear missions. The first problem occurred in 1978 when a Soviet spacecraft (Cosmos 954) operating in LEO could not be boosted to a high disposal orbit as planned. As a result, the reactor system aboard the spacecraft reentered and dispersed radioactive debris over an uninhabited region of Canada. The Canadian Atomic Energy Control Board led an air and ground search and recovery operation called *Operation Morning Light*. Debris, including a number of large highly radioactive fragments, was found over a 600 km track. A photograph of a large recovered fragment, often referred to as the antlers, is shown in Figure 3.4. Small fuel particles were found over a territory covering more than 100,000 km^2. It is believed that all radioactive fragments of significant size were recovered [6]. Though no significant environmental effects remained after the cleanup operation, the reactor system did not respond in a manner consistent with current safety standards.

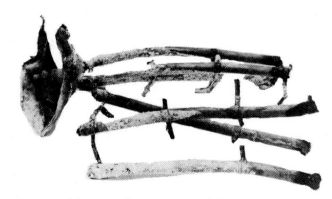

Figure 3.4 COSMOS 954 debris (*antlers*) recovered during Operation Morning Light [8].

As a consequence of the Cosmos 954 incident, the Soviets redesigned their nuclear systems to assure complete high-altitude dispersal in the event of an accidental reentry. Upon completion of its operational life, the spacecraft is designed to separate into several pieces. One piece consists of the reactor and a small kick stage. After separation, the reactor is boosted to a higher orbit; and the reactor core is ejected to facilitate burnup during eventual reentry. This kick-stage approach was used successfully until 1982, when the kick stage aboard the COSMOS 1402 failed to separate from the spacecraft. The reactor core was ejected, however, and is reported to have completely burned up during reentry (as designed) over the southern Atlantic Ocean [9].

In 1988, the Soviet Union reported that their COSMOS 1900, with an onboard reactor, failed to boost to a disposal orbit. This incident was resolved without consequence, and the spacecraft was properly disposed [8]. The last Russian incident with a nuclear system onboard occurred in 1996. The Russian Mars 96 spacecraft, with a plutonium radioisotope source onboard, was successfully placed in a circular 160 km orbit. The second burn, however, was unsuccessful, resulting in reentry. Russia reports that the radioisotope source reentered intact and is now submerged in the Pacific Ocean. [10].

References

1. *Webster's New World College Dictionary Third Edition*, V. Neufelt, Ed., Macmillan, Hudson, OH, 1996.
2. Atomic Energy Commission, *AEC Establishes Aerospace Nuclear Safety Board*, AEC public release, (from DOE archives), Nov. 1959.
3. Marshall, A. C., R. A. Bari, N. W. Brown, H. S. Cullingford, A. C. Hardy, J. H. Lee, W. H. McCulloch, K. Remp, G. F. Niederauer, J. W. Rice, J. C. Sawyer, and J. A Sholtis, Nuclear Safety Policy Working Group Recommendations, NASA Technical Memorandum 105705, Apr. 1993.
4. *NEP Space Test Program Preliminary Nuclear Safety Assessment*, SDI Report, Nov. 1992.
5. Polansky, G. F., R. F. Rochow, N. A. Gunther, and C. H. Bixler, "A Bimodal Spacecraft Bus Based on a Cermet Fueled Heat Pipe Reactor." Proceedings of the 31st AIAA/ASME/SAE/ASEE Joint Propulsion Conference, San Diego CA, July 10–12, 1995.
6. Angelo, J. A. Jr. and D. Buden, *Space Nuclear Power*. Orbit Book Co., Malabar, FL, 1985.
7. U.S. Department of Energy, *Atomic Power in Space*. DOE/NE/321174, National Technical Information Service, Springfield, VA, May 1987.
8. Gummer, W. R. et al., *COSMOS 954, the Occurrence and Nature of Recovered Debris*. Canadian Government Publishing Center, Catalogue No. CC 172-2/1980E, May 1980.
9. Bennett, G. L., "A Look at the Soviet Space Nuclear Power Program." Proceedings of the24th Intersociety Energy Conference, Washington, DC, Aug. 9, 1989, 2:1187–1194.
10. Lissov, I., *What Really Happened with Mars-96*. http://www.fas.org/spp/eprint/mars96, Nov. 1996.

Chapter 4

Radiation Protection

Albert C. Marshall and F. Eric Haskin

The objective of this chapter is to provide the reader with an understanding of radiation safety concepts, health effects, and analysis methods. Space nuclear radiation guidelines and regulations are reviewed. Sources of radiation are discussed and radiation exposures from normal operation of space nuclear systems are compared to exposure from natural radiation sources. Radiation shielding principles and simple analysis methods are presented.

1.0 Radiation Concepts and Units

All nuclear safety issues ultimately relate to the possibility of exposure to ionizing radiation and potential health effects from exposure. Familiarity with radiation concepts and units is essential to the study of space nuclear safety. Both standard international (SI) units and older radiation units are discussed in this chapter.

1.1 Source Strength

The rate of particle emission from a radioisotope material is called the *source strength* s_R (in particles/s) for radiation type R. The concept of source strength is appropriate for characterizing isotropic radiation from a radioactive source and is directly related to *activity* $\mathcal{A}$ (discussed in Chapter 1). Activity, however, gives the rate of nuclear decay in disintegrations per second or becquerel (becquerel = Bq = 1 dis/s), or in curies (1 Ci = 3.7×10^{10} dis/s). Source strength describes the emission rate of nuclear particles, in particles/s, resulting from nuclear disintegration. (Source strength s_R should not be confused with the volumetric source strength $\bar{S}$ or the surface source strength S discussed in Section 5.) Source strength depends on the composition and quantity of radioisotopes in the source and the number of each type of particle emitted per disintegration from each radioisotope. If one radioisotope species emits only one particle per disintegration, and essentially all radiation escapes the source, then $s_R = \mathcal{A}\,(3.7 \times 10^{10})$, where $\mathcal{A}$ is given in curies. This simple relationship of source strength to activity does not apply if more than one type of particle is emitted and the relationship does not distinguish between source strengths of particles of the same type that are emitted at different energies. Source strength is occasionally used to roughly characterize radiation from reactors in order to carry out simple shielding calculations. For operating reactors, however, source strength is not related to the decay of radioactive materials; it is, instead, an approximate method for characterizing escaping neutron and gamma radiation produced by nuclear fission.

1.2 Particle Flux

Source strength gives the rate of emission of nuclear particles in a quantity of radioactive material; it is not a measure of the radiation incident upon living tissue, equipment, or some material of interest. *Particle flux* ϕ_R, in units of particles/cm²·s, is used to express the incident radiation at a particular location. The particle flux incident on a particular organ will depend on whether one is exposed externally to direct radiation from a source or internally to radiation emitted by radioactive material that has been inhaled or ingested.

The particle flux generally depends on the distance from the source. For a point source of radiation type R in a vacuum, the particle flux decreases due to geometric attenuation as the inverse square of the distance r from the source; i.e.,

$$\phi_R(r) = \frac{s_R}{4\pi r^2}. \qquad (4.1)$$

Absorption, scattering, and other types of particle interactions within a real source (as opposed to a point source) can alter the relationship of source strength to particle flux. Furthermore, the magnitude of the incident particle flux will also depend on whether materials between the source and the individual (or object) absorb or scatter some of the incident radiation. Radiation absorbing materials are referred to as shielding materials when they are used to protect individuals and equipment from radiation exposure. Shielding is discussed in Section 6.

1.3 Dose Concepts

Although the particle flux can be used to express the magnitude of the incident radiation, it does not directly provide the dose absorbed by an individual, organ, or component. Dose is a quantitative measure of the effect due to radiation exposure. A number of dose concepts are used in radiation safety analysis, regulation, and monitoring. The most important concepts and units are discussed here.

Absorbed Dose

When radiation interacts with matter it can cause damage to the material. The extent of damage induced in a given mass of material is roughly proportional to the energy absorbed, called the *absorbed dose D*. The SI unit for absorbed dose is the *gray* (Gy). One gray is equal to an absorbed radiation dose of 1 joule per kilogram (1 J/kg). An older unit of radiation dose is the *rad* (radiation absorbed dose). One rad is an absorbed radiation dose of 100 ergs per gram. Since 1 J = 10^7 ergs and 1 kg = 1000 g, 1 Gy = 100 rad. Although the gray is the newer unit, the rad continues to be widely used.

Biological Effectiveness

The biological effect due to radiation depends on the type and energy of the radiation as well as the magnitude of the absorbed dose. In order to account for the effect of different forms of radiation, a quantity called the *relative biological effectiveness* (RBE) is used. The RBE is defined as the ratio of the absorbed energy from 200-keV X-rays required to produce a given biological effect to the absorbed energy from another radiation to produce the same effect. The RBE of any specific radiation depends on the exact biological effect being considered. A conservative upper limit of the RBE for humans is used as a normalizing factor for adding absorbed doses from various types of radiation. The U.S. Nuclear Regulatory Commission (NRC) calls this normalizing factor the *quality factor Q*. The International Committee on Radiation Protection (ICRP) refers to this normalizing factor as the *radiation weighting factor* W_R. Values assigned to Q by the NRC and to W_R by the ICRP are given in Table 4.1.

An important consideration for determining the biological effects of radiation is the rate of energy loss per unit distance traversed by the radiation. The higher this rate of linear energy transfer (LET), the more effective the radiation is in causing biological damage. Heavy charged particles such as alpha particles and protons have much higher LET than photons or electrons. Neutrons, which create highly charged recoil nuclei when they scatter, are also classified as high-LET radiation. Although alpha particles have a high LET, internal organs are protected from external alpha radiation because alpha particles produced by nuclear decay do not penetrate the dead layer on the surface of human skin. On the

Table 4.1   Quality and Radiation Weighting Factors for Various Radiations

Type of Radiation	Quality Factor, Q	Radiation Weighting Factor, W_R
Xrays, gamma rays, beta particles	1	1
Neutrons		
thermal energy	2	5
0.01 MeV	2.5	10
0.1 MeV	7.5	10
0.5 MeV	11	20
>0.5 to 2 MeV		20
>2 to 20 MeV		5
Unknown energy	10	
High energy protons	10	5
Alpha particles, fission fragments, heavy nuclei	20	20

Source: Adapted from 10 CFR 20 (Q) and ICRP 60 (W_R).

other hand, if alpha-emitting material is ingested or inhaled, alpha radiation can produce cell damage in the living tissue of internal organs with potentially serious consequences.

Dose Equivalent

Radiation-weighted doses H incorporate the relative biological effectiveness of different forms of radiation. Radiation-weighted dose is used for radiological design criteria, safety analysis, and regulatory purposes. The ICRP calls the radiation-weighted dose the *dose equivalent* and uses the *sievert* (Sv) as the unit of measure. The dose equivalent H_T in tissue or organ T is defined by the ICRP as

$$H_T = \sum_R W_R \, D_{T,R}, \tag{4.2}$$

where $D_{T,R}$ is the absorbed dose in gray due to radiation R averaged over the tissue or organ T. If the tissue or organ is irradiated by only one type of radiation, the preceding equation becomes $H_T = W_R \, D_{T,R}$.

According to the values for W_R in Table 4.1, an absorbed dose of 1 Gy of X, beta, or gamma radiation corresponds to an equivalent dose of 1 Sv, while an absorbed dose of 1 Gy of 0.1 MeV neutrons gives an equivalent dose of 10 Sv. In U.S. regulations, the radiation-weighted dose unit is the *rem* (one rem = 0.01 Sv). The preceding equations still hold when H_T is expressed in rem, $D_{T,R}$ is expressed in rad, and W_R is replaced by the quality factor Q_R. For brevity, the generic term *dose* is frequently used for the dose equivalent and is expressed in units of sieverts or rem.

Effectance

The probability of a detrimental effect in any organ is generally taken to be proportional to the dose equivalent in that tissue or organ. Because of differences in sensitivity, however, the proportionality factors differ from organ to organ. The relative sensitivity to detrimental effects is expressed as a tissue-weighting factor W_T. Table 4.2 provides the tissue weighting factors recommended in ICRP 26 [1] and ICRP 60 [2].

If different organs receive different doses, the weighting factors in Table 4.2 are used to calculate an effective dose equivalent or *effectance* H_E as

$$H_E = \sum_T W_T H_T. \tag{4.3}$$

Table 4.2 Tissue Weighting Factors[a], W_T

Tissue or Organ	W_T (ICRP 26)	W_T (ICRP 60)
Gonads	0.25	0.20
Bone marrow (red)	0.12	0.12
Colon	—	0.12
Lung	0.12	0.12
Stomach	—	0.12
Bladder	—	0.05
Breast	0.15	0.05
Liver	—	0.05
Esophagus	—	0.05
Thyroid	0.03	0.05
Skin	—	0.01
Bone surface	0.03	0.01
Remainder[b]	0.30	0.05

[a] The values are based on a reference population of equal numbers of both sexes and a wide range of ages. In the definition of effective dose, they apply to workers, to the whole population, and to either sex.

[b] For purposes of calculation, the remainder is composed of the following additional tissues and organs: adrenals, brain, upper large intestine, small intestine, kidney, muscle, pancreas, spleen, thymus, and uterus. The list includes organs which are likely to be selectively irradiated. Some organs in the list are known to be susceptible to cancer induction. If other tissues and organs subsequently become identified as having a significant risk of induced cancer, they will be included either with a specific W_T or in this additional list constituting the remainder. The latter may also include other tissues or organs selectively irradiated. In those exceptional cases in which a single one of the remainder tissues or organs receives an equivalent dose in excess of the highest dose in any of the 12 organs for which a weighting factor is specified, a weighting factor of 0.025 should be applied to that tissue or organ and a weighting factor of 0.025 to the average dose in the rest of the remainder as defined above.

Combining Eqs. 4.2 and 4.3 we obtain

$$H_E = \sum_T W_R \sum_R W_T D_{T,R}. \tag{4.4}$$

Dose Rate and Committed Dose

Absorbed dose rate $\dot{D}$ and dose equivalent rate $\dot{H}_T$ are the absorbed dose and dose equivalent received per unit time, with units Gy/s and Sv/s, respectively. The dose rate may be time dependent due to changes in source strength or changes in the relative position of the source, exposed organs, and shielding materials. Following ingestion or inhalation of radioactive materials, the dose rate will be time dependent. The time integral of the equivalent dose rate $\dot{H}_T$ over a specified period τ is called the *committed dose equivalent*:

$$H_T = \int_\tau \dot{H}_T(t)dt. \tag{4.5}$$

Similarly a committed effectance is defined as

$$H_E = \int_\tau \dot{H}_E(t)dt. \tag{4.6}$$

Collective Dose Equivalent and Effectance

A *collective dose equivalent* S_T for a specific organ T, and a *collective effectance* S_E have been used to quantify the committed dose equivalents and effectance for exposed groups or populations; i.e.,

$$S_T = N\bar{H}_T, \tag{4.7}$$

and
$$S_E = N\bar{H}_E. \tag{4.8}$$

Here, $\bar{H}_T$ is the time-averaged and population-averaged committed dose equivalent for organ T, $\bar{H}_E$ is the averaged committed effectance, and N is the number of exposed individuals.

1.4 Dose-Flux Relationships

Alpha and beta particle flux can be predicted using decay chain and particle attenuation calculations, and the dose can be calculated from the particle flux. However, the external dose due to alpha radiation is of little consequence, and only intense beta radiation doses to the skin result in significant injury. Neutron and gamma radiation constitute the principal external radiation hazard from space nuclear systems. For space reactors, transport theory or Monte Carlo computer calculations are used to provide neutron and gamma fluxes at a particular location. Approximate methods for computing neutron and gamma fluxes are provided in Section 5 of this chapter. Once the neutron and gamma fluxes have been determined for a particular location, the following relationships can be used to obtain dose rates.

Gamma Radiation

For gamma exposure, the dose rate in Gy/s or Sv/s can be computed using

$$\dot{H}_T(t) = \dot{D}_T(t) = \frac{\phi_{\gamma,T}(t)E_\gamma\mu_\gamma}{\varsigma_T}(1.603 \times 10^{-10}), \tag{4.9}$$

where $\phi_{\gamma,T}$ is the gamma flux (photons/cm^{2}s) incident on organ T, at time t. The parameter E_γ is the energy of the incident photons in MeV/photon, μ_γ is the linear absorption coefficient with units of cm^{-1}, and ς_T is the tissue density of organ T in g/cm^3. The linear absorption coefficient is discussed in Chapter 1. The quantity $\phi_{\gamma,T}E_\gamma$ gives the energy flux in MeV/cm^{2}s, the product $\phi_{\gamma,T}E_\gamma\mu_\gamma$ is the rate of energy absorption per cm^3, and $\phi_{\gamma,T}E_\gamma\mu_\gamma/\varsigma_T$ gives the energy absorption rate per gram of tissue (MeV/g·s). The quantity (1.603×10^{-10}) converts units from MeV/g·s to Gy/s. Because the radiation weighting factor for gamma rays is 1.0, the computed dose rate in Gy/s is numerically equal to the dose equivalent rate $\dot{H}_T$, in Sv/s.

Neutron Radiation

The relationship of flux to dose is more complex for neutrons than for gamma rays; consequently, the relationship of dose rate to neutron flux [(mrem/hr)/(n/cm^{2}s)] is provided in graphical form as a function of energy in Figure 4.1.

Example 4.1

Calculate the dose rate, individual dose, effectance, and collective dose effectance for the following conditions:

- 30 people receive radiation to the thyroid from a 10–millicurie point source of Co-60.
- Each disintegration emits two photons, $\gamma_1 = 1.17$ MeV and $\gamma_2 = 1.33$ MeV.
- The distance from the source is 15 cm and irradiation is for a period of 3 hours.
- Neglect air absorption and assume that only the thyroid is irradiated.
- Use $\mu_\gamma/\varsigma_T = 0.06$ cm^2/g for both photon energies.

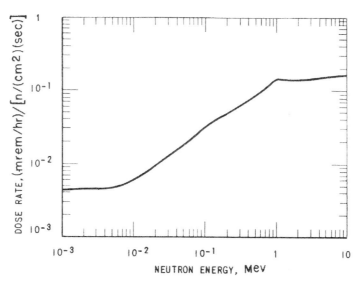

Figure 4.1 Plot for conversion between neutron flux and dose as a function of energy. *Courtesy of Argonne National Laboratory* [3].

Solution:

From Eq. 4.1, we obtain the gamma radiation flux for each photon

$$\phi_{\gamma 1} = \phi_{\gamma 2} = \frac{S}{4\pi r^2} = \frac{10^{-2}\text{Ci}(3.7 \times 10^{10} \text{ photon/s/Ci})}{4\pi(15\text{cm})^2} = 1.31 \times 10^5 \text{ photons/cm}^2\text{s}.$$

Using Eq. 4.9, we obtain

$$\dot{H}_{T,\gamma 1} = (1.31 \times 10^5 \text{ photons/cm}^2\text{s})(0.06 \text{ cm}^2/\text{g})(1.17 \text{ MeV/photon})(1.6 \times 10^{-10} \text{ Sv/MeV/g})$$
$$= 1.47 \times 10^{-6} \text{ Sv/s},$$

and $\dot{H}_{T,\gamma 2} = (1.31 \times 10^5 \text{ photons/cm}^2\text{s})(0.06 \text{ cm}^2/\text{g})(1.33 \text{ MeV/photon})(1.6 \times 10^{-10} \text{ Sv/MeV/g})$
$= 1.67 \times 10^{-6} \text{ Sv/s},.$

The total dose equivalent rate is

$$\dot{H}_T = (1.47 \times 10^{-6} + 1.67 \times 10^{-6}) = (3.14 \times 10^{-6} \text{ Sv/s})(1000 \text{ mSv/Sv})(3600 \text{ s/hr}) = 11.3 \text{ mSv/hr}.$$

For the brief irradiation period, the dose rate is constant; consequently, the individual dose equivalent is simply

$$H_T = (11.3 \text{ mSv/hr})(3\text{hr}) = 33.9 \text{ mSv}.$$

From Table 4.2, $W_T = 0.05$, and only the thyroid is irradiated; thus, the effectance for an individual is

$$H_E = 33.9 \times 0.05 = 1.7 \text{ mSv}.$$

Finally, the collective dose is

$$S_E = 30 \times 1.7 \text{ mSv} = 51 \text{ mSv}.$$

2.0 Radiation Health Effects

Radiation sources are often categorized as either *external* sources, originating outside the body, or *internal* sources, resulting from inhalation or ingestion of radioactive substances. Exposure to either external or internal radiation can cause both *somatic* effects (in which injury is inflicted on the irradiated individual) and *hereditary* effects (genetic effects passed on to the children of an individual receiving germ cell radiation damage). Radiation injuries can be further classified as either *deterministic* or *stochastic* health effects. A deterministic effect is one for which (1) a certain minimum dose must be exceeded before the effect is observed; (2) the magnitude of the effect increases with the size of the dose; and (3) there is a clear, unambiguous causal relationship between the exposure and the observed effect. For example, sunburn is a deterministic effect of overexposure to sunlight. A stochastic effect is one that is characterized by probabilities; it occurs among unexposed as well as exposed individuals. Radiation-induced cancers and genetic effects are stochastic.

Radiation-induced health effects result from either direct or indirect effects. Direct effects result from ionization and excitation of atoms in a cell. Indirect effects result from chemical processes initiated by ionization of water within a cell. Ionization of water molecules leads to the production of the free radicals H•, OH•, and the oxidizing agent H_2O_2. These highly reactive chemical products can interact with organic molecules within the cell. The resulting damage may cause early death of the cell or prevention of cell division. Furthermore, radiation damage can result in molecular changes at specific DNA sites within a cell. Altered DNA that does not result in cell death and does not lead to a loss of ability to divide can result in cancer induction. However, a subsequent sequence of events is required for a mutation to develop into cancer. Birth defects can result from radiation damage to the DNA in germ cells (spermatozoa or egg) that are passed on to progeny. Both radiation-induced cancer and radiation-induced birth defects are stochastic effects. The probability of stochastic effects increases with increasing received dose.

Much effort has been made over the years to determine the effects of radiation on the human body. Based on these efforts considerable information is found on the effects of large, *acute* (short-term) doses in excess of 0.1 to 0.2 Sv. On the other hand, very limited information exists regarding the effects of smaller acute doses or *chronic* (delivered over a long time) doses of millirems per year. In discussing health effects caused by acute doses, *early effects* (evident within 60 days of the exposure) are commonly distinguished from *late health effects* (which occur after 60 days). Early effects are generally deterministic in nature. Late effects arise from both stochastic and deterministic processes.

2.1 Large Acute Doses—Early Health Effects

Large doses received over short time periods threaten both the short and long term health of exposed individuals. If exposures are sufficiently intense, exposed organs are damaged causing radiation sickness or death within days or months. Radiation sickness includes vomiting, diarrhea, loss of hair, nausea, hemorrhaging, fever, loss of appetite, and general malaise. Deaths can be caused by failures of the lungs, small intestine, or blood-forming bone marrow. Barring death or complications, recovery from radiation sickness occurs in a few weeks to a year depending on the dose received. Exposed individuals who survive radiation sickness are still subject to increased risk of late effects.

Radiation-induced sickness and death are deterministic effects. Such effects are not observed until the dose received is greater than the associated threshold dose D_{th}. Once the threshold dose has been exceeded, the fraction of the exposed population in which the health effect occurs (the health effect's incidence) rises rapidly with increasing dose until the effect appears in all of the exposed individuals. Estimates of the threshold for radiation sickness caused by whole body γ-ray exposure range from 0.5 to 1 Gy. The threshold for early deaths due to whole body γ-ray exposures is about 2.5 Gy. The acute dose that leads to a deterministic effect of half of the exposed population within t days is called the ED_{50}/t dose (LD_{50}/t if the dose is lethal). Without supportive medical treatment, about 50% of the people who receive a whole-body-ray dose of 3 Gy would die within 60 days ($LD_{50}/60 = 3$ Gy). The $LD_{50}/60$ dose has been estimated to increase to 4.5 Gy with supportive medical treatment [4].

In the range of doses from about 1 to 10 Gy, the most important effects are those to the blood-forming organs, especially the red bone marrow. The resulting *hematopoietic syndrome* weakens the body's defenses to infection and prevents normal clotting of the blood, which in severe cases may cause hemorrhage and internal bleeding. Medical treatment for patients suffering from hematopoietic syndrome includes isolation in a sterile environment and the administration of antibiotics. Bone-marrow transplants have been attempted with little success. The use of growth agents has been proposed and offers considerable promise based on animal test data.

An acute whole body dose between 10 and 50 Gy leads to what is known as the *gastrointestinal syndrome*. The dominant effect and ultimate cause of death is failure of the intestinal wall. The individual remains in satisfactory condition for a few days while the existing wall cells continue to function, but as these cells are sloughed off, the patient succumbs to infection and usually dies within 2 weeks. Persons who receive doses in excess of 50 Gy die within a few hours of exposure. The cause of death is not entirely clear, but probably involves the rapid accumulation of fluid in the brain. The symptoms are known as the *central nervous system syndrome*. If only a portion of the body is exposed, the resulting early effects depend upon which organs are irradiated. Based on animal data, exposure to the lungs alone can cause death due to *respiratory impairment* for low-LET doses exceeding 10 Gy. Lung impairment is, therefore, a factor in deaths attributed to the hematopoietic and gastrointestinal syndromes.

2.2 Late Health Effects

Populations receiving doses or dose rates insufficient to cause early fatalities are subject to increased risk of cancer or other health effects in 2 to 30 years. Cancers may also result from chronic exposures to low levels of radiation. Experimental data suggest that the probability of incurring health effects due to exposure of a population at low doses and low dose rates can be two-to-ten times smaller than the probability for the same collective dose at high doses and high dose rates [5]. A brief discussion of potential late or chronic effects is provided here.

Cancer: Data on human exposure to acute radiation doses above 1 Sv indicate that the excess incidence of cancer increases approximately linearly with dose for both low-LET and high-LET radiation. A plot of the excess cancer incidence versus dose, called the *dose-response curve* is therefore a straight line at high doses, as shown in Figure 4.2. The cancer risk posed by small acute doses or moderately

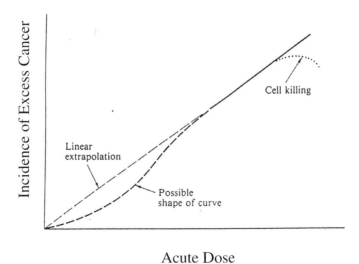

Figure 4.2 Dose-response curves for acute radiation doses.

large doses received at low dose rates (e.g., long term exposure to low levels of ground contamination) are more controversial. Statistically significant associations of radiation doses with cancer have not been demonstrated below about 0.05 to 0.1 Sv. The traditional practice has been to extrapolate the dose-response curve linearly to zero dose as indicated in Figure 4.2. This procedure is referred to as the *linear hypothesis*. It is generally agreed that such an extrapolation from high doses is proper for high-LET radiation. For low-LET radiation, on the other hand, there is evidence to suggest that linear extrapolation overestimates the cancers attributable to chronic or small acute doses because it ignores the capability of biological systems to repair themselves. That is, the evidence suggests that the actual dose response curve lies below the curve linearly extrapolated from high doses, as depicted in Figure 4.2. A factor-of-two reduction in the incidence-per-dose relationship is commonly used for low doses or low dose rates. Using this reduction factor, the ICRP estimates that a collective dose of 100 person-Sv (0.1 Sv to 1000 people, 10^{-3} Sv to 100,000 people, etc.) will result in five radiation-induced cancer fatalities in an affected general population [6].

Thyroid Nodules: Thyroid nodules can result from direct radiation or inhalation of radioiodine. This effect is noteworthy because recent events indicate that the latency period for thyroid cancers appears to be quite short. Increased incidence rates of thyroid cancer in children and adolescents following the Chernobyl nuclear power plant accident of April 1986 were observable within 6 years of the accident. Such incidences continue to be monitored. Thyroid abnormalities are treatable, so fatalities from thyroid damage are rare.

Cataracts: Radiation-induced cataracts, which impair vision by clouding the lenses of the eyes, may appear from 10 months to 35 years after exposure. This is a deterministic effect. The threshold occurs in the 2–5 Gy range for low-LET radiation. Populations exposed to whole body doses this large may have few survivors with radiation-induced cataracts. The threshold is lower for high-LET radiation, 0.75 to 1 Gy for fission neutrons.

Sterility: The threshold for brief γ-induced sterility in humans, both male and female, is about 1.5 Gy to the gonads. Doses exceeding of 2.5 Gy can cause sterility for 1 to 2 years. Permanent sterility can be caused by doses exceeding 5 Gy and is virtually certain for doses exceeding 8 Gy.

Degenerative Effects: Radiation exposure can also cause degenerative conditions in various body organs due to the failure of the exposed tissues to regenerate properly. This leads to permanent though not necessarily debilitating impairment of organ functions.

Life Shortening: The overall effect of radiation exposure may be seen in its influence on life span. It is generally believed, however, that life shortening from radiation exposure is due solely to the appearance of radiation-induced cancers.

Mutations: Genetic effects of radiation exposure in humans have not been demonstrated at the present time. The risk of such effects has been estimated from data on laboratory mice. The dominant cause of genetic effects in mice is damage to male spermatozoa. Studies indicate that high dose rates produce many more mutations per unit dose than low dose rates. This implies the damage is clearly repairable. The transmission of genetic damage from acute doses of radiation can therefore be reduced by delaying conception until new sperm cells have matured from cells in a less sensitive stage at the time of exposure. Quantitative estimates of the risk of radiation-induced genetic effects in humans are complicated and highly uncertain.

Mental Retardation: Studies of children who were irradiated in utero and survived the nuclear bombings in Japan show a dose-related increase in mental retardation. The effect is especially pronounced for those exposed during the 8th through the 17th week of pregnancy when the number of neurons in the developing brain increases rapidly. Six out of 9 individuals who received doses in excess of

1 Gy during the 8th to 17th week of pregnancy developed mental retardation, and the effect is clearly observable for doses as low as 0.1 Gy [7].

3.0 Radiological Regulations and Guidelines

Normal operations of nuclear systems deployed in space have no radiological effect on Earth's biosphere; however, astronauts in the vicinity of operating systems may be exposed to radiation. Astronauts are also exposed to the natural radiation environment of space. A number of ground activities are carried out in support of space nuclear programs, such as fabrication, testing, and transportation of nuclear materials and systems. Radiation workers, and possibly the public, may be exposed to radiation from these activities. Space nuclear programs must follow the established policy to assure that radiation exposure to astronauts, radiation workers, and the public are kept as low as reasonably achievable (ALARA). In addition, specific radiation protection guidelines have been established to protect astronauts both from natural radiation and from radiations produced by on-board nuclear systems. Radiation regulations are also established for radiation workers and for the general public for all ground activities preceding launch.

3.1 Radiological Guidelines for Astronauts

In the United States, NASA has established radiological regulations for astronauts. These regulations include OSHA 29 CFR 1910.96 [8] and the recommendations of the National Council on Radiation Protection and Measurements (NCRP 98) [9]. Federal regulation 20 CFR 1910.96 applies to astronaut exposure to radiation from on-board nuclear systems. These regulations are equivalent to 10 CFR 20 [10] worker dose limits, presented in Table 4.3. The NCRP 98 regulations apply to astronaut exposure to both on-board nuclear systems and to exposure to the natural radiation environment of space. The NCRP recommends the guidelines given in Table 4.4 for the total of both natural and on-board radiation sources. These guidelines are recommended for all space activities except for . . . *exceptional exploratory circumstances in space (e.g., Mars mission, or some such)*. . . . Radiation protection from natural radiation sources is beyond the scope of this book; however, a brief summary of natural radiation sources is given in Section 4.2 to provide perspective. Note that both 29CFR1910.96 and NCRP 98 must be followed.

3.2 Radiological Guidelines for Ground Operations

Regulations for ground operations, given in Table 4.3, are governed by Department of Energy Order 5480 [11]; these regulations are essentially the same standards as Code 10 of Federal Regulations, Part 20 (10 CFR 20). Radiation regulations for the general public are presented in Table 4.5. Addi-

Table 4.3 NASA Guideline 29 CFR 1910.6 Astronaut Exposure Guidelines for On-board Nuclear Sources and DOE Order 5480 Radiation Worker Guidelines for Ground Activities

- Occupational exposures (annual)
  1. Effectance (stochastic effects) — 50 mSv
  2. Dose equivalent limits for tissues and organs (non-stochastic effects)
     a. Lens of eye — 150 mSv
     b. All others — 500 mSv
        (e.g., red bone marrow, breast, lung, gonads, skin, and extremities)
  3. Guidance: Cumulative exposure — 10 mSv × age

- Embryofetus exposures*
  1. Total dose equivalent limit — 5 mSv
  2. Dose equivalent limit in a month — 0.5 mSv

* Sum of internal and external exposures

Table 4.4 NASA Guideline NCRP 98 Astronaut Total Exposure Guidelines for All Radiation Sources (Natural and Onboard)

(a) Career Whole-body Dose Equivalent Limit (Sv) for a Lifetime Excess Risk of Fatal Cancer of 3% as a Function of Age at Exposure

Age	25	35	45	55
Male	1.5	2.5	3.25	4.0
Female	1.0	1.75	2.5	3.0

(b) Recommended Organ Dose Equivalent Limits (Sv) for All Ages

	Blood-forming Organs	Eye	Skin
Career	See Table 4.4 (a)	4.0	6.0
Annual	0.50	2.0	3.0
30 Days	0.25	1.0	1.5

tional radiation regulations are provided in 10 CFR 20 for doses from airborne radioactive materials and radioactive contaminants in water. These regulations include exposure limits for radiation workers and for releases into the environment. Inhalation or ingestion of radioactive materials will result in internal radiation exposure. The chemical form of the inhaled or ingested radioactive material, particularly its solubility in body fluids, will determine its deposition within the body. Typically, one critical organ receives most of the radiation damage from a particular radioactive substance; for example, bone and the thyroid are the principal recipients of radiation from strontium and iodine, respectively. Regulations for radioactive concentrations in air and water are given for both soluble and insoluble forms of the radioisotope.

Radiation regulations for transportation of nuclear materials are provided in 10 CFR 71 [12]. For small quantities of radioactive materials, a Type-A shipping container is acceptable. Type-A shipping containers are not required to meet strict standards for containment of materials. Radioactive materials in quantities exceeding limits for Type-A packages require a Type-B package for shipment. Very rigid standards are specified for Type-B packages. The development and qualification of a large Type-B shipping package can be very expensive. Transportation of radioactive materials to the launch site is an important consideration in the design of space nuclear systems. This consideration is especially important for large space reactor systems.

Radiological protection also includes operational practices during fabrication and testing. These standards and practices are well established for most ground activities for space nuclear systems. Future missions may require a prelaunch criticality test at the launch site. No precedence has been established for a launch site criticality test in the United States; consequently, if this option is desired, the regulatory requirements and facility modifications needed for launch site criticality testing must be addressed.

Table 4.5 DOE, NCRP, ICRP Public Radiological Exposure Regulations Applicable to Ground Activities

- Public exposures (annual)
  1. Effectance limit, continuous or frequent exposure   1 mSv
  2. Effectance limit, infrequent exposure   5 mSv
- Public exposures (in any one hour)
  Effectance limit, continuous or frequent exposure   0.02 mSv
  - Negligible individual risk level (annual)
    Effectance per source or practice   0.01 mSv

Table 4.6   U.S. Reporting and Review Classification for the Launch of Radioisotopes

Categories	Radiotoxicity Groups			
	I	II	III	IV
Category A NASC Staff Review	≤ 20 Ci	≤ 20 Ci	≤ 200 Ci	≤ 200 Ci
Category B Agency Approval NASC Quarterly Report	≤ 1.0 mCi	≤ 50 mCi	≤ 3 Ci	≤ 20 Ci
Category C No Reports	See Appendix B of Reference [13]			

3.3 Radiological Guidelines for Launch Review

INSRP approval is not required for the launch of space systems with minor quantities of radioactive sources onboard. The approval process for minor quantities of radioisotope sources is specified in a National Aeronautics and Space Council (NASC) report [13]. Three levels of reporting are required for minor radioisotope sources. The levels are determined by the total activity of the sources and the radiotoxicity of the radioisotopes as shown in Table 4.6. Radioisotopes are grouped into four groups of radiotoxicity in Appendix A of Reference [13]. Group I includes the most hazardous radioisotopes such as the plutonium isotopes. Group IV includes the least hazardous radioisotopes such as Mo-99. Category A quantities of radioactive materials requires submission of a report by the user agency for review by NASC staff prior to each launch. Launch of Category B material requires user-agency approval and submission of a quarterly report to NASC listing nuclear materials for planned launches. No reporting is required for Category C quantities of radioactive materials.

4.0 Natural Radiation Sources

Radiation exposure can result from either radiations produced by space nuclear systems or from natural radiation sources. In this text, discussion of radiation protection methods and analysis is limited to on-board radiation sources. Nonetheless, natural sources are briefly discussed to provide a reference for comparing astronaut exposure due to on-board sources.

4.1 Earth Radiation Sources

Exposures from natural radiation on Earth can come from a variety of sources. As indicated in Figure 4.3, an average individual receives about 0.4 mSv (40 millirems) each year from radionuclides in soil, rock, and building materials that surround us. On the East Coast, the average dose is about 0.2 mSv; near the Rocky Mountains, the dose is about 0.9 mSv. Earth's atmosphere shields the biosphere from the intense radiation sources found in space; nonetheless, cosmic rays account for an annual dose of about 0.3 mSv to individuals at sea level. The dose increases to 1.6 mSv for persons in the mile-high city of Denver. Radionuclides in food, water, and the human body account for an additional 0.4 mSv per year. The average annual dose from radon gas is about 2.0 mSv. For the United States, the average annual dose from all sources is about 3.7 mSv (370 mrem); however, differences in location, medical procedures, airline flights, and other factors can result in significant variations about this average.

4.2 Space Radiation Sources

Natural radiation in near-Earth space (up to geosynchronous orbit) has three primary components: solar radiation (particularly solar flare particles), galactic cosmic radiation, and radiation produced by trapped particles (Van Allen belts). Space radiation levels vary substantially both with time and with distance from the Earth because all three components are influenced by solar activity and the Earth's mag-

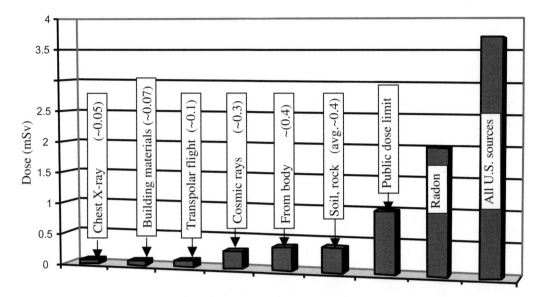

- Chest X-ray and transpolar flight are single event doses; all other are annual doses.
- Soil, rock, cosmic ray, and radon doses are U.S. averages.
- Public dose limit is for continuous or frequent exposure to radiation from nuclear systems.

Figure 4.3 Comparison of various dose sources with public dose limit.

netic field. Other celestial bodies may also possess unique radiation fields. Spatial and temporal fluctuations must be taken into account while planning space missions [14].

Solar Wind, Flares, and Proton Events

In addition to visible and ultraviolet radiation, a steady stream of charged particles (mostly electrons, protons, and alpha particles) flows outward from the Sun in what is termed the *solar wind*. The solar wind blows outward past the planets, filling the solar system with charged particles. Usually the speed past Earth is about 500 km/s with a density of about 100 particles/cm^3. Solar activity varies in intensity with an 11-year cycle, and the solar wind increases in both speed and density during periods of high solar activity. Even during periods of high solar activity, the exposure to the solar wind does not pose a significant threat to astronauts or spacecraft. The electrons and protons in the solar wind are slow enough that they can be stopped by spacesuits.

A more significant threat to astronauts is posed by protons expelled into space during solar flares. Flares may last from a few minutes to a few hours. Even small flares may be larger than Earth, releasing as much energy as a billion hydrogen bombs. The occurrence-frequency for flares follows the 11-year sunspot cycle. As the number of sunspots increases, so does the number of flares. The largest flares, however, often occur after the sunspot cycle reaches it maximum. Bursts of visible light, X-rays, ultraviolet light, and radio noise are observed with flares. In addition, some charged particles, mostly electrons and protons, are accelerated to high speeds, exceed escape velocity from the Sun, and are propelled into space. Occasionally, during very intense flares, protons reach speeds up to one-fourth the speed of light and fit into the classification of *solar cosmic rays*. A solar cosmic event, also referred to as a *proton event* or *flare*, may last from several hours to a day. Protons with energies above 10 MeV can penetrate spacesuits (of current design), and protons with energies above 30 MeV are capable of penetrating the protective shell of a typical satellite.

Galactic Cosmic Radiation

Galactic cosmic radiation (GCR) originates outside our solar system. It is comprised of high-energy (greater than 0.1 GeV) protons, electrons, and other heavy charged particles. When first discovered in the early 1900s, such particles were thought to be electromagnetic radiation and were dubbed *cosmic rays*. The composition of cosmic radiation is 85 to 90% protons, 10 to 12% alpha particles, about 1% electrons, and 1% nuclei of heavy atoms such as oxygen, nitrogen, iron, and neon. Some of these particles may have originated shortly after the "big bang." Others come from proton events on distant stars and from exploding dying stars called supernovas. These particles travel through space and arrive at earth from all directions.

Galactic cosmic radiation is biologically important despite its low density. Many galactic cosmic rays travel at such high speed that they pass right through the spacecraft at a rate of perhaps two per cm² per hour. Doses received from cosmic radiation particles tend to be proportional to the square of the charge on the interacting particle. For spacecraft, both the altitude of the vehicle and the inclination of the orbit are important in determining the radiation dose rate that would be received due to galactic cosmic radiation. Spatial variations in galactic cosmic radiation flux are produced by variations in source location, the Earth's magnetic field, atmospheric shielding, and altitude. Low-altitude, low-inclination orbits experience much smaller GCR doses due to strong shielding produced by the combined effects of the atmosphere and the geomagnetic field. Doses for such orbits tend to be associated with trapped radiation belts. The most important temporal variation in flux is associated with the 11-year solar cycle.

Trapped Radiation Belts

The Van Allen belts are two geomagnetically trapped belts of radiation that surround the Earth. The inner belt consists of both protons and electrons and extends out to about 12,000 km. The peak of this belt ranges between 2,000 and 5,000 km. The outer belt, which contains mainly electrons, extends from about 16,000 to 36,000 km with a peak at an altitude of about 20,000 km. The two Van Allen belts are separated by a region of relatively low intensity. The most hazardous regions within the belts occur at the maximum densities of the most energetic particles. Energetic protons trapped in the inner belt are the major source of radiation for Earth orbiting spacecraft above 500 km. The amount of radiation varies with latitude and longitude (the inner belt extends to about 45° latitude). The inner belt proton population is also susceptible to solar induced variations. Particle densities vary out of phase with the 11-year solar cycle so that the inner belt is most concentrated during the solar minimum.

The outer belt is asymmetric, with the nightside being elongated and the dayside flattened. Generally, particle energy and outer boundary location vary with the 11-year solar cycle. During solar maximum, the outer boundary of the electron belt is closer to the Earth and contains higher energy particles. At solar minimum, the outer boundary moves outward and contains fewer energetic electrons. Outer belt electron densities undergo order of magnitude changes over time scales of weeks. The short-term variations can produce significant radiation dose variations and are related to the level of geophysical activity. During or shortly after very active periods, the high-energy electron density increases in the outer belt increasing the radiation hazard substantially. Diurnal variations in radiation dose inside a spacecraft in high-altitude circular orbit can occur when the trajectory crosses the asymmetric outer electron belt.

4.3 Perspective on Operational Doses

As stated in Section 3.1, the radiation dose limit to astronauts from space nuclear systems (equivalent to 10 CFR 20) and dose limits for natural sources plus onboard sources (NCRP 98) apply to astronauts

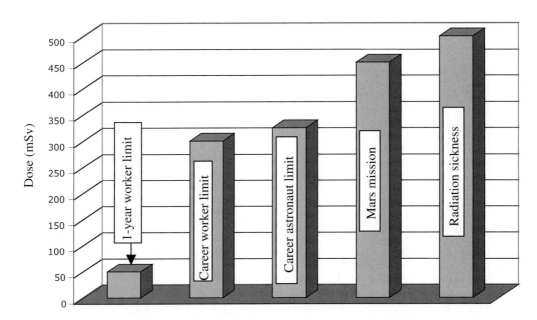

- Worker limit applies to terrestrial radiation workers and dose to astronauts due to space nuclear systems.
- Astronaut career limit is the total dose permitted from all radiation sources for a 45-year-old male astronaut.
- Mars mission dose to astronauts is due to natural radiation only.
- Radiation sickness dose is the lower limit of possibility.

Figure 4.4 Comparison of astronaut dose limits with radiation worker limits, health effects, and dose from natural sources for a Mars mission.

on space missions. The NCRP guidelines allow much higher exposure levels from natural radiation sources than from on-board nuclear systems. Figure 4.4 compares astronaut dose limits to doses from natural radiation sources for a projected Mars mission and to levels required for identifiable health effects. For this projected mission, the dose predicted from natural radiation sources (450 mSv) is much greater than that permitted from nuclear power systems (50 mSv). In fact, the dose level exceeds the highest career exposure limit recommended by the NCRP. Approximate values for sources of radiation during the proposed Mars mission are as follows:

• Passage through belts	50 mSv
• Galactic cosmic radiation (10 months in transit)	380 mSv
• Solar proton events	(cannot be predicted)
• On Mars surface	20 mSv
Total	450 mSv

Solar proton events cannot be predicted; consequently, doses received from solar proton events are not provided. Special *storm shelters* may be required to protect the crew during solar proton events.

The NCRP guidelines do permit exposures beyond the numerical values given in Table 4.4 for exceptional exploratory circumstances. Nonetheless, lower exposures may be achievable if a high-specific impulse nuclear propulsion system is used to reduce trip time. Thus, the use of a space nuclear system may significantly reduce radiation exposure for astronauts.

5.0 Radiation from Space Nuclear Systems

Radiation exposure from space nuclear systems must be considered for both normal operation and for postulated accident conditions. Radiation exposure in space can be a consideration if astronauts are in the vicinity of a space nuclear system or in an area contaminated as a result of an accident. Astronaut exposure for normal operation is limited by occupational radiological regulations. Exposure is controlled by providing radiation shielding, requiring a minimum separation distance between personnel and the source, maintaining an exclusion area, limiting exposure time, or combinations of these measures.

Radiation exposure on Earth must be considered for all mission phases including manufacture, testing, transport, launch, deployment, operation, and disposal. Normal operation of nuclear systems in space does not present a risk to Earth's environment or populace. Radiation hazards on Earth, for the launch through disposal phases, are only a consideration for postulated accidents in which systems either fail to properly deploy in space or undergoe an unplanned reentry of the system following deployment. During the launch and deployment phases the safety focus is on assuring very low consequences for postulated accidents. During the operational and disposal phases, the focus is generally on assuring a low probability of accidental reentry.

5.1 Radiation from Radioisotope Sources

Shielding requirements for radioisotope sources will depend on the radioisotope used, the quantity of radioactive materials, and the proximity to workers and astronauts. Only modest shielding is needed to protect astronauts from radiation produced by Pu-238 radioisotope sources. Radiological data for Pu-238 is summarized in Table 2.3 in Chapter 2. A Pu-238 radioisotope source produces heat by α-emission resulting from nuclear decay. The charged α particle cannot penetrate encapsulating materials; however, spontaneous fission and α-n reactions with source materials (e.g., oxygen in oxide fuels) can produce a small fast neutron flux. Gamma rays are also produced during spontaneous fission and during neutron interactions with source materials. Although the neutron and gamma radiation from a typical PuO_2 source is very small, neutron and gamma radiation can penetrate thin shielding. As a consequence, shielding requirements for radioisotope sources are dictated by secondary neutron and gamma radiations rather than from primary radiation due to alpha decay. The neutron production rate due to spontaneous fission in Pu-238 is 3420 n/s per gram of Pu metal and the rate due to α-n reaction with oxygen in PuO_2 is 1.4×10^4 n/s per gram of PuO_2. Measured neutron and gamma dose equivalents for GPHS-RTGs are presented as a function of distance from the source in Figure 4.5 (a) and (b), respectively. Other choices for radioisotope fuels are possible, and some choices may include beta radiation as well.

The most significant issue associated with radioisotope sources is their very large inventory of radioactive material during and prior to launch. For Pu-238, the specific activity is 17.56 Ci/g. The total activity for the Transit power source was about 25,000 Ci, and an 18-module GPHS contains 130,000 curies of plutonium. The principal radiological concern in this case is the possibility of inhalation or ingestion of alpha-emitting materials following a postulated disruptive accident. For radioisotope sources, these radioactive materials are present during prelaunch, launch, and deployment, as well as during operation and disposal. Radioisotope sources are designed and tested to assure that containment barriers cannot be breached for credible reentry accidents.

For some missions, radioisotope sources are left in a very high Earth orbit so that radioactive materials will have decayed to low levels before reentry can occur. In order to determine a Sufficiently High Orbit (SHO), one must consider all radioisotopes in the radioisotope source, as well as daughter products. From Table 2.3 we find that a typical source also contains Pu-239, Pu-240, Pu-241, and

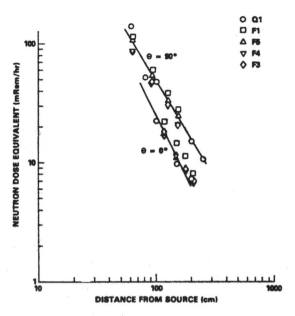

(a) Neutron radiation measurements

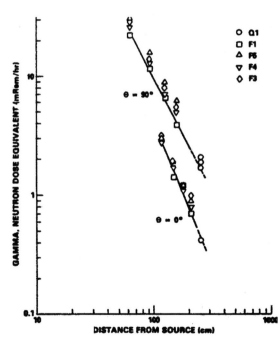

(b) Gamma radiation measurements

Figure 4.5   Radiation measurements for GPHS-RTGs [15].

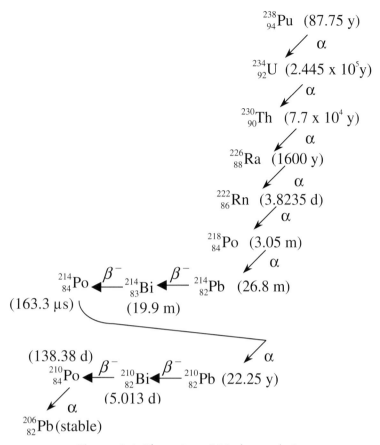

Figure 4.6 Plutonium-238 decay chain.

Pu-242. Each of these radioisotopes decays into another radionuclide. The full decay chain for Pu-238 is illustrated in Figure 4.6. As discussed in Chapter 1, the rate of decay Pu-238 is given by

$$\frac{dN_1(t)}{dt} = -\lambda_1 N_1(t), \tag{4.10}$$

with the solution

$$N_1(t) = N_1(0)\, e^{-\lambda_1 t}. \tag{4.11}$$

Here, N_1 and λ_1 are the atom density and decay constant for Pu-238. The half-life for Pu-238 is 87.75 years which gives a decay constant of $\lambda_1 = \ln2/87.5\ y = 7.9 \times 10^{-3}\ y^{-1}$. Plutonium-238 decays into its daughter product U-234, and the rate of change in the atom density N_2 of U-234 is given by

$$\frac{dN_2(t)}{dt} = \lambda_1 N_1(t) - \lambda_2 N_2(t), \tag{4.12}$$

The atom density of U-234 at time t is

$$N_2(t) = N_1(0)\frac{\lambda_1}{\lambda_2 - \lambda_1}\left(e^{-\lambda_1 t} - e^{-\lambda_2 t}\right). \tag{4.13}$$

The process can be continued, usually by using computer programs, to obtain the time dependent atom densities of all radionuclides in the chain. The atom densities are then used to determine the time dependent activity from

$$A(t) = V \sum_{n=1}^{12} \lambda_n N_n,$$ (4.14)

where V is the source volume.

Example 4.2

Determine the time-dependent activity of Pu-238 and the combined activities of Pu-238 and U-234 for a 60 g source of PuO_2. Assume a PuO_2 density of 10 g/cm^3 and assume that the isotopic content of the source is 90% Pu-238. For this simple example we ignore the activity due to other plutonium isotopes and other daughter products.

Solution: The initial atom density for Pu-238 is given by

$$N_1(0) = \varsigma_{PuO_2} \frac{N_A}{MW_{PuO_2}}.$$

thus $\qquad N_1(0) = 10 \text{ g/cm}^3 (0.9) \dfrac{6.022 \times 10^{23} \text{ atoms/gmole}}{[238 + 2(16)] \text{ gmole}} = 2.0 \times 10^{22} \text{ atoms/cm}^3.$

From Fig. 4.6 $\quad ^{238}\lambda = 0.693/(87.75 \text{ y}) = 7.9 \times 10^{-3} \text{ y}^{-1},$

and $\qquad\qquad ^{234}\lambda = 0.693/(2.445 \times 10^5 \text{ y}) = 2.83 \times 10^{-6} \text{ y}^{-1}.$

From Eqs. 4.11 and 4.13

$$N_1(t) = (2.0 \times 10^{22}) e^{-(7.9 \times 10^{-3}) t},$$

and $\qquad N_2(t) = (2.0 \times 10^{22}) \dfrac{7.9 \times 10^{-3}}{(2.83 \times 10^{-6}) - (7.9 \times 10^{-3})} (e^{-7.9 \times 10^{-3} t} - e^{-2.83 \times 10^{-6} t}).$

From Eq. 4.14 the calculated activities are given by

$$A_1(t) = \frac{60 \text{ g}}{10 \text{ g/cm}^3} \frac{(7.9 \times 10^{-3} \text{y}^{-1}) N_1(t)}{(3.7 \times 10^{10} \text{ dis/s/Ci})(3.15 \times 10^7 \text{ s/y})},$$

and $\qquad A_{(1+2)}(t) = \dfrac{60 \text{ g}}{10 \text{ g/cm}^3} \dfrac{[(7.9 \times 10^{-3} \text{ y}^{-1}) N_1(t) + (2.83 \times 10^{-6} \text{ y}^{-1}) N_1(t)]}{(3.7 \times 10^{10} \text{ dis/s/Ci})(3.15 \times 10^7 \text{ s/y})},$

The calculated activity vs. time is presented in Figure 4.7.

5.2 Radiation from Reactor Systems—Normal Operation

The radiological considerations for space reactors differ from those for radioisotope sources. Whereas the radiological inventory of radioisotope sources can be very high at launch, a reactor that contains fresh fuel at launch emits very little radiation. On the other hand, nuclear fission in reactors produces

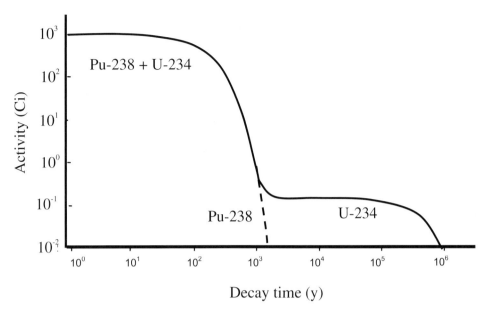

Figure 4.7 Activity due to decay of a 60 g plutonium-238 source and its daughter product uranium-234 from Example 4.2.

intense neutron and gamma radiation during operation. If astronauts are in the vicinity of an operating space reactor, radiation shielding is needed for protection from emitted neutron and gamma radiation. Radiation protection will also be required for operational staff during ground testing of space reactor prototypes and criticality testing of space reactors before launch. For the purpose of understanding radiation protection issues and mitigation methods, the following simple approximations can be used to estimate neutron and gamma source strengths.

Neutron Source Strength

We can define an average volumetric fission source strength $\bar{S}_f$ (using 200 MeV per fission) as

$$\bar{S}_f = \left[\frac{6.25 \times 10^{12}\ \text{MeV/watt·s}}{200\ \text{MeV/fiss}}\right]\left[\frac{P(\text{watt})}{V(\text{cm}^3)}\right] = (3.1 \times 10^{10})\frac{P}{V}\ \text{fiss/cm}^3\text{·s},\qquad (4.15)$$

where P is the thermal power in watts and V is the volume of the reactor in cm^3. The reactor volume includes all reactor components, including the fuel, moderator, structure, coolant, voids, and reflectors. Alternatively, the volume in Eq. 4.15 can be limited to the core, and the reflectors can be treated as part of the radiation shield. Although the latter is a better approximation, the first approach is simpler and will be used in our discussions. The average volumetric neutron source strength $\bar{S}_n$ can be computed from the reactor power density by multiplying Eq. 4.15 by the neutron yield per fission (approximately 2.5), to obtain

$$\bar{S}_n = 7.8 \times 10^{10}\frac{P}{V}\ \text{n/cm}^3\text{s}.\qquad (4.16)$$

Glasstone and Sesonske [16] have shown that the average volumetric neutron source strength can be used to estimate an equivalent surface strength for neutrons S_n at the surface of a reactor as

$$S_n \approx \frac{\bar{S}_n}{\mu_{r,n}}\ \text{n/cm}^2\text{·s},\qquad (4.17)$$

where $\mu_{r,n}$ is the reactor attenuation coefficient for neutrons, with units of cm^{-1}. The neutron attenuation coefficient is approximately equal to the macroscopic neutron removal cross section, discussed in Section 4.2, for the reactor core. The reactor attenuation coefficient approximately accounts for the attenuation of the fast neutron flux within the reactor volume. This effective attenuation coefficient includes the attenuation due to all reactor components, such as fuel, moderator, structure, and the reflector. The components used in the computation of $\mu_{r,n}$ must be consistent with the components included in the reactor volume in Eq. 4.16. The equivalent surface source strength given by Eq. 4.17 is based on the assumption of a uniform volumetric source strength. This approximation tends to overestimate the surface source strength because the reactor fast neutron flux is usually lower at the outer region of the reactor.

Example 4.3

A moderated space reactor will operate at 100 kWe, with a net efficiency of 22%. Calculate the average volumetric neutron source strength at full power operation and estimate the neutron flux at the reactor surface. The reactor is a cylinder with a radius of 30 cm and a height of 60 cm. Assume a reactor neutron attenuation coefficient of 0.09 cm^{-1}.

Solution:

The reactor volume is $V = \pi(30\text{cm})^2(60\text{cm}) = 1.7 \times 10^5$ cm^3.

We compute the average volumetric neutron source strength using Eq. 4.16,

$$\bar{S}_n = 7.8 \times 10^{10} \frac{100 \times 10^3/0.22}{1.7 \times 10^5} = 2.09 \times 10^{11} \text{ n/cm}^3\text{s}.$$

Using Eq. 4.17, the surface source strength is approximately

$$S_n \approx \frac{2.09 \times 10^{11} \text{ n/cm}^3 s}{(0.09 \text{ cm}^{-1})} = 2.3 \times 10^{12} \text{ n/cm}^2\text{s}.$$

Gamma Source Strength

Gamma rays are released during the fission process and during the decay of fission products. Gamma radiation is also produced by inelastic scattering of neutrons, by neutron capture, and by neutron activation of materials. The energy range for reactor-produced gamma photons is fairly broad. For dose calculations it is common to divide the spectrum of gamma radiation into a number of energy groups, such that each group is represented by a single gamma ray energy. The spectrum for each energy group is provided in Table 4.7 for gamma radiation due to both fission and fission product decay. The fission product gamma spectrum corresponds to equilibrium fission product concentrations. Spectra are given in terms of MeV of gamma energy per fission, for both prompt and decay sources, at photon energies of 1, 2, 4, and 6 Mev.

Table 4.7 Four-Energy-Group Gamma Ray Energy per Fission for Uranium-235 Prompt Fission and Fission Product Gamma Rays [3]

Gamma Source	Photon Energy (MeV)			
	1	2	4	6
Prompt Fission (MeV/fiss)	3.45	3.09	1.04	0.26
Fission Products (MeV/Fiss)	5.16	1.74	0.32	—

In order to estimate the equivalent surface energy source strength (MeV/cm²·s) for gamma rays S_γ for a reactor we can use an equation similar to Eq. 4.17; i.e.,

$$S_\gamma = E\kappa_\gamma \frac{\bar{S}_f}{\mu_{r,\gamma}}, \qquad (4.18)$$

where $\mu_{r,\gamma}$ is the reactor attenuation coefficient for gamma rays, with units of cm^{-1}. The parameter κ_γ is the gamma production rate in photons per fission, and E is the photon energy in MeV. The gamma ray attenuation coefficient depends on the reactor composition and on the energy of the gamma photons under consideration. In order to obtain the surface source strength in photons/cm²s, the right-hand side of Eq. 4.18 must be divided by the energy per photon E. As discussed subsequently, the approximation given by Eq. 4.18 does not account for gamma flux buildup due to multiple scattering in the reactor.

Example 4.4

Estimate the 2 MeV volumetric gamma source strength for the case described in Example 4.3, and compute the equivalent surface source strength for gamma rays in MeV/cm²s and photons/cm²s. Assume an effective reactor gamma ray attenuation coefficient of 0.08 cm^{-1}. (Ignore multiple scattering, capture gamma rays, and activation effects.)

Solution:

From Eq. 4.15 and Example 4.3 we obtain

$$\bar{S}_f = 3.1 \times 10^{10} \frac{100 \times 10^3/0.22}{1.7 \times 10^5} = 8.3 \times 10^{10} \text{ fiss/cm}^3\text{s}.$$

From Table 4.7, an average of 4.83 MeV per fission ($E\kappa_\gamma$) is due to 2 MeV photons; thus,

$$S_\gamma = \frac{4.83 \text{ MeV/fiss} \times 8.3 \times 10^{10} \text{ fiss/cm}^3\text{s}}{(0.08 \text{ cm}^{-1})} = 5.01 \times 10^{12} \text{ MeV/cm}^2\text{s},$$

or in terms of photons,

$$S_\gamma = \frac{5.01 \times 10^{12} \text{ MeV/cm}^2\text{s}}{2 \text{ MeV/photon}} = 2.5 \times 10^{12} \text{ photons/cm}^2\text{s}.$$

Capture and Activation Radiation

Neutron capture by reactor and shielding materials, nearby components, and the surrounding environment can result in significant radiation due to capture gamma radiation and activated materials. Gamma ray spectra due to neutron capture by various materials are presented in Table 4.8 in units of MeV per neutron capture. Activated components typically decay by emitting beta particles and gamma rays, contributing to the total gamma dose received during normal operation. Gamma radiation characteristics due to activation of possible reactor and shield materials are presented in Table 4.9. For reactors based on the surface of a planet or moon, activation of the surrounding environment may be a consideration. If the planet sustains an atmosphere, the potential for distributing activated soil by wind must be considered. Potential gamma-emitting activation products for Martian soil are presented in Table 4.10, along with their isotopic abundance, the half-life of the activation products, and their emitted radiations.

Table 4.8 Capture Gamma Ray Spectra Adapted from Reference [17]

Photon Energy (MeV)

Material	1	2	4	6	8	10
			(MeV /Neutron Capture)			
Hydrogen	—	2.2	—	—	—	—
Beryllium	0.2	0.6	1.1	4.5	—	—
Carbon	0.4	—	4.5	—	—	—
Sodium	0.2	—	—	1.3	—	—
Aluminum	—	0.3	1.4	—	2.0	—
Iron	0.1	0.2	0.1	1.1	4.2	0.3
Zinc	0.2	—	—	0.55	1.31	—

Table 4.9 Some Potential Reactor Materials Activation Products [3]

Element	Isotope Mass No.	Abund. %	Activ. x-sec. (b)*	Activ. Product	Half-life	Radiation Types	Max γ Energy (MeV)
Na	23	100	0.53	Na-24	14.9 hr	β^-,γ	2.76
Cr	50	4.3	14	Cr-51	28 d	γ	0.3
Mn	55	100	13.4	Mn-56	2.6 hr	β^-,γ	3.0
Co	59	100	36	Co-60	5.3 y	β^-,γ	1.33
Cu	63	69	4	Cu-64	12.8 hr	β^-,β^+,γ	1.34
Zr	94	7.4	0.1	Zr-95	65 d	β^-,γ	0.75
W	186	28.4	34	W-187	24 h	β^-,γ	0.686

* Activation x-sec. are 2200 m/s cross-sections.

Table 4.10 Some Potential Martian Soil Activation Products [3, 18]

Element in Soil	Target Isotope	Abund. %	Activ. x-sec. (b)*	Activ. Product	Half-life	Radiation Types	Max γ Energy (MeV)
O	O-16	99.8	2×10^{-5}*	N-16	7.4 s	β^-,γ	7.1
	O-18	0.20	2×10^{-4}	O-19	29.4 s	β^-,γ	1.6
Mg	Mg-26	11.3	0.02	Mg-27	9.5 min	β^-,γ	1.0
Al	Al-27	100	0.2	Al-28	2.3 min	β^-,γ	1.8
Si	Si-30	3.1	0.11	Si-31	2.7 h	β^-,γ	1.26
S	S-36	0.02	0.14	S-37	5.04 min	β^-,γ	3.12
Cl	Cl-37	24.5	0.56	Cl-38	37.5 min	β^-,γ	2.15
K	K-41	6.9	1.15	K-42	12.5 h	β^-,γ	1.53
Fe	Fe-58	0.3	0.9	Fe-59	45 d	β^-,γ	1.29

* Activation x-sec. for O-16, the cross section is for the fast energy range (*n,p*) averaged over the fission spectrum; all others are 2200 m/s cross sections.

Other Operational Radiations

Most other types of radiation produced during normal operation will have an insignificant radiological effect. One possible exception is the potential for positron annihilation in the vicinity of astronauts. As discussed in Chapter 1, gamma rays passing in the vicinity of a nucleus can create positron-electron pairs. Thus, positrons are produced by the intense gamma flux in an operating reactor. Because positrons quickly annihilate when they encounter an electron in dense reactor and shield materials, positron emission is not a consideration for terrestrial reactors. However, for space-based systems, positrons and electrons produced by pair production near an unshielded reactor surface can escape into the vacuum of space. Compton scattering by gamma photons provides another mechanism for ejection of electrons from reactor materials. For some orbits, the charged positrons and electrons can become trapped in Earth's magnetic field. The trapped positrons and electrons spiral back and forth along Earth's magnetic field lines between the Northern and Southern poles, creating an artificial radiation field. In principle, it is possible for positrons to annihilate on spacecraft components, producing two 0.51 MeV gamma rays for each annihilation interaction. Electrons may also produce X-rays from bremsstrahlung interactions with the spacecraft. The potential for significant crew dose due to positron annihilation and bremsstrahlung seems remote; nonetheless, the possibility should be examined to assure that a unique feature of the mission or system does not create conditions for positron and electron induced dose effects.

5.3 Radiation from Reactor Systems—Accident Conditions

In addition to radiological protection during normal operation, reactor safety analysis must assess the potential risk associated with radiation exposure or radiological contamination due to postulated accident conditions. These considerations include the possibility of inadvertent criticality and supercriticality accidents and the potential for dispersal of fuel materials during a postulated launch accident. Once a reactor has operated at high power levels, significant radiological inventories will build up in the form of fission products and activation products. The potential radiological hazard associated with these materials must be assessed for a variety of postulated accidents.

Radiologically Cold Fuel

The radiological inventory of a reactor that has never been operated is limited to the relatively low-level natural decay of the fuel's uranium isotopes, typically referred to as *radiologically cold fuel*. Because most space reactor designs use highly enriched uranium fuel, most of the activity of the fresh (unirradiated) fuel is associated with alpha emission from U-234. Enriching uranium in U-235 also increases the fraction of U-234 during the enrichment process; and the half-life of uranium-234 is 2.44×10^5 years, compared to 7.04×10^8 years for uranium-235. Consequently, even though the fraction of uranium-234 in enriched fuel is still relatively small, its shorter half-life results in substantially higher activity. A plot of specific activity as a function of uranium-235 enrichment in weight percent (wt%) is presented in Figure 4.8. For 93% enriched fuel, the specific activity is about 7×10^{-5} curies/g. Prior to

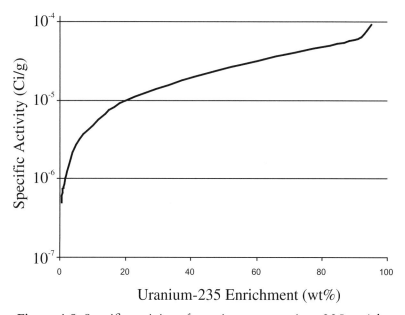

Figure 4.8 Specific activity of uranium vs. uranium-235 enrichment [12].

launch, a brief criticality test may be performed to assure that neutronic characteristics meet design criteria. Some fission and activation products will be generated during testing. If the test is performed at low power and sufficient time is allowed for decay of fission and activation products, the radiological contribution from testing will not significantly increase the fuel activity relative to the activity of fresh fuel. Thus, the specific activity for a reactor at launch will be very small, and the extremely stringent containment requirements for radioisotope fuels are not necessary for a reactor.

Reactivity Excursions

A severe reactivity excursion accident may result in an excessive direct neutron and gamma dose to workers or astronauts. If the total energy E produced during the excursion is known (see Chapter 9), the time-integrated volumetric neutron source strength can be obtained by replacing P (power) in Eq. 4.16 with E. The time-integrated surface source strength can be estimated by using the time-integrated neutron source strength in Eq. 4.17. The integrated gamma flux can be obtained using the same basic approach.

Fission and Activation Products

Once a reactor has been operated at high power, significant quantities of fission products will be generated in the fuel and activation products can be produced in reactor components. The decay of fission and activation products results in continued emission of gamma radiation from the reactor core after shutdown. If we postulate an accident scenario in which a reactor returns to Earth following operation, potential exposure to radiation from fission and activation products must be considered. If we also assume that the core is disrupted, then the release of fission products and dispersion of fuel or activated particulate becomes a potential radiological contamination hazard. The calculated radioisotope inventory due to fission products is shown in Figure 4.9 as a function of time after shutdown of the SNAP-10A reactor. Figure 4.9 provides the total fission product activity, the contributions of important fission product isotopes, and the activity due to the actinides.

Radiation from fission products is primarily beta and gamma radiation. A reasonable approximation of the beta and gamma activity from fission products has been obtained [16] for light water cooled reactors; using

$$\mathcal{A}_\beta = 1.4 P_0 \left[(t - t_0)^{-0.2} - t^{-0.2} \right], \qquad (4.19)$$

and

$$\mathcal{A}_\gamma = 0.7 P_0 \left[(t - t_0)^{-0.2} - t^{-0.2} \right]. \qquad (4.20)$$

Here, $\mathcal{A}$ is given in curies, P_0 is the operational power level in watts, t_0 is the operational time in days, and t is the operational time t_0 plus the time after shutdown, in days. Although obtained for water moderated reactors, Eqs. 4.19 and 4.20 have been found to provide reasonable approximations of fission product activity for other types of reactors.

Example 4.5

For a reactor operating at 100 kWe for 1 year, estimate the beta activity following shutdown for 1 day, 1 month, and 1 year. Assume a net system efficiency of 4%.

Solution:

From Eq. 4.19 we obtain

$$\mathcal{A}_\beta (1 \text{ day}) = 1.4 \, \frac{100 \times 10^3}{0.04} \left[(366 - 365)^{-0.2} - 366^{-0.2} \right] = 2.4 \times 10^6 \text{ Ci,}$$

$$\mathcal{A}_\beta \text{ (1 month)} = 1.4 \frac{100 \times 10^3}{0.04} \left[(31)^{-0.2} - 396^{-0.2}\right] = 7.0 \times 10^5 \text{ Ci},$$

and
$$\mathcal{A}_\beta \text{ (1 year)} = 1.4 \frac{100 \times 10^3}{0.04} \left[(365)^{-0.2} - 730^{-0.2}\right] = 1.4 \times 10^5 \text{ Ci}.$$

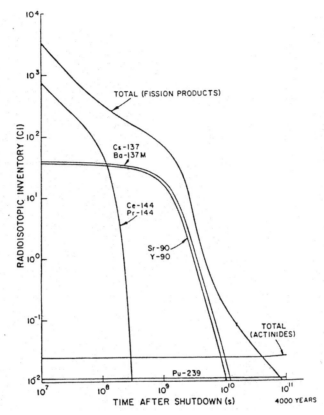

Figure 4.9 Calculated radionuclide activity following operation of the SNAP-10A reactor as a function of time after shutdown [19].

The radiological source term due to fission products and activation products will depend on the nature of the postulated reactor accident. For example, in a pin-type reactor, volatile fission products present in the free volume within the cladding may be released if the pressure vessel and cladding are disrupted during a postulated reentry accident. The fuel and activated components may subsequently fragment creating radioactive particulate. If fuel melting occurs, volatile fission products trapped in the fuel will be released. If we postulate an intact reentry of a space reactor, anyone approaching a reactor may be exposed to gamma radiation from fission products contained within the core and from activation products. It must be understood, however, for some mission types, a reactor disruption accident in space may have no safety impact on Earth. Even if we postulate radiologically hot reentry, the orbital decay time may be months or years for some missions. If the system is shutdown, either intentionally or as a result of a disruption accident, radioactive decay can reduce the potential source term significantly during the gradual orbital decay period.

6.0 Radiation Shielding

During normal operation, radiation protection from an operating nuclear system is usually provided by a combination of shielding materials and exclusion areas. Protection from low level radiation escaping a radioisotope source can be provided by encapsulating materials, light shielding, and limiting close contact with the source. For missions that involve reactor systems and astronauts, radiation protection can be provided by a radiation shield and by separating the crew quarters from the reactor system. Even for missions that do not involve astronauts, radiation shielding is typically required to protect electronic equipment.

6.1 Requirements

Basic shielding principles developed for ground-based reactor systems also apply to space nuclear systems. Space and mass constraints for space nuclear systems, however, suggest shielding strategies different from those used for terrestrial reactor systems. The driving requirement for space reactor radiation protection is to obtain a system that provides adequate radiation protection while minimizing system mass and volume. These goals are achieved by finding an optimal combination of shielding materials, shield configuration, and methods for providing a separation distance between the radiation source and the crew or electronic equipment. Shielding materials for an operating reactor perform several necessary functions; i.e., slowing down of fast neutrons, capture of slow neutrons, and attenuation of gamma radiation. The source of neutron radiation is from fission events in the nuclear fuel of an operating reactor. Gamma radiation results from fission, fission product decay, radiative capture, and neutron activation of components. Materials that are most effective for attenuating neutrons can differ from the materials used for attenuating gamma radiation; consequently, some shielding materials are called neutron shields and other shield materials are called gamma shields. Nonetheless, neutron shields also attenuate gamma radiation and gamma shields attenuate neutron radiation.

6.2 Neutron Shielding

Neutrons from the high end of the fission energy spectrum are the most penetrating; consequently, neutron shielding focuses on methods for slowing down high-energy neutrons. Figure 1.13 in Chapter 1 shows that a fraction of neutrons emitted during fission will have energies greater than 1 MeV. Although low atomic weight nuclei are effective for slowing down neutrons by elastic collisions, the cross section for scattering is relatively small for energies in excess of 1 MeV. On the other hand, heavy elements, such as tungsten, have appreciable cross sections for inelastic scattering and neutron capture in the high-energy range. In order to effectively attenuate fast neutrons, radiation shielding typically includes a region containing a heavy element material, such as tungsten, followed by a region containing a light element material, such as lithium hydride. The heavy element region slows down or captures neutrons from the high-energy end of the spectrum, such that most of the neutrons transmitted into the light element region have energies less than 1 MeV. Neutrons are then slowed to thermal energies by elastic scattering with light nuclei, and are subsequently captured by neutron-absorbing materials. Shield designs often incorporate neutron-absorbing materials, such as boron, within the neutron shield region containing the light element materials.

Collimated Neutron Beam

In Chapter 1, Section 5.2, it was shown that a narrow beam of gamma rays passing through a material will be attenuated with an exponential dependence on the product of the material thickness and its attenuation coefficient (Eq. 1.25). Similarly, the attenuation of the neutron flux from a collimated beam of neutrons has an approximate exponential behavior, given by

$$\phi(x) = \phi_0 \, e^{-\Sigma_R x}. \tag{4.21}$$

Here, ϕ is the neutron flux, x is the distance into the shield, ϕ_0 is the neutron flux incident at the surface of the shield, and Σ_R is an attenuation cross section for fast neutrons, commonly referred to as a *removal cross section*. Removal cross sections are primarily associated with inelastic scattering of high-

Table 4.11   Neutron Removal Cross Sections for Various Materials [20]

Material	σ_R (barns)	Atom Density at 273 K (Atoms/b·cm)	Σ_R (cm^{-1})
H	1.00	Gas	—
Li	1.01	0.046	0.046
Be	1.07	0.120	0.132
B	0.97	0.139	0.135
C	0.81	0.113	0.065
Al	1.31	0.060	0.079
Fe	1.98	0.084	0.168
Ni	1.89	0.091	0.173
Zr	2.36	0.042	0.101
W	3.13	0.063	0.198
Pb	3.53	0.033	0.116
U	3.60	0.047	0.170

energy fission neutrons (6 to 8 Mev). The numerical value of the removal cross section is based on the assumption that the fast neutron shield material is followed by a hydrogenous material. Removal cross sections can be interpreted as a measure of the probability that a fast neutron will be slowed to less than 1 MeV, where it can then be further slowed to thermal energies and captured. Historically, removal cross sections have been determined empirically using Eq. 4.21. Removal cross sections for a number of materials are presented in Table 4.11.

Example 4.6

For a collimated beam of high-energy (~7 MeV) neutrons, estimate the required thickness of a tungsten shield to attenuate the neutron flux by a factor of 100.

Solution:

From Eq. 4.21 we obtain

$$\frac{\phi(x)}{\phi_0} = e^{-\Sigma_R x}.$$

Using Table 4.11, and rearranging this expression gives

$$x = \frac{1}{\Sigma_R} \ln\left(\frac{\phi_0}{\phi(x)}\right) = \frac{1}{0.198} \ln(100) = 23.3 \text{ cm}.$$

Reactor Neutron Source

For a point source surrounded by a spherical radiation shield of radius R_s, illustrated in Figure 4.10 (a), the neutron flux at distance R is approximated by

$$\phi(R) = s_n \frac{e^{-\Sigma_R R_S}}{4\pi R^2}. \tag{4.22}$$

Equation 4.22 assumes that the region beyond the shield is a vacuum.

We are more generally interested in calculating shielding requirements for volumetric sources, such as an operating reactor. A simplification of a common space reactor/shield geometry is shown in Figure 4.10 (b). The reactor is represented as a cylinder with a slab neutron shield separating the reactor

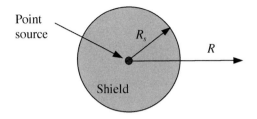

(a) Point source surrounded by a radiation shield.

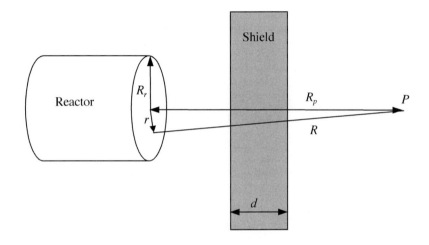

(b) Cylindrical reactor represented as a disk source with
a slab radiation shield.

Figure 4.10 Nomenclature for a point source and cylindrical reactor with a slab shield.

from the payload at point P. Here, we approximate the reactor neutron source as an equivalent disk surface source, and we use Eq. 4.22 to define a *point kernel* $G(R)$. In order to determine the flux at a point P due to an equivalent surface source at a point on the reactor surface; we use the point kernel,

$$G(R) = \frac{e^{-\Sigma_R d \sec\theta}}{4\pi R^2} = \frac{e^{-\Sigma_R d\, R/R_p}}{4\pi R^2}. \tag{4.23}$$

As shown in Figure 4.10 (b), R is the distance from a point on the reactor surface to point P at the payload. We obtain the flux at point P, due to neutrons emitted from all points on the reactor surface, by integrating the point kernel over the reactor surface. Thus, treating the reactor as an equivalent surface source and using Figure 4.10 (b), we can write

$$\phi(R_p) = 2\pi S_n \int_{R_p}^{\sqrt{R_p^2+R_r^2}} G(R)R\,dR. \tag{4.24}$$

Using Eq. 4.23 in Eq. 4.24, gives

$$\phi(R) = 2\pi\,S_n \int\limits_{R_P}^{\sqrt{R_P^2+R_r^2}} \frac{e^{-\Sigma_R d\ R/R_P}}{4\pi R}\,dR. \tag{4.25}$$

Solving Eq. 4.25 we obtain

$$\phi(R_p) = \frac{S_n}{2}\left[\mathrm{E}_1(\Sigma_R d) - \mathrm{E}_1\left(\Sigma_R d\,\frac{\sqrt{R_P^2+R_r^2}}{R_P}\right)\right], \tag{4.26}$$

with units n/cm^2·s. The quantity $\mathrm{E}_1(\Sigma_R x)$ is the exponential integral function defined by

$$\mathrm{E}_1(\Sigma_R d) = \int_{\Sigma_R x}^{\infty} \frac{\exp(-q)}{q}\,dq. \tag{4.27}$$

A plot of the exponential integral function is presented in Figure 4.11.

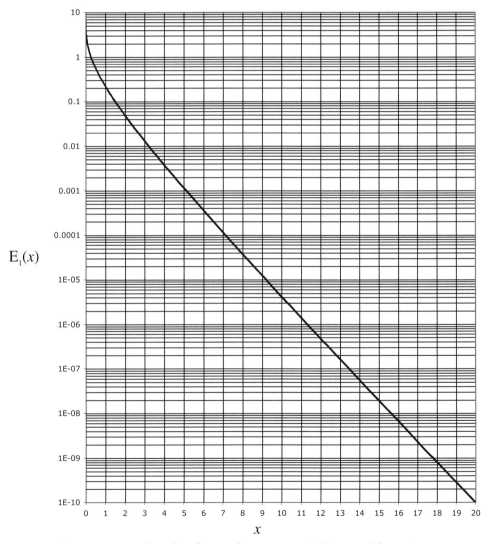

Figure 4.11 Values for first-order exponential integral function.

For the simple case of no shielding between the reactor and point P, Eq. 4.25 becomes

$$\phi(R_p) = \frac{S_n}{2} \int_{R_R}^{\sqrt{R_P^2+R_r^2}} \frac{1}{R^2} R dR,$$

(4.28)

or

$$\phi(R_p) = \frac{S_n}{2} \ln\left(\frac{\sqrt{R_P^2+R_r^2}}{R_r}\right).$$

(4.29)

6.3 Gamma Shielding

The basic approach for neutron shielding described in the preceding discussion also applies to gamma radiation shielding. However, as discussed in Section 5, it is necessary to explicitly account for the energy dependence of the incident gamma rays. Furthermore, the approximate exponential dependence on shield thickness does not account for multiple scattering of photons in thick shields.

Collimated Gamma Ray Beam

The effect of multiple scattering is illustrated in Figure 4.12. The contribution due to multiple scattering events is included by a so-called *buildup factor* $B(\mu_\gamma x)$. The equation for attenuation of the gamma flux for a collimated beam is

$$\phi_\gamma(x) = B(\mu_\gamma x)\phi_{\gamma 0}\, e^{-\mu_\gamma x}.$$

(4.30)

Buildup factors for various materials have been computed, assuming an isotropic point source; and are presented in Table 4.12. Values for the linear attenuation coefficients were provided in Chapter 1, Table 1.4.

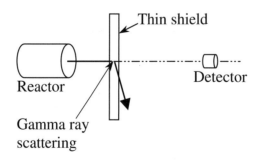

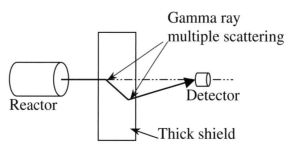

Figure 4.12 Illustration of gamma dose buildup effect.

Table 4.12 Dose Buildup Factors for a Point Isotropic Source [3]

Gamma Ray Energy (MeV)

Material	μx	0.5	1.0	2.0	3.0	4.0	6.0	8.0	10.0
Water	1	2.52	2.13	1.83	1.69	1.58	1.46	1.38	1.33
	2	5.14	3.71	2.77	2.42	2.17	1.91	1.74	1.63
	4	14.3	7.68	4.88	3.91	3.34	2.76	2.40	2.19
	7	33.8	16.2	8.46	6.23	5.13	3.99	3.34	2.97
	10	77.6	27.1	12.4	8.63	6.94	5.18	4.25	3.72
	15	178	50.4	19.5	12.8	9.97	7.09	5.66	4.90
	20	334	82.2	27.7	17.0	12.9	8.85	6.95	5.98
Iron	1	1.98	1.87	1.76	1.55	1.45	1.34	1.27	1.20
	2	3.09	2.89	2.43	2.15	1.94	1.72	1.56	1.42
	4	5.98	5.39	4.13	3.51	3.03	2.58	2.23	1.95
	7	11.7	10.2	7.25	5.85	4.91	4.14	3.49	2.99
	10	19.2	16.2	10.9	8.51	7.11	6.02	5.07	4.35
	15	35.4	28.3	17.6	13.5	11.2	9.89	8.50	7.54
	20	55.6	42.7	25.1	19.1	16.0	14.7	13.0	12.4
Tungsten	1	1.28	1.44	1.42	1.36	1.29	1.20	1.14	1.11
	2	1.50	1.83	1.85	1.74	1.62	1.43	1.32	1.25
	4	1.84	2.57	2.72	2.59	2.41	2.07	1.81	1.64
	7	2.24	3.62	4.09	4.00	4.03	3.60	3.05	2.62
	10	2.61	4.64	5.27	5.92	6.27	6.29	5.40	4.65
	15	3.12	6.25	8.07	9.66	12.0	15.7	15.2	14.0
	20	—	7.35	10.6	14.1	20.9	36.3	41.9	39.3
Lead	1	1.24	1.37	1.39	1.34	1.27	1.18	1.14	1.11
	2	1.42	1.69	1.76	1.68	1.56	1.40	1.30	1.23
	4	1.69	2.26	2.51	2.43	2.25	1.97	1.74	1.58
	7	2.00	3.02	3.66	3.75	3.61	3.34	2.89	2.52
	10	2.27	3.74	4.84	5.30	5.44	5.69	5.07	4.34
	15	2.65	4.81	6.87	8.44	9.80	13.8	14.1	12.5
	20	2.73	5.86	9.00	12.3	16.3	32.7	44.6	39.2

Example 4.7

Assume that a point source emits 8 MeV gamma rays with a source strength of 10^6 photons/s. A 3.85-cm thick lead spherical shield surrounds the source with an inner diameter of 25 cm. Compute the gamma particle flux at the outer surface of the shield.

Solution:

From Table 1.4, we obtain an attenuation coefficient of 0.521 for 8 MeV gamma rays in lead.

Thus,
$$\mu_\gamma x = (0.521 \text{ cm}^{-1})\,(3.85 \text{ cm}) = 2.0,$$

and from Table 4.12 the buildup factor for 8 MeV gamma rays in lead $= B(2.0) = 1.3$.

Using Eq. 4.30 and accounting for $1/R_d^2$, we compute the transmitted flux from a point source as

$$\phi_\gamma = (10^6 \text{ photons/s})\frac{1.30\,e^{-2}}{4\pi\left(\dfrac{25}{2} + 3.85\right)^2} = 52.4 \text{ photons/cm}^2\cdot\text{s}.$$

Reactor Source

The approximate gamma flux, at location R_p in a shield from a cylindrical reactor, can be estimated using the same basic approach used for neutrons. We write the buildup factor in the form $B(\mu_\gamma d) = 1 + b\mu_\gamma d$. Using this expression with the point kernel for the gamma flux, the solution is

$$\phi_\gamma(R_p) = \frac{S_\gamma}{2}\left[E_1(\mu_\gamma d) - E_1\left(\mu_\gamma d\,\frac{\sqrt{R_P^2+R_r^2}}{R_P}\right)\right][1 + b\exp(-\mu_\gamma d)], \tag{4.31}$$

with units MeV/cm^2·s.

6.4 Shielding Materials

Angelo and Buden [20] list seven factors for the selection of shielding materials:

1. Ability to attenuate radiation
2. Minimum mass
3. Resistance to radiation-induced thermophysical damage
4. Stability at elevated operating temperatures
5. Ease of fabrication
6. Availability
7. Cost

The ability to attenuate gamma radiation and fast neutrons can be gauged by comparing the values of gamma attenuation coefficients and neutron removal cross sections of various materials in Tables 1.4 and 4.11, respectively. From these tables, tungsten is observed to exhibit high attenuation coefficients for both fast neutrons and gamma rays, and the attenuation per gram of tungsten is excellent. High temperatures can result from radiation induced heating in the shield or from the passage of high-temperature working fluid pipes through or around the shield. Tungsten is stable in both high temperature and intense radiation environments. Although lead and uranium also exhibit high attenuation coefficients, they do not exhibit tungsten's high temperature stability. Lead is also toxic, and depleted uranium is an alpha emitter. The cost of tungsten is high, but for the relatively small quantities required for space reactor shielding, the shielding advantages of tungsten outweigh the added material cost. For these reasons, tungsten is frequently chosen as a primary shielding material for space reactors. Boron carbide (B_4C) also exhibits excellent thermal stability and reasonably good shielding properties.

As mentioned previously, a low atomic mass material is also needed to slow neutrons to the thermal energy range where they are captured by neutron absorbing materials. Hydrogenous materials are typically used to perform this function; however, other materials can be used. Although high-density materials are more effective than hydrogenous materials for attenuating neutrons with energies > 6 MeV, hydrogenous shield materials can also contribute to slowing down high-energy neutrons. Lithium hydride is often selected as the low atomic mass neutron shield, because it possesses a high hydrogen atom density, has a low mass density, produces little secondary radiation, and is relatively stable at fairly high temperatures. Shields can be designed to allow sufficient heating of a lithium hydride shield to permit annealing of radiation damage in the shield while maintaining temperatures below a level that would result in significant thermal dissociation of hydrogen. Lithium hydride shield designs may employ depleted lithium to reduce helium generation by n,α reactions with lithium-6. Excessive helium generation can result in high pressures within the shielding canister. Important lithium hydride properties are presented in Table 4.13. The relationship of neutron attenuation to lithium hydride thickness is presented in Figure 4.13.

Table 4.13   Lithium Hydride (LiH) Properties [20]

Density (g/cm³)	0.775
Molecular weight	7.95
Hydrogen content (atoms H/cm³)	0.585×10^{23}
(weight %)	12.68
Melt temperature (K)	960
Volume change on melting (%)	25
Crystal structure	fcc
Heat of fusion (kJ/mole)	21.77
Molar volume (cm³/mole)	10.254

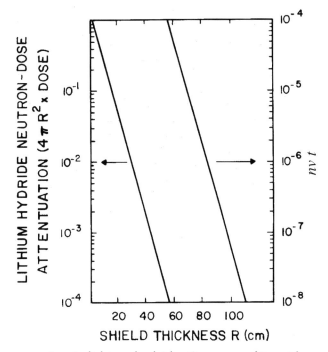

Figure 4.13 Neutron attenuation in lithium hydride. *Courtesy of Los Alamos National Laboratory.*

Shielding Strategies

Although launch considerations place space and mass constraints on the design of radiation shielding, the remote location of space nuclear systems can reduce shielding requirements. For example, the *shadow shield* approach illustrated in Figure 4.14 requires radiation shielding on only one side of the reactor; i.e., the side facing the crew quarters or payload. Shadow shields are possible for orbiting systems because gamma and neutron radiations in the vacuum of space follow straight-line trajectories. For this approach, astronaut movement (space walks) beyond the protective "shadow" of the shield is prohibited. In order to reduce shield mass, most orbiting reactor designs include a boom to separate the spacecraft from the reactor. For fairly large distances, the attenuation due to separation has a *1/R²* dependence. The ability to reduce shield and system mass by increasing separation distance is generally limited by the increase in boom mass and the mass of power and instrumentation cables from the reactor system to the spacecraft. System designers select an optimum separation distance, so that the combination of the shield, boom, and cables masses yield a minimum total mass. Other factors, such as flexing in a long boom, can limit the separation between the spacecraft and the reactor. The shadow shield is typically conical in shape to minimize mass while covering all straight-line radiation trajectories between the reactor system and the spacecraft. Note that the radiation shield diameter covers radiation contributions due to scattering off

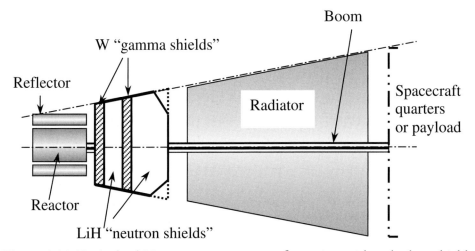

Figure 4.14 Typical orbiting reactor system configuration with a shadow shield.

the reflectors. Furthermore, the radiator is contained within the shadow of the radiation shield to prevent scatter of unshielded radiation from the radiator to the payload. For missions requiring astronaut activities beyond the zone easily protected by a shadow shield, other shield geometries may be required. Some of these alternative geometries are illustrated in Figure 4.15.

As shown in Figure 4.14, tungsten, or another heavy-atom shield material, is typically placed as close to the reactor as possible. This first shield performs two functions: (1) it slows down high-energy neutrons escaping from the core by inelastic scattering collisions, and (2) it attenuates primary gamma rays from the reactor. By placing the heavy-atom shield as close to the core as possible, the high-density

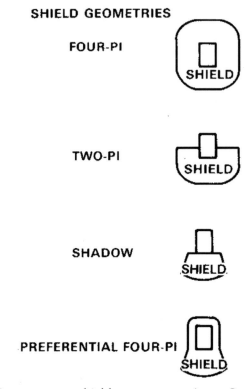

Figure 4.15 Space reactor shield geometry options. *Courtesy of NASA.*

material will be located where it will have the smallest radius and mass. The next shield zone consists of a low atomic mass material, such as lithium hydride, to slow down and capture transmitted neutrons. Although the first shield attenuates gamma radiation, neutron interactions in the first heavy-atom shield result in the production of secondary gamma radiation. In order to attenuate this secondary gamma radiation, without producing substantial additional secondary gamma radiation from neutron interactions, a second gamma shield is usually placed within the light-atom shield region where the high-energy neutron flux is greatly reduced. Even though the first heavy-atom shield is used to attenuate fast neutrons as well as gamma rays, both heavy-atom shields are commonly called *gamma shields*; and the low atomic weight shield region is called the *neutron shield*.

If the reactor power system is located on the surface of a moon or planet, a different set of strategies may be used. Several shielding options for surface missions are illustrated in Figure 4.16. One low shield mass approach enforces an exclusion area with a habitat and work area located at a safe distance

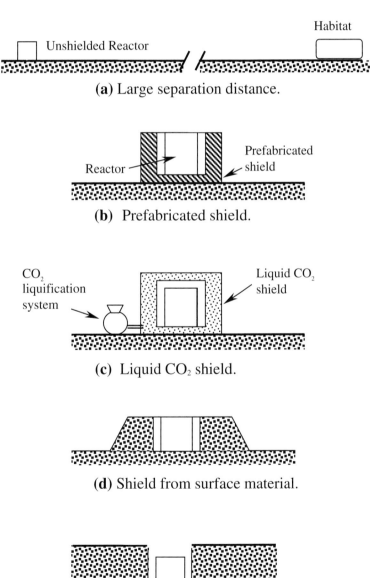

(a) Large separation distance.

(b) Prefabricated shield.

(c) Liquid CO_2 shield.

(d) Shield from surface material.

(e) Shield by excavated pit.

Figure 4.16 Shield geometry options for planet or moon surface.

from the operating system, as shown in Figure 4.16 (a). Some shielding will be required to protect instrumentation and control hardware; activation of surface material may be an issue. This approach requires long power cables and careful monitoring to prevent astronaut exposure. Another approach, similar to shielding for orbiting systems, is shown in Figure 4.16 (b). This approach requires launching and deploying the reactor system with prefabricated shields installed. The advantages of prefabricated shields include minimal restriction on astronaut movement, no extended site preparation, and minimal activation of the surface environment. The disadvantage of prefabricated shields is the high shield mass at launch. Figure 4.16 (c) illustrates a novel approach to reduce shield mass at launch without large exclusion areas or extensive site preparation. For this approach, a cryogenic cooling system is used to liquefy the atmosphere. The liquefied atmosphere is stored in tanks surrounding the reactor to serve as a radiation shield. This idea was suggested for a Mars base, using the Martian CO_2 atmosphere [21]. The liquefied atmosphere method is limited to celestial bodies that support an atmosphere and requires the development of a reliable cryogenic system. Other issues, such as the potential for leaks, must be addressed. Two other methods for reducing system shield mass at launch are illustrated in Figure 4.16 (d) and Figure 4.16 (e). For these approaches, the reactor system is either surrounded by excavated surface materials; or the system is placed in an excavated pit. Shielding by excavation requires extensive site preparation, and activation of surface material may be an issue.

Computer Methods

The actual design of a space reactor radiation shield is more complex than a simple conical shape. As shown in Figure 4.14 the neutron shield mass is reduced, relative to a simple conical design, by tapering the shield at locations where a smaller thickness provides adequate radiation attenuation. The radiation shield for the SP-100 space reactor (Figure 2.34, in Chapter 2) is penetrated by safety rods and is surrounded by coolant lines and other hardware that can scatter radiation. The shield assembly also contains materials that are not for the purpose of radiation attenuation, such as heat conduction plates and materials for structural support. These other materials may provide additional attenuation, and they may alter the radiation field by scattering or by the production of secondary gamma rays. The reactor is far more complex than a simple spherical source. The neutron flux from the core exhibits a spectrum of energies and is spatially non-uniform. Reactor gamma ray sources include gamma radiation from fission, fission products, neutron capture, and activation products. As stated previously, gamma rays also originate from neutron capture and neutron activation of shield materials. Radiation streaming through voids and scattering off reflectors and structure further complicates the analysis.

Given the potential complexity of an actual space reactor shield design, the simple calculational techniques described in the foregoing cannot be used for design or safety analysis beyond first-order estimates. Radiation shielding design and safety analysis is usually performed using either Monte Carlo or transport theory computer codes. Monte Carlo and transport theory computer methods are described in Chapter 8. Monte Carlo neutronic analysis tools, such as the MCNP computer code [22], use probabilistic methods to track the interactions and individual histories of many neutrons and gamma rays. Transport theory methods, such as TWODANT [23] use a deterministic approach to predict the neutron and gamma flux in the reactor and shield. The transport equation is an adaptation of the Boltzmann equation to neutron and gamma transport. In general, Monte Carlo methods are better suited for shielding analysis of complex geometries. The neutron fluence and gamma dose data presented in Figure 4.17 were obtained using computer methods for an unshielded 1.6 MWt reactor. The dose has been time-integrated over a 7-year period of operation.

7.0 Summary

The rate of particle emission from a radioisotope material is called the *source strength*. Source strength is occasionally used to roughly characterize radiation from reactors. Radiation incident on a particular

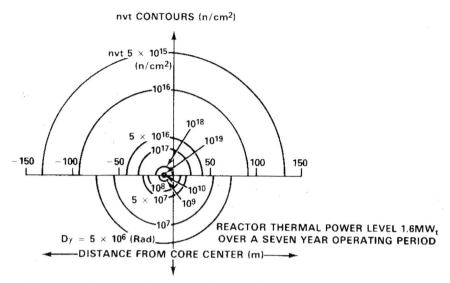

Figure 4.17 Neutron fluence and gamma dose around an unshielded reactor. *Courtesy of National Aeronautics and Space Administration.*

body organ can be described by the *particle flux*. The extent of damage induced by the incident particle flux is roughly proportional to the energy absorbed, called the *absorbed dose D*. A *radiation weighting factor* W_R (or Q) is used to account for the biological effect of different types of radiation. The *dose equivalent H* incorporates the relative biological effectiveness of different forms of radiation as the product of W_R and D. The relative sensitivity of different organs to detrimental effects of radiation is expressed by a *tissue weighting factor* W_T. Tissue weighting factors are used to calculate an effective dose equivalent or *effectance* H_E. The time integral of the dose rate and the effectance rate are called the *committed dose equivalent* and the *committed effectance*, respectively. A *collective dose equivalent* and a *collective effectance* have been established to quantify the committed dose equivalents and effectance for exposed groups or populations.

Radiation sources are often categorized as either external sources, originating outside the body, or internal sources, resulting from inhalation or ingestion of radioactive substances. Exposure to either external or internal radiation can cause both somatic effects in which injury is inflicted on the irradiated individual, and hereditary effects which are passed on to the children of an individual receiving germ cell radiation damage. Radiation injuries can be further classified as either deterministic or stochastic health effects. Large doses received over short time periods threaten both the short and long term health of exposed individuals. If exposures are sufficiently intense, exposed organs are damaged causing radiation sickness or death within days or months. Radiation-induced sickness and death are deterministic effects. Such effects are not observed until the dose received is greater than an associated threshold dose D_{th}. Populations receiving doses or dose rates insufficient to cause early fatalities are subject to increased risk of cancer or other health effects in 2 to 30 years. Birth defects can result from radiation damage to the DNA in germ cells that are passed on to progeny. Both radiation-induced cancer and radiation-induced birth defects are stochastic effects. The probability of stochastic effects increases with increasing received dose. Cancers and birth defects may also result from chronic exposures to low levels of radiation. The probability of incurring health effects due to exposure of a population at low doses and low dose rates, is often smaller than the probability for the same collective dose at high doses and high dose rates.

Space nuclear programs follow established policy to assure that radiation exposure to astronauts, radiation workers, and the public is kept as low as reasonably achievable. In addition, specific radiation protection guidelines have been established to protect astronauts from both natural radiation and radiations produced by on-board nuclear systems. Normal operation of nuclear systems in space does not present a radiological risk to Earth's environment or populace. For space nuclear systems, radiation hazards on Earth are only a consideration for postulated accidents that either fail to properly deploy the nuclear system in space or that result in an unplanned reentry of the system following deployment. The most significant issue associated with relatively large radioisotope sources is their high inventory of radioactive material during and prior to launch. Radioisotope sources are designed and tested to assure that containment barriers cannot be breached for credible reentry accidents.

Mission plans typically require launch of space reactors in the radiologically cold condition. Once deployed and operating, however, nuclear fission in reactors produces intense neutron and gamma radiation. Gamma rays are released during the fission process and during the decay of fission products. Gamma radiation is also produced by neutron capture, inelastic scattering processes, and activated materials. Radiation exposure in space can be a consideration if astronauts are in the vicinity of a space nuclear system. Exposure is controlled by providing radiation shielding and maintaining an exclusion area.

Shielding requirements for radioisotope sources are dictated by the secondary neutron and gamma radiations rather than by the primary radiation due to alpha decay. Radiation shields for an operating reactor typically include a region containing a heavy element material, followed by a region containing a light element material. The heavy element region attenuates gamma radiation and slows down or captures neutrons from the high-energy end of the spectrum. Neutrons are then slowed to thermal energies by elastic scattering with light nuclei, and are subsequently captured by neutron absorbing materials. In shielding materials, radiation is attenuated with an approximately exponential dependence on the product of the shield thickness and the attenuation coefficient for the shield material.

Symbols

$\mathcal{A}$	activity
b	barn
b	buildup factor coefficient
B	buildup factor
Ci	curie
d	shield thicness
eV	electron volt
$D_{T,R}$	absorbed dose tissue T from radiation R
D	dose rate
E_1	first order exponential function
E	gamma ray energy
ED_{50}/t	dose effect, 50% of population in t days
$G(R)$	point kernel
Gy	gray
H_E	effectance or committed effectance
$\dot{H}_E$	effectance rate
$\bar{H}_E$	average effectance
H_T	dose equivalent
$\dot{H}_T$	dose equivalent rate
$\bar{H}_T$	average dose equivalent
LD_{50}/t	lethal dose, 50% of population in t days
LET	linear energy transfer
N	atom density
N_A	Avogadro's number
n	neutron
P	power

Q	quality factor
r	radial distance
R	distance from surface to point P
R_r	reactor radius
R_P	payload separation distance
R_s	shield radius
rad	radiation absorbed dose
RBE	relative biological effectiveness
rem	roentgen equivalent man
s	second
s_R	source strength, radiation type R
S_E	collective effectance
S_H	collective dose equivalent
$\bar{S}_f$	average volumetric fission source strength
$\bar{S}_n$	average volumetric n source strength
S_T	collective dose equivalence
S_γ	surface γ source strength
S_n	surface n source strength
Sv	sievert
t	time
T	tissue or organ type
V	volume
W_R	radiation weighting factor
W_T	tissue weighting factor
x	distance

α alpha particle
β beta particle
ϕ_R particle flux for radiation type R
ϕ neutron flux
ϕ_0 neutron flux incident at surface
γ gamma ray
κ_γ photons/fission

λ decay constant
$\mu_{r,\gamma}$ reactor γ-attenuation coefficient
$\mu_{r,n}$ reactor n-attenuation coefficient
μ_γ linear attenuation coefficient for γ
σ_R microscopic removal cross section
Σ_R macroscopic removal cross section
ς_T tissue density

References

1. *ICRP Publication 26, Recommendations of the ICRP.* Annals of the ICRP **1** (3) 1987.
2. *ICRP Publication 60, Recommendations of the ICRP.* Pergamon Press, UK, 1991.
3. Argonne National Laboratory, *Reactor Physics Constants.* ANL-5800 (2nd ed.), 1963.
4. Abrahamson, S., M. A. Bender, S. Book, C. Buncher, C. Denniston, E. Gilbert, F. Hahn, V. Hertzberg, H. Maxon, B. Scott, W. Schull, and S. Thomas, *Health Effects Models for Nuclear Power Plant Accident Consequence Analysis, Low-LET Radiation, Part II: Scientific Bases for Health Effect Models.* NUREG/CR-4214, SAND-7185, Rev. 1, May 1989.
5. *ICRP Publication 90, Recommendations of the ICRP.* Annals of the ICRP **1** 1990.
6. *ICRP Recommendations of the International Commission of Radiological Protection,* Ann. ICRP **1**, No. 3, 1991.
7. Abrahamson, S., M. A. Bender, B.B. Boecker, E.S. Gilbert, and B.R. Scott, *Health Effects Models for Nuclear Power Plant Accident Consequence Analysis, Modifications of Models Resulting from Recent Reports on Health Effects of Ionizing Radiation, Low-LET Radiation, Part II Scientific Bases for Health Effects Models.* Prepared for U.S. Nuclear Regulatory Commission, NUREG/CR-4214, Rev. 1, Part II Addendum 1, LMF-132, Aug. 1991.
8. U.S. Occupational Safety and Health Administration, Department of Labor, *U.S. Code of Federal Regulations, 29 CFR 1910.96: Ionizing Radiation.* Office of the Federal Register, National Archives and Records, July 1, 1989.
9. National Council on Radiation Protection and Measurements, *Guidance On Radiation Received in Space Activities.* NCRP Report Number 098, 1989.
10. U.S. Nuclear Regulatory Commission, *U.S. Code of Federal Regulations, 10 CFR 20: Standards for Protection against Radiation,* Office of the Federal Register, National Archives and Records, Jan. 1, 1991.
11. U.S. Department of Energy, *DOE Order 5480.4: Environmental Protection, Safety, and Health Protection Standards.* May 15, 1984.
12. U.S. Nuclear Regulatory Commission, *U.S. Code of Federal Regulations, 10 CFR 71: Packaging and Transportation of Radioactive Material.* Office of the Federal Register, National Archives and Records, Jan. 1, 1991.
13. Executive Office of the President, National Aeronautics and Space Council, *Nuclear Safety Review and Approval Procedures for Minor Radioactive Sources for Space Operations.* 1970.
14. Tascione, T. F., *Introduction to the Space Environment.* Orbit Book Company. Malabar, FL, 1988.
15. Bennett, G. L., J. J. Lombardo, R. J. Hemler, and J. R. Peterson, "The General Purpose Heat Source Radioisotope Thermoelectric Generator: Power for the Galileo and Ulysses Missions." Proceedings of the 21st Intersociety Energy Conversion Engineering Conference, San Diego, CA, Aug. 25–29, 1986, p. 1999–2001 v.3.
16. Glasstone, S. and A. Sesonske, *Nuclear Reactor Engineering.* Van Nostrand Reinhold Co., New York, 1967.
17. Reedy, R. C. and S, C. Frankle. "Neutron Capture Gamma-Ray Data for Obtaining Elemental Abundances from Planetary Spectra." Lunar and Planetary Science XXXII, Houston, TX, Mar. 12–16, 2001.
18. Alexander, M. (Ed.), *Mars Transportation Environment Definition Document,* Huntsville AL, National Aeronautics and Space Administration, NASA/TM-2001-210935, Mar. 2001.
19. Buden, D. and G. Bennett, "On the Use of Nuclear Reactors in Space." *Physics Bulletin* **33:12**, 1982, p. 432.
20. Angelo, J. A., Jr. and D. Buden, *Space Nuclear Power.* Orbit Book Co., Malabar, FL, 1985.
21. Houts, M. G., D. I. Poston, H. R. Trellue, J. A. Baca, and R. J. Lipinski, "Planetary Surface Reactor Shield Using Indigenous Materials." Space Technologies and Applications International Forum-1999, AIP Conference Proceeding No. 458, Albuquerque, NM, 1999.

22. Briesmeister, J. F. (Ed.), "MCNP-A General Monte Carlo Code for Neutron and Photon Transport, Version 3A." LA-7396-M Rev. 2, Los Alamos, NM, Los Alamos National Laboratory, 1986.
23. Alcouffe, R. E., "User's Guide for TWODANT: A Code Package for Two-Dimensional, Diffusion Accelerated, Neutral Particle Transport." LA-10049-M Rev. 1, Los Alamos, NM, Los Alamos National Laboratory, 1984.

Student Exercises

1. (a) A radioisotope source design uses a 100 g sphere of PuO_2 fuel, consisting of 90% ^{238}Pu, 9% ^{239}Pu, and 1% other plutonium isotopes. Spontaneous fission constitutes 1.8×10^{-7} percent of all ^{238}Pu decays and constitutes 4.4×10^{-10} percent of all ^{239}Pu decays. Assume that about 2.9 neutrons are released per fission and assume that all neutrons escape the source. Using a half-life of 87.75 years for Pu-238 and a half-life of 24,131 years for Pu-239, compute the neutron source strength of the fuel due to spontaneous fission. Ignore the contribution due to other Pu isotopes. (b) Using a specific source strength of 1.4×10^4 n/s·g for α-n reactions in $^{238}PuO_2$, estimate the specific source strength for α-n reactions in $^{239}PuO_2$. Assume that the cross sections for α-n reactions for α particles emitted from ^{238}Pu and ^{239}Pu are approximately equal. Compare the spontaneous fission neutron source strength for ^{238}Pu plus ^{239}Pu to the source strengths due to α-n reactions with oxygen. (c) For ^{238}Pu and ^{239}Pu, determine the neutron source strength due to spontaneous fission and the neutron source strength due to α-n reactions (assuming no loss of oxygen) after 1000 years. (d) Compare the sum of all neutron source strengths (both isotopes and both decay modes) at beginning of mission to the sum of all neutron source strengths after 1000 years. Also compare the total neutron source strength (spontaneous fission plus α-n) of Pu-238 to Pu-239 at beginning-of-mission, and make the same comparison for after 1000 years. Comment on your results.

2. Consider a cylindrical liquid metal-cooled reactor having a length of 50 cm and a 53 cm diameter. The reactor produces 150 kW_e, and the power conversion efficiency is 23%. (a) Calculate the average volumetric neutron source strength during full power operation. Assuming a reactor attenuation coefficient of 0.06 cm^{-1} for neutrons, estimate the average surface source strength. (b) Estimate the neutron flux at 15 m from the core center. (c) Assuming that most of the emitted neutrons have an energy of ≥ 1 MeV, compute the unshielded dose rate in Sv/hr at 15 m from the core center.

3. For the reactor in Exercise 2, compute the gamma ray flux for each energy group at 15 m from the core center in MeV/cm^2·s and photons/cm^2·s. Use the gamma rays spectrum presented in Table 4.7 and assume a reactor attenuation coefficient of 0.075 cm^{-1} for all gamma rays.

4. Using results from Exercises 2 and 3, carry out the following calculations. (a) Compute the unshielded gamma dose rate for each energy group in Sv/hr at 15 m from the core center. (b) Using the neutron dose rate from Exercise 2 and the gamma dose rate from part (a) of this exercise, determine the length of time an astronaut can work at 15 m from the reactor without shielding and without exceeding the annual worker dose limit for on-board nuclear systems. (The allowed time should be very small.)

5. A 1-cm diameter spherical rock on the surface of Mars receives an average thermal neutron flux of 3×10^9 n/cm^2·s from a reactor operating on the surface. The density of the rock is 1.6 g/cm^3 and a mass fraction for iron is 0.13. Using Table 4.8 and an average thermal capture cross section for iron of 2.4 b, compute the gamma ray source strength in photons/s due to neutron capture. Assume no attenuation of the neutron flux within the rock and assume that all gamma radiation escapes the rock.

6. For a reactor operating at 250 kWe with an efficiency of 8%, estimate the gamma activity following 1, 5, and 10 years of operation at 1 week, 1 year, 25 years, 100 years, 300 years, and 1000 years after shutdown.

7. The radioisotope source in Exercise 1 will be used on the surface of the Moon inside a human habitat. Approximate the radioisotope source as a point source and assume no scattering by other materials in the habitat. (a) Calculate the unshielded neutron dose for six months of continuous operation at a point 50 cm from the surface. (b) Assume that a 3-cm thick spherical lead shield surrounds the source, and the lead shield is surrounded by 17 cm of LiH such that all neutrons exiting the lead shield with energies < 1 MeV will be reduced to thermal energies and captured. Estimate the shielded neutron dose for the conditions in part (a). Use the removal cross section for lead from Table 4.11 and estimate the neutron attenuation by LiH using Figure 4.13.

8. Assume that a 10-cm thick tungsten radiation shield is adjacent to the reactor in Exercise 4. Using the results from in Exercise 4, estimate the shielded gamma dose for 30 days of operation at 15 m from the reactor. Use Tables 1.4 and 4.1 and interpolate data as needed.

Chapter 5

Explosions and Fires

Albert C. Marshall, Mehdi Eliassi, and David L. Y. Louie

This chapter introduces the reader to environments resulting from postulated propellant fire and explosion accidents. Simple methods are provided for characterizing fire and explosion environments and for estimating the possible effect of these environments on onboard space nuclear systems. A brief discussion is provided of typical computer methods used for detailed analysis.

1.0 Definitions, Scenarios, and Issues

The proximity of space nuclear systems to propellants raises the possibility of nuclear system damage and radiological release in the event of a propellant explosion or fire. In order to facilitate the discussion of scenarios and issues, we begin with a preliminary definition of terms and a discussion of categories of explosions.

1.1 Definitions and Categories

A *fire* is the rapid oxidation of a substance accompanied by the production of heat and light. A *blast wave* is a wave in the surrounding air produced by an explosion, and an *explosion* is defined as a sudden violent expansion of a gas. In general, explosions can arise from either physical or chemical origins. *Physical explosions* originate from strictly physical phenomena, such as the rupture of a pressurized tank. For example, a physical explosion can result from the violent boiling of liquid hydrogen or liquid oxygen. This phenomenon is often referred to as the boiling liquid expanding vapor explosion (BLEVE). For *chemical explosions*, the process is more complicated in that the chemical reaction can result in slow burning, a deflagration, or detonation. Also, the reaction may progress from a deflagration to a detonation. For liquid propellants, vapor cloud explosions are possible.

BLEVE

A boiling liquid expanding vapor explosion can occur when there is a sudden loss of integrity of a vessel containing a superheated liquid or liquefied gas. The primary cause of BLEVE is usually an external heat source (e.g., a fire) or a high velocity fragment striking and rupturing a pressure vessel. Potentially any mechanism that can result in a sudden containment vessel failure, thereby allowing a superheated liquid to flash, could lead to a BLEVE. If the released fluid is flammable, ignition of the vapor can result in a fireball.

Chemical Explosions

A chemical explosion originates from a rapid chemical reaction, such as the combustion of hydrogen in air. In a chemical explosion, an uncontrolled combustion quickly converts the released chemical energy into mechanical energy and heat. The mechanical energy is the main source of a blast wave. Chemical reactions can be classified as either *uniform* or *propagating reactions*. For uniform reactions the conditions are usually such that the reaction can occur uniformly throughout the volume of the reactants.

For a propagating reaction, the chemical reaction initiates at a specific location within the propellant mass and propagates from that point as a reaction front throughout the unreacted material. For a propagating reaction to continue under steady-state conditions, energy from the reaction zone must be able to reach the unreacted material.

Propagating Reactions

Fires involving a fuel that has not premixed with an oxidant burn relatively slowly at the interface between the fuel and oxidant. When the fuel and oxidant are premixed, however, combustion can propagate rapidly through the combustible mixture. Propagating reactions are classified as either *deflagrations* or *detonations* depending upon the mechanism for energy transfer from the reaction zone to the unreacted material. *Deflagrations* (fast burning) are propagating reactions in which the energy transfer from the reaction zone to the unreacted material occurs through ordinary transport processes such as heat and mass transfer. *Detonations,* on the other hand, are propagating reactions in which the energy transfer from the reaction zone to the unburned material occurs via a reactive shock wave; i.e., the detonation velocity exceeds the sound velocity of the unburned fuel-oxidant mixture. In general, explosions can range from fading detonations and deflagrations for lower energy propellants to high-order detonations for high-energy propellants. Figure 5.1 schematically illustrates the behavior of fires, deflagrations, and detonations.

Vapor Cloud Explosions

Vapor cloud explosions (VCEs) are unique to liquid propellants. A VCE can occur when accidental spillage of a propellant is followed by mixing of the propellant fuel with an oxidant, formation of a vapor cloud, and subsequent ignition. The oxidant may be the on-board liquid oxidant or oxygen in the atmosphere. In general, a liquid fuel that has not formed a vapor cloud and mixed with an oxidant will burn rather than explode on ignition. Once the vapor cloud is ignited, the resulting events can range from deflagration to detonation.

1.2 Scenarios

The characteristics of postulated fire and explosion environments will depend on the type of launch vehicle, the type and quantity of the propellant, and the specific environment at the time of the postulated accident. The principal initiators for postulated fire and explosion accidents depend on the mission phase. Explosion and fire accidents can be postulated to occur during the prelaunch, launch, and deployment mission phases. Fires or explosions occurring at or near launch involve larger quantities of propellants than for accidents occurring later in the deployment phase. The proximity of the nuclear system to the propellant and the confinement of the propellant will affect the accident environment and the consequences of the postulated accident.

Prelaunch and Launch Phase

Prelaunch fires and explosions may result from propellant spillage or accidental propellant mixing during launch vehicle fueling. During the early launch phase, commonly postulated fire and explosion initiators include fallback and tipover on the launch pad, as illustrated in Figure 5.2. A fallback accident could occur if a launch vehicle suffers an accidental loss of thrust just after takeoff, resulting in a fall back onto the launch pad. Such an accident is especially a concern for vehicles with liquid propellant systems, where a fallback could cause propellant tanks to rupture. The unconfined liquid fuel could burn or explode in the atmosphere. If both the fuel and oxidizer tanks rupture, the fuel and oxidizer could mix and explode. A tipover accident, caused by failure of one of multiple rocket engines, could also result in propellant tank ruptures and a propellant explosion.

For the Space Shuttle, failure of both solid rocket motors has been postulated to result in a pushover accident, illustrated in Figure 5.3. For this scenario, the external tank directly impacts the launch pad.

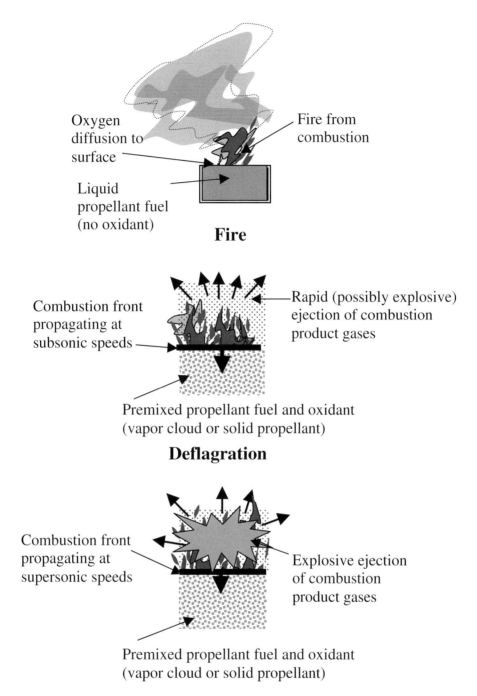

Fire

Deflagration

Detonation

Figure 5.1 Schematic comparison of fires, deflagrations, and detonations.

The impact may rupture the liquid oxygen (LOX) tank, resulting in spilling of the contents into the pit, followed by a liquid hydrogen (LH$_2$) tank rupture and mixing of large quantities of LOX and LH$_2$. A resulting LOX/LH$_2$ explosion has been estimated to have the yield equivalent to ~91,000 kg of TNT [1]. Broken solid rocket propellant pieces lying in a liquid propellant mixture are subject to very high dynamic pressures when the mixture explodes. The high dynamic pressures can cause a subsequent explosion of the solid propellant pieces. Other possible prelaunch propellant explosion initiators include propellant tank failure due to overpressure and an accidental command destruct signal.

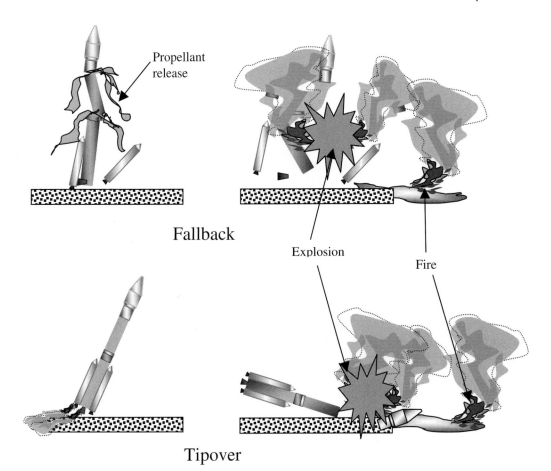

Figure 5.2 Illustration of fallback and tipover accidents.

Ascent Phase Explosion

During the early ascent phase, a premature thrust termination, guidance failure, or structural failure of the launch vehicle could result in ground impact. If the engines are still thrusting during a guidance failure, high-speed ground impact of the launch vehicle is possible. Ground impact from any of these scenarios could initiate a propellant explosion. Structural failure of the launch vehicle and overpressurization failure of propellant tanks during ascent could initiate an explosion while in flight. These potential initiators also apply to the late ascent phase. In addition, a failure of the guidance system will most likely result in a command-destruct signal, resulting in a propellant explosion. Nonpropellant types of explosions are also possible, such as explosions resulting from overheating of onboard pressure vessels or explosions relating to a liquid metal reactor coolant.

1.3 Issues

For radioisotope sources, the principal issue is the potential for disruption of protective barriers and the release and dispersion of the radioisotope fuel into the biosphere. A propellant fire accident may overheat and cause a failure of protective barriers by burning, melting, or weakening barrier materials. If a failure of barriers is predicted, the fire may, depending on the type of radioisotope fuel, alter the chemistry of the nuclear fuel material or create fuel particle aerosols. Updrafts produced by the fire can carry fuel particulate to high altitudes, potentially resulting in broad dispersal of radioactive fuel particles. Released fuel particulate may contaminate the environment and the possibility of inhalation or ingestion of radioactive particles presents a potential radiological hazard. Consequences for scenarios of this type are discussed in Chapter 11.

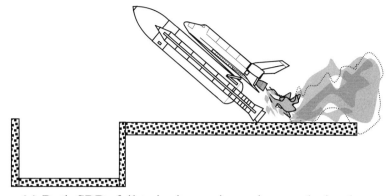

(a) Both SRBs fail to ignite, main engines push shuttle over.

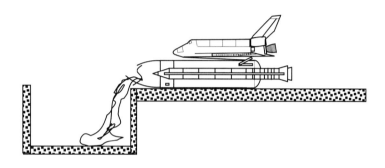

(b) ET ruptures, LH_2 and LOX released into pit.

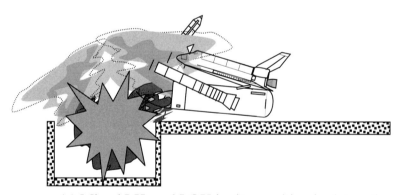

(c) Mixed LH_2 and LOX ignite resulting in detonation.

Figure 5.3 Illustration of Shuttle pushover accident.

Propellant explosions can also threaten radioisotope containment barriers by propelling the source, resulting in high-speed impact of the source against a hard surface. Propellant explosions can also create and propel fragments that may strike the source. In either case, impact may breach containment barriers, fragment the fuel, and release radioactive material into the environment. Release of radioactive fuel is prevented or mitigated by careful design of containment barriers and by rigorously testing these systems under credible propellant fire and explosion accident conditions. Sequential accidents must be assessed as well. For example, a postulated accident might include a propellant explosion followed by impact and a fire.

For reactors, the principal issue is the possibility of an accidental criticality. A fire or explosion accident can be postulated to result in an inadvertent accidental criticality by altering the reactor geometry or moving reactor shutdown devices or reflectors. For a well-designed system, however, an inadvertent criticality due to these potential mechanisms is an extremely unlikely event. A reactor immersion accident may be a more plausible inadvertent criticality scenario; i.e., one may postulate that subsequent to a fire or explosion, the reactor is immersed in a moderating or reflecting fluid, such as water. As discussed in Chapter 8, if reactor barriers have been breached by a fire, explosion, or impact, water filling the core may increase neutron moderation and result in an inadvertent criticality. If reactor reflectors or external shutdown devices are ejected, additional neutron reflection due to water immersion may result in an unplanned criticality. Accidents of this type are discussed in greater detail in Chapter 8. Although the radioactive content of fresh reactor fuel is very low, the potential for dispersal of the reactor fuel must be assessed as well.

1.4 Considerations and Perspective

A summary of the basic types and progression of potential fire and explosion accidents is presented in Figure 5.4. This figure is, of course, an oversimplification. For launch vehicle accidents, the explosion from one propellant source may induce fires or subsequent explosions in other onboard propellants. For example, explosion-generated high-speed fragments can impact other nearby propellants, or the initial explosion may cause a hard-surface impact of the pressure vessels for other propellants. In addition, the intense air shock wave from the first explosion can place a dynamic load on other propellants. All of these insults from the initial explosion can result in subsequent deflagrations or detonations in other propellants. For some multiple explosions that are sufficiently far apart, the separate explosions may interfere such that the blast waves subdue or cancel each other, reducing the overall blast effect.

In order to assure that nuclear safety issues relating to fires and explosions are properly assessed, the following basic questions must be addressed:

- What scenarios can lead to propellant fires and explosions?
- Will these scenarios result in a fire, deflagration, detonation, or some combination of these outcomes?
- For an explosion, what are the ranges of possible overpressures and impulses that the nuclear system may experience?
- For fires, what is the temperature history and what is the possible spatial extent of the fire?
- How does the system respond to the range of conditions presented by these postulated accidents?

A general discussion of scenarios was provided in this section. The following sections of this chapter will provide an introduction into the methods used to address the remaining questions.

2.0 Launch Vehicles and Propellants

The launch and deployment of space nuclear systems require that the systems must be placed in close proximity to large quantities of propellants. The launch vehicle geometry and the quantity and type of propellants used are important considerations. Launch vehicles may employ liquid propellants, solid propellants, or a combination of both. As discussed in Chapter 2, the specific impulse I_{sp} for rockets can be used as a measure of rocket performance. Often, specific impulse is expressed in English units of (pounds force)/(pound mass/s). At sea level, (pound force) = (pound mass) and I_{sp} can be expressed with units of seconds (s). The specific impulse is defined as the propellant thrust (force)/mass flow rate of the propellant. The specific impulse for a rocket can also be expressed as

$$I_{sp} = A_P C_F \sqrt{T_{ch}/M_r},$$

(5.1)

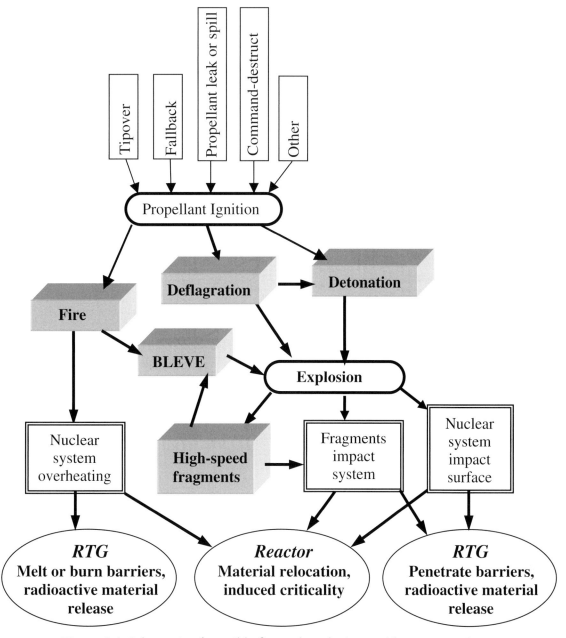

Figure 5.4 Schematic of possible fire and explosion accident progressions.

where A_P is a propellant performance factor, C_F is a coefficient that is a function of nozzle design, T_{cb} is the chamber temperature (K), and M_r is the relative molecular mass of the propellant (u). Thus, high combustion temperatures and low propellant relative molecular mass are desired.

2.1 Launch Vehicles

In the United States, the Space Shuttle, Atlas, and Delta launch vehicles are most likely to be used for deploying nuclear systems into space. The Titan class launch vehicles have also been used for launching space nuclear systems; however, the Titan has been phased out. The Space Shuttle, the Atlas V, Delta II, and the Russian Proton launch vehicles are briefly described here.

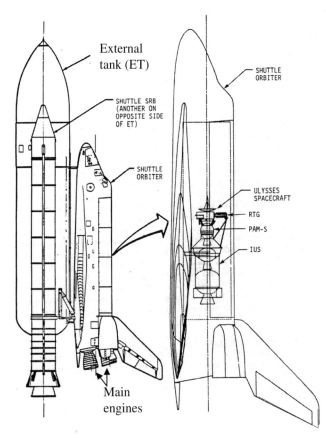

Figure 5.5 Space Shuttle configuration at launch with onboard Ulysses spacecraft. *Courtesy of U.S. Department of Energy.*

Space Shuttle

As shown in Figure 5.5, the Space Shuttle consists of four main components, the orbiter, two solid rocket boosters (SRBs), an external tank (ET), and the Shuttle's three main engines. The propellant for the main engines is liquid oxygen/liquid hydrogen (LOX/LH$_2$). The solid rocket motors use the propellant TD-H1148 HB polymer (16% aluminum, 70% ammonium perchlorate). For some missions an inertial upper stage (IUS) is used to boost the payload to a high Earth orbit. The propellant for the inertial upper stage is HTPB UTP-19360A. Figure 5.5 also shows the Ulysses spacecraft in the Shuttle cargo bay with the IUS and RTG power supply. The principal specifications for the Space Shuttle are presented in Table 5.1. The Space Shuttle is unique in that it is a reusable, piloted craft with a crew of up to eight astronauts. The Shuttle can deliver an 18 m long payload with a mass of up to 26,000 kg to LEO, or a 2,270 kg payload to GEO. At end-of mission, the orbiter is deorbited and the craft is maneuvered to a runway landing.

Following launch, within the first 2 minutes, the SRBs are jettisoned and are recovered for reuse on a subsequent mission. About 6 minutes later, the main engines are shut down and the ET is separated and falls back to Earth. For safety purposes, each of the SRBs and the ET are equipped with command decoders/receivers that can execute destruct signals received from the ground station. The ET carries a receiver and each SRB carries two receivers. The entire system is connected so that a signal received by each SRB can initiate the destruction of the other two systems as well. For the ET, one linear shaped-charge is attached to the LOX tank and another to the LH$_2$ tank. The destruct charges are designed to split the case, allowing the chamber pressure to sharply drop and thus terminate thrust. The linear

Table 5.1   Space Shuttle Specifications [2]

	Length (m)	Diameter (m)	Launch Mass (kg)	Thrust (kN)	Burn Time (s)	Propellant	Deployment Capability
Shuttle	56.14	23.79	2.50×10^6	34,622 SL	—	—	—
Orbiter	37.24	23.79 (WS)	8.2×10^4	—	—	—	LEO
Max Payload	18.3	4.57	2.6×10^4	—	—	—	LEO
External Tank: Dry (LOX/LH$_2$)	46.88	8.4	2.6×10^4 7.3×10^5	—	—	—	—
2 Solid Rocket Motors	38.47	3.71 (1)	1.14×10^6 (2)	29,360 SL (2)	124	TB-H1148 HB polymer	—
3 Main Engines	4.24	2.39	9.54×10^4 (3)	5,262 SL (3)	510	LOX/LH$_2$	—
Inertial Upper Stage (2 Stages)	5.17	2.9	1.47×10^4	185 Vac. S1 78 Vac. S2	152 S1 103 S2	HTPB UTP-19360A	2.27×10^3 kg to GEO

GEO = Geosynchronous Earth orbit; LEO = Low Earth orbit; LOX = Liquid oxygen; LH$_2$ = Liquid hydrogen; SL = Sea level; S = Stage; Vac. = Vacuum; WS = Wing span; HTPB UTP-19360A = Aluminum, ammonium perchlorate, and hydroxyl-terminated polybutadiene binder

shape charges for each SRB are approximately 24 meters long, divided to roughly six 4-meter sections. The destruct charges are placed in the cable trays on the outside of the booster case, but on the side that is away from the ET.

The IUS is comprised of two stages, with the first stage including the interstage structure, the large solid rocket motor, and the avionics [2]. Note that the payload is contained within the Shuttle bay area adjacent to the SRMs and the LOX/LH$_2$ propellant in the ET. This arrangement places the space nuclear system in closer proximity to the propellants than for typical expendable launch vehicles (ELVs), where the space nuclear system is contained within the faring at the top of the launch vehicle.

Atlas V

The Atlas series of expendable launch vehicles has been used since the early 1960s. Over the years, Atlas vehicles have taken part in many government, military, and civilian space missions. Illustrations of the Atlas V 400 and 500 series launch vehicles are shown in Figure 5.6. The Atlas V incorporates design features developed for Atlas III. Both the 400 and 500 series use a common core booster (CCB), a Centaur upper stage (CIII), and a payload faring. The propellant for the CCB is LOX/RP-1. The 500 series can also use up to five strap-on SRBs (using hydroxyl-terminated polybutadiene propellant). The first digit in the three digit naming convention gives the payload faring diameter in meters; thus, all 400 series Atlas V vehicles have a 4 m diameter faring and all 500 series Atlas V vehicles have a 5 m diameter faring. The second digit gives the number of strap-on boosters, and the last digit identifies the number of Centaur engines (one or two). The specifications for an Atlas V 552 series are shown in Table 5.2.

A typical ascent profile to geosynchronous transfer orbit (GTO) is described in Reference [3] for the Atlas V 501 launch vehicle. At liftoff, the CCB and all strap-on boosters provide thrust. At about 99 seconds into the flight, the first three SRBs are jettisoned, and the last two are jettisoned at 100 seconds. The payload faring is jettisoned at 212 seconds. The CCB is shut down, then jettisoned at an altitude of 155 km at 257 seconds into the flight, and the Centaur main engine is ignited. After the Centaur first-burn main engine cutoff (MECO1) at 801 seconds, the Centaur and the spacecraft enter a coast period. At 1307 seconds, following vehicle alignment, the second burn of the main engine (MES2) is initiated. After reaching the target position, the main engines are shut down (MECO2) and the Centaur aligns the spacecraft and separates.

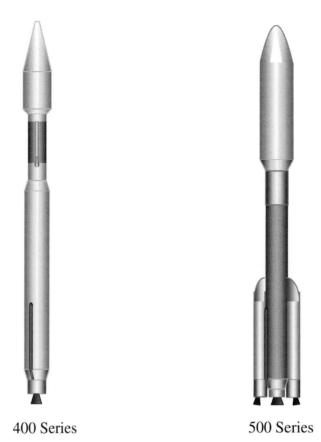

400 Series   500 Series

Figure 5.6 Illustration of Atlas V 400 and 500 series launch vehicles.

Delta II

The Delta II is a series of ELVs that can be configured as two or three-stage vehicles. Delta II launch configurations are illustrated in Figure 5.7. The first stage for the 7000 series is powered by an RS-27A main engine and two LR-101-NA vernier engines, using refined kerosene (RP1) and liquid oxygen as the propellant. Additional boost can be provided using three, four, or nine strap-on solid rocket motors. The solid rocket motors are graphite-epoxy motors (GEMs) and use hydroxyl-terminated polybutadiene (HTPB) as solid propellant. An Aerojet AJ10-118K powers the second stage using a mixture of hydrazine (N_2H_4) and unsymmetrical dimethyl hydrazine (UDMH, $C_2H_8N_2$) liquid propellants. A Thiokol Star is used for a third stage, fueled also with HTPB solid propellant. A four-digit designator is used to identify Delta configurations. The first digit identifies the first stage type and the second digit identifies

Table 5.2 Atlas V 552 Specifications [3]

	Length (m)	Diameter (m)	Launch Mass (kg)	Thrust (kN)	Burn Time (s)	Propellant
Atlas V 552	59.7	5.4	5.41×10^5	9,490 (Launch)	—	—
Faring	20.7	5.4	4.09×10^3	—	—	—
Max Payload to LEO	20.7	5.4	1.70×10^4	—	—	—
Common Core Booster:	32.4	3.8	2.84×10^5	3,820 SL	236	LOX/RP-1
5 Strap-On Boosters	17.7	1.55 (1)	2.1×10^5 (5)	5,670 SL (5)	94	HTPB
Centaur Upper Stage (DEC)	11.7	3.0	2.3×10^4	198 Vac.	429	LOX/LH$_2$

DEC = Dual Engine Centaur; GEO = Geosynchronous Earth orbit; LOX = Liquid oxygen; LH$_2$ = Liquid hydrogen; L = Sea level; Vac. = Vacuum; HTPB = Aluminum, ammonium perchlorate, and hydroxyl-terminated polybutadiene binder

Figure 5.7 Illustration of Delta II launch vehicle.

the number of strap-on solid rocket motors. The third and fourth digits identify the type of second and third stage, respectively. A zero for the second or third digits indicates no second or third stage, respectively. The specifications for the Delta II 7925 are presented in Table 5.3.

A two-stage Delta II rocket is typically used to deploy payloads to LEO, and a three-stage is used to deploy to GTO. The Delta II can deliver up to 5.8 metric tons to GEO and up to 2 metric tons to GTO. The flight sequence for the Thor III mission launch begins with a first stage burn for 4 minutes and 23 seconds. At launch, six of the nine strap-on boosters fire for 1 minute. The first six strap-on

Table 5.3   Delta II Specifications [4-6]

	Length (m)	Diameter (m)	Launch Mass (kg)	Thrust (kN)	Burn Time (s)	Propellant
Delta II 7925	38.32	2.9	2.32×10^5	3,748 (Launch*)		
Faring		2.9		—	—	—
Max Payload to LEO			5.2×10^3	—	—	—
Core Booster:	26.1	2.4	1.0×10^5	890 SL	240	RP1/LOX
9 Strap-On Boosters	13.0	1.0	1.1×10^5 (9)	4,042 SL (9)	60	HTPB
Second Stage	6.0	2.4	5.9×10^3	44 Vac.	422	UDMH/N_2O_4
Third Stage	2.0	1.25	2.0×10^3	66.4 Vac.	87.2	HTPB

*6 Boosters thrusting at launch
LOX = Liquid oxygen; SL = Sea level; Vac. = Vacuum; UDMH = Unsymmetrical dimethyl hydrazine; HTPB = Aluminum, ammonium perchlorate, and hydroxyl-terminated polybutadiene binder

boosters then separate and the remaining three are fired for 1 minute. After the first stage burn is complete, the first stage separates and the second stage burns for about 5 minutes and 20 seconds. After 22 minutes and 13 seconds into the flight, the second stage burns again for about 50 seconds. After 69 minutes and 30.2 seconds the second stage burns again, for the last time, for about 52 seconds. The second stage then separates and the third stage ignites about 72 minutes after liftoff and burns for 87.2 seconds. Finally, after 90 minutes from liftoff, the spacecraft separates [4-7].

Proton

The Russian Proton launch vehicle uses all liquid fuel propellants and is available in three or four stage vehicles. A four-stage Proton 8K82K/DM1 is illustrated in Figure 5.8. In this figure the outer shell is rendered semitransparent to illustrate the major internal structure, including fuel and oxidizer tanks. For the Proton M configuration, all four stages use a N_2O_4/UDMH propellant. The specifications for the Proton M configuration are given in Table 5.4. The first stage consists of a 4.2 m diameter central oxidizer tank, surrounded by six 1.6 m diameter fuel tanks, each with one RD-253 engine. The second stage is 17.1 m long and uses four RD-0210 engines. The third stage is 4.1 m long and uses one RD-0210 engine. For the Proton M, the fourth stage, required for deployment in GTO, uses an R2000 engine. The Proton can deploy more than 20,000 kg into LEO, and the four-stage Proton M is capable of deploying 5,500 kg into GTO [8,9].

2.2 Liquid Propellants

In a liquid propellant rocket, the fuel and oxidizer are stored in separate tanks, and are fed through a system of pipes, valves, and turbo pumps to a combustion chamber where they are combined and

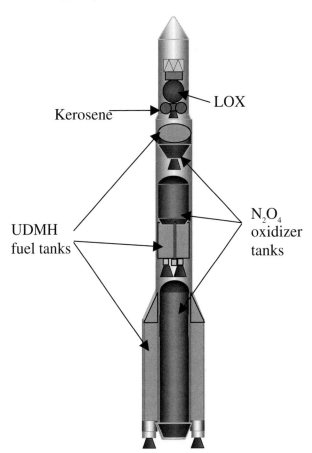

Figure 5.8 Illustration of Proton 8K82K/DM1 launch vehicle (semitransparent, revealing major internal structure).

Table 5.4   Proton Specifications [8,9]

	Length (m)	Diameter (m)	Launch Mass (kg)	Thrust (kN)	Burn Time (s)	Propellant
Proton M	59	7.4	7.1×10^5	1,745 SL		
Max Payload to LEO			2.1×10^4			
Stage 1	21.2	7.4	4.5×10^5	1,745 SL	150	N_2O_4/UDMH
Stage 2	17.1	4.1	1.18×10^5	582 Vac.	200	N_2O_4/UDMH
Stage 3	4.1	4.1	5.2×10^4	582 Vac.	250	N_2O_4/UDMH
Stage 4	2.6	4.0	2.1×10^4	19.6 Vac.	Up to 3,150	N_2O_4/UDMH

SL = Sea level; Vac. = Vacuum; UDMH = Unsymmetrical dimethyl hydrazine

burned to produce thrust. Liquid propellant engines are more complex than their solid propellant counterparts; however, they offer several advantages. By controlling the flow of propellant to the combustion chamber, the engine can be throttled, stopped, and restarted.

As stated previously, a high specific impulse is desired for propellants. From Eq. 5.1 we found that a high specific impulse is achieved for propellants with high combustion temperatures and for exhaust gases with small molecular weights. Several other considerations are important in the selection of propellants, such as propellant density, toxicity, and corrosiveness. The toxicity concern relates to safety considerations when handling, transporting, or storing propellants. Corrosive propellants may require special materials, and low-density propellants require large onboard storage tanks that add weight to the launch vehicle.

Liquid propellants used by NASA and in commercial launch vehicles can be classified into three types: cryogenic, hypergolic, and petroleum propellants. Properties of common liquid propellants are summarized in the following and compared in Table 5.5.

Cryogenic Propellants

Cryogenic propellants, such as liquid hydrogen and liquid oxygen, are liquefied gases stored at very low temperatures. LH_2 serves as the fuel and LOX is the oxidizer. LH_2 remains liquid at temperatures of 20 K and LOX remains in a liquid state at temperatures of 90 K. Because of the low temperatures of cryogenic propellants, they are difficult to store over long periods of time. For this reason, they are less desirable for use in military rockets which must be kept launch ready for months at a time. Also,

Table 5.5   Properties of Liquid Rocket Propellants [10]*

Compound	Chemical Formula	Molecular Mass (g/mole)	Density (g/cc)	Melting Point (K)	Boiling Point (K)
Liquid Oxygen	O_2	32.00	1.141	54.35	90.15
Nitrogen Tetroxide	N_2O_4	92.01	1.45	263.85	294.3
Nitric Acid	HNO_3	63.01	1.55	231.55	356.15
Liquid Hydrogen	H_2	2.016	0.071	13.85	20.25
Hydrazine	N_2H_4	32.05	1.004	274.55	386.65
Methyl Hydrazine	$CH_3N_2H_3$	46.07	0.866	220.75	360.65
Unsymmetrical Dimethyl Hydrazine (UDMH)	$(CH_3)_2N_2H_2$	60.10	0.791	215.15	337.05
Dodecane (Kerosene)	$C_{12}H_{26}$	170.34	0.749	263.55	489.45

* Notes: (1) Nitrogen tetroxide and nitric acid are hypergolic with hydrazine, MMH and UDMH. Oxygen is not hypergolic with any commonly used fuel. (2) Chemically, kerosene is a mixture of hydrocarbons; the chemical composition depends on its source; however, it usually consists of approximately 10 different hydrocarbons in which each contains from 10 to 16 carbon atoms per molecular. The constituents include n-dodecane, alkyl benzenes, and naphthalene and its derivatives.

liquid hydrogen has a very low density (~0.071 g/cm^3) and, therefore, requires a storage volume many times greater than other fuels. Despite these drawbacks, the high I_{sp} of liquid hydrogen/liquid oxygen makes these problems worth coping with when reaction time and storability are not too critical. Liquid hydrogen delivers a specific impulse about 40% higher than other rocket fuels. Liquid hydrogen and liquid oxygen are used as the propellant in the high efficiency main engines of the Space Shuttle. LH$_2$/LOX also powered the upper stages of the Saturn V and Saturn lB rockets as well as the second stage of the Atlas/Centaur launch vehicle, the first U.S. LH$_2$/LOX rocket, launched in 1962.

Hypergolic Propellants

Hypergolic propellants are fuels and oxidizers, which ignite spontaneously on contact with each other and require no ignition source. The easy start and restart capability of hypergolics makes them ideal for spacecraft maneuvering systems. Also, since hypergolics remain liquid at normal temperatures, they do not pose the storage problems of cryogenic propellants. Hypergolics are highly toxic and must be handled with extreme care.

Hypergolic fuels commonly include hydrazine, monomethyl hydrazine (MMH), and unsymmetrical dimethyl hydrazine (UDMH). The oxidizer is typically nitrogen tetroxide (N$_2$O$_4$) or nitric acid (HNO$_3$). UDMH is used in many Russian, European, and Chinese rockets while MMH is used in the orbital maneuvering system and reaction control system of the Space Shuttle orbiter. The Titan family of launch vehicles and the second stage of the Delta use a fuel called Aerozine 50, a mixture of 50% UDMH and 50% hydrazine.

Hydrazine is also frequently used as a monopropellant in *catalytic decomposition engines*. In these engines, a liquid fuel decomposes into hot gas in the presence of a catalyst. The decomposition of hydrazine produces temperatures of about 1200 K and a specific impulse of about 230 or 240 seconds.

Petroleum Propellants

Petroleum fuels are refined from crude oil and are a mixture of complex hydrocarbons, i.e., organic compounds containing only carbon and hydrogen. The rocket fuel RP-1 is a highly refined kerosene that is used in combination with liquid oxygen as the oxidizer. RP-1 and liquid oxygen are used as the propellant in the first-stage boosters of the Atlas/Centaur and Delta launch vehicles. This combination also powered the first stages of the Saturn 1B and Saturn V rockets. RP-1 delivers a specific impulse considerably less than cryogenic fuels.

2.3 Solid Propellants

Small solid propellant motors are often used to power the final stage of a launch vehicle or attach to payloads to boost them to higher orbits. Larger solid propellant motors such as the payload assist module and the inertial upper stage provide the boost required to place satellites into geosynchronous orbit or on planetary trajectories. Both the Delta and the Space Shuttle launch vehicles use strap-on solid propellant motors to provide added thrust at liftoff.

Unlike liquid propellants in which the fuel and oxidizer are stored separately, solid propellant motors usually consist of a casing filled with a mixture of solid fuel and oxidizer compounds. The solid fuel and oxidizer burn at a rapid rate and expel the hot product gases from a nozzle to produce thrust. Once the ignition is achieved, a solid propellant burns from the center out toward the sides of the casing. The shape of the center channel determines the pattern and rate of the burn, thus providing a means to control thrust. The two families of solid propellants include homogeneous and composite (heterogeneous) propellants. Both types of solid propellants are dense, stable at ordinary temperatures, and easily storable.

Table 5.6   Combustion Properties of Selected Elements [10]

Chemical Element	Combustion Temperature (K)	Heating Value (kJ/kg)
Hydrogen	3000.00	120.95
Lithium	2611.11	41.87
Beryllium	4388.89	65.13
Boron	3000.00	60.48
Carbon	2055.56	32.56
Sodium	2111.11	9.30
Magnesium	3388.89	25.59
Aluminum	3888.89	30.24

Homogenous Solid Propellants

Homogeneous solid propellants can be grouped as either simple or double base. A simple base propellant contains a single compound, usually nitrocellulose. Double base propellants are usually composed of nitrocellulose and nitroglycerine, to which a plasticizer is added. Homogenous propellants generally do not have a specific impulse greater than 210 seconds under a normal condition, and their main use is in tactical weapons because they do not produce traceable fumes. They are also used in subsidiary functions such as jettisoning spent parts or separating one stage from another.

Composite Solid Propellants

The solid propellant motors in the Delta and the Space Shuttle launch vehicles are made up of composite solid propellants. Composite solid propellants are heterogeneous mixtures of fuel and oxidizer powders and a binder. Black powder is the oldest composite solid propellant. Powdered metals are often used as the fuel, and aluminum is used extensively in propellant formulations because of its heating value (30 kJ/kg) and stable burning characteristics. The combustion temperatures and heating values of potential fuel materials are listed in Table 5.6. Between 60% and 90% of the propellant mass is an oxidizer consisting of crystallized or finely ground mineral salt. Ammonium perchlorate is the most widely used oxidizer today. Ammonium perchlorate contains 34% oxygen by weight and its density is 1.9 g/cm^3. Other oxidizers such as potassium perchlorate and ammonium nitrate are used for fast burning rates and slower burning rates, respectively.

Hydroxyl-terminated polybutadiene (HTPB) is used as a binder in solid propellant motors for Delta launch vehicles. Polybutadiene acrylic acid, a new type of binder based on long-chained polybutadiene, is used extensively in NASA propulsion programs. A polybutadiene binder enables a propellant to withstand high strain rates, such as those encountered on ignition of a large-diameter rocket motor. The SRB of the Space Shuttle uses polybutadiene acrylic acid acrylonitrile as the binder.

3.0 Propellant Explosions

This section begins with a discussion of the important considerations relating to possible propellant explosions. The basic theory of explosions is then addressed, followed by a discussion of the application of the theory to space nuclear missions.

3.1 Propellant Explosion Considerations

The effect of the explosions on the nuclear payload depends upon where the explosion initiates. Also an explosion at one location may induce a secondary explosion that may be more severe than the primary explosion. Some considerations will depend on whether the propellant is a liquid or solid propellant.

Liquid Propellants

For liquid propellants, the fuel and oxidizer are contained in separate tanks. Hence, the bulkhead separating the fuel and oxidizer or the fuel tank itself must be breached for an explosion to occur inside a propellant tank. Subsequent, multisource explosions behave quite differently from the single-source explosion. Each explosive source must be considered, including interactions through multiple blast-wave reflections. When analyzing liquid propellant explosions, a number of factors must be kept in mind. These include initial vehicle geometry, propellant tank types and their configurations, the tank's length-to-diameter ratio, whether the tanks share a common bulkhead, single or multiple fuel and oxidizer tanks, relative tank locations, propellant type, the timing of ignition, the quantity of propellant, and the concentrations of fuel and oxidants. The relative confinement of spilled liquid propellants is also an important consideration. The PYRO project [11] generalized the following four basic categories of spilled propellant accident environments:

- *Unconfined*: Propellants are ejected into the atmosphere during flight.
- *Confined by Ground Surface (CBGS)*: Propellants are spilled and spread over a large area, with minimal confinement. Typical examples for such a situation may be launch vehicle fallback or tipover to the ground with subsequent tank rupture and ignition.
- *Confined by Missile (CBM)*: Propellants are spilled but are confined by the launch vehicle or are internally mixed within the tank (e.g., common bulkhead failure).
- *High Velocity Impact (HVI)*: Propellants are released and mixed due to a HVI on a ground surface. Here, however, the type of surface becomes important. For impact on a hard surface the propellants tend to spread over a large area (i.e., similar to the CBGS case). For the case of an impact on a soft surface, a higher degree of mixing and confinement may result if the fuel and oxidizer collect in the impact crater.

Solid Propellants

Unlike liquid propellants, wherein the fuel and oxidizer are stored in separate tanks, solid propellants are premixed with optimum levels of fuel and oxidizer. The explosive nature of solid propellants is characteristically similar to that of common solid explosives, such as TNT. Although solid propellants are often designed to be less sensitive than solid explosives (e.g., requires high ignition temperatures), solid propellants can still be classified as explosive materials. Erdahl et al. [12] identified the following considerations to assess the likelihood of solid propellant explosion:

- Propellant toughness
- Motor design (core configuration, diameter, length to diameter ratio, chamber pressure, case bonding technique, and propellant residual strain)
- Propellant critical diameter and geometry
- Propellant granular bed characteristics (pyrolysis and ignition)
- Propellant response to direct shock and delayed reduced shock
- Solid rocket impact velocity and the type of surface impacted

3.2 Vapor Explosions

The requirements for and consequences of vapor explosions will depend on chemical considerations and whether the explosion is confined or unconfined. Conditions can lead to deflagrations or detonations or deflagrations transitioning to detonations. For a deflagration, the speed of the combustion wave is limited by the process of heat and mass transfer into the unburned propellant mixture. For a detonation, the reaction generates a shock wave causing an increase in the temperature in the unburned propellant and subsequent ignition in the unburned gas mixture. Thus, for a detonation, the combustion wave propagates at supersonic speeds.

Chemical Considerations

The energy released during propellant explosions results from the rapid exothermic chemical reaction between the propellant fuel and an oxidant. For example, consider a propellant consisting of LH_2 and LOX. The energy release from the explosive reaction of LH_2 and LOX can be expressed by the chemical equation

$$2H_2 + O_2 \rightarrow 2H_2O + E_d. \qquad (5.2)$$

For an explosion, the parameter E_d is the total heat of detonation or heat of reaction, typically expressed in kJ. The heat of detonation is directly proportional to the total mass of the explosive m_e and the specific heat of reaction $\tilde{H}_e$ (kJ/g). Tables 5.6 and 5.7 list the heat of the reactions for selected chemical elements (in kJ/kg) and liquid propellants (in kJ/mole), respectively. Equation 5.2 is given for a stochiometric composition of fuel and oxidant. The stochiometric composition of the vapor cloud is the composition in which the fuel and oxidant are in balance such that no excess fuel or oxidant remains when the reaction has been completed. In general, the stochiometric composition is the composition that yields the highest explosion pressure.

A vapor cloud will not ignite, however, if the fuel concentration in a fuel/oxidant mixture is above the upper flammability limit or below the lower flammability limit. For hydrogen, the flammability range is very large. At 300 K, the flammability concentration for hydrogen in air ranges from a few volume percent hydrogen to about 75% hydrogen. As shown in Figure 5.9, the flammability range widens

Table 5.7 Stochiometric Equations and Heat of Reactions for Selected Liquid Propellants

Propellant Combination	Stochiometric Equation	Adiabatic Flame Temp. (K)	ΔH_r,* kJ/mole
LH_2/LO_2	$H_{2(l)} + 0.5O_{2(l)} \Rightarrow H_2O_{(g)}$	3007	-226
LH_2/N_2O_4	$4H_{2(l)} + N_2O_{4(l)} \Rightarrow 4H_2O_{(g)} + N_{2(g)}$	2891	-912
MMH/LO_2	$CH_3N_2H_{3(l)} + 2.5O_{2(l)} \Rightarrow$ $3H_2O_{(g)} + N_{2(g)} + CO_{2(g)}$	2971	-1141
MMH/N_2O_4	$4CH_3N_2H_{3(l)} + 5N_2O_{4(l)} \Rightarrow 4CO_{2(g)}$ $+ 12 H_2O_{(v)} + 9N_{2(g)}$	2903	-4598
$RP\text{-}1/LO_2$**	$C_{12.1}H_{23.5(l)} + 17.975O_{2(l)} \Rightarrow$ $12.1CO_{2(g)} + 11.75H_2O_{(v)}$	3079	-7073
$UDMH/N_2O4$	$(CH_3)_2N_2H_{2(l)} + 2N_2O_{4(l)} \Rightarrow$ $2CO_{2(g)} + 4 H_2O_{(v)} + 3N_{2(g)}$	3415	-1765

* The following heat of formation values were used for these calculations:
$MMH_{(l)}$ = 54.836 kJ/mole [13]
$CO_{2(g)}$ = -393.533 kJ/mole [13]
$N_2O_{4(l)}$ = -19.564 kJ/mole [13]
Jet A, $C_{12}H_{23(l)}$ = -303.403 kJ/mole [14]
$H_2O_{2(l)}$ = -187.78 kJ/mole [14]
$H_{2(l)}$ = -9.012 kJ/mole (at 20.27 K) [14]
$O_{2(l)}$ = -12.979 kJ/mole (at 90.17 K) [14]
$H_2O_{(v)}$ = -241.826 kJ/mole [13]
$C_2H_8N_{2(l)}$ = 49.7896 kJ/mole [obtained from http//www.dunnspace.com/isp.html]
RP-1, H/C ratio of 1.9423 = -24.7177 kJ/mole [4], so to estimate the heat of formation for RP-1, -24.7177 kJ/mole was multiplied by 12 carbons, -296.6 kJ/mole.

** References list RP-1 as $C_{12.069321}H_{25.56698}$. This results in a H/C ratio of 2.11. Ref [14] was used to obtain the heat of formation for RP-1. However, Ref [14] lists the H/C ratio of RP-1 as 1.9423. In order to obtain an estimate for the heat of reaction, the molecular formula for RP-1 was estimated based on Ref [14]'s H/C ratio.

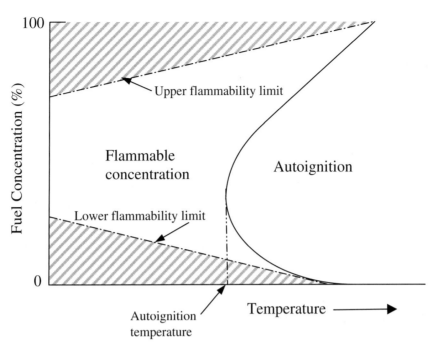

Figure 5.9 Illustration of fuel/oxidant concentration flammability and autoignition ranges as a function of temperature.

with increasing temperature. At lower temperatures, an ignition source with an ignition energy of sufficient strength is required in order to ignite a vapor cloud. The minimum ignition energy depends on the type of propellant fuel and the fuel/oxidant concentration in the vapor cloud. The minimum ignition energy for hydrogen is very low. Above some temperature, called the minimum autoignition temperature, a chemical reaction will take place without an external ignition source. For hydrogen, the minimum autoignition temperature is ~800 K.

Confined Vapor Explosions

Chapman (in 1899) and Jouguet (in 1905) independently developed a simple one-dimensional model for a combustion wave and a supersonic shock wave. The model, now called the Chapman-Jouguet (C-J) model, makes the following assumptions: (1) laminar one-dimensional flow, (2) instantaneous chemical reaction at the combustion wave front, and (3) thermodynamic equilibrium. The combustion products emerging from the combustion front are described by steady state conditions; however, the flow following this point can be time-dependent.

Here the C-J model will be used to explore explosions within a confining cylinder. The basic geometry and velocities are shown schematically in Figure 5.10. An analysis of the simple cylindrical structure illustrates some of the effects of confinement by the launch vehicle structure. For simplicity, the combustion wave will be used as the reference frame to develop the C-J model. In the combustion wave frame, the speed of the unburned gas v_u is equal to the combustion wave speed U. Momentum and energy balance requires, respectively,

$$p_u v_u + \varsigma_u v_u^2 = p_b v_b + \varsigma_b v_b^2, \tag{5.3}$$

and

$$c_p T_u + \frac{1}{2} v_u^2 + \tilde{q} = c_p T_b + \frac{1}{2} v_b^2. \tag{5.4}$$

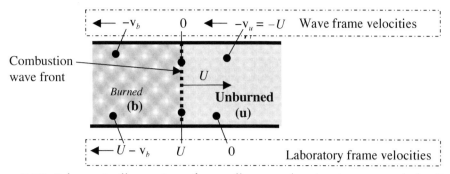

Figure 5.10 Schematic illustration of propellant combustion in open-ended cylinder.

The parameters p, v, and ς are the gas pressure, speed, and density, respectively. The subscripts b and u correspond to the burned and unburned propellant gases. Equation 5.3 is Bernoulli's equation. In Eq. 5.4, c_p, T, and $\tilde{q}$ are the specific heat at constant pressure, the gas temperature, and the heat released by combustion per unit mass, respectively. For simplicity, the specific heats for the unburned and burned propellant are assumed to be equal. Also, the continuity equation requires

$$\varsigma_u v_u = \varsigma_b v_b, \tag{5.5}$$

and assuming the ideal gas law, the equation of state for the unburned and burned gasses are, respectively,

$$p_u = \varsigma_u R T_u, \tag{5.6}$$

and

$$p_b = \varsigma_b R T_b. \tag{5.7}$$

Here, R is the universal gas constant.

Using Eq. 5.3 with Eq. 5.5 gives

$$v_u = U = \frac{1}{\varsigma_u} \sqrt{(p_b - p_u)\bigg/\left(\frac{1}{\varsigma_u} - \frac{1}{\varsigma_b}\right)}, \tag{5.8}$$

and

$$v_b = \frac{1}{\varsigma_b} \sqrt{(p_b - p_u)\bigg/\left(\frac{1}{\varsigma_u} - \frac{1}{\varsigma_b}\right)}. \tag{5.9}$$

For constant U, Eq. 5.8, defines the *Rayleigh line* in a p_b vs. $1/\varsigma_b$ plot. Similarly, for constant v_b Eq. 5.9 gives the curve of constant speed for the burned gas. The speed of sound in the unburned gas c_u is given by $c_u = \sqrt{\gamma R_u T_u} = \sqrt{\gamma p_u/\varsigma_u}$, where $\gamma = c_p/c_v$ and c_p and c_v are the specific heats at constant pressure and constant volume, respectively. Thus, from Eq. 5.8 we obtain

$$M_u = \sqrt{\frac{1}{\gamma}\left(\frac{p_b}{p_u} - 1\right)\bigg/\left(1 - \frac{1/\varsigma_b}{1/\varsigma_u}\right)}, \tag{5.10}$$

where $M_u = v_u/c_u$ is the Mach number for the unburned propellant. We can also obtain the Mach number for the burned propellant gas as,

$$M_b = \sqrt{\frac{1}{\gamma}\left(1 - \frac{p_u}{p_b}\right) \Big/ \left(\frac{1/\varsigma_u}{1/\varsigma_b} - 1\right)}. \tag{5.11}$$

From the preceding equations, and using the relationship for the specific heat $c_p = R[\gamma/(\gamma - 1)]$, we can rewrite Eq. 5.4 as,

$$\frac{\gamma}{\gamma - 1}\left(\frac{p_b}{\varsigma_b} - \frac{p_u}{\varsigma_u}\right) - \frac{1}{2}(p_b - p_u)\left(\frac{1}{\varsigma_u} + \frac{1}{\varsigma_b}\right) = \tilde{q}. \tag{5.12}$$

Equation 5.12 provides the so-called Hugoniot curve illustrated in Figure 5.11. The Hugoniot curve describes the pressure in the burned propellant plotted as a function of $1/\varsigma_b$. The point O is the pressure/density point for the unburned gas prior to ignition. For zone D, $p_b \gg p_u$ and $\gamma \sim 1.4$ such that $M_u > 1$. Thus, for zone D the combustion wave is supersonic and combustion for this zone corresponds to a detonation. For zone E, no real solution exists; consequently, combustion cannot yield parameters in this zone. For zone F, $p_u > p$ and $1/\varsigma_b \gg 1/\varsigma_u$ such that $0 < M_u < 1$; thus, the combustion wave is subsonic, corresponding to a deflagration.

Steady state solutions to the C-J equations are obtained as the intersection of the Rayleigh line and the Hugoniot curve. The so-called C-J points shown in Figure 5.11 correspond to the Rayleigh line for $M_u = 1$. For subsonic wave speeds, Figure 5.11 shows that the Rayleigh line does not intersect the Hugoniot curve; thus, no solutions are possible for U subsonic. For $M_u > 1$, two detonation solutions are obtained. Rayleigh lines intersecting the Hugoniot curve in the deflagration region also provide two solutions. The upper and lower C-J points mark the boundaries between strong and weak detonations

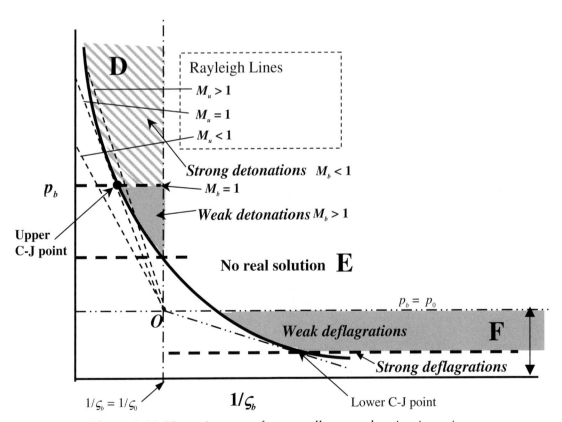

Figure 5.11 Hugoniot curve for propellant combustion in a pipe.

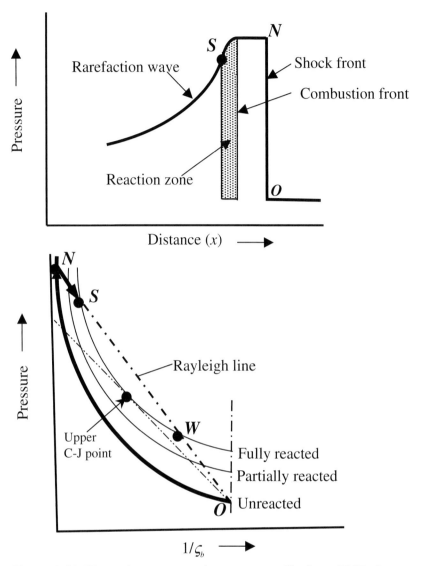

Figure 5.12 Hugoniot curves and pressure profile from ZND theory.

and deflagrations, respectively. It can be shown that above the upper C-J point the speed of the burned gas v_b is subsonic, and below the upper C-J point, in zone D, v_b is supersonic. For deflagrations below the lower C-J point, the burned gases are subsonic, and above the lower C-J point, in zone F, the burned gases are supersonic.

During World War II, Zeldovich, von Neumann, and Doring (ZND) developed an improvement on the Chapman-Jouguet model [15]. Whereas the C-J model assumes that the combustion front is a discontinuity with an infinite reaction rate, the ZND model takes the reaction rate into account and provides a reaction zone of finite thickness. For the ZND theory, the $\tilde{q}$-term in Eq. 5.12 is multiplied by a reaction fraction λ. The ZND model can be used to provide Hugoniot curves for the unreacted ($\lambda = 0$), partially reacted ($0 < \lambda < 1$), and fully reacted ($\lambda = 1$) states, as shown in Figure 5.12. The Rayleigh line is also shown in Figure 5.12. The pressure increases from its initial value and rises sharply at the shock front from point O to N along the unreacted Hugoniot curve. The burned gas then expands to its final steady state value at point S (strong solution). A pressure trace across the combustion front is provided in the top portion of Figure 5.12, with the corresponding points O, N, and S shown. Using the theory described in the preceding discussion, the basic pressure vs. distance profile can be obtained

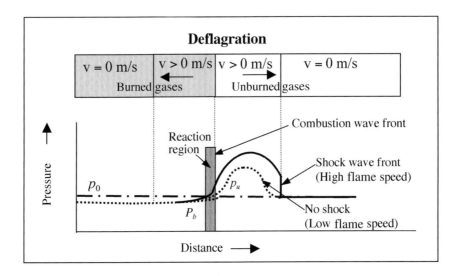

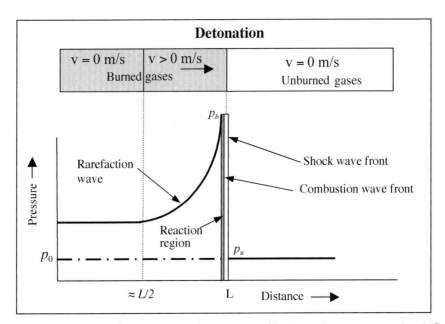

Figure 5.13 Schematic illustrations of pressure vs. distance profile for a detonation and a deflagration in a cylinder closed at one end.

for both a detonation and a deflagration, as shown in the schematic illustration in Figure 5.13. A schematic illustration of the gas temperature in and near the reaction zone is presented in Figure 5.14 for deflagrations and detonations.

Although the preceding discussion suggests that a range of strong and weak detonations and deflagrations are possible, not all of these possible outcomes yield steady state solutions. For strong detonations, heat losses, turbulence, and friction at the cylindrical wall produce sonic expansion waves that catch up with the shock wave. The expansion waves weaken the shock, reducing the pressure, and eventually the shock front evolves to a sonic wave at the C-J point. According to the relatively simple ZND theory, weak detonations are physically unrealizable because all of the chemical energy of the unburned gas was expended in moving from point N to point S (Figure 5.12). From these arguments it can be concluded that the only steady state solution for the detonation region is at the upper C-J point,

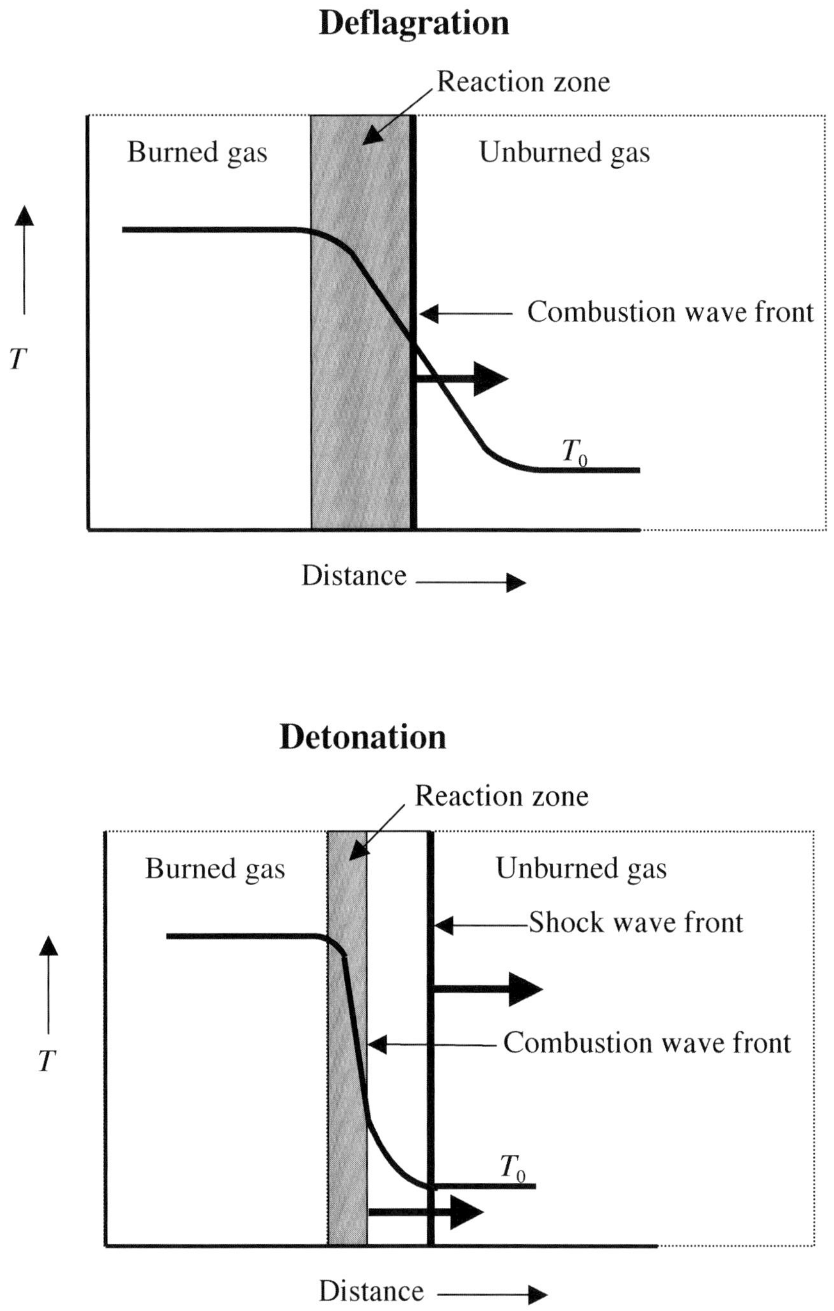

Deflagration

Detonation

Figure 5.14 Schematic illustrations of temperature vs. distance for a detonation and a deflagration.

and weak detonations cannot be achieved. However, more elaborate detonation models have shown that for some conditions weak detonations are possible [15]. Strong deflagrations can be ruled out based on arguments from compressible flow theory, but the entire range of weak deflagrations is permitted. A summary of typical parameters is provided in Table 5.8.

During a deflagration, the pressure generated by the flame can propagate away from the combustion front. The principal mechanism for flame acceleration is turbulence at the interface of the burned and unburned gas resulting in an increased burning rate. An accelerating deflagration can suddenly transition to a detonation. The transition is called the deflagration to detonation transition (DDT). At the point of DDT, very high pressures are generated. The mechanism for DDT is not fully understood.

Table 5.8 Parameter Ratio Ranges for Detonations and Deflagrations [16,17]

Parameter	Detonation Strong	Weak Deflagration
U (km/s)	1.5–3	0.001–0.5
v_b/c_u	5–10	0.0001–0.03
p_{max} (MPa)	1.4–4	0.1–0.3
p_b/p_u	13–55	0.98–0.976
T_{max} (K)	2500–3700	1200–3500
T_b/T_0	8–21	4–16
ς_b/ς_0	1.4–2.6	0.06–0.25

Vapor Cloud Explosions

A vapor cloud explosion (VCE) can result when an accident causes the release of the liquid propellant followed by the vaporization and subsequent ignition of the propellant. An explosive mixture can result by mixing of the propellant fuel and oxidizer vapors or by mixing of the propellant fuel with oxygen in the atmosphere. As for a confined vapor, ignition could result in either a deflagration or a detonation; however, only detonations propagate at high velocity without vapor confinement. The flame acceleration depends on local conditions, such as the presence of an obstacle, whether the release of the combustible gas occurs in laminar or turbulent flow, and the density of the propellant in the vapor cloud. When obstacles are present, the flame speed approaches that of a solid explosive. In an open space and in the absence of any obstacles, however, the explosion speed slows, approaching only a fraction of the solid explosives' speeds. Combustion wave speeds for VCEs can range from subsonic (deflagration) to sonic and supersonic (detonation). An illustration of the propagation of an unconfined VCE is shown in Figure 5.15.

A VCE can be analytically described by considering the energy addition within an expanding spherical flow (gas) field. The conservation equations for mass, momentum, and energy in one-dimensional spherical geometry yield a set of nonlinear partial differential equations, requiring numerical methods for their complete solution. By making some simplifying assumptions, however, it can be shown [18] that these equations can be solved to yield both the velocity and maximum overpressure; i.e.,

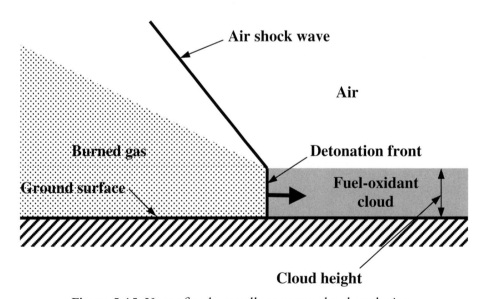

Figure 5.15 Unconfined propellant vapor cloud explosion.

$$\frac{\mathrm{v}}{c} = \frac{\alpha - 1}{\alpha} \, M_f^3 \left[\frac{c^2 t^2}{r^2} - 1 \right], \tag{5.13}$$

and
$$\frac{p - p_0}{p_0} = 2\gamma \left[\frac{\alpha - 1}{\alpha} \right] M_f^3 \left[\frac{ct}{r} - 1 \right]. \tag{5.14}$$

Here, r is the radial coordinate, t is time, p_0 is the ambient pressure, α is the isobaric expansion ratio, and M_f is the flame Mach number.

3.3 Solid Propellant Explosions

Detonations in solid propellants differ in several ways from vapor detonations. These differences are generally associated with the incompressibility and higher density of solid propellants relative to vapor clouds. The higher densities typically result in higher detonation velocities in solid propellants (up to 9 km/s). Nonetheless, the simple methods used to analyze vapor explosions discussed in the previous section can also be used to analyze solid propellant explosions.

Solid propellant explosions can be accidentally initiated by a shock or as a result of heating (cook-off). Shock initiation may result from solid propellant impact or from an air shock wave produced by the detonation of another propellant source. The mechanism of impact-shock initiation of a solid propellant is illustrated in Figure 5.16. High-speed impact can fracture a solid propellant creating a highly porous zone near the point of impact. The fractured region then becomes recompressed, producing a compression wave and initiating a detonation.

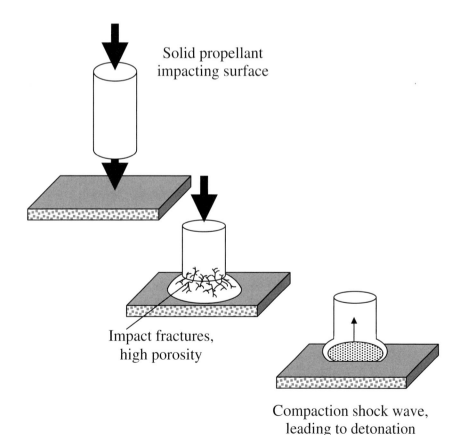

Figure 5.16 Schematic illustrations of impact induced detonation of a solid propellant.

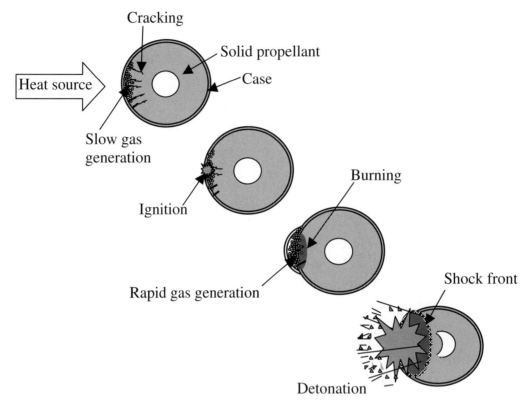

Figure 5.17 Schematic of a fast cook-off detonation of a solid rocket propellant. *Adapted from [16].*

The term *cook-off* refers to solid propellant detonation as a result of heating. The two categories of accidental cook-off include *fast cook-off* and *slow cook-off*. Slow cook-off events are due to very slow heating of the propellant and detonation, if induced, may take many hours to initiate. The slow cook-off scenario is generally not a principal concern for launch or prelaunch accidents with onboard nuclear systems. Fast cook-offs, however, can be postulated for some accident scenarios. We assume that an accident has resulted in a propellant fire (or another heat source such as an onboard liquid metal fire). It is also assumed that a solid rocket motor has either landed within or close to the fire. If the solid propellant is still confined by the motor housing, the basic mechanisms illustrated in Figure 5.17 have been postulated for fast cook-off [16]. The heat source or fire causes slow evolution of gas forming pockets near the propellant-case interface. At sufficiently high temperatures, ignition occurs at the hot spot resulting in burning at the hot surface. Rapid production of gaseous combustion products causes the pressure to rise, and if the casing is strong, the rapid pressure increase can initiate a detonation.

3.4 BLEVE

As noted in Section 1.0, one of the phenomena associated with liquid propellant explosion is the possible rupture of a vessel containing a pressurized liquid, or BLEVE. This phenomenon usually results from boiling of cryogenic propellants such as LH_2/LOX, liquid N_2O_4, and liquid combustible hydrocarbons. The magnitude of the explosion due to a BLEVE depends on the liquid temperature and internal pressure of the vessel as well as the ambient pressure at the moment of vessel failure. The BLEVE may result from any initiator that causes rupture of the pressurized tank. Examples include an external fire, mechanical impact (such as a fragment impact), excessive internal pressure (e.g., heating from an accident involving a propellant fire), or metallurgical failure. Failures of this type are possible for launch vehicles during the prelaunch and early launch stages.

The explosive energy E_{ex} from a BLEVE can be expressed as

$$E_{ex} = \tilde{E}_{ex} m_1, \tag{5.15}$$

where $\tilde{E}_{ex}$ is the specific work done by the expanding fluid (J/kg), and m_1 is the mass of the fluid released. Specific work is defined as the initial internal specific energy minus the final internal specific energy; i.e.,

$$\tilde{E}_{ex} = \tilde{U}_1 - \tilde{U}_2, \tag{5.16}$$

where $\tilde{U}_1$ and $\tilde{U}_2$ are the specific internal energies of the fluid before and after failure, respectively. The specific internal energy is given by

$$\tilde{U}_1 = \tilde{H}_1 - \frac{p_1}{\varsigma_1}, \tag{5.17}$$

where $\tilde{H}$ is specific enthalpy (J/kg). The specific internal energy at state 2 is given by

$$\tilde{U}_2 = (1 - X)\tilde{H}_{sl} + X\tilde{H}_g - (1 - X)\frac{p_0}{\varsigma_{sl}} - X\frac{p_0}{\varsigma_g} \tag{5.18}$$

where X is vapor ratio $(\tilde{S}_1 - \tilde{S}_{sl}/\tilde{S}_g - \tilde{S}_{sl})$, and $\tilde{S}$ is specific entropy (J/kg). The subscript sl refers to saturated liquid state at ambient pressure, and g refers to saturated vapor state at ambient pressure.

Example 5.1

During a prelaunch accident, a tank containing LOX is ruptured. The initial pressure of the tank is 1.5 atm and the ambient pressure is 1 atm. What is the explosive energy of this BLEVE if the mass of LOX is 50,000 kg? LOX has the following saturation properties:

At 1.5 atm, $\tilde{H}_1 = -1.2641 \times 10^5$ J/kg, $\varsigma_1 = 1120.7$ kg/m^3, and $\tilde{S}_1 = 3.0168$ J/g·K.

At 1.0 atm, $\tilde{H}_{sl} = -1.3337 \times 10^5$ J/kg, $\varsigma_{sl} = 1141.2$ kg/m^3, $\tilde{H}_g = 7.9688 \times 10^4$ J/kg,

$\varsigma_g = 4.467$ l kg/m^3, $\tilde{S}_{sl} = 2.9418$ J/g·K, and $\tilde{S}_g = 5.3042$ J/g·K.

Solution:

Using Eq. 5.17,

$$\tilde{U}_1 = -1.2641 \times 10^5 \text{ J/kg} - \frac{(1.5 \text{ atm})(1.013 \times 10^5 \text{ Pa/atm})}{1120.7 \text{ kg/m}^3}\frac{\text{J/m}^3}{\text{Pa}} = -1.2655 \times 10^5 \text{ J/kg}$$

Using Eq. 5.18, with $X = \dfrac{3.0168 - 2.9418}{5.3042 - 2.9418} = 0.03175$; thus,

$$\tilde{U}_2 = (1 - 0.03175)(-1.3337 \times 10^5) + (0.03175)(7.9688 \times 10^4)$$

$$-(1 - 0.03175)\frac{1.013 \times 10^5}{1141.2} - 0.03175\frac{1.013 \times 10^5}{4.4671} = -1.2741 \times 10^5 \text{ J/kg}$$

Finally, applying Eq. 5.16, the explosive energy due to a BLEVE is

$$E_{ex} = \left[-1.2655 \times 10^5 \frac{\text{J}}{\text{kg}} - \left(-1.2741 \times 10^5 \frac{\text{J}}{\text{kg}}\right)\right](5 \times 10^4 \text{ kg}) = 4.3 \times 10^7 \text{ J}$$

4.0 Explosion Effects

The discussion of explosion effects in this chapter is limited to blast effects and fragment generation. The discussion of blast wave effects focuses on the characterization of the effect of an explosion on the surrounding air. Fragment impact effects are discussed in Chapter 7.

4.1 Blast Fundamentals

Because the explosive properties of propellants are similar to conventional explosives, many of the techniques developed to analyze the blast phenomenon for high explosives also apply to propellants. Blast wave properties determine the potential effect of an explosion. The propellant configuration and environment are also important factors. For example, for an unconfined liquid propellant or a solid propellant with broken casing, the detonation wave propagates directly into the surrounding air. If the propellant is encased, the detonation wave must first breach the casing material before driving the propellant outward. For a postulated propellant explosion, the blast wave can reflect from surfaces (e.g., the ground surface) and interact with structures near the onboard space nuclear system. These interactions can cause the pressure, density, temperature, and particle velocity to change rapidly with position and time.

Blast wave properties are often characterized by the undisturbed wave as it propagates through the air. Figure 5.18 presents a depiction of some of these properties for an ideal blast wave as a function of time. Before the shock front arrives (i.e., when $t < t_a$), the pressure is at ambient pressure p_0. At arrival time t_a the pressure p rises abruptly to a peak value of $\hat{p} = \Delta p^+ + p_0$, then decays to the ambient pressure when $t = t_a + t_d^+$. Following time $t_a + t_d^+$, the pressure decreases to a partial vacuum with minimum pressure equal to $p_0 - \Delta p^-$, and the pressure eventually returns to p_0 after a time of $t = t_a + t_d^+ + t_d^-$. Here, Δp^+ is defined as the peak overpressure (Pa). The portion of the time history above initial p_0 is called the positive phase and spans from $t = t_a$ to $t_a + t_d^+$. Conversely, the portion that is below p_0, for the time period from $t_a + t_d^+$ to $t = t_a + t_d^+ + t_d^-$, is called the negative phase. For simplicity, we

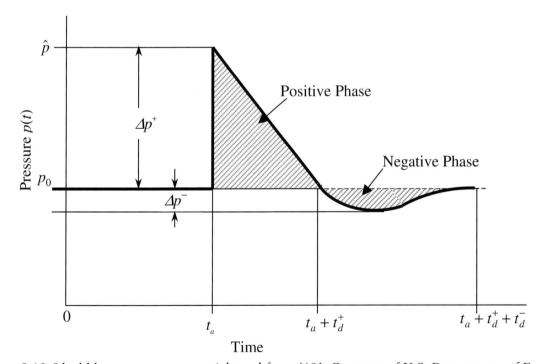

Figure 5.18 Ideal blast wave structure. *Adapted from [19]. Courtesy of U.S. Department of Energy.*

only focus on the positive phase because $\Delta p^+ >> \Delta p^-$. The positive impulse I_s^+ resulting from the blast is given by the time integral of the pressure increase over the positive phase; i.e.,

$$I_s^+ = \int_{t_a}^{t_a+t_d^+} [p(t) - p_0]\, dt. \tag{5.19}$$

At the shock front, in free air and away from reflecting surfaces, a number of wave properties are interrelated through the Rankine-Hugoniot equations. Using Eqs. 5.3 through 5.7 and eliminating the combustion energy term, these equations can be rewritten as

$$\hat{\varsigma}(U - \hat{v}) = \varsigma_0 U, \tag{5.20}$$

$$\hat{p} - p_o = \varsigma_0 \hat{v} U, \tag{5.21}$$

and

$$\hat{\varsigma}(U - \hat{v})^2 + \hat{p} = \varsigma_0^2 U^2 + p_0. \tag{5.22}$$

Here, U is the shock front velocity and the hat for $\hat{\varsigma}$, $\hat{p}$, and $\hat{v}$ indicates the peak values behind the shock front. The parameter ς_0 is the ambient air density, and the peak pressure (also referred to as the side-on pressure) is

$$\hat{p} = \Delta p^+ + p_0. \tag{5.23}$$

Equations 5.20 through 5.22 consist of three equations with unknowns U, $\hat{\varsigma}$, and $\hat{v}$, which describe the properties across a normal shock front. The ambient air properties (i.e., ς_0 and p_0) are usually tabulated as a function of altitude, as presented in Table 5.9. In order to simplify this discussion of these relations, we can focus on a specific working fluid; i.e., constant molecular weight (combustion prod-

Table 5.9 Physical Properties of the Standard Atmosphere [18]

Altitude (m)	Temperature (ΔK)	Pressure (Pa)	Density (kg/m³)
0	288.2	1.013×10^5	1.225
1,000	281.7	8.988×10^4	1.112
2,000	275.2	7.950×10^4	1.007
3,000	268.7	7.012×10^4	9.093×10^{-1}
4,000	262.2	6.166×10^4	8.194×10^{-1}
5,000	255.7	5.405×10^4	7.364×10^{-1}
6,000	249.2	4.722×10^4	6.601×10^{-1}
7,000	242.7	4.111×10^4	5.900×10^{-1}
8,000	236.2	3.565×10^4	5.258×10^{-1}
9,000	229.7	3.080×10^4	4.671×10^{-1}
10,000	223.3	2.650×10^4	4.135×10^{-1}
15,000	216.7	1.211×10^4	1.948×10^{-1}
20,000	216.7	5.529×10^4	8.891×10^{-2}
30,000	226.5	1.197×10^3	1.841×10^{-2}
40,000	250.4	2.871×10^2	3.996×10^{-3}
50,000	270.7	7.978×10^1	1.027×10^{-3}
60,000	255.8	2.246×10^1	3.059×10^{-4}
70,000	219.7	5.520	8.754×10^{-5}
80,000	180.7	1.037	1.999×10^{-5}
90,000	180.7	1.644×10^{-1}	3.170×10^{-6}

uct gases not included) polytropic gases. For a polytropic gas the relation between the pressure and density is

$$\frac{p}{p_0} = \left(\frac{\varsigma}{\varsigma_0}\right)^{\gamma}.$$

(5.24)

Using these equations and the relationship for the speed of sound, the shock front velocity relationship is obtained as

$$U = c_0 \left[1 + \left(\frac{\gamma + 1}{2\gamma}\right)\left(\frac{\hat{p}}{p_0}\right)\right]^{1/2},$$

(5.25)

where c_0 is the ambient speed of sound. We also have

$$\hat{\varsigma} = \varsigma_0 \frac{2\gamma p_0 + (\gamma + 1)\hat{p}}{2\gamma p_0 + (\gamma - 1)\hat{p}},$$

(5.26)

and the peak particle velocity behind the shock front is

$$\hat{v} = \frac{c_0 \hat{p}}{\gamma p_0}\left[1 + \frac{\gamma + 1}{2\gamma}\left(\frac{\hat{p}}{p_0}\right)\right]^{-1/2},$$

(5.27)

Another blast wave property of interest is the dynamic pressure, because of its importance in describing wind effect, explosion-driven fragments, and drag. Dynamic pressure p_{dy} is the pressure due to fluid motion, given by

$$p_{dy} = \frac{1}{2}\hat{\varsigma}\hat{v}^2.$$

(5.28)

By substituting the density and particle velocity from Eqs. 5.26 and 5.27 into Eq. 5.28 and using $\gamma = 1.4$ for air, the dynamic pressure for air becomes

$$p_{dy} = \frac{5}{2}\left(\frac{\hat{p}^2}{7p_0 + \hat{p}}\right).$$

(5.29)

For explosions close to structures or surfaces, the shock wave of a blast can reflect from the structure, such that the blast load can become larger than that of side-on pressure. This situation may be important during prelaunch stage where the launch vehicle is on the ground. Figure 5.19 shows the hypothetical progress of a blast wave, where at time t_3, the blast wave has reflected from the surface. The overpressure due to the reflected wave can exceed the overpressure from the incident wave, because the reflected wave can travel at a higher velocity in the high density atmosphere behind the incident shock wave. The peak pressure in normally reflected waves is designated as $p_r(t)$. Given the reflected pressure history $p_r(t)$ the reflected specific impulse I_r can be written as

$$I_r = \int_{t_a}^{t_a + t_d} [p_r(t) - p_0]dt.$$

(5.30)

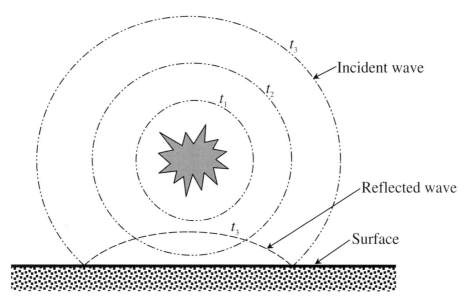

Figure 5.19 Reflection of blast wave at the ground surface. *Adapted from [20]. Courtesy of U.S. Department of Defense.*

Note that Eq. 5.30 is similar to Eq. 5.19. The parameter p_r in Eq. 5.30 can be shown [19] to be related to the peak pressure defined in Eq. 5.23 by

$$\Delta p_r = \hat{p} \frac{\left(2 + \dfrac{1}{v^2}\right)\left(\dfrac{\hat{p}}{p_0}\right) - 1}{\dfrac{1}{v^2} + \dfrac{\hat{p}}{p_0}} - p_0, \tag{5.31}$$

and

$$\frac{1}{v^2} = \frac{\gamma + 1}{\gamma - 1}. \tag{5.32}$$

Note that Eq. 5.31 represents the highest peak pressure for a given blast that is far away from other reflective surfaces, and is used to derive an empirical curve relating to the side-on pressure for a blast of a TNT explosive (see Section 4.2).

4.2 Blast Scaling

Blast effects are frequently analyzed using established scaling laws; i.e., correlations have been developed using scaling laws to determine the blast wave properties of explosion. Some of the most common scaling methods include the cube root method, the *TNT equivalency method*, and the *multi-energy method*.

Cube Root, Sachs, and Pressure-Time Scaling

Cube Root (Hopkinson-Cranz): The most common form of blast scaling is the *Hopkinson-Cranz* or *cube root* scaling law. The Hopkinson-Cranz scaling law states that when two explosive charges of similar geometry and type, but of two different sizes, are detonated under the same atmospheric conditions they produce similar blast waves at identical scaled distances. The scaling distance, which is a function of the explosive mass, is given by

$$Z = \frac{r}{m_{ex}^{1/3}}, \tag{5.33}$$

where Z is the scaled distance (m/kg$^{1/3}$), r is the distance from the center of the explosion (m), and m_{ex} is the total mass (kg) of the explosive. An explosive scaling factor λ is also defined as

$$\lambda = \left(\frac{m_{exr}}{m_{ex}}\right)^{1/3}, \tag{5.34}$$

where m_{exr} is a reference mass. Thus,

$$r = \lambda r_r, \tag{5.35}$$

where r_r is the reference distance that gives the same blast effect that mass m_{ex} gives at location r. The cube root scaling also applies to time; i.e.,

$$t_a = \lambda t_{ar}, \tag{5.36}$$

and
$$t_d = \lambda t_{dr}. \tag{5.37}$$

Sachs: Another popular blast scaling rule is Sachs scaling, which is almost exclusively used to predict blast waves properties from explosions at high altitude. The Sachs law states that dimensionless overpressure and impulse can be expressed as unique functions of a dimensionless scaling distance, where the ambient atmospheric conditions prior to the explosion are used to provide dimensionless quantities. For instance, Sachs scaled pressure and impulse can be defined as

$$\underline{\Delta p} = \Delta p/p_o, \tag{5.38}$$

and
$$\underline{I} = \frac{I c_o}{E_{ex}^{1/3} p_o^{2/3}}, \tag{5.39}$$

where $\underline{\Delta p}$ and $\underline{I}$ are the dimensionless peak overpressure and impulse, respectively. The dimensionless Sachs scaled distance $\underline{r}$ is

$$\underline{r} = r\left(\frac{p_o}{E_{ex}}\right)^{1/3}. \tag{5.40}$$

Both Hopkinson-Cranz and Sachs scaling laws can also be applied to reflected blast wave parameters as well as side-on parameters. Also note that for a reflected surface, such as a ground reflection in a pre-launch situation, E_{ex} in Eq. 5.40 needs to be multiplied by a factor of two.

Pressure-Time: The time dependence of the pressure profile can be accurately represented by

$$\Delta p(t) = \Delta p_0\left(\frac{t - t_0}{t_d}\right)\exp\left(-\alpha_w \frac{t - t_0}{t_d}\right), \tag{5.41}$$

where α_w is a dimensionless waveform parameter and t the time from the instant the shock front arrives. Equation 5.41 is generally valid, regardless of the source of the explosion.

TNT Equivalency and Multi-energy Methods

Because of the large amount of experimental data available for TNT explosions, explosive characteristics of different propellants and pressurized containers are commonly scaled to TNT blast data. The available data include spatial and temporal variations of the blast properties, including reflected pressures, at various measurement locations and orientations (e.g., side-on or face-on). The data are provided as a function of charge size, shape, orientation, and location (i.e., in air, on the surface, or buried). These data are used in the *TNT equivalency models* that include a *free air model* and a *hemispherical model*. The free air model assumes that the explosion is far away from surfaces, such that the reflected pressure is not important. The hemispherical model is assumed to have explosion close to the ground surface, wherein the reflecting pressure is important. The resulting pressure predicted by the hemispherical model is about twice that predicted using the free air model.

The TNT equivalency model relates the amount of flammable material to an equivalent amount of TNT, based on the relative heat of combustion as

$$m_{TNT} = \eta\, m_{ex} \left(\frac{\mathcal{H}_c}{\mathcal{H}_{cTNT}} \right), \tag{5.42}$$

where m_{TNT} is the equivalent mass of TNT (kg), η is a dimensionless empirical explosion yield factor, m_{ex} is the mass of explosive (kg), $\mathcal{H}_c$ is the net heat of combustion of the explosive (J/kg), and $\mathcal{H}_{cTNT}$ is the heat of combustion of TNT (~ 46.556 MJ/kg). For gases, η is ~ 0.01 to 0.2; e.g., for hydrogen gas in an unconfined space, η is 0.03, and for a solid propellant, η is unity. The TNT fractional equivalency χ is defined as $\eta(\mathcal{H}_c/\mathcal{H}_{cTNT})$. If χ is known, Eq. 5.42 can be simplified to

$$m_{TNT} = m_{ex}\chi. \tag{5.43}$$

Table 5.10 lists empirically determined values of χ for a number of liquid propellants, as a function of mass.

Reference [19] provides an empirical relationship between the reflected and side-on pressure. Converting the expression to metric units gives the following equation:

Table 5.10 TNT Equivalency Data for Selected Liquid Propellants [21]

Total Propellant Mass (kg)	Maximum Reported χ		
	LH$_2$/LO$_2$	LO$_2$/RP-1	N$_2$O$_4$/Aerozine 50[a]
m < 5	0.95	1.3042	<0.01
5 < m ≤ 45	0.60	0.45	0.05
45 < m ≤ 450	1.35[b] (0.504)	0.85 (0.59)	0.90c (0.402)[d]
450 < m ≤ 4,500	0.04 (0.355)[e]	0.01	No reported data
4,500 < m ≤ 45,000	0.10	0.40	0.015
m > 45,000	0.01	0.07	0.02

[a]Aerozine 50 is comprised of 50% hydrazine and 50% unsymmetrical dimethyl hydrazine.

[b]TNT equivalency of 1.35 was recorded for a high velocity impact mixing which meant the propellants were mixed at an impact velocity of > 45.72 m/s. Next closest value (non-HVI) was 1.05.

[c]This was a peak reported value. Next closest value was 0.60.

[d]WSTF tested N$_2$O$_4$ and MMH.

[e]HOVI Test 9 resulted in a 0.355 TNT equivalency based on impulse and airburst TNT data. Next closest value was 0.057.

$$\frac{\Delta p_r}{\Delta p} = \frac{3.864 \times 10^{-7} \Delta p}{1 + 1.125 \times 10^{-7} \Delta p + 2.796 \times 10^{-13} \Delta p^2} + 2$$

$$+ \frac{4.218 \times 10^{-3} + 7.021 \times 10^{-6} \Delta p + 9.979 \times 10^{-10} \Delta p^2}{1 + 1.164 \times 10^{-6} \Delta p + 5.595 \times 10^{-10} \Delta p^2}. \qquad (5.44)$$

For explosion of combustible gases in free air, once m_{TNT} is known, the distance to a given overpressure can be calculated using the empirical correlation

$$r = 0.3048 \, (2.2046 \, m_{TNT})^{1/3} \, e^{\omega(\Delta p)}, \quad \text{where} \qquad (5.45)$$

$$\omega(\Delta p) \equiv 3.5031 - 0.7241[\ln(1.4554 \times 10^{-4} \, \Delta p)] + 0.0398[\ln(1.4554 \times 10^{-4} \, \Delta p)]^2. \quad (5.46)$$

The assumptions and limitations of the TNT equivalency model are:

- The source of the explosion is assumed to be a point. This assumption poorly represents the conditions for a vapor cloud explosion.
- Decay of the overpressure with distance is assumed to be similar to that associated with high explosives (i.e., TNT).
- The model overpredicts overpressures at locations near the source of the explosion.
- The model does not account for the effects of terrain, buildings, or obstacles.
- Finally, this model is established at the ambient condition at sea level. In order to apply the model correctly at various altitudes, a correction factor must be applied to the results. Pressures should be multiplied by (p/p_0) and impulse should be multiplied by $(p/p_0)^{2/3}(c_0/c)$.

The multi-energy model is often used to estimate vapor cloud explosion blast effects. The model assumes that a vapor cloud explosion can only occur within regions of the vapor cloud that is partially confined, and treats the vapor as a hemispherical volume. The hemispherical volume is equal to either the volume of the partially confined region (if the vapor cloud is larger than the region), or the volume of the vapor cloud (if the cloud is smaller than the partially confined region). Nondimensional charts have been developed for predicting the maximum blast overpressure and duration that can be used with the multi-energy method. For example, the peak dimensionless overpressure as a function of the Sachs scaled distance is provided in Figure 5.20 for a hydrocarbon gas and air mixture. The curves numbered 1 through 10 represent the relative blast strength. A blast strength of 10 is the detonative strength, and a strength of 1 is a very low blast strength. These curves can be used to estimate peak overpressures by:

1. Multiplying the partially confined volume by the energy density of the mixture,
2. Computing the energy-scaled distance $\underline{r}$ for the distance of interest r,
3. Selecting the blast strength, and obtaining the scaled overpressure from the curve, and
4. Multiplying the $\underline{\Delta p}$ by the ambient atmospheric pressure p_0 to obtain Δp.
5. As for the TNT equivalency model, a correction factor should be applied for the altitude under consideration.

The preceding discussion provides methods for estimating the potential blast environment that may be produced by a postulated propellant explosion accident. The overpressure from the blast is unlikely to result in a safety concern for most system designs. The dynamic pressure and blast wind, however, may propel the system at high velocity against hard surfaces or objects causing system disruption or reconfiguration. This possibility is discussed in Chapter 7. Shock damage may also result from the blast wave.

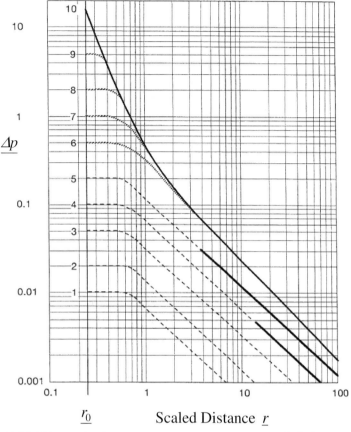

Figure 5.20 Dimensionless overpressure vs. energy-scaled distance [22].

Example 5.2

Compute the TNT equivalent mass for an explosion due to a leak in a fuel tank containing LH_2 during the ascent phase of a mission. The propellant mass is 53,000 kg and the ambient pressure is at 0.25 atm.

Solution:

Using Eq. 5.43 and $\chi = 0.01$ from Table 5.10,

$$m_{TNT} = 5.3 \times 10^4 (0.01) = 530 \text{ kg}.$$

4.3 Explosion-Induced Fragments

In addition to the damaging effect of an explosion-induced blast wave, a propellant explosion typically generates high-velocity fragments that can impact and damage an onboard nuclear system. In order to predict the potential effect of explosion-induced fragments on onboard nuclear power systems, the fragment properties must be characterized (velocity, mass, shape, size, and trajectory). These properties, however, depend largely on the initial geometry, structural material, and configuration of the propellant tank, as well as the severity of the explosion. Data from experiments, such as those conducted as a part of the PYRO project (see Chapter 7), have been used to quantify fragment environment and to develop fragmentation models. These models are used to predict the flux of fragments and the spectrum of fragment sizes and speeds at various locations. These characteristics are particularly important when assessing the consequences of liquid propellant tank explosions.

Fragment Size and Speed Distributions

Fragment size d (m) and speed v (m/s) can be estimated using the simple empirical relationships,

$$d = d_0 \mathcal{Z}^a, \tag{5.47}$$

and

$$v = v_0 \mathcal{Z}^b, \tag{5.48}$$

respectively. Here, d_0, v_0, a, and b are constants determined from test data, and $\mathcal{Z}$ is the scaled distance Z (m/kg$^{1/3}$), defined by Eq. 5.33, divided by the cube root of the TNT fractional equivalency χ; i.e.,

$$\mathcal{Z} = \frac{Z}{\chi^{1/3}}. \tag{5.49}$$

As a part of the Space Shuttle safety program, the four parameters d_0, v_0, a, and b, were obtained using fragment data from Saturn (S-IV) test series. By fitting the test data, the values $d_0 = 0.769$ (m), $v_0 = 128.9$ (m/s), $a = 0.9$, and $b = -1.3$ were obtained [1].

Liquid Propellant Tank Fragmentation

In the absence of air resistance, the laws of motion predict that a projectile propelled at a 45° angle has a range of v^2/g, where v is the projectile velocity. For instance, a projectile having a v = 150 m/s ($g = 9.8$ m/s^2) can travel 2.3 km. However, this prediction assumes that gravitational forces dominate the projectile's movement. Drag forces are more important than the gravitational forces, particularly for lighter fragments. Neglecting the effect of gravity, the drag effect can be examined using Newton's second law; i.e.,

$$m_F \frac{dv}{dt} = F_D, \tag{5.50}$$

where m_F is the fragment mass (kg), t is time (s), and F_D represents the drag forces (N). The drag force is given by

$$F_D = -\frac{1}{2} C_D \varsigma_g A_F v^2, \tag{5.51}$$

where C_D is the drag coefficient, ς_g is the air or gas density behind the blast front (kg/m^3), and A_F is the fragment's front face cross-sectional area (m^2). Substitution of Eq. 5.51 into Eq. 5.50 and solving yields the equation for the fragment velocity as a function of position

$$v(r) = v_0 e^{-(r/\xi)} \tag{5.52}$$

where r is the fragment range or the distance traveled from the explosion source (m), v_0 is the initial fragment velocity (m/s), equal to the velocity that is imparted by the explosive gases to the fragment, and $\xi = 2(\beta/\varsigma_g)$ is a characteristic length (m) and is a direct function of the ballistic coefficient, $\beta \equiv m_F/A_F C_D$ of the object. Thus, objects with larger ballistic coefficents tend to decelerate more slowly than lower β projectiles. Anderson and Owing [23], using fragment data from the PYRO project, were able to confirm that Eq. 5.52 can adequately predict the fragment velocity distribution for Shuttle accidents.

The initial fragment velocity is often not known *a priori*. Project PYRO data has been used to develop empirical relations for determining the initial fragment velocity and range; i.e.,

$$v_0 = 73.96 \, \psi^{0.43}, \tag{5.53}$$

and
$$r = 95.93 \, \psi^{0.28}, \tag{5.54}$$

where ψ is defined as

$$\psi = \chi \frac{m_p}{m_F}. \tag{5.55}$$

Here m_p and m_F are the total mass of the propellant (fuel plus oxidant) and the total fragment mass, respectively.

Example 5.3

Estimate the fragment size and velocity for an accidental explosion of the Space Shuttle external tank, for a distance of 30 m from the center of explosion (COE).

Solution:

From Eq. 5.33 and Table 5.1, the scaled distance is a function of Z is

$$Z = \frac{30 \text{ m}}{(7.3 \times 10^5 \text{ kg})^{1/3}} = 0.33 \text{ m/kg}^{1/3},$$

and from Table 5.10 $\chi = 0.01$. Thus,

$$\mathcal{Z} = \frac{Z}{\chi^{1/3}} = \frac{0.33}{(0.01)^{1/3}} = 1.5 \text{ (m/kg}^{1/3})$$

Using Eqs. 5.47 and 5.48, we find fragment size and velocity to be

$$d = d_0 \mathcal{Z}^a = 0.769 \times (1.5)^{0.9} = 1.11 \text{ (m)}$$

$$v = v_0 \mathcal{Z}^b = 128.9 \times 1.5^{-1.3} = 76 \text{ (m/s)}$$

5.0 Propellant Fires

The principal types of liquid propellant fires include *pool fires, vapor cloud fires,* and *fireballs.* The formation of a fireball subsequent to a propellant explosion depends strongly on the quantity and configuration of the propellant at the time of ignition. These considerations, in turn, depend on the time and mission phase when the accident is postulated to occur. For example, during the prelaunch phase the amount of propellant spilled will affect the duration of the fire and the potential magnitude of damage. At the time of liftoff, the entire propellant mass is available to support a fire. During ascent, the available propellant decreases as the propellant is consumed. The possibility of a ground fire is eliminated once the predicted impact point for the vehicle clears land. Although both liquid and solid propellant fires share many characteristics, processes involved in solid propellant fires can be more complex than those of liquid propellants. As a consequence, the discussion of solid propellant fires at the end of this section is very rudimentary.

5.1 Pool Fires and Vapor Cloud Fires

Pool fires often result when liquid propellants are accidentally spilled on the ground. Spilled liquids generally spread laterally on the ground and fill small-scale crevices for rough surfaces or larger-scale depressions in an undulating terrain. While spilled cryogenic propellants will boil violently due to heat transfer from the relatively warm ground, non-cryogenic liquid propellants evaporate more slowly. Ignition of a pool fire produces a flame that spreads along with the liquid film and the flame height is determined by the evaporation rate of the fuel. The burning rate depends on the rate at which air and fuel mix, as the combustion process proceeds, and burning will continue until the fire exhausts all available fuel.

One consequence of liquid propellant accidents is the potential for formation of vapor cloud fires. Unlike fireballs that are a direct consequence of liquid propellant explosions (see Section 5.2), vapor cloud fires can occur due to non-explosive burning of vapor clouds. The ignition source can be simply the result of friction of structures with the ground, electrical sparks, or an existing open fire. In general, large amounts of vapor could be produced and then transported by the prevailing winds to form vapor clouds. These can occur, for example, when the liquid pool, due to lack of an ignition source, does not immediately ignite. Also, certain propellants (e.g., LH_2) have a high vapor pressure, which boil due to heat transfer from the surrounding environment and not necessarily from fire. For these cases, the resulting cloud has an elongated shape, which is often referred to as a *plume*. A plume's leading edge advances with the wind, while its trailing edge is formed at the vapors' source. As the leading edge moves further downwind, ambient air is entrained in the cloud, thus increasing its volume and decreasing the vapor concentration. As the cloud dispersion continues, the vapor concentration may drop below the flammable limit prior to encountering an ignition source. In this case the fire hazard may be avoided altogether. Thermal radiation is the main hazard associated with pool and vapor cloud fires. The effect of thermal radiation depends on the flame height, the flame's emissive power, local atmospheric conditions, vapor cloud size, and the distance from the cloud. Although empirical relations are available, numerical solutions of the Navier-Stokes and flame chemistry equations are often used to analyze vapor and pool fires.

5.2 Fireballs

Fireballs are generally a direct result of the explosion of liquid propellant containers (BLEVE). During a BLEVE, the liquid propellant can instantaneously boil, followed by ignition of the vaporized propellant. The most visible characteristic of a fireball is that it quickly rises from the ignition source and disperses into the surrounding atmosphere. By nature, fireballs can yield extremely high temperatures and thus large amounts of radiated energies. For the purpose of this discussion the fireball is assumed to be comprised of two phases; i.e., first the combustion phase and second the entrainment phase. During the combustion phase, both the fireball temperature and size increase as the propellant burns. Radiative and convective heat losses from the fireball directly influence the ambient environment as well as any structures within the fireball, e.g., refer to Dobranich et al. [24].

When an unconfined propellant fire is initiated, a series of events contribute to the formation and growth of a fireball (see Figure 5.21). If the fireball starts on a surface, it initially grows as a hemisphere bounded by the surface (e.g., Earth). Subsequently, the bulk of the propellants rapidly burn on or near the confining surface. However, if the fireball occurs above a confining surface (e.g., in the atmosphere), it will take on a fully spherical shape from the start. Explosions resulting from the liquid propellants usually involve either a detonation or rapid deflagration in the early stages of the fireball development. Such high-rate reactions produce shock waves at the outer fireball boundaries coupled with low pres-

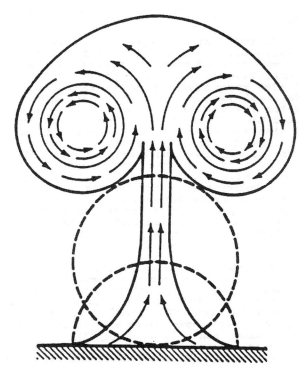

Figure 5.21 Illustration of fireball development [25].

sures behind the shock wave in the center of the fireball. These phenomena regulate the initial lateral growth pattern, air infusion into the fireball, and the overall fireball upward motion. Following the dissipation of the shock wave and when the pressure gradient across the fireball has diminished, the initial growth phase is complete. At this point, the fireball dynamics are controlled by buoyant forces that are a result of the high temperature and low gas density created in the first stage of the fireball development. The fireball rises as a result of the reduced gas density.

When the upward velocity of the fireball gases exceeds its growth rate, the fireball begins to change shape from a hemisphere to a sphere truncated on the bottom by the ground. Liftoff occurs when essentially all of the fuel is consumed and the lateral growth rate is slower than the vertical movement of the fireball due to buoyant forces. As the fireball continues to rise, it creates a vortex motion, which along with natural convective forces causes the surrounding ambient air and any remaining propellant to be drawn into the fireball from below, forming a stem. The fireball continues to grow as it rises due to air entrainment and residual combustion. When combustion is complete, the fireball begins to rapidly cool, due to continued air entrainment and radiation from the fireball surface. During this period, the induced vortex motion changes the fireball shape to an oblate spheroid and then to a toroid. Eventually, the cloud cools sufficiently to eliminate buoyant forces, and the fireball attains some maximum height. The following key fireball characteristics, which define its thermal effects, must be known to assess potential damage to an onboard space nuclear system:

- fireball size and duration,
- liftoff time,
- adiabatic flame temperature,
- average fireball temperature,
- propellant burning rate, and
- radiation and convection fluxes associated with the fireball.

Fireball Size and Duration

Based on data from a test series, a number of empirical relationships have been developed for fireball size and time dependence. Some of the more important relationships are provided here. At liftoff, the maximum fireball radius r_{LF} (m) can be approximated [12, 26, 27] using

$$r_{LF} = 1.44 \, (m_P)^{1/3}, \tag{5.56}$$

and the maximum fireball size after liftoff is approximately

$$r_{FB} = 1.88 \, (m_P)^{0.325}. \tag{5.57}$$

Here, m_p is the propellant mass (kg), including both the fuel and oxidant.

Liftoff time t_{LF} (time from propellant ignition to liftoff, in seconds) can be estimated using

$$t_{LF} \approx 0.65 \, (m_p)^{1/6}, \tag{5.58}$$

and the stem liftoff time, t_{LS} is approximated as 1.5 times t_{LF}. In general, liftoff may not occur if the mass of the propellant participating in the burning is less than 90 kg. The fireball duration t_{FB} is the length of time that heating occurs within the fireball. The fireball duration can be estimated using

$$t_{FB} \approx 0.26 \, (m_P)^{0.349}. \tag{5.59}$$

The predictions given by Eqs. 5.56 through 5.59 represent a worst-case scenario.

Fireball Thermal Properties

A time-dependent form of the energy balance for a fireball can be written as (e.g., see [24]),

$$\frac{d(n_{FB}H'_{FB})}{dt} = H'_{re}\frac{dn_{re}}{dt} + H'_a\frac{dn_a}{dt} - L, \tag{5.60}$$

where n_{re} and n_a are the molar quantities of the reactants and the entrained air in the fireball, respectively. The parameter $n_{FB} = n_{cp} + n_a$, where n_{cp} is the molar quantity of the combustion products. The parameters H'_{FB}, H'_{re}, and H'_a are the molar enthalpies (J/mol) of the fireball, reactants, and the ambient air, respectively. The parameter L is the heat loss rate (W) due to energy loss in the fireball by radiation, convection, etc. In Eq. 5.60, the left-hand side represents the rate of energy change in the fireball, and the first and second terms on the right-hand side are the rates of change in energy of the reactants and the entrained air, respectively. Using the chain rule, the production rate to the reactant combustion can be written as

$$\frac{dn_{cp}}{dt} = \left(\frac{dn_{cp}}{dn_{re}}\right)\frac{dn_{re}}{dt} = y\,\frac{dn_{re}}{dt}, \tag{5.61}$$

where the ratio $y = dn_{cp}/dn_{re}$ is the change in the molar quantity of the combustion products per mole of reactant. Substituting Eq. 5.61 into Eq. 5.60 gives

$$n_{FB}\frac{dH'_{FB}}{dt} + \left(y\,\frac{dn_{re}}{dt} + \frac{dn_a}{dt}\right)H'_{FB} = H'_{re}\frac{dn_{re}}{dt} + H'_a\frac{dn_a}{dt} - L. \tag{5.62}$$

The loss term can be written as

$$L = \varepsilon\sigma A_{FB}(T_{FB}^4 - T_a^4) + \varepsilon_s\sigma A_s(T_{FB}^4 - T_s^4) + h_s A_s(T_{FB} - T_s) + q_s, \qquad (5.63)$$

where A_{FB} is the spherical fireball surface area (m²), T_a is the temperature of the ambient air (K), ε_s is the emissivity of the launch vehicle structure within the fireball, σ is the Stephan-Boltzmann constant, T_s is the structure temperature (K), h_s is the heat transfer coefficient for the convective heat transfer between the structure and the fireball (W/m²·K), and q_s is the added heat due to the combustion of the structure (W).

Equation 5.63 is typically solved numerically using, for example, the *Fireball Integrated Code Package* [24]. The Fireball Integrated Code Package is a fairly comprehensive tool, including the combustion chemistry, fireball dynamics, effects of turbulent motion, soot and dirt addition on the fireball, and the effect of fire damage to structure on the onboard nuclear system (e.g., PuO_2 vaporization). The fireball code package is well suited for parametric or statistical studies, where a variety of accident scenarios can be considered. A brief description of this code package is provided at the end of this chapter.

Semi-empirical Model for Fireball Temperature

As described in [12], by simply assuming that radiation is the dominant mode of heat transfer within the fireball, the fireball energy equation may be written as

$$V_{FB}(t)\varsigma_{FB}(t)\frac{d[c_{cp}T_{FB}(t)]}{dt} = \dot{H}'_{r-p} - \varepsilon\sigma A(t)T_{FB}^4(t), \qquad (5.64)$$

where V_{FB} is the fireball volume (m³), ς_{FB} is the fireball density (kg/m³), c_{cp} is the specific heat of the fireball gases (J/kg·K). The left-hand side of Eq. 5.64 is the rate of enthalpy change within the fireball. The first term on the right-hand side $\dot{H}'_{r-p} = \dot{H}'_{re} - \dot{H}'_{cp}$ is the difference between the rate of enthalpy change between the reactants and products (J/s). The second term on the right-hand size gives the rate of energy loss by radiation.

Several empirical relationships can be used to provide a more useful form of Eq. 5.64. For example, the fireball mass (the product of the fireball volume and density $V_{FB}\varsigma_{FB}$) is found [12] to be equal to the product of propellant consumption rate R_a and time t; i.e.,

$$V_{FB}\varsigma_{FB} = R_a t. \qquad (5.65)$$

Furthermore, the fireball surface area, $A = 4\pi r_{FB}^2$, can be used along with empirical relationships for the fireball time-dependent radius and density, to rewrite Eq. 5.64 as

$$t\frac{d[c_{cp}T_{FB}(t)]}{dt} = \tilde{H}_{r-p} - k_0[tT_{FB}^7(t)]^{2/3}, \qquad (5.66)$$

where k_0 is defined as

$$k_0 \equiv 4\pi\varepsilon\sigma R_a^{-1/3}\left(\frac{3}{4\pi}\frac{z\mathrm{R}}{p_a M_{rp}}\right)^{2/3}. \qquad (5.67)$$

Here, $\tilde{H}_{r-p}$ is the specific enthalpy difference between the reactants and products (J/kg), z is the gas compressibility factor, R is the universal gas constant (J/mole·K), p_a is the atmospheric pressure (Pa), and

M_{rp} is the relative molecular mass (kg/mole) of the fireball products. Equation 5.66 is an initial value problem for the fireball temperature as a function of bulk propellant properties, which can be integrated numerically using standard techniques. This equation is generally valid provided that $t \leq t_{LF}$. Following fireball liftoff ($t > t_{LF}$), $\tilde{H}_{r-p}$ is set equal to zero and the temperature at $t = t_{LF}$ is used as the initial temperature.

Illustrative Examples of Fireball History

Figure 5.22 shows the time-dependent radiative heat flux, resulting from a Space Shuttle fireball, for two different models. The thermo-chemical model is based on Eqs. 5.66 and 5.67, described by Williams [28], and the experimental data are derived from the Saturn fireball experiments. The maximum radiation flux for the thermo-chemical model is typically 2.5 times larger than that of the experimental data. The thermo-chemical model also shows a slope change at fireball liftoff time and at stem liftoff; whereas, the experimental data continuously decreases until stem liftoff. The corresponding temperature histories for the experimental and thermo-chemical models are presented in Figure 5.23.

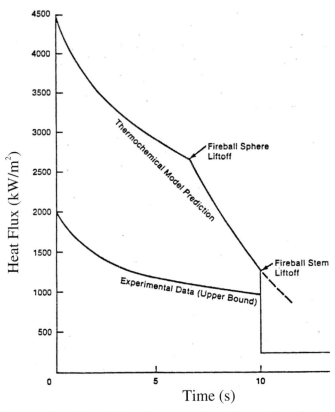

Figure 5.22 Fireball heat flux as a function of time for postulated Space Shuttle accident, based on the thermo-chemical model prediction and experimental data [25].

Example 5.4

A guidance failure is postulated to cause an Earth impact of a Proton launch vehicle with 191,000 kg of propellant mass remaining. Assuming that all of the remaining propellant explodes, calculate: (a) maximum fireball size and the diameter at the liftoff, (b) fireball duration and the time to liftoff, and stem liftoff time.

Solution:

a) From Eqs. 5.57 and 5.56, the maximum fireball diameter and the diameter at the time of liftoff, respectively, become:

$$D_{FB} = (2 \times 1.88)\ (191000)^{0.325} = 195.67 \text{ m, and}$$

$$D_{LF} = (2 \times 1.44)\ (191000)^{1/3} = 165.9 \text{ m.}$$

Note that, since $D_{LF} < D_{FB}$, the fireball continues to expand following liftoff.

b) Fireball duration and the time to liftoff can be respectively calculated using Eqs. 5.59 and 5.58: i.e.,

$$t_{FB} = 0.26\ (191000)^{0.349} = 18.1 \text{ s,}$$

$$t_{LF} = 0.65\ (191000)^{1/6} = 4.9 \text{ s, and}$$

$$t_{LS} = 1.5 t_{LF} = 7.35 \text{ s}$$

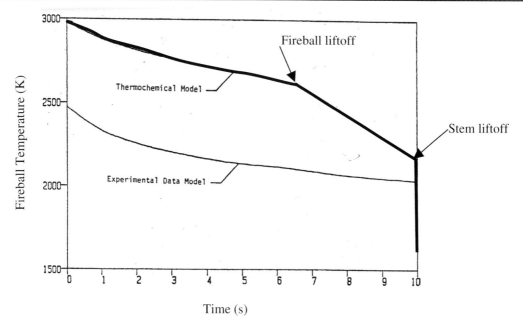

Figure 5.23 Fireball temperature as a function of time for a hypothetical Space Shuttle LOX/LH$_2$ spill, using a thermochemical model and experimental data [25].

5.3 Solid Propellant Fires

Although the explosive characteristics of solid propellants are similar to that of solid high explosives, solid propellants for rocket application are designed to burn or deflagrate in a steady state manner. Burning of solid propellants is generally more complex than that of liquid propellants because for solid propellants the burning rate is critically dependent on the available burning surface.

Burning Rate

Burning rate is a basic characteristic of solid propellants. Fry [29] states the burning rates generally depend on the

- nature of energetic materials (basic ingredients and mixture ratio),
- chemical composition (catalysts, additives, modifiers, etc.),
- physical effects (particle and/or grain size distribution),

- the way the propellant was manufactured, and
- operating conditions and the mode of operations.

Operating conditions determine the initial temperature, pressure, heat losses, gas flow parallel to the burning surface, acceleration, etc. The mode of operations refers to whether the propellant is burning under steady or transient conditions. Under steady operation, the most common relationship for the burning rate is written as

$$R_b = a_b p_c^n,\qquad(5.68)$$

where R_b is the burning rate (cm/s), a_b is the pressure coefficient for the steady burning, p_c is the chamber pressure (MPa), and n is the pressure exponent for the steady burning process. Both a_b and n are determined experimentally. Equation 5.68 is known as the Vieille or de Saint Robert law. Although Eq. 5.68 is widely accepted, variations in pressure result in significant uncertainties associated with the solid propellant burning. For this reason, other forms of Eq. 5.68 have been developed. Fry [29] has reviewed 31 different methods for estimating the burning rate, either based on thickness/time or on mass balance.

Experiments based on steady burning of strands of solid propellants have shown that, on a log-log plot, the relation between R_b and p_c is nearly linear, with $n\sim1$. In 1942, Zeldovich [30] proposed an exponential form for the burning rate; i.e.,

$$R_b = \frac{\hat{R}_s}{\varsigma_c}\exp\left(-\frac{E_z}{RT_s}\right).\qquad(5.69)$$

Here $\hat{R}_s$ is the maximum or asymptotic mass burning rate, ς_c is the chamber density, E_z/R is the activation temperature, and T_s is the solid temperature. For nitrocellulose and double base propellants, it is found that $\hat{R}_s = 18000$ kg/m²·s and $E_z/R = 5000$ K. If the motor pressure remains constant during the burn, the burning is known as *neutral burning*. However, all motors experience some transient burning, which could either be *progressive* or *regressive* [29, 31]. During progressive burning the chamber pressure (or thrust) increases beyond some average value with time. For regressive burning, the chamber pressure decreases with time.

Combustion Process

While a solid propellant's burning process is by nature three-dimensional, many researchers have used an idealized one-dimensional model to distinguish the various combustion regimes. Figure 5.24 presents a schematic illustration of the temperature profiles within each combustion zone for such a one-dimensional idealization for a typical double-based propellant. The temperature within the solid zone of the propellant is at an initial temperature T_{s0}. A foam zone near the burning surface involves exothermic processes and partial solid propellant gasification [29], but the true nature of this zone is not well known [31]. The temperature across the foam zone slightly increases from T_{s0} and the gradient is fairly small. A dark zone [29] (called the *induction zone* in [31]) is located just beyond the burning surface. The material in this zone is predominantly in gas phase. The actual process within the dark or induction zone is far from one-dimensional. In this zone, large quantities of nitric oxide are usually present in the fuel and oxidizer mix as a result of both turbulent and diffusion processes. The dark zone can contain a *fizz reaction zone* where incomplete burning can occur. Finally, the flame zone is the so-called *luminous flame* [29] or *hot reaction* [31] regime, where the highest temperatures exist. For pressures below 0.15 MPa, a flame zone may not form.

Different zones are mathematically described by a combination of energy balances and the chemical processes involved. The simplest zone is the solid zone, where the heat conduction equation applies. For instance, the transient form of the energy equation in the solid phase can be stated as [32]

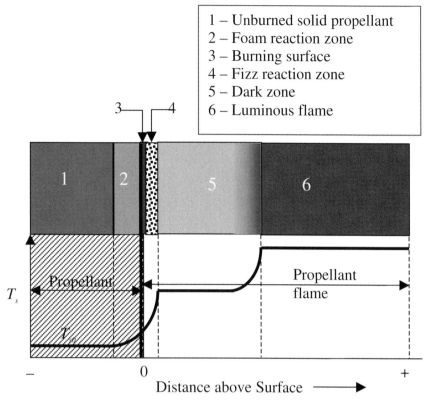

1 – Unburned solid propellant
2 – Foam reaction zone
3 – Burning surface
4 – Fizz reaction zone
5 – Dark zone
6 – Luminous flame

Figure 5.24 Schematic representation of the propellant and flame temperature profiles for a double-based solid propellant.

$$\frac{\partial T_s}{\partial t} + R_b \frac{\partial T_s}{\partial x} - \alpha_c \frac{\partial^2 T_s}{\partial x^2} = \frac{1}{\varsigma_c C_c}(\dot{Q} + \dot{Q}_r), \tag{5.70}$$

where x is distance in the solid phase, subscript c refers to the condensed or solid phase, $\dot{Q}$ is the volumetric heat generation rate due to surface chemical reactions, and $\dot{Q}_r$ is the volumetric heat generation rate due to radiation.

Solid Propellant Cook-off Modeling

Although a relatively simple one-dimensional model can be used to analyze the solid propellant burning, a more complex model is often used to predict fast cook-off behavior. A fast cook-off is often modeled using coupled codes for thermal behavior (heat transfer, decomposition, ignition, and burning), hydrodynamics (case expansion velocity, detonation, fragmentation), and structural or mechanical failure (propellant liner, etc.) [33]. In recent years, such coupled numerical approaches have become mature enough that detailed analysis is quite possible. Sandia National Laboratories researchers have coupled several existing computer codes to simulate fast cook-off experiments. The codes include COYOTE (thermal-chemical analysis) [34], JAS3D (quasi-static mechanics) [35], ALEGRA (dynamic mechanics) [36], and CTH (shock physics) [37].

5.4 Propellant Fire Experiments and Analysis

To date, project PYRO has provided the most comprehensive set of data related to liquid propellant fires. General Electric Corporation also performed a test of liquid propellant fires in the early 1970s, as a part of the Multi Hundred Watt safety test program. The test included the immersion of a fuel capsule in burning Aerozine-50 for 20 minutes. The measured fire temperature ranged from 1090 to 1200 K. For this temperature range, the test did not show any thermochemical degradation of the fuel [38].

As a part of the Ulysses program, a test was conducted to assess the ability of the radioisotope cladding to prevent fuel release during a postulated solid propellant fire accident. The test articles were preheated to a temperature of about 1360 K and were placed next to a cube (0.9 m³) of UTP-3001 solid rocket propellant. The solid propellant cube (protected on five sides) was electrically ignited and burned with great intensity for 10.5 min. According to Tate and Land [39], once the burning started, a violent flame zone extended ~4.6 m from the unprotected side. Thermocouples measured temperatures as high as 2330 K at roughly 1.8 m from the fire source. Post experiment analysis of the cladding revealed dome thinning of the bare fueled clad, but showed that solid fires would not result in any fuel release [40]. In addition to the test, a series of numerical simulations were performed to determine the response of RTG units exposed to fires. Detailed analysis also considered liquid propellants and exposure to a fireball. The predicted results were consistent with the experiments, and no overall loss of cladding was detected.

The simple equations for analysis of liquid propellants fires, discussed in this section, only provide basic features of the fires. For instance, in the case of fireballs we may be able to estimate the size, liftoff time, rise velocity, maximum radiation flux, etc. However, the full description of fire environment requires solving the flow equations (i.e., mass, momentum, and energy conservation equations) along with the chemical species equations (i.e., fire chemistry), and combustion thermodynamics. A detailed analysis of the aerosol physics associated with the fire is also required; i.e., entrainment of sand and dirt into the fire and soot generation, interactions within the fire, sources of turbulence, and dispersion within the atmosphere. Such sets of coupled equations, describing fire dynamics are, in general, highly nonlinear in nature and can only be solved numerically. On the other hand, fast-running computational tools capable of predicting the first-order effects of various launch-abort fires are quite useful; especially, when parametric and stochastic analyses are required. One such model for fireball analysis, developed at Sandia National Laboratories, is the *Fireball Integrated Code Package* (FICP), also known as the Sandia fireball model (SFM) [24]. The FICP captures many of the dominant physical and chemical processes that take place in a fireball, where the ultimate objective of this model is to quantify the amount and size distribution of the plutonium-bearing particles in a fireball. The FICP is a fast-running, fully integrated code package that is well suited for parametric launch-abort studies. A brief description of the FICP underlying physics is provided in the following.

As shown in Figure 5.25, the FICP is comprised of two basic modules describing the fireball and aerosol physics. The fireball physics is based on the numerical solution of an energy equation for the fireball temperature, within a single uniformly mixed control volume (CV). However, the CV's size changes as the combustion, air entrainment, and heat loss continue. The aerosol physics within the FICP is based on the Maeros2 code [41], which divides the continuous distribution of particle sizes into a finite number of size bins. The model assumes that the aerosol within the fireball consists of the following five components:

1. plutonium dioxide (PuO_2) debris generated from the initial explosion,
2. condensed PuO_2 particles,
3. carbon or soot particles,
4. aluminum oxide (Al_2O_3), where Al_2O_3 is assumed to be predominant alloy product of burning aluminum from the rocket structure, and
5. entrained dirt.

All of the particles listed above can agglomerate to form other particles. The code models agglomeration due to gravitational settling, Brownian motion, and turbulent diffusion. To account for the quantity and composition of the combustion products, the FICP assumes the thermodynamics within the fireball is at an equilibrium state. This is a reasonable assumption because of the high temperatures

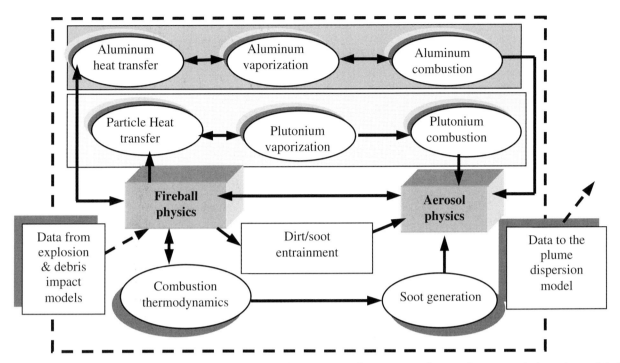

Figure 5.25 Flowchart of physics modules within the FICP code. Arrows depict the general direction of information exchange among the various submodels [25].

and high reaction rates within the fireball. Because different abort scenarios can result in different rates of propellant introduction into the fireball, the FICP allows the user to choose the combustion rate and pressure for each reactant mix. A more detailed discussion of the underlying physics can be found in References [24] and [41].

6.0 Summary

The proximity of space nuclear systems to propellants raises the possibility of nuclear system damage and radiological release in the event of a propellant explosion or fire. In general, explosions can arise from either physical or chemical origins. Physical explosions originate from strictly physical phenomena, such as the rupture of a pressurized tank. The violent boiling of liquid hydrogen or liquid oxygen is a physical explosion referred to as the boiling liquid expanding vapor explosion (BLEVE). A chemical reaction, however, can result in slow burning, a deflagration, or detonation. For a deflagration, the speed of the combustion wave is limited by the process of heat and mass transfer into the unburned propellant mixture; consequently, the speed of a deflagration combustion wave is subsonic. For a detonation, the reaction generates a shock wave causing an increase in the temperature in the unburned propellant and subsequent ignition in the unburned gas mixture. Thus, for a detonation, the combustion wave propagates at supersonic speeds. An accelerating deflagration can suddenly transition to a detonation. The transition is called the deflagration to detonation transition. A simple one-dimensional analysis approach, called the Chapman-Jouguet (C-J) model, describes the conditions required for deflagrations and detonations and the characteristics of possible deflagrations and detonations. Whereas the C-J model assumes that the combustion front is a discontinuity with an infinite reaction rate, the ZND model takes the reaction rate into account and includes a reaction zone of finite thickness.

For liquid propellants, vapor cloud explosions are also possible. A vapor cloud explosion can result when an accident causes the release of the liquid propellant followed by the vaporization of the propellant and subsequent ignition. As for a confined vapor, ignition could result in either a deflagration or a detonation; however, only detonations propagate at high velocity without vapor confinement. When obstacles are present, the flame speed approaches that of a solid explosive. In an open space and in the absence of any obstacles, however, the explosion speed slows, approaching only a fraction of the speed of the combustion front for a solid explosive. A vapor cloud will not ignite if the fuel-oxidant concentration is above the upper flammability limit or below the lower flammability limit. Above the minimum autoignition temperature, a chemical reaction will take place without an external ignition source. The higher densities for solid propellants typically result in higher detonation velocities than for liquid propellants. Solid propellant explosions can be accidentally initiated by a shock or as a result of heating (cook-off). Shock initiation may result from solid propellant impact or from an air shock wave produced by the detonation of another propellant source.

Blast wave properties determine the potential effect of an explosion. Blast wave properties are often characterized by the undisturbed wave as it propagates through the air. For explosions close to structures or surfaces, the shock wave of a blast can reflect from the structure, such that the blast load can become larger than that of side-on pressure. Blast effects are frequently analyzed using established scaling laws; i.e., correlations have been developed using scaling laws to determine the blast wave properties of explosion. Some of the most common scaling methods include the cube root method, the TNT equivalency method, and the multi-energy method. The overpressure from the blast is unlikely to result in a safety concern for most system designs. The dynamic pressure and blast wind, however, may propel the system at high velocity against hard surfaces or objects causing system disruption or reconfiguration. In addition to the damaging effect of an explosion-induced blast wave, a propellant explosion typically generates high-velocity fragments that can impact and damage an onboard nuclear system. In order to predict the potential effects of explosion-induced fragments on onboard nuclear power systems, the fragment properties must be characterized (velocity, mass, shape, size, and trajectory).

Both liquid and solid propellant fires can present potential hazards for postulated accidents involving space nuclear systems. The principal types of liquid propellant fires include pool fires, fireballs, and vapor cloud fires. Fire environments depend strongly on the quantity and configuration of the propellant at the time of ignition. Ignition of a pool fire produces a flame that spreads along with the liquid film and the flame height is determined by the evaporation rate of the fuel. The burning rate depends on the rate at which air and fuel mix. Fireballs can be a direct result of the explosion of liquid propellant containers (BLEVE). The most visible characteristic of a fireball is that it quickly rises from the ignition source and disperses into the surrounding atmosphere. Fireballs can yield extremely high temperatures and radiate large amounts of energy. Once the initial growth phase is complete, the fireball dynamics are controlled by buoyant forces that are a result of the high temperatures and low densities created in the first stage of the fireball development. Unlike fireballs that are a direct consequence of liquid propellant explosions, vapor cloud fires can occur due to non-explosive burning of vapor clouds. Solid propellants for rocket application are designed to burn or deflagrate in a steady state manner. Simple one-dimensional models are often used to characterize solid propellant fires; however, transient and fast cook-off is often modeled using coupled thermal, hydrodynamic, and structural or mechanical codes.

Symbols

a	parameter in Eq. 5.47	b	parameter in Eq. 5.48
a_b	parameter in Eq. 5.68	c_u	sound speed in unburned gas
A_F	fragment projected area	c_0	sound speed in ambient air
A_{FB}	fireball surface area	c_p	specific heat, constant pressure
A_P	propellant performance factor	c_v	specific heat, constant volume

C_D	dimensionless drag coefficient	p	pressure
C_F	nozzle coefficient	$\tilde{q}$	combustion heat release/g
d	fragment size	q_s	heat of structure combustion
d_0	parameter in Eq. 5.48	Q	volumetric heat generation rate
D	diameter	r	distance, range, or radius
E_d	heat of detonation	R	universal gas constant
E_z/R	activation temperature	R_a	consumption rate
F_D	drag force	R_b	burning rate
g	acceleration due to gravity	$\hat{R}_s$	maximum mass burn rate
E	energy	$\underline{r}$	dimensionless distance
$\mathcal{H}_c$	heat of combustion	$\tilde{\bar{S}}$	specific entropy
H'	molar enthalpy	t	time
$\tilde{H}$	specific enthalpy	T	temperature
h_s	structure heat transfer coefficient	T_{ch}	chamber temperature
I_{sp}	propulsion specific impulse	U	combustion wave speed
$\underline{I}$	dimensionless impulse	$\tilde{U}$	specific internal energy
I_s^+	shock impulse, + phase	V	volume
k_0	defined by Eq. 5.67	v_0	parameter in Eq. 5.48
L	fireball heat loss term (Eq. 5.60)	v	speed
m	mass	x	position
m_{TNT}	equivalent TNT mass	X	vapor ratio
M	Mach number	y	d_{np}/dn_{re}
M_r	relative molecular mass	Z	scaled distance
M_{rp}	M_r for products	z	compressibility factor
n	molar quantity	$\mathcal{Z}$	fragment scaled distance
n	pressure exponent (Eq. 5.68)		
α_w	wave form parameter	λ	explosive scaling factor
α	isobaric expansion ratio	σ	Stephan-Boltzmann constant
β	ballistic coefficient	ω	defined by Eq. 5.46
Δp^+	overpressure	ς	density
Δp^-	negative phase pressure drop	ξ	characteristic length
$\underline{\Delta p}$	dimensionless peak overpressure	υ	defined by equation 5.32
ε	emissivity	χ	fractional equivalence
γ	specific heat ratio	Ψ	defined by Eq. 5.55
η	yield factor		

Special Subscripts/Superscripts

0	initial	LS	stem liftoff
a	arrival time	p	propellant
a	ambient	r	reflected
b	burned gases	re	reactant
c	condensed phase	r-p	reactants, products difference
cp	combustion products	s	structure or side-on
d	duration	sl	saturated liquid
dy	dynamic	T	total
ex	explosive	u	unburned
f	flame	z	Zeldovich param., Eq. 5.69
F	fragment	TNT	TNT equivalent
FB	fireball	^	peak
g	saturated vapor	~	specific (per mass)
LF	fireball liftoff		

References

1. Vedder, J. D. and R. H. Brown, *Space Shuttle Data for Nuclear Safety Analysis*, Rev.1. JSC-16087, NASA Johnson Space Center, July1982.
2. http://users.commkey.net/Braeunig/space/specs/orbiter
3. International Launch Services, *Atlas Launch System Mission Planner's Guide Atlas V Addendum (AVMPG) Revision 8*. San Diego CA, Dec. 1999.
4. http:www.fas.org/pp/military/ptogram/nssrm/initiatives/detail
5. http://stardust.jpl.nasa.gov/spacecraftt/sd-luanch.html
6. http://www.kevinforsyth.net/delta/vehicle.htm
7. http://www.floridatoday.com/space/explore/stories
8. http://www.spaceandtech.com/spacedata/elvs/proton_specs.shtml
9. http://www.astronautix.com/lvs/pron8k82
10. Lide, D. R. (Ed.), *Handbook of Chemistry & Physics 1996–1997*. 77th ed.: CRC Press, 1996.
11. Willoughby, A. B., C. Wilton, and J. Mansfield, *Liquid Propellant Explosion Hazards, Volume 3, Prediction Methods*. URS 652-35, AFRPL-TR-68-92, URS Research Co., Burlingame, CA, Dec. 1968.
12. Erdahl, D. C., D. W. Banning, and E. D. Simon, *Space Propulsion Hazards Analysis Manual (SPHAM), Volume 1*. Prepared for the Air Force Astronautics Laboratory, AFAL-TR-88-096, 1988.
13. Baker D. L., D. D. Davis, L. A. Dee, C. H. Liddell, G. Greene, and S. Woods, *Fire, Explosion, Compatibility, and Safety Hazards of Nitrogen Tetroxide*. RD-WSTF-0017, NASA White Sands Test Facility, Las Cruces, NM, June 30, 1995.
14. McBride, B. and S. Gordon, Chemical Equilibrium and Applications (CEA) Code, NASA-Glenn Chemical Equilibrium Program, Oct. 17, 2000.
15. Fickett, W. and W. C. Davis, *Detonation: Theory and Experiment*, Dover Publications, Minerola, NY, 2000.
16. Beckstead, M., *Lectures on Combustion and Detonation*. From a presentation at BYU, Apr. 2000.
17. Raun, R. L., A. G. Butcher, D. J. Caldwell, and M. W. Beckstead, "An Approach for Predicting Cookoff Reaction Time and Reaction Severity." *JANNAF Propulsion Systems Hazards Meeting*, 1, 407-504, 1992.
18. Center of Chemical Process Safety, *Guidelines for Evaluating the Characteristics of Vapor Cloud Explosions, Flash Fires and BLEVEs*. American Institute of Chemical Engineers, New York, 1994.
19. U.S. Department of Energy, *A Manual for the Prediction of Blast and Fragment Loadings on Structures*. DOE/TIC-11268, DOE Albuquerque Office, July 1992.
20. Glasstone, S. and P. J. Dolan, *The Effects of Nuclear Weapons*. U.S. DoD and ERDA, 1977.
21. National Aeronautics and Space Administration, *Test Report—Correlation of Liquid Propellants NASA Headquarters RTOP*. NASA Lyndon B. Johnson Space Center, White Sands Test Facility.
22. Ozog, H. and G. A. Melhelm, "Facility Sitting—Case Study Demonstrating Benefit of Analyzing Blast Dynamics." Proceedings of International Conference and Workshop on Process Safety Management and Inherently Safe Processes, AIChE/CCPS, 293–315, Oct. 1996.
23. Anderson, D. C. and W. D. Owing, *An Improved Fragment Model for Internal Explosion of LO2/LH2 Propellant Tanks*. Teledyne Energy Systems, TES-16018-5, Jan. 1981.
24. Dobranich, D., D. A. Powers, and F. T. Harper, *The Fireball Integrated Code Package*. Sandia National Laboratories Report, SAND97-1585, Albuquerque NM, July 1997.
25. Final Safety Analysis Report for the Galileo Mission and Ulysses Mission, Vol. II (Book 2), Accident Model Document-Appendices, GESP 7200, General Electric Spacecraft Operations, Oct. 8, 1985.
26. Gayle, J. B. and J. W. Bradford, *Size and Duration of Fireballs from Propellant Explosions*. NASA-TM-X-53314, George C. Marshall Space Flight Center, Huntsville, AL, Aug. 1965.
27. Merrifield, R. and R. Wharton, "Measurement of the Size, Duration, and Thermal Output of Fireballs Produced by a Range of Propellants." *Prop., Explosions, and Pyrotechnics*, 25, 179–185, 2000.
28. Williams, D. C., *Vaporization of Radioisotope Fuels in Launch Vehicle Abort Fires*. SC-RR-71-0118, Sandia National Laboratories, Albuquerque, NM, Dec. 1971.
29. Fry, R. S., *Solid Propellant Subscale Burning Rate Analysis Methods for U.S. and Selected NATO Facilities*. Chemical Propulsion Information Agency, Johns Hopkins University, CPTR 75, Jan. 2002.
30. Zeldovich, Ya. B. "On the Combustion Theory of Powder and Explosive." *J. Exp. Theor. Phys.*, Vol. 12, p. 498–510, 1942.
31. Strehlow, R. A., *Fundamentals of Combustion*. International Textbook Company, Scranton, PA, 1968.

32. Kuo, K. K., J. P. Gore, and M. Summerfield, "Transient Burning of Solid Propellants." *Fundamentals of Solid Propellant Combustion*. K. K. Kuo and M. Summerfield (Eds.), Progress in Astronautics and Aeronautics, AIAA, New York, 1984.

33. Cocchiaro, J. E., *Subscale Fast Cookoff Testing and Modeling for the Hazard Assessment of Large Rocket Motors*. Chemical Propulsion Information Agency, The Johns Hopkins University, CPTR 72, Mar. 2001.

34. Gartling, D. K. and R. E. Hogan, *Coyote II – A Finite Element Computer Program for Nonlinear Heat Conduction Problems, Part I- Theoretical background*. Sandia Report, SAND94-1173, 1994.

35. Blanford, M. L., M. W. Heinstein, and S. W. Key, *JAS3D – A Multi-Strategy Iterative Code for Solid Mechanics Analysis, Users' Instruction Release*. Sandia National Laboratories, 2001.

36. Boucheron, E. A. et al., *ALEGRA: User Input and Physics Descriptions, Version 4.2*. Sandia Report, SAND2002-2775, 2002.

37. McGlaun, J. M., S. L. Thompson, and M. G. Elrick, "CTH-A Three-Dimensional Shock-Wave Physics Code." *Int. Journal Impact Eng.*, **10**, 351, 1990.

38. General Electric, Final Report, Safety Test No. S-2, *Liquid Propellant Thermochemical Effects*. GEMS-413, General Electric Co., King of Prussia, PA, 1973.

39. Tate, R. E. and C. C. Land, *Environmental Safety Analysis Tests on the Light Weight Radioisotope Heater Unit* (LWRHU). LA-10352-MS, Los Alamos National Laboratories, Los Alamos, NM, 1985.

40. Bronisz, S. E., *Space Nuclear Safety Program*. LA-9934-PR, Los Alamos National Laboratories, Los Alamos, NM, 1983.

41. Gelbard, F., *Maeros Users Manual*. NUREG/CR-139, Sandia Report, SAND80-822, 1980.

Student Exercises

1. Using the results from Example 5.1, estimate the overpressure if the nuclear payload is located 10 m from the BLEVE.

2. During the ascent phase of a mission, a tank containing LH_2 is ruptured. If the initial pressure of the tank is at 3 atm, the ambient pressure is at 0.25 atm, and the propellant mass is 53,000 kg, what is the explosive energy due to this BLEVE? LH_2 has the following saturation properties:

   - at 3 atm, $\tilde{H}_1 = 5.084 \times 10^4$ J/kg, $\varsigma_1 = 65.033$ kg/m^3, and $\tilde{S}_1 = 2.1275$ J/g·K.
   - at 0.25 atm, $\tilde{H}_{sl} = -3.3862 \times 10^4$ J/kg, $\varsigma_{sl} = 74.777$ kg/m^3, $\tilde{H}_g = 4.18 \times 10^5$ J/kg, $\varsigma_g = 0.3901$ kg/m^3, $\tilde{S}_{sl} = -1.7894$ J/g·K, and $\tilde{S}_g = 25.839$ J/g·K.

3. For Exercise 2, what is the overpressure at a distance of 10 m?

4. Consider a chemical explosion of the propellant mass in Exercise 2. (a) Compute the heat of combustion. (b.) Assuming a yield fraction of 0.01, compute the explosive energy release. (c) Estimate the overpressure due to the explosion. (d) Estimate the reflected pressure.

5. Consider the shrapnel velocity as a result of a hypothetical Atlas Centaur tipover accident. The typical dimension for steel bolts following such an accident is estimated to be roughly 0.0254 m by 0.00635 m and having a mass of 7.7 gm. Assuming a maximum (initial) velocity of 75 m/s and a drag coefficient of 0.7, plot the fragment velocity as a function of distance. What are the fragment velocities if the pieces travel 0.1 m, 1 m, and 10 m from explosion source to the nuclear payload?

6. Assuming a perfectly spherical fireball (e.g., barely attached to the ground surface), use Eq.5.65 to derive a relationship for the fireball radius as a function of time. Consider a catastrophic launch pad accident of the Space Shuttle external tank, where the entire LOX/LH_2 inventory results in a fireball. Plot the fireball diameter as a function of time up to the stem liftoff. Assume a constant fireball density of $\varsigma_{FB} = 0.1$ kg/m^3 and refer to Example 5.4 for an empirical relation for R_a. How long does it take for the all-liquid fuel to be consumed by the fireball?

7. Using Eq. 5.66, derive an expression for the fireball temperature history following the stem liftoff (hint: for simplicity assume the fireball specific heat is a constant). Assume at the time of stem liftoff, the fireball maintains a uniform temperature at the maximum flame temperature and can be treated as an ideal gas with $\gamma = 1.2$, $M_r = 11.8$ g/mole. Plot the temperature and radiation flux per surface area time histories for 30 s after the stem liftoff, using $\varepsilon = 1$ and 0.5. Discuss the results.

Chapter 6

Reentry

Leonard W. Connell

This chapter will address issues associated with postulated accident scenarios in which space nuclear systems reenter Earth's biosphere. Fundamental principles of orbital mechanics are discussed to permit computation of the initial reentry speed and angle for postulated failures during orbital maneuvering or trajectory changes. Approximate methods are provided for analyzing reentry characteristics and reentry heating.

1.0 Scenarios and Issues

The objective of reentry (return-entry) analysis is to select mission parameters and system designs that will either make the probability of accidental reentry very small or will assure that the consequences of accidental reentry are manageable. Several accident scenarios can be postulated for space nuclear systems that result in reentry from Earth orbit or reentry during the launch and deployment phase.

One category of postulated accident scenarios includes reentry of intact nuclear systems during launch and deployment or during orbital maneuvering. These scenarios include (1) launch vehicle failure during launch and deployment, (2) failure to transfer the nuclear payload from low Earth orbit (LEO) to its working orbit, (3) failure to transfer the nuclear payload to a disposal orbit (if operated in LEO), (4) failure to transfer the nuclear payload to an Earth-escape trajectory (for lunar or planetary missions), and (5) trajectory failures for nuclear rockets. The characteristics of the subsequent Earth atmospheric reentry depend on the type of failure. For example, the rocket may explode, fail to ignite, or fail to burn for the correct length of time. In each case the reentry angle of the nuclear payload (relative to the local horizontal) and the reentry speed will be different. These different initial reentry conditions have an effect on the severity of the reentry heating rate and, consequently, will influence the design of the payload.

The other category of premature reentry scenarios includes postulated in-orbit accidents that disrupt the integrity of the space nuclear system. Disruption may result from meteoroid or debris impact or from postulated system-generated accidents such as a reactivity initiated reactor accident. The disrupted system or debris generated by disruption may have a ballistic coefficient that differs from the original nuclear system; consequently, the disrupted nuclear system or radioactive debris may prematurely reenter Earth's biosphere.

The consequences of a postulated reentry accident depend on the system involved, the nature of the failure, the location of reentry, meteorological conditions, and other factors. Possible consequences include radiological exposure from atmospheric dispersal; environmental contamination of air, land, and water bodies; and direct exposure of individuals to radiation from fully or partially intact systems or from debris generated by Earth impact. For some scenarios involving space reactor systems, the possibility of criticality accidents induced by Earth impact must be considered.

2.0 Orbital Mechanics

As discussed in the preceding section, some space nuclear missions require changes in orbit or trajectories during or following operation. For example, nuclear propulsion systems may undergo repeated orbital maneuvers that lead to changes in the trajectory of the spacecraft. One important maneuver involves the transfer of a spacecraft from LEO to a flight trajectory toward its ultimate destination such as the Moon or Mars. If the engine thrust is misaligned (a highly unlikely event), the spacecraft may thrust into a trajectory that intersects Earth's atmosphere and result in an inadvertent Earth reentry. Failure of the rocket used to perform orbital transfer may also result in reentry of the nuclear power system. Some understanding of basic orbital mechanics is required to permit computation of the initial reentry speed and angle for postulated failures during orbital maneuvering. Orbital mechanics is briefly reviewed in this section. A more comprehensive treatment of basic orbital mechanics is provided in Reference [1].

2.1 Motion about a Celestial Body

The basic equations for motion about a celestial body are derived in this section. We begin with a derivation of the basic equations for the classic two-body problem.

Two-Body Problem

From Newton's law of gravitation the force, $\mathbf{F}_m$ on a body of mass m in Earth's gravitational field is

$$\mathbf{F}_m = -\frac{Gm_e m}{r^2}\left(\frac{\mathbf{r}}{r}\right), \tag{6.1}$$

where G is the gravitational constant, m_e is the mass of the Earth, $\mathbf{r}$ is the position vector, and r is the scalar distance between Earth and the body (see Figure 6.1).

Defining $\mu \equiv Gm_e = 3.99 \times 10^5 \text{ km}^3/\text{s}^2$, then

$$\mathbf{F}_m = -\mu\frac{m}{r^3}\mathbf{r}. \tag{6.2}$$

From Newton's second law of motion $\mathbf{F} = m\ddot{\mathbf{r}}$; thus,

$$\ddot{\mathbf{r}} + \frac{\mu}{r^3}\mathbf{r} = 0. \tag{6.3}$$

Equation 6.3 is the two-body problem equation of motion.

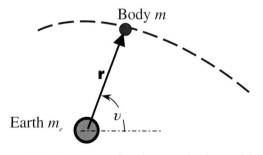

Figure 6.1 Geometry for the two-body problem.

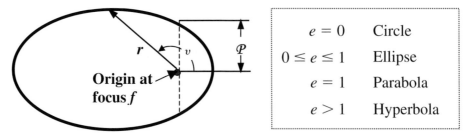

Figure 6.2 Polar coordinate parameters for conic sections.

The Trajectory Equation: Conic Sections

Solution of the two-body problem equation yields an expression for r. In polar coordinates, the expression is

$$r = \frac{\mathcal{P}}{1 + e\cos(v)}.\qquad (6.4)$$

Equation 6.4 is the equation for a conic section in polar coordinates, where $\mathcal{P}$ the semi-latus rectum, e is the eccentricity, and v is the polar angle, shown in Figure 6.2. The three types of conic sections are presented in Figures 6.3 (a) through 6.3 (c).

Ellipse: An ellipse, shown in Figure 6.3 (a), is defined as the locus of points such that the sum of the distances from two fixed points, f and f' is a constant. The points f and f' are called the foci of the ellipse, and c is defined as the half distance between f and f'. For celestial orbital applications, the main gravitational body is located at one of the foci. The parameters a and b are the semi-major axis and the semi-minor axis, respectively. The relationship between these parameters is given by, $a^2 = b^2 + c^2$. The eccentricity is defined as $e = c/a$; and for an ellipse, $0 \leq e \leq 1$. Note that a circle is just a special case of an ellipse where the eccentricity is zero. In rectangular coordinates the equation for a circle is $x^2 + y^2 = a^2$. A body on an elliptical orbit has insufficient kinetic energy to escape the gravitational attraction of the celestial body.

Parabola: For a parabola, shown in Figure 6.3 (b), both c and a are infinite and the eccentricity e is unity. A parabola can be thought of as an ellipse in which the second focal point f' is at infinity. A space vehicle on a parabolic trajectory has exactly the amount of kinetic energy needed to escape the gravitational pull of a celestial body located at f.

Hyperbola: For a hyperbola, $e > 1.0$, where $e = (c/a)$ and the parameters a and c are identified in Figure 6.3 (c). The kinetic energy of a body having a hyperbolic trajectory is greater than the energy required to escape the gravitational attraction of the celestial body at focus f.

Useful Relationships

In Figure 6.2, the polar radius r at angle $v = 0$ is called the periapsis radius r_p (or perigee for an Earth trajectory). The radius at $v = \pi$ it is called apoapsis radius r_a (or apogee for an Earth trajectory). Using Eq. 6.4 for a conic section in polar coordinates, the following relationships are obtained:

$$r_a - r_p = 2c, \quad r_p = \frac{\mathcal{P}}{1 + e}, \quad r_a = \frac{\mathcal{P}}{1 - e}, \quad \text{and} \quad a = \frac{r_p + r_a}{2}.\qquad (6.5)$$

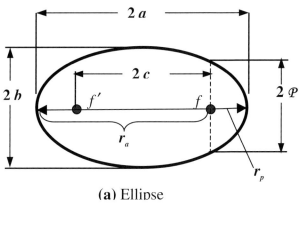

(a) Ellipse

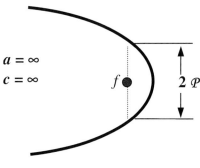

(b) Parabola

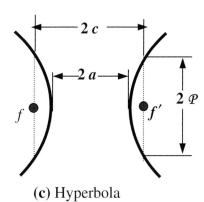

(c) Hyperbola

Figure 6.3 Conic section geometries identifying parameters a, b, and c.

The relationship $e = c/a$ for the eccentricity of an ellipse or hyperbola is not applicable for the parabola ($e = 1$). For a parabola, both c and a are infinite, resulting in an indeterminate form. Consequently, the following equations (employing the relationship $e = c/a$) do not apply for a parabola,

$$e = \frac{c}{a}, \qquad e = \frac{r_a - r_p}{r_a + r_p}, \qquad \mathcal{P} = a(1 - e^2), \quad r_P = a(1 - e), \quad \text{and} \quad r_a = a(1 + e). \qquad (6.6)$$

The period τ of an elliptical orbit (derived from angular momentum considerations) is

$$\tau = \frac{2\pi}{\sqrt{\mu}}\, a^{3/2}. \tag{6.7}$$

Equation 6.7 is known as Kepler's third law.

2.2 Conservation of Energy and Angular Momentum

The conservation laws of energy and momentum and the relations derived above provide the methodology needed to determine the initial conditions for postulated reentry accident scenarios.

Conservation of Energy

Since gravity is a conservative vector force field, the total mechanical energy E of the orbiting satellite is conserved. The total mechanical energy consists of the sum of the kinetic energy $E_k = mv^2/2$ and the gravitational potential energy $U = -m\mu/r$. Thus,

$$E = \frac{1}{2}mv^2 - \frac{m\mu}{r}, \tag{6.8}$$

where m is the mass of the object; v is its speed relative to the celestial body; and r is the polar radius. The negative sign associated with the gravitational potential energy is an artifact of how it is defined; i.e., as the work done by an external applied force, equal but opposite to the gravitational force, in moving a mass from $r = \infty$ to position r. However, only changes in potential energy are important and the equation for potential energy gives a positive change as r increases and more energy is stored in the gravitational field. Dividing E by the satellite mass, we obtain the energy per unit mass or the specific energy $\tilde{E}$,

$$\tilde{E} \equiv \frac{E}{m} = \frac{v^2}{2} - \frac{\mu}{r}. \tag{6.9}$$

Equation 6.9 is generally a more useful relationship than Eq. 6.8 because it does not require the space vehicle mass.

Angular Momentum

The gravitational force acts along the radius vector. In the absence of any force external to this system, no torque will be generated about the location of the celestial body (the focal point f in Figure 6.4). The angular momentum $\mathbf{L}$ about f is conserved, and

$$\mathbf{L} = \mathbf{r} \times m\mathbf{v}. \tag{6.10}$$

As for energy, it is more useful to work with specific angular momentum,

$$\tilde{\mathbf{L}} = \mathbf{r} \times \mathbf{v}, \tag{6.11}$$

and

$$\tilde{L} = (rv)\cos\omega, \tag{6.12}$$

where ω is the flight path angle shown in Figure 6.4. The semi-latus rectum can be expressed in terms of the specific angular momentum; i.e.,

$$\mathcal{P} = \frac{(\tilde{L})^2}{\mu}. \tag{6.13}$$

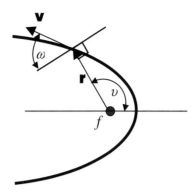

Figure 6.4 Parameters relationship for angular momentum about a celestial gravitational body.

Using the equations for $\tilde{L}$ evaluated at periapsis and apoapsis and the above relationship between $\mathcal{P}$ and $\tilde{L}$, gives

$$\tilde{L} = r_p v_p = r_a v_a. \tag{6.14}$$

From the relationships developed in this section, we can derive the following expression for the specific energy:

$$\tilde{E} = -\frac{\mu}{2a}. \tag{6.15}$$

2.3 Basic Orbital Maneuvers and the Rocket Equation

Relatively simple equations can be developed for basic orbital maneuvering by assuming that the propulsion force required for maneuvering occurs instantaneously. In the following discussion the propulsive force is treated as an impulse applied to the spacecraft such that the spacecraft's initial orbital parameters are assumed to instantaneously change to a new set.

The Hohmann Transfer

Referring to Figure 6.5, consider a spacecraft in circular orbit about a celestial body. For a circular orbit, the circular orbital speed v_{cs} is obtained from the Eqs. 6.9 and 6.15 as

$$v_{cs} = \sqrt{\frac{\mu}{r}}. \tag{6.16}$$

Suppose, as shown in the figure, a transfer to a new orbit of larger radius r_2 is desired. The transfer is accomplished by thrusting the spacecraft into a *transfer ellipse* which is just tangent to the initial and final circular orbits. The specific energy required to change the orbit from r_1 to r_2 is

$$\tilde{E}_t = -\frac{\mu}{2a_t}, \qquad \text{where} \quad a_t = \frac{r_1 + r_2}{2}, \tag{6.17}$$

and $\tilde{E}_t$ is the specific energy of the transfer ellipse. The parameters r_1 and r_2 are known; consequently, $\tilde{E}_t$ can be readily computed. Also, $\tilde{E}_t = [(v_1^2/2) - (\mu/r_1)]$; thus, the velocity v_1 can be computed from

$$v_1 = \sqrt{2\left(\tilde{E}_t + \frac{\mu}{r_1}\right)}. \tag{6.18}$$

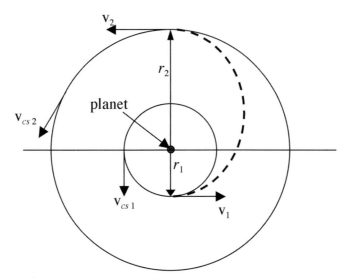

Figure 6.5 Diagram for a Hohmann orbital transfer maneuver.

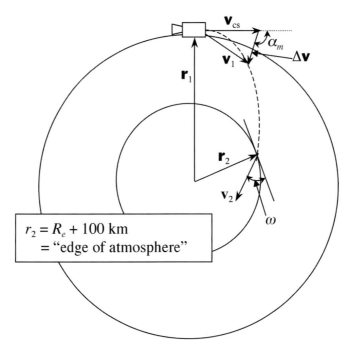

Figure 6.6 Illustration of trajectory for a misaligned thrust vector for Example 6.1.

The Δv needed to get onto the transfer ellipse is given by,

$$\Delta v_1 = v_1 - v_{cs1}. \tag{6.19}$$

Similarly, when at apoapsis on the transfer ellipse another Δv is needed to convert to the circular orbit of radius r_2; thus,

$$v_2 = \sqrt{2\left(\tilde{E}_t + \frac{\mu}{r_2}\right)}. \tag{6.19}$$

and
$$\Delta v_2 = v_2 - v_{cs2}. \tag{6.21}$$

The total velocity change is $\Delta v_{tot} = \Delta v_1 - \Delta v_2$.

Example 6.1

A space nuclear system is in a circular orbit around the Earth at an altitude of 600 km. It is to be transferred onto a lunar trajectory and delivered to a lunar outpost for use as an electric power supply for exploration missions. The payload assist transfer rocket motor attached to the reactor applies a net Δv of 2 km/s. The thrust is to be applied parallel to the circular orbit velocity vector. However, the thrust vector becomes misaligned by $\alpha_m = 95$ degrees relative to the circular orbit velocity vector in the plane of the orbit and toward Earth (see Figure 6.6). Compute the atmospheric reentry conditions for the reactor payload.

Use: $R_e = 6378$ km and $\mu = 3.99 \times 10^5$ km^3/s^2.

Solution:

The initial circular orbit speed, v_1, is obtained using Eq. 6.16,

$$v_{cs1} = \sqrt{\mu/r_1} = \sqrt{(3.99 \times 10^5 \text{ km}^3/\text{s}^2)/(6378 + 600 \text{ km})} = 7.56 \text{ km/s}.$$

Next, adding (vectorially, using the law of cosines) the payload assist rocket $\Delta \mathbf{v}$ to $\mathbf{v}_{cs1}$, and we obtain transfer orbit speed v_1,

$$v_1 = \sqrt{v_{cs1}^2 + \Delta v^2 - 2v_{cs1}\,\Delta v \cos\left(180° - 95°\right)} = 7.65 \text{ km/s}.$$

Using Eq. 6.9, the specific energy of the object on its transfer ellipse is

$$\tilde{E}_t = \frac{v^2}{2} - \frac{\mu}{r} = \frac{(7.65)^2}{2} - \frac{3.99 \times 10^5}{6378 + 600} = -27.92 \text{ km}^2/\text{s}^2.$$

Because the energy is negative, the transfer orbit is an ellipse. We will also need to compute the angular momentum. From Figure 6.6 and the law of sines,

$$\frac{v_1}{\sin(85°)} = \frac{\Delta v}{\sin\omega_1}; \quad \text{or} \quad \sin\omega_1 = \frac{\Delta v_1}{v_1}\sin(85°) = 0.2604; \quad \text{thus, } \omega_1 = 15.09°.$$

From Eq. 6.12, the specific angular momentum at point 1 is,

$$\tilde{L}_1 = r_1 v_1 \cos\omega_1 = (6978 \text{ km})(7.65 \text{ km/s}) \cos(15.09°) = 5.15 \times 10^4 \text{ km}^2/\text{s}.$$

Assuming that the edge of the sensible atmosphere begins at 100 km, the speed and flight path angle can be computed when the object reaches this altitude. Using the conservation laws for energy and momentum,

$$v_2 = \sqrt{2\left(\tilde{E}_t + \frac{\mu}{r_2}\right)} = \sqrt{2\left(-27.92 + \frac{3.99 \times 10^5}{6478}\right)} = 8.21 \text{ km/s}; \text{ also}$$

$$\cos\omega_2 = \frac{\tilde{L}}{r_2 v_2} = \frac{5.15 \times 10^4 \text{ km}^2/\text{s}}{(6478 \text{ km})(8.21 \text{ km/s})} = 0.968, \quad \text{and } \omega_2 = 14.4°.$$

Consequently, the nuclear payload will enter Earth's atmosphere at 8.21 km/s at an angle of 14.4° relative to the local horizontal. These parameters are the reentry conditions (i.e., speed and angle) needed to perform reentry heating and breakup analysis.

Example 6.2

A satellite with an RTG onboard is in a stable, circular orbit at 5000 km altitude when a hydrogen fuel cell explosion results in an impulse Δv of 0.22 km/s being applied in direct opposition to the circular orbit speed. Calculate the new orbit parameters, apogee, perigee, speeds at these points, and the new eccentricity of the orbit.

Solution

Using Eq. 6.16, the initial circular orbit speed at 5000 km altitude is

$$v_{cs1} = \sqrt{\mu/r_1} = \sqrt{(3.99 \times 10^5 \text{ km}^3/\text{s}^2)/(6378 + 5000 \text{ km})} = 5.92 \text{ km/s}.$$

The new speed at point 1 thus becomes,

$$v_1 = v_{cs1} - \Delta v = 5.92 - 0.22 = 5.70 \text{ km/s}.$$

Point 1 is the apogee point for new elliptic orbit, which of course coincides with the Hohmann transfer orbit. Thus, apogee occurs at an altitude of 5000 km and the spacecraft speed at this point is 5.70 km/s. Perigee will be chosen as point 2. From the Hohmann relations,

$$\tilde{E}_t = \frac{v_1^2}{2} - \frac{\mu}{r_1} = \frac{(5.70 \text{ km/s})^2}{2} - \frac{3.99 \times 10^5 \text{ km}^3/\text{s}^2}{(6378 + 5000 \text{ km})} = -18.82 \text{ km}^2/\text{s}^2. \text{ Also,}$$

$$\tilde{E}_t = -\frac{\mu}{r_1 + r_2} = -\frac{3.99 \times 10^5}{(6378 + 5000) + r_2}.$$

Solving the above equation for r_2,

$$r_2 = \frac{3.99 \times 10^5 \text{ km}^3/\text{s}^2}{18.82 \text{ km}^2/\text{s}^2} - (6378 + 5000 \text{ km}) = 9822 \text{ km}.$$

The altitude at perigee is

$$y = 9822 - 6378 = 3444 \text{ km.}$$

The speed at point 2 can be obtained from the total specific energy equation,

$$\tilde{E}_t = -18.82 = \frac{v_2^2}{2} - \frac{\mu}{r_2}, \text{ and solving for } v_2 \text{ we obtain}$$

$$v_2 = 6.60 \text{ km/s.}$$

Finally, using the relations in Eqs. 6.6 we can obtain the eccentricity of the ellipse,

$$e = \frac{r_a - r_p}{r_a + r_p} = \frac{11378 - 9822}{11378 + 9822} = 0.0734.$$

Note that, the negative Δv applied at point 1 causes the orbit to become slightly elliptical with perigee at a lower altitude than the initial circular orbit altitude of 5000 km. Note also that the speed at perigee, 6.60 km/s exceeds the speed at apogee. These conditions are simply the result of energy conservation; i.e., the loss in potential energy by dropping in altitude is converted to kinetic energy.

The Rocket Equation and Fuel Expenditure Calculations

In the previous section, equations were derived for the Δv needed to perform various orbital maneuvers; however, in order to change orbit, energy must be expended to produce the net force and do the work required to change the spacecraft's momentum. The required energy is provided by the chemical energy stored in the propellant (rocket fuel). The rocket equation relates the Δv required to the mass and performance of the rocket fuel and to the mass of the rocket and its payload. A simple form of the rocket equation,

$$m_{tot} = m_{RP} \exp\left(\frac{\Delta v}{g I_{sp}}\right), \qquad (6.22)$$

is applicable to a single stage rocket motor operating in the vacuum of space. The parameters m_{tot} and m_{RP} are, respectively, the total rocket system mass (rocket, payload, and fuel) and the rocket and payload mass (without fuel). The quantity g is the acceleration due to Earth's gravity (9.8 m/s²) and I_{sp} is the specific impulse of the rocket fuel. The specific impulse is the thrust provided by the rocket fuel per unit mass flow rate of the fuel from the rocket nozzle (see Chapter 2). The rocket equation is useful for quantifying the requirements for a particular orbit maneuver. Reference [2] is recommended for a more complete discussion and derivation of the rocket equation.

3.0 Atmospheric Reentry Analysis

In addition to a reentry caused by a misfire of a rocket thruster, some orbital decay scenarios can lead to premature reentry. For example, failure of the orbital transfer rocket to ignite could leave a spacecraft, intended for high-orbit disposal, stranded in a LEO with subsequent premature orbital-decay reentry. The failure may be a result of a defect in the rocket or the failure may be induced by a collision with space debris causing damage to thrusters or other systems. Also, collisions with space debris or an internally initiated accident (e.g., reactor loss-of-coolant) could break up the nuclear payload into smaller pieces that subsequently undergo orbital decay reentry.

Predicting atmospheric reentry of a space vehicle is normally an iterative process. The trajectory is first calculated assuming no ablation or shape change during reentry. The trajectory calculation provides the altitude and velocity of the object as a function of time. Normally, a point-mass (three-degrees-of-freedom) trajectory computer code provides an adequate level of detail for safety assessments. This first trajectory estimate is used to compute atmospheric heating and thermal response. A boundary layer computer code is then used to estimate the flow field about the reentry object and resulting aerodynamic heating to the body. Using a thermal response computer code, an analysis of surface melting, ablation, and conduction is carried out to complete the first iteration of the analysis. If the thermal analysis determines that significant changes have occurred to the object's ballistic characteristics, then another trajectory calculation is performed with the altered ballistic parameters. Reentry analysis involves a suite of sophisticated computer codes. In this section, some of the theory that forms the basis of these codes will be presented. Simple closed-form analytical models are provided for estimating reentry behavior of basic geometric shapes (e.g., spheres and cylinders).

3.1 Orbital Lifetimes

This section provides a discussion of orbital lifetime as determined by orbital decay. Factors affecting orbital decay include fluid flow regimes, drag coefficients, and ballistic coefficients.

Flow Regimes

Three flow regimes are identified for reentry analysis [3]; i.e., free molecular flow, transition flow, and continuum flow. The dimensionless Knudsen number Kn is often used to define the boundaries of these regimes, where

$$Kn = \frac{\lambda}{d_c}. \qquad (6.23)$$

Here, λ is the mean free path of air molecules and d_c is a characteristic dimension of the reentry body. For a sphere, the characteristic dimension is the diameter; while for a cylinder either length or diameter is used, whichever is larger. The flow regime boundaries are defined by

Free molecular flow: Kn > 10

Transition flow: $0.001 \le Kn \le 10$

Continuum flow: Kn < 0.001 .

Figure 6.7 presents a plot of the mean free path of air molecules versus altitude above Earth's surface. As an object enters the Earth's atmosphere from high altitude, it will move through all three flow regimes, starting with free molecule flow, then transition flow, and finally continuum flow. The altitude at which these three flow regimes interface depends upon the characteristic dimension of the object.

Drag Coefficients

The drag coefficient and heat transfer coefficients will be dependent on the relevant flow regime. The drag force F_D is given by

$$F_D = C_D A_{ref}\left(\frac{1}{2} \varsigma_\infty v_\infty^2\right), \qquad (6.24)$$

where A_{ref} is a reference (cross-sectional) area, C_D is the drag coefficient, ς_∞ is the free stream air density, and v_∞ is the free stream air speed relative to the body. The reference area for a sphere of diameter D is given by its cross-sectional area, $\pi D^2/4$, while for a cylindrical body the product of the diameter and the length L is used. Although C_D is normally a function of the Mach and Reynolds numbers, under

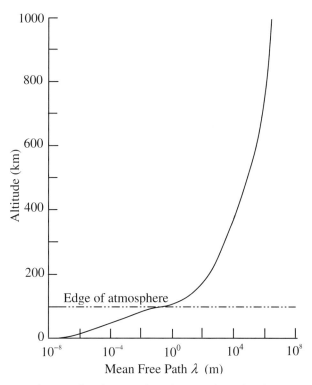

Figure 6.7 Mean free path of air molecules vs. altitude above Earth's surface.

hypersonic flow (Mach number greater than 6) the Reynolds number is nearly constant, and the drag coefficients are given by the values in Table 6.1 [4]. Note from this table that for free molecular flow, the drag coefficient for spheres and broadside cylinders is 2.0. When C_D is not known for a given object, a drag coefficient of 2.0 is usually a good first order estimate.

Table 6.1   Drag Coefficients for Cylinders and Spheres [4]

		Flow Regime Drag Coefficients	
Geometry	Case	Free Molecular	Laminar Continuum
Cylinder	Broadside	2.0	$0.667\,(2-\kappa)$
	End-on	$1.57\,\dfrac{D}{L}$	$0.714\,\dfrac{D}{L}\,(2-\kappa)$
	End-over-end tumble	$1.273 + \dfrac{D}{L}$	$\left(0.283 + 0.303\,\dfrac{D}{L}\right)(2-\kappa)$
	Random tumble	$1.57 + 0.785\,\dfrac{D}{L}$	$\left(0.393 + 0.178\,\dfrac{D}{L}\right)(2-\kappa)$
Sphere	Random spin	2.0	1.0

D = diameter, L = length, $\kappa = 1.67$ for our applications

Ballistic Coefficients

The ballistic coefficient β is used in the analysis of orbital decay and orbital lifetime, and it is used in the derivation of a simple reentry trajectory equation. The ballistic coefficient is defined as

$$\beta = \frac{m}{C_D\,A_{ref}}, \quad \text{or alternatively as} \quad \beta_F = \frac{m\,g}{C_D\,A_{ref}}. \tag{6.25}$$

Both forms of the ballistic coefficient are used. The first version has units of kg/m^2 and the second version β_F has units of force per unit area (N/m^2). Table 6.2 gives approximate values for β for several space nuclear system objects.

Table 6.2   Ballistic Coefficients for Representative Objects

Object	Ballistic Coefficient β (kg/m^2)
Nuclear rocket engine	1000
Space reactor system	100
Space reactor core	100
NERVA fuel rod	50
PBR fuel particles	5

Orbital Lifetime

Example 6.2 showed that a negative Δv applied to an initially circular orbit resulted in an elliptic orbit with a perigee less than the radius of the initial circular orbit. A perturbation analysis performed on the Hohmann transfer equations [5] shows that

$$\frac{\Delta r}{r} = -2\left\{1 + \frac{1}{[(\Delta v/v) + 1]^2 - 2}\right\}. \tag{6.26}$$

In Eq. 6.26, Δv is a small perturbation in the initial circular orbit speed v, and Δr is the resulting change in the orbit radius which results in an elliptic orbit with apogee given by the original orbit radius r and perigee given by $r + \Delta r$. If the values for $\Delta v/v$ in Eq. 6.26 are negative, the result is a negative $\Delta r/r$; and if the values are positive, the result is a positive $\Delta r/r$. In other words, any sudden change in the circular orbit speed will change the orbit into an ellipse; and a positive Δv will increase perigee relative to r while a negative Δv will decrease it. The Hohmann relations assume an instantaneous application of the Δv produced, for example, a short impulse thrust from a rocket engine. The drag forces exerted on an orbiting spacecraft are of course continuously applied. Nonetheless, this force can be represented as a series of small negative impulses (resulting in a series of negative Δvs). The result is a continuous decrease in the orbit altitude; i.e., a spiraling orbital decay ultimately ending in an atmospheric reentry.

The primary source of drag on an orbiting spacecraft comes from the planet's atmosphere. Figure 6.8 presents the Earth atmospheric density predicted by various models [6]. The solar flux goes through an 11-year cycle. Increases and decreases in the solar flux cycle cause coincident expansions and contractions in the atmosphere as evidenced in changes in the atmospheric density. The atmospheric density differences for the active Sun and quiet Sun boundaries are shown in Figure 6.8 (based on the Jacchia model). From the previous discussion on drag force affecting orbital decay, it is evident that the predicted orbital lifetime will depend on the orbit altitude, the atmospheric density model used, and on the ballistic coefficient of the body. Figure 6.9 presents estimates of lifetime for various object's ballistic coefficients in circular orbit [7]. The calculations for the lifetime estimates used the standard, average Sun atmosphere model. Note the initial orbital altitudes and lifetimes for COSMOS 954 and SNAP-10A.

In addition to the effect of atmospheric drag, Earth-Moon perturbations and solar radiation pressure can influence the orbital lifetime of an object. Solar radiation pressure can be a dominant influence on the orbital lifetime at altitudes $\geq$ few thousand kilometers for particles with dimensions $\leq$100 mm. For particles in this range, resonance-induced oscillations in the orbit eccentricity can be caused by the solar radiation pressure. Solar radiation pressure can cause the orbit's perigee to fall into regions where

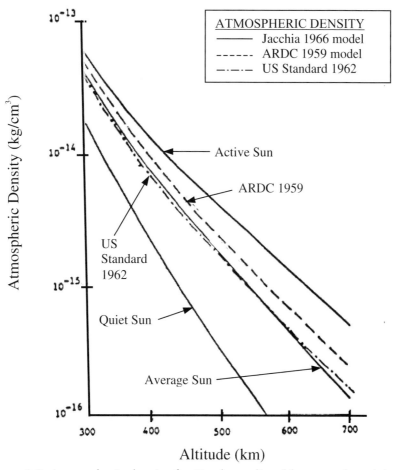

Figure 6.8 Atmospheric density for Earth predicted by several models [6].

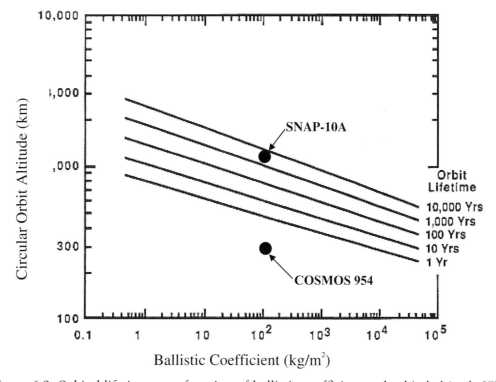

Figure 6.9 Orbital lifetimes as a function of ballistic coefficient and orbital altitude [7].

atmospheric drag can operate on the particle, thus affecting its lifetime. Solar resonance effects appear to be important in a narrow band centered at approximately 6000 km [6].

Simple Method for Estimating Trajectories

Three flow regimes are identified for reentry analysis; i.e., free molecular flow, transition flow, and continuum flow. The objective of trajectory analysis is to estimate the altitude and speed of the vehicle versus time. This information is needed to compute the reentry heating and thermal response of the vehicle. Detailed trajectory analysis is typically performed on a computer using sophisticated models for gravity, atmospheric density, and the drag coefficient. A simple analytic model [8] for the trajectory of a reentry object is derived in the following.

Consider the free-body diagram of a reentry body shown in Figure 6.10. Beginning with Newton's second law and summing forces in the horizontal (x-direction), we obtain

$$-C_D A_{ref} \varsigma_\infty \frac{v_\infty^2}{2} \cos\omega_0 = m \frac{dv_x}{dt}, \tag{6.27}$$

where the ∞ subscript stands for free-stream conditions. In the vertical (y-direction)

$$-mg + C_D A_{ref} \varsigma_\infty \frac{v_\infty^2}{2} \sin\omega_0 = m \frac{dv_y}{dt}, \tag{6.28}$$

and $v_x^2 + v_y^2 = v_\infty^2$.

Dividing each term in Eq. 6.28 by mg and using the definition of the ballistic coefficient gives

$$\left(\frac{\varsigma_\infty v_\infty^2}{2\beta} \sin\omega \right) - 1 = \frac{1}{g} \frac{dv_y}{dt}. \tag{6.29}$$

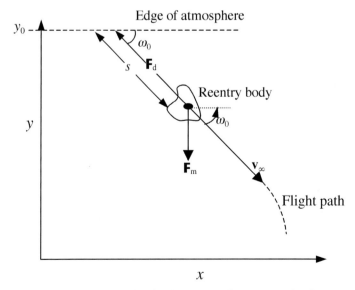

Figure 6.10 Free-body diagram of a reentry body.

The ballistic coefficient is an indicator of the rate of deceleration an object will experience during reentry. The smaller the ballistic coefficient, the greater the deceleration will be. Like the drag coefficient, the ballistic coefficient is normally a function of Mach and Reynolds numbers; but under hypersonic flight, its value is fairly constant.

Under typical hypersonic flight conditions, the drag force is much larger than the force of gravity $\mathbf{F}_m$; consequently, $(\varsigma_\infty v_\infty^2 / 2\beta) \gg 1$, and Eqs. 6.27 and 6.28 become

$$\frac{1}{g}\frac{dv_x}{dt} = -\frac{\varsigma v_\infty^2}{2\beta}\cos\omega_0, \tag{6.30}$$

and

$$\frac{1}{g}\frac{dv_y}{dt} = \frac{\varsigma v_\infty^2}{2\beta}\sin\omega_0. \tag{6.31}$$

Thus, with the effects of gravity being negligible compared to the much larger drag force, the vehicle flies along a straight line. The straight path is inclined at a fixed angle ω_0 to the local horizontal, where ω_0 is the initial flight path angle at first reentry into the atmosphere (as shown in Figure 6.10).

In order to derive the governing equation for motion along the straight line, we can apply Newton's second law to obtain

$$\frac{1}{g}\frac{dv_\infty}{dt} = -\frac{\varsigma_\infty v_\infty^2}{2\beta}. \tag{6.32}$$

The negative sign in Eq. 6.32 indicates a deceleration. To reduce clutter, the ∞ subscript is dropped; i.e., it is implicitly assumed throughout. By the chain rule we have,

$$\frac{dv}{dt} = v\frac{dv}{ds}. \tag{6.33}$$

Here, s is the distance traveled by the reentry body (from the edge of the atmosphere). The distance s shown in Figure 6.10. By the geometry of the flight path,

$$\frac{dv}{ds} = -\sin\omega_0\frac{dv}{dy}; \tag{6.34}$$

thus,

$$\sin\omega_0\frac{v}{g}\frac{dv}{dy} = \frac{\varsigma v^2}{2\beta}. \tag{6.35}$$

From Eq. 6.35, we obtain

$$\ln\frac{v_0}{v} = \frac{g}{2\beta\sin\omega_0}\int_y^{y_0}\varsigma(y)dy. \tag{6.36}$$

Here, v_0 and y_0 are the speed and altitude, respectively, at the edge of the atmosphere. In order to perform the integral on the right-hand side of Eq. 6.36, an analytic model is needed for the variation of atmospheric density with altitude.

Atmospheric Density Model

In the altitude band where significant reentry heating occurs, ς_∞ can be approximated by a simple exponential [8]; i.e.,

$$\varsigma = \varsigma_0 \exp\left[-\frac{y}{Y_0}\right]. \tag{6.37}$$

Here ς_0 is the reference density equal to 1.75 kg/m^3, y is the altitude, and Y_0 is called the scale height, equal to 6700 meters. Using this relationship, the integral in Eq. 6.36 is solved. The solution includes the factor $[\exp(y/Y_0) - \exp(y_0/Y_0)]$. The second exponential term is very small and has no significant effect on the predicted speed; consequently, the second exponential term is set to zero and we obtain

$$v = v_0 \exp\left[-\frac{\varsigma_0\, g\, Y_0}{2\beta\, \sin\omega_0} \exp(y/Y_0)\right]. \tag{6.38}$$

Figure 6.11 compares the predicted speed as a function of altitude with and without the effect of gravity [8].

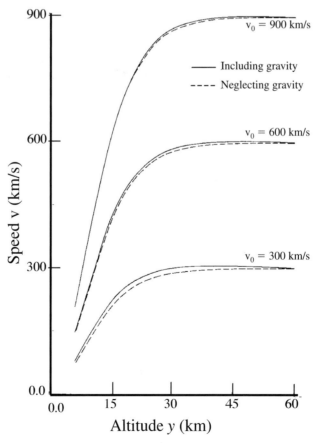

Figure 6.11 Comparison of trajectory model for several v_0 (for a specific case) with and without including the effect of gravity. *Courtesy of the U.S. National Aeronautics and Space Administration.*

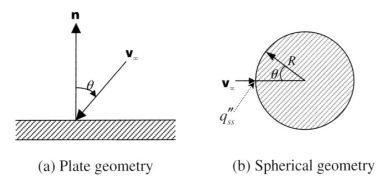

(a) Plate geometry   (b) Spherical geometry

Figure 6.12 Angle θ for flat plate and spherical geometries.

3.2 Reentry Heating for Free Molecular Flow

In the preceding section the discussion focused on the conditions leading to reentry and on simple methods for estimating trajectories. In this section, methods are provided for estimating reentry heating. For simplicity, free molecular flow heating is discussed here for the basic geometries of a flat plate, a sphere, and a cylinder.

Flat Plate

For a flat plate, the heat flux for angle θ [see Figure 6.12 (a)] is given by

$$q''(\theta) = \frac{1}{2}\alpha\varsigma_\infty \, v_\infty^3 \cos\theta, \qquad (6.39)$$

where $q''(\theta)$ has units of W/m². The quantity α is called the accommodation coefficient. The accommodation coefficient expresses the fraction of the incident particle's kinetic energy converted to thermal energy. A typical value for α is about 0.9. The parameters ς_∞ and v_∞ represent the free stream air density and speed, respectively.

Sphere

For a sphere, the local heat flux at angular location θ [see Figure 6.12 (b)] is also given by Eq. 6.39. The sphere and flat plate equations are identical because the incidence of the atmosphere at location θ on a sphere is equivalent to incidence on a flat plate at angle θ relative to $\mathbf{v}_\infty$. The average heat flux over the surface area A of the sphere is obtained from

$$\bar{q}'' = \frac{1}{A}\int q''(\theta) \, dA = \frac{1}{4\pi R^2}\int_0^{\pi/2} 2\pi r^2 \, q''(\theta)\sin\theta d\theta = \left(\frac{1}{4}\right)\left(\frac{\alpha\varsigma_\infty \, v_\infty^3}{2}\right) = F_{fm}q''_\perp, \qquad (6.40)$$

where F_{fm} is the free molecular flow average heating factor equal to 1/4 for the case of the sphere and $q''_\perp$ is the free molecular flow heat flux to a flat plate at normal incidence. The same definitions will be used to obtain the average heat flux to a cylinder. The average heat flux is a useful quantity because in many cases slight imperfections in an entering spherical object will cause it to randomly tumble and spin. Thus the heat flux average shown above represents a time average heat flux to any surface point on the sphere.

Cylinder

The average heat flux to a cylinder is also given by Eq. 6.40. However, several different values for the heating ratio F_{fm} apply depending on the type of motion experienced by the cylinder. For a sphere, only random spinning applies; however, for a cylinder, four types of rigid body motion must be considered:

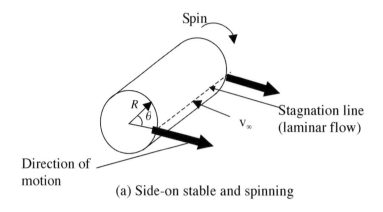

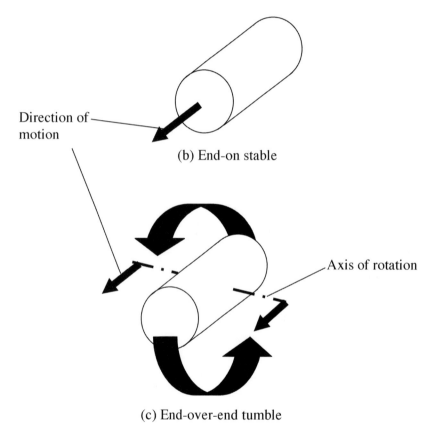

Figure 6.13 Cylinder reentry modes: (a) side-on stable, (b) end-on stable, and (c) end-over-end tumble.

(a) side-on spinning, (b) end-on stable, (c) end-over-end tumble, and (d) random tumble and spin. Cases (a), (b), and (c) are shown in Figure 6.13. A different value of F_{fm} applies for each of these motions. Column 3 of Table 6.3 along with Figures 6.14 and 6.15 (for parameters Y and Z, respectively) provide the method for computing F_{fm} [4].

3.3 Laminar Continuum Flow Heating

Free molecular flow transitions into laminar continuum flow and may then transition to turbulent continuum flow. The form of the heating equation is the same for both laminar and turbulent flow. Because

Table 6.3 Heating Ratios for Spherical and Cylindrical Geometries for Free Molecular Flow F_{fm} and Laminar Continuum Flow F_C [4]

Case	Location	Free Molecular Flow (F_{fm})*	Laminar Continuum Flow (F_C)**
Sphere:			$\dfrac{0.438}{\sqrt{R}}$
Random Spin	Whole body	0.25	
Cylinder:			
Side-on Spinning	Ends	Z	$\dfrac{0.147}{\sqrt{R}}$
	Sides	Y	$\dfrac{0.269}{\sqrt{R}}$
End-on	Ends	1.0	$\dfrac{0.613}{\sqrt{R}}$
	Sides	Z	$\dfrac{B}{\sqrt{R}}$
End-over-end Tumble and Spin	Ends	0.322	$\dfrac{0.329}{\sqrt{R}}$
	Sides	$0.637(Y+Z)$	$\dfrac{0.134 + (0.5)\,B}{\sqrt{R}}$
Random Tumble and Spin	Ends	0.255	$\dfrac{0.323}{\sqrt{R}}$
	Sides	$(0.785)\,Y + (0.5)\,Z$	$\dfrac{0.179 + (0.333)\,B}{\sqrt{R}}$

*Use Y from Fig. 6.14 and Z from Fig. 6.15.
**Use B from Fig. 6.17.

most reentry breakup phenomena occur in the laminar flow regime and above it in the free molecular flow regime, our discussion includes only the laminar and free molecular flow regimes. A semi-empirical methodology [4], developed and used during 1960s and 1970s, will be used to compute the reentry heat flux for the simple shapes of spheres, cylinders, and plates. These formulas provide rough estimates for reentry heating for more complex shapes such as the sphere-conic shapes of space capsules. Calculations of this type should be performed to gain a feel for the problem and to check the validity of detailed computer models. This simple methodology was sufficiently accurate to allow U.S. astronauts to safely travel to the moon and back.

Sphere: Stagnation Point Heat Flux

We begin with a spherical geometry and a formula for determining the heat flux to the stagnation point of the sphere. The stagnation point is a point of symmetry ($\theta = 0$) where the incoming streamline terminates just outside the surface of the sphere (see Figure 6.16). The general form of the heat transfer equation to the stagnation point is

$$q_{ss}'' = \varsigma_\infty v_\infty \, \text{St} \, (\tilde{H}_0 - \tilde{H}_w). \tag{6.41}$$

As for free molecular flow, the ∞ subscript stands for free stream conditions; i.e., the conditions of the air stream outside the boundary layer (see Figure 6.16). The coefficient St is the dimensionless heat transfer coefficient also known as the Stanton number. The quantity $\tilde{H}_0$ is the specific enthalpy (with

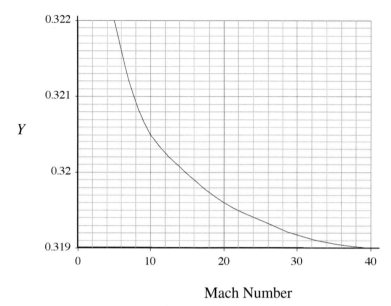

Figure 6.14 *Y*-value vs. Mach number for computing F_{fm} for free molecular flow [4]. *Courtesy of Sandia National Laboratories.*

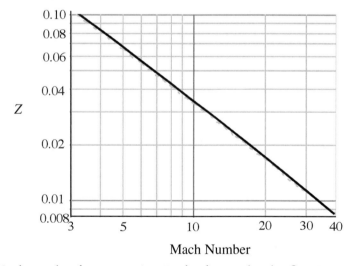

Figure 6.15 *Z*-value vs. Mach number for computing F_{fm} for free molecular flow [4]. *Courtesy of Sandia National Laboratories.*

units of J/kg) of the air that is brought to stagnation (the kinetic energy converted to internal energy). The quantity $\tilde{H}_w$ is the specific enthalpy of the air that is in contact with the wall surface and is assumed to equal the wall temperature. Most engineers are familiar with heat transfer rates expressed in terms of temperature difference. However, when dealing with air temperatures where significant dissociation and ionization of the air occur (the regime that we are investigating), it is more convenient to express the driving potential for heat transfer in terms of specific enthalpy. Specific enthalpy and temperature are related through the constant pressure specific heat at constant pressure c_p,

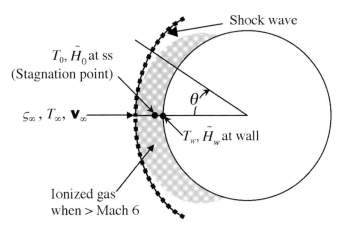

Figure 6.16 Heat transfer parameters for a sphere in the laminar flow regime.

$$(\tilde{H}_0 - \tilde{H}_w) = c_p(T_0 - T_w), \tag{6.42}$$

where $\tilde{H}$ has units of J/kg and c_p is in J/(kg·K). Therefore, we could also write Eq. 6.41 as $q''_{ss} = \varsigma_\infty v_\infty \, \text{St} \, c_p(T_0 - T_w)$. From the first law of thermodynamics (i.e. energy conservation), we have

$$\tilde{H}_0 = \tilde{H}_\infty + \frac{v_\infty^2}{2}. \tag{6.43}$$

For the first iteration, we assume that the wall enthalpy (or temperature) is equal to the free stream air temperature; thus,

$$q''_{ss} = \text{St} \, \varsigma_\infty \frac{v_\infty^2}{2}. \tag{6.44}$$

The parameter q''_{ss} is called the *cold-wall stagnation point heat flux* because of the assumption that the temperature of the reentry body surface is assumed to equal the free stream air temperature.

Combining Eq. 6.44 with an empirical correlation for St, an expression is obtained for the cold wall sphere stagnation point heat flux in terms of a reference speed and atmospheric density [9]; i.e.,

$$q''_{ss} = \left(\frac{C}{\sqrt{R}}\right)\left(\frac{\varsigma_\infty}{\varsigma_{ref}}\right)^{1/2}\left(\frac{v_\infty}{v_{ref}}\right)^{3.15}, \tag{6.45}$$

where $\varsigma_{ref} = 1.2$ kg/m³, $v_{ref} = 7926$ m/s, and $C \equiv 1.10 \times 10^8$ W/m³/². The radius of the sphere R is in meters and q''_{ss} is given in W/m².

Example 6.3

Consider a spherical pellet of space nuclear reactor fuel of radius 1 cm released during an inadvertent atmospheric reentry of a space nuclear reactor. Calculate the cold wall stagnation point heat flux at an altitude of 50 km and an entry speed of 7500 m/s.

Solution:

Using the exponential-atmosphere model (Eq. 6.37), we obtain $\varsigma_\infty = 1.005 \times 10^{-3}$ kg/m³.

Then using Eq. 6.45,

$$q''_{ss} = \left(\frac{1.1 \times 10^8}{\sqrt{0.01}}\right)\left(\frac{1.005 \times 10^{-3}}{1.2}\right)^{1/2}\left(\frac{7.5}{7.93}\right)^{3.15} = 26.7 \text{ MW/m}^2.$$

To provide some perspective, a typical propane blowtorch produces a cold wall heat flux of about 1 MW/m².

Sphere: Average Heat Flux

The heat flux distribution over the surface of a sphere can be expressed as

$$q''_{sph}(\theta) = q''_{ss} f(\theta),\tag{6.46}$$

where $f_{cl}(\theta)$ is the heat flux distribution over the surface of the sphere.

For a sphere that is rapidly spinning in a random fashion, an element on its surface will receive an average heat flux given by

$$\bar{q}'' = \frac{1}{A}\int q''dA,\tag{6.47}$$

or

$$\bar{q}'' = \frac{1}{4\pi R^2}\int_0^\pi q''_{ss} f(\theta)2\pi R^2 \sin\theta \, d\theta.\tag{6.48}$$

Using a semi-empirical expression for $f_{cl}(\theta)$ [10], numerical integration of Eq. 6.48 results in

$$\bar{q}'' = (0.438)q''_{ss}.\tag{6.49}$$

Equation 6.49 can in turn be rewritten in terms of a reference heat flux. For free molecular flow, the heat flux to a flat plate at normal incidence was used as the reference heat flux; for continuum flow, the stagnation point heat flux to a reference sphere will be used. The conventional reference standard is a sphere of unit radius. Thus q''_{ref} is readily obtained from Eq. 6.45 by setting $R = 1.0$. Using this definition of q''_{ref}, the equation for the average cold wall heat flux for a randomly spinning sphere of radius R is

$$\bar{q}'' = \left(\frac{0.438}{\sqrt{R}}\right)q''_{ref} = F_c q''_{ref},\tag{6.50}$$

where F_c is defined as the heating ratio for laminar continuum flow. This value for F_c is shown in column 4 of Table 6.3. Values of F_c for laminar continuum flow for other cases are also provided in Table 6.3. Our general expression for continuum flow average aerodynamic heating then becomes

$$\bar{q}'' = F_c C\left(\frac{S_\infty}{S_{ref}}\right)^{1/2}\left(\frac{v_\infty}{v_{ref}}\right)^{3.15}.\tag{6.51}$$

Cylinder

The average heat flux on the cylinder is obtained using our previously established approach; i.e., we first compute a stagnation heat flux on a reference body assumed to be traveling along the same trajectory

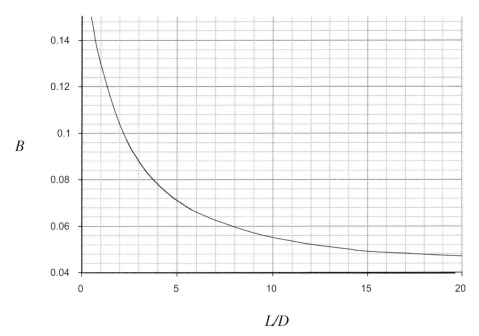

Figure 6.17 *B*-value vs. length-to-diameter ratio for computing F_c for laminar continuum flow [4]. *Courtesy of Sandia National Laboratories.*

and with the same speed as our body of interest. The average heating ratio F_c is then obtained for the real object. From a table of empirical data, the average heat flux on the real object is computed as the product of F_c (using column 4 of Table 6.3 and Figure 6.17 for the *B*-parameter for the tumble and spin cases) and the stagnation heat flux on the reference object. The reference condition for the cylinder will again be the stagnation point heat flux to a sphere of unit radius.

Figure 6.13 (a) shows the geometry for a cylinder entering the atmosphere side-on stable and spinning. For cylindrical geometry, a stagnation line is used rather than a single stagnation point, as for the case of a sphere. The stagnation line heat flux on the cylinder is a fixed fraction of the stagnation point heat flux to a sphere of the same radius [4]; i.e.,

$$q''_{sl} = 0.747 \, q''_{ss}. \tag{6.52}$$

Furthermore, the heat flux distribution $f_{cl}(\theta)$ over the cylinder when normalized by q''_{sl} is nearly identical to the distribution over the sphere when normalized by q''_{sp}; i.e.,

$$\frac{q''_{cyl}(\theta)}{q''_{sl}} = \frac{q''_{sph}(\theta)}{q''_{ss}} = f(\theta). \tag{6.53}$$

The average heat flux on a cylinder is calculated using the expression

$$\bar{q}''_{cyl} = \frac{1}{A} \int q''_{cyl} dA. \tag{6.54}$$

Equation 6.54 can be normalized to the stagnation heat flux on a sphere of equal radius. For a cylinder of radius R and length L, this becomes

$$\frac{\bar{q}''_{cyl}}{q''_{ss}} = \frac{q''_{sl}}{q''_{ss}} \frac{2}{2\pi RL} \int_0^\pi f(\theta) RL \, d\theta = \frac{0.747}{\pi} \int_0^\pi f(\theta) \, d\theta = 0.269. \qquad (6.55)$$

Finally, using our reference condition, the stagnation point heat flux to a sphere of unit radius, Eq. 6.55 becomes

$$\bar{q}''_{cyl} = F_c \bar{q}''_{ref}, \qquad (6.56)$$

with

$$F_c = \frac{0.269}{\sqrt{R}} \qquad .$$

The other values presented in the last column of Table 6.3 are obtained using this same basic approach.

4.0 Thermal Response

Given a reentry object's trajectory, boundary layer fluid flow, and heat flux as input, determination of the temperature history of the object normally requires numerical solution of the general heat conduction equation. For these calculations, the surface heat flux represents a time-dependent convective boundary condition. The problem becomes even more complex when the outer surface of the reentry body begins to lose mass via chemical reactions with the very high temperature air flow in the boundary layer or due to melting, stripping, vaporization, or mechanical erosion. This mass loss process, regardless of whether it is driven by chemical reactions or more simple physical processes, is called *ablation*. The ablating material flows into and perturbs the boundary layer, consequently affecting the heat flux.

Determining the body's temperature and ablation versus time is a complex problem. The analysis requires coupled equations for the boundary layer flow, the surface chemistry, and the subsequent heat transfer to the surface and into the body. In this section, some simple analytic models are provided for estimating the temperature-time history for special cases. We will consider two bounding cases: (1) a steep reentry angle where convective aerodynamic heating dominates over radiation heat transfer and (2) a shallow reentry angle (e.g., orbital decay reentry) where the body achieves an equilibrium surface temperature governed by a balance between the convective aerodynamic heat input and the radiation losses. For these two special cases, we will use two simple thermal response models; i.e., the *lumped parameter model* (or *thin skin approximation*) and the *radiation equilibrium model*.

4.1 Lumped Parameter Analysis, Steep Reentry Angle

Consider the reentry object in Figure 6.18. An approximate (order of magnitude) heat balance equation can be used to relate the convective heat flux at the skin surface to the conductive heat flux into the skin. The temperature gradient within the body provides the mechanism for heat conduction into the body (for the moment the radiant heat flux is ignored); thus,

$$q'' = h\Delta T_{bl} = \frac{k}{\delta_s} \Delta T_{body}. \qquad (6.57)$$

In this equation h represents the bulk convective heat transfer coefficient; i.e., $h = \varsigma_\infty v_\infty St \, c_p$, with units (W/m^2·s). Here $\Delta T_{bl} = (T_0 - T_w)$ is the temperature difference across the boundary layer and δ_s is the skin thickness (cm). Rearranging Eq. 6.57, we can express the ratio of the temperature drop across the body to that across the boundary layer as

$$\frac{\Delta T_{body}}{\Delta T_{bl}} = \frac{h\delta_s}{k}, \qquad (6.58)$$

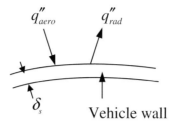

Figure 6.18 Heat transfer for the thin skin model.

where k is the thermal conductivity of the material. The dimensionless quantity $h\delta_s/k$ is the Nusselt number Nu. If Nu is small, the difference across the body is small relative to the distance across the boundary layer. An object with a high thermal conductivity (e.g., a metal object) and a small δ_s, has a relatively small Nu. Generally, if Nu < 0.1, the temperature difference within the body can be ignored and treated as one uniform lumped temperature; thus, the term *lumped parameter analysis*. The thin metal skin of a hypersonic missile or aircraft fits this model; consequently, the method is also called the *thin skin approximation*. The lumped parameter model may be applicable to several elements of a space nuclear reactor undergoing accidental entry. Possible elements for lumped parameter analyses are the reactor vessel wall, fuel element cladding, and individual fuel pellets. Reentry thermal response analysis of clad fuel elements and pellets is required for successive failure of the reactor vessel and cladding, respectively; i.e., these components are assumed to be released and independent from the rest of the reactor.

We can derive a differential equation for the reentry object (refer to Figure 6.18), which expresses the law of energy conservation,

$$mc_p \frac{dT}{dt} = q_{aero} - q_{rad},\qquad(6.59)$$

where m is the mass of the entry object, T is the lumped object's bulk temperature, t is time, and q represents the total heat transfer rate (q'' times surface area). The subscripts *aero* and *rad* refer to aerodynamic heat input and radiation loss, respectively. For fairly steep reentry angles (i.e., $\omega >$ about 10 degrees), the radiation heat loss from the object's surface is small relative to the aerodynamic heat input (with the exception of the stagnation point where radiation losses can be high). Thus, for a randomly tumbling object the radiation losses can be ignored. The thin skin equation can then be manipulated to obtain

$$\varsigma c_p \delta_s \frac{dT}{dt} = q''_{aero}.\qquad(6.60)$$

Previous sections provided methods for calculating the average aerodynamic heat flux depending on the flow type (free molecule or continuum), body type (sphere or cylinder), and the flight dynamics (stable, or various random motions of the object). For the current problem, continuum flow is assumed. The energy balance equation then becomes

$$\varsigma \delta_s \frac{dT}{dt} = F_c C \left(\frac{\varsigma_\infty}{\varsigma_{ref}}\right)^{1/2}\left(\frac{v_\infty}{v_{ref}}\right)^{3.15}\left(\frac{T_0 - T}{T_0}\right),\qquad(6.61)$$

where the right-hand side of Eq. 6.61 is readily derived from Eqs. 6.41 through 6.56. T_0 is the stagnation point air temperature, which can be derived from the definition of stagnation enthalpy (Eq. 6.42)

$$T_0 = \frac{v_\infty^2}{2c_{p(air)}}. \tag{6.62}$$

The temperature ratio term in Eq. 6.61 converts the cold-wall heat flux term to the true hot-wall heat flux.

In order to solve Eq. 6.61, we use the expression for the atmospheric density given by Eq. 6.37. Assuming a fixed flight path angle and constant reentry speed v_∞ we use the simple trajectory equation,

$$y(t) = y_o - v_\infty \sin(\omega_o t). \tag{6.63}$$

Combining Eqs. 6.37, 6.61, and 6.63 gives

$$\varsigma c_p \delta_s \frac{dT}{dt} = K \exp\left(-\frac{y_o - v_\infty t \sin\omega_0}{2Y_o}\right)(T_0 - T), \tag{6.64}$$

where

$$K \equiv \left(\frac{F_c C}{v_\infty^2/2c_{p(air)}}\right)\left(\frac{\varsigma_o}{\varsigma_{ref}}\right)^{1/2}\left(\frac{v_\infty}{v_{ref}}\right)^{3.15}. \tag{6.65}$$

Solution of Eq. 6.64 is straightforward provided v_∞ can be treated as a constant. This assumption is usually reasonable for objects with large ballistic coefficients and for reentry altitudes above roughly 20 km (see Figure 6.11). An intact reactor vessel (with internal core) fits this description; however, an individual fuel pellet or particle, with low ballistic coefficient, would undergo rapid deceleration. For low ballistic coefficients, coupled differential equations must be solved to obtain the trajectory and the thermal response. Here we assume a constant reentry speed v_∞, and solving Eq. 6.64 we obtain

$$T(t) = T_0 - (T_0 - T_i) \exp\left\{-\left(\frac{2Y_0 K \exp(-y_0/2Y_0)}{\varsigma c_p \delta_s v_\infty \sin\omega_0}\right)\exp\left(\frac{v_\infty t \sin\omega_0}{2Y_0}\right)\right\}. \tag{6.66}$$

4.2 Radiation Equilibrium Model, Shallow Reentry Angle

Consider a body possessing a thin skin of low heat capacity material reentering at a shallow angle (an orbital decay scenario). Under these conditions the thin skin of the reentry object will quickly establish equilibrium conditions between the aerodynamic heat input and the thermal radiation losses. Therefore, we can use Eq. 6.59 to calculate the equilibrium temperature at each point along the trajectory by setting the dT/dt equal to zero; hence, $q_{aero} = q_{rad}$. Using $\varsigma_\infty(y) = \xi_0 \exp(-y/Y_0)$, Eq. 6.37, and $\bar{q}''_{aero} = F_c q_{ref} = \bar{q}''$ as given in Eq. 6.51, we obtain

$$F_c C \left(\frac{v_\infty}{v_{ref}}\right)^{3.15}\left(\frac{\varsigma_0}{\varsigma_{ref}}\right)^{1/2}\exp\left(\frac{-y}{2Y_0}\right)\left(\frac{T_0 - T}{T_0}\right) = \varepsilon\sigma T^4. \tag{6.67}$$

Here ε is the surface emissivity and σ is the Stefan-Boltzmann constant, $5.669 \times 10^{-8}\,W/(m^2K^4)$.

5.0 Reentry Analysis Methodology and Computer Modeling

Reentry analysis by mission safety specialists generally uses the following procedure:

1. Compute the ballistic coefficient of the body.
2. Compute the trajectory; i.e., the flight path angle, altitude, and speed as a function of time.

3. At selected trajectory points, compute the heat transfer rate to selected points on the body. The selected points are locations on the body that are judged to be most likely to fail first.

4. Using the heat transfer rates, compute the time dependent surface ablation rates and temperature profiles.

5. Predict the disassembly of the system based on a chosen failure criterion. For metal components, failure is assumed to occur at or near the point where the aerodynamic induced stresses in the metal exceed the yield stress.

6. Return to step 1 and repeat the process for the new configuration. Repeat the analysis process until complete disassembly or until aerothermal effects cease to drive disassembly.

The output from these calculations is a picture of the breakup process that includes, as a minimum, the altitude points along the trajectory at which important nuclear system components break free. For a reactor system, the breakup would include separation of the thermal radiator, radiation shield, reflector system, pressure vessel, core, fuel elements, and fuel pellets. The predicted reentry breakup of the Enisey reactor system is shown in Figure 6.19. The analysis results also provides the thermal history of key components including a calculation of the ablation mass loss. The ablation data can be used as input to atmospheric dispersal codes to perform radiological consequence calculations.

Breakup of the non-nuclear components of a reactor system does not usually have nuclear safety consequences, and partial or complete reactor breakup during reentry may not have significant safety consequences. The safety significance of reactor system breakup will depend on the system design, mission characteristics, the mission phase, and other considerations. Thus, for some conditions (particularly when the reactor has not been operated), prediction of reactor breakup will not require redesign of the reactor system. Prediction of the release of significant nuclear material from radioisotope sources, however, is usually a more serious matter. If predictions, supported by testing, indicate release of significant quantities of radioactive material for credible postulated reentry accidents, a redesign of the source protective barriers (or other modifications) is needed.

5.1 Reentry Trajectory Codes

The predecessor to NASA, the National Advisory Committee for Aeronautics (NACA), published one of the first concise papers on the subject of reentry trajectory analysis [8]. This paper provided simple

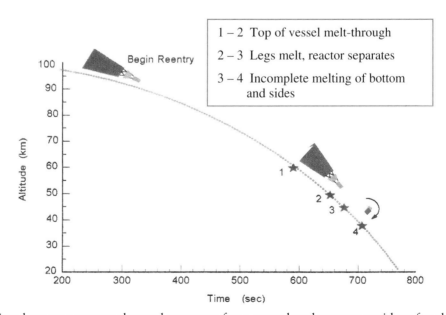

Figure 6.19 Predicted reactor system thermal response for a postulated reentry accident for the Enisey space reactor system [11].

analytical models for calculating the trajectory and bulk heating to a reentering body. The basic methodology provided by the paper was used in Section 3.1 of this chapter. Advancements in computing that came along in the 1960s made it possible to integrate the more exact equations of motion and to obtain accurate predictions of reentry trajectories.

Although 6-degree-of-freedom codes are available, a point-mass, 3-degree-of-freedom trajectory code normally provides the level of detail necessary for reentry breakup analysis. As discussed above, a point-mass trajectory computation is performed for each segment of the reentry. The desired output from this computation is the altitude and speed of the object versus time. The required input includes the initial conditions (speed, altitude, flight path angle) and the ballistic coefficient. If the ballistic coefficient data are not already available, a flow field code will be needed to determine the pressure distribution about the body; and from the pressure distribution, the drag coefficient and the ballistic coefficient are obtained.

Trajectory analysis tools are very well developed and trajectory analysis is fairly routine. Many trajectory codes are available. One of the best written and documented is the Trajectory Simulation and Analysis Program, TSAP [12]. Predicting the reentry trajectory of a space nuclear power system is more complicated than that of a space capsule that was designed for reentry. For reactors, the intact nuclear power system, with its large conical radiator, generally exhibits static aerodynamic stability. Once the radiator breaks away, however, the reactor proper is usually unstable during reentry and will begin a random tumble and spin (RTS) motion. Predictions for RTS reentry must rely on experimental data and analytical simplifications (discussed in Section 3) to compute the average drag and heat transfer coefficients.

5.2 Aerodynamic Heating Codes

The foundation for computing reentry aerodynamic heating was formed in the late 1950s by the efforts of a small group of researchers [13]. This group developed theoretically based formulas that closely matched experimental data. The formulas allowed computation of the reentry aerodynamic heat flux to blunt-nosed, axisymmetric bodies (e.g., spherically tipped cones) in stable flight for both laminar and turbulent flow. Although these correlations represented closed-form solutions, they were complex and required knowledge of free stream and boundary layer fluid properties integrated over the surface of the body. Initial computer programs written to compute the heat flux employed clever simplifying techniques to obtain the correct fluid properties at the location of interest on the body. Present day codes solve the boundary layer equations. If greater accuracy is required, the parabolized Navier-Stokes equations are used. The output from this analysis is the time dependent heat transfer rate (or heat flux) at selected locations on the vehicle surface.

Although the above techniques work well for an axisymmetric body in stable flight, as previously mentioned, space nuclear systems are likely to exhibit RTS reentry; and conventional methods are not applicable. Approximate methods for computing the surface averaged heat flux to a randomly tumbling and spinning cylinder, sphere, or flat plate were developed in support of the SNAP aerospace nuclear safety program. The Klett report [4] presents simple correlations for the drag coefficient and heat transfer coefficient for right circular cylinders in four modes of flight: (1) side-on stable, (2) end-on stable, (3) end-over-end tumble, and (4) random tumble and spin. The data cover the flow regimes of free molecular, transition, and laminar continuum flow. The HANDI Code (Heating Analysis Done Interactively) [14] uses the Klett correlations, streamlines the heating calculations, and generates the output in a format acceptable to the standard thermal response codes. These approximate methods continue to offer the only practical means of computing the heat flux and drag to various reactor components in RTS reentry. Determination of the time dependent flow field surrounding such objects is beyond the computational power of present day computers and is likely to remain out of reach for the near future.

5.3 Thermal Response Codes

Prior to the development of advanced computers, the simple analytical methods discussed in Section 4 were used to obtain the temperature history of a reentry body. The *thin skin approximation* was used to calculate the lumped temperature of a thermally thin body such as a reactor pressure vessel wall. Failure of the body was predicted using the assumption that failure occurs when the body approached the melt temperature. For thick objects, however, such as a reactor core or beryllium control drum, this approach could not be used. Extensive testing was performed to obtain data on in-depth temperature and surface mass loss. Scale models of the reentry object were tested in rocket nozzle exhausts, shock tubes, and plasma arc-jet test facilities.

With the development of advanced computers and the Charring Material Ablation (CMA) code [15], accurate simulation of the ablation and in-depth thermal response of reentry bodies could be performed. The CMA code computes the one-dimensional time dependent temperature profile in the reentry body and the surface recession rate caused by various ablation mechanisms (vaporization, melting and stripping, mechanical removal, and heterogeneous chemical reactions). Any material can be modeled as long as the thermo-physical and thermo-chemical properties are available. Given the time dependent heat transfer coefficient obtained from the aero-heating analysis, CMA performs a surface energy balance and determines how the incoming aerodynamic heat flux is apportioned into thermal radiation, ablation, and in-depth conduction. The two-dimensional code Axi-Symmetric Transient Heat Conduction and Material Ablation (ASTHMA) [16] was also developed; but, because of its complexity, the code is not as widely used. A companion code, Aerotherm Chemical Equilibrium (ACE), is often needed to characterize the chemical composition of the gas at the surface of the body from chemical reactions occurring between the air and the body.

The availability of these codes has not obviated the usefulness and importance of ablation testing. Many cases arise where the surface chemistry is not well understood and ablation tests must be performed. These tests provide empirical factors such as the heat of ablation and ablation temperature which are used in CMA in lieu of the ACE data. Although most of the methodology was developed in the early 1960s, very little improvement has been made on reentry analysis tools. New thermal response codes have been written that utilize finite element methods but do not incorporate the capability of CMA to perform detailed ablation of a chemically decomposing heat shield material.

6.0 Evolution of Reentry Space Nuclear Safety Practices

The purpose of the reentry analysis is to estimate the fate of the object; i.e., whether the object will reenter essentially intact or whether it will break up under the intense heating and dynamic pressures of reentry. The ability to accurately predict the fate of a space nuclear payload during reentry is important for predicting the radiological consequences of reentry. The analytical and experimental means to calculate the sequential disassembly of a space nuclear payload undergoing an atmospheric reentry were developed in the late 1950s and early 1960s. These approaches resulted from the efforts of the U.S. defense establishment to develop ballistic missiles and from the growth of the civilian space program. Both of these programs were accelerated by the successful launch of Sputnik I in 1957. The manned space program and the intercontinental ballistic missiles program had to solve the problem of ensuring the survival of an object returning to the Earth from space. The reentry problem was one of the most difficult problems confronting aerospace engineers of that time [17]. The aero-thermal environment created by the movement of a body through the atmosphere at near orbital speeds is so severe that structural materials cannot survive without an innovative thermal protection system. The thermal protection system or reentry heat shield design must be lightweight, sturdy construction, and capable of being securely bonded to the substructure. The method chosen was to employ an ablative material [13] that melts, vaporizes, or out-gases, thereby carrying away the reentry heat from the surface of the body and limiting the deep conduction into the structure.

From the beginning of the space nuclear age, the need for special precaution during reentry was recognized. The reentry safety approach for RTGs underwent an evolutionary process due in part to the changes made in RTG design. On the SNAP-3 RTG used for the Transit launches, the design approach for reentry protection called for complete breakup of the RTG and ablation of the plutonium-238 fuel at high altitude. Beginning with the development of SNAP-19, used on the Nimbus program, the practice changed to intact-through-impact reentry. This change was brought on by analysis which demonstrated that the degree of ablation and dispersal of the RTG fuel was scenario dependent, and that complete breakup and dispersal at high altitude was not a certainty as was previously believed [18]. Furthermore, the later RTG programs incorporated much larger amounts of plutonium-238 (45,000 Ci for SNAP-27 versus 1,800 Ci for SNAP-3); and the fuel form was changed from metal to oxide which is much more resistant to reentry ablation. As a consequence, the intact reentry approach became the more preferred safety option.

Space nuclear reactors offer certain safety advantages over RTGs because the radiological activity of unfissioned uranium is very small compared to significant quantities of Pu-238. Although the trend for reactor missions has moved from aerothermal dispersal to intact reentry, the reentry safety practice for reactors is less well established. The SNAP-10A space reactor, shown in Figure 2.2, was placed into a 1300 km circular polar orbit by an Atlas-Agena launch vehicle on April 3, 1965. The entire SNAP-10A reactor power system weighs 440 kg and measures 3.4 meters in length and 1.3 meters in diameter (at the base of the radiator). The cone of the thermal radiator has a half-angle of 7.5 degrees. The reactor contains approximately 5 kg of highly enriched (93%) uranium-zirconium-hydride ($UZrH_{1.6}$). SNAP-10A employed operational constraints to preclude the reentry of a hot reactor system. The operating orbit of 1300 km (4000 years in orbit) was selected to assure the decay of the fission and activation products to approximately the activity level of the actinides in the fuel. In addition, start-up of the reactor was not permitted until a stable orbit was confirmed. Under these restrictions, reentry accidents occurring during launch and at the end of orbital life do not pose a serious radiological hazard because, in both cases, the radioactivity of the system at the time of reentry is quite low.

A thorough examination of the reentry heating and breakup of SNAP-10A was performed at both the analytical and experimental levels, including a reentry test of a full-scale model [19]. The conclusion drawn from this extensive program was that the reactor vessel would melt and release the core materials. Uncertainties existed, however, as to the fate of the fuel elements; complete ablation and dispersal of the fuel could not be guaranteed. Although complete dispersal of the core was the desired objective, no attempts were made to actively disperse the core, such as with an explosive charge. Calculations demonstrated that the risks to the public from postulated reentry accidents for SNAP-10A were very low. The operational constraints on orbital lifetime and reactor start-up were the key elements in establishing reentry safety.

The SP-100 space reactor, shown in Figures 2.21 and 2.34 in Chapter 2, was under development during the early 1980s as a joint DOE/DOD/NASA effort. Budget constraints and the absence of a clear mission and customer caused the termination of the program. The SP-100 was a fast reactor design using uranium nitride fuel and a lithium liquid metal coolant. The major subsystems included a beryllium sliding reflector, radiation shield, primary heat transport system, thermoelectric electromagnetic pumps, thermoelectric power converters, heat pipe radiators, and a carbon-carbon reentry heat shield. The SP-100 was designed to support a variety of space missions encompassing both national defense and space science objectives. For SP-100, a reentry safety approach was needed that could satisfy a large range of missions, including low Earth orbit and deep space operations. For some possible missions, reentry ablation calculations demonstrated that high altitude dispersal was not feasible for the SP-100 design; and safeguard concerns for postulated partial breakup was an issue [20]. Consequently, intact reentry was selected and implemented by shrouding the reactor with a reentry heat shield. Other reentry safety features of SP-100 included reactor start-up interlocks to prevent nuclear operations until the required stable orbit was achieved, and in-core safety rods that ensured the reactor would remain subcritical if submerged in water or buried in moist sand or soil.

7.0 Summary

Reentry accidents can result from failures during launch and from failures while in Earth orbit. Once a satellite is in orbit, reentry can occur either by an orbital decay or by a sudden change in orbital speed resulting in a transfer orbit that intersects the atmosphere. Reentry heating may disrupt the nuclear system leading to a variety of possible consequences. Reentry consequences can include radiological exposure from atmospheric dispersal; environmental contamination of air, land, and water bodies; and direct exposure of individuals to radiation from fully or partially intact systems or from debris following Earth impact. For some scenarios involving space reactor systems, the possibility of criticality accidents induced by Earth impact must be considered.

A ballistic coefficient (object mass divided by the drag coefficient and cross section area) is used in the analysis of orbital decay and orbital lifetime. The orbital lifetime of a satellite is strongly dependent on its orbital altitude as well as its ballistic coefficient. For reentry, three flow regimes are defined to facilitate analysis. The regimes include: free molecular flow, transition flow, and continuum flow. A reentry analysis begins by first computing the altitude and speed of the reentry object as a function of time. This information is then used to compute reentry heating and the thermal response of the body. An analysis of surface melting, ablation, and conduction is carried out to complete the first iteration of the analysis. If the thermal analysis determines that significant changes have occurred to the object's ballistic characteristics, then another trajectory calculation is performed with the altered ballistic parameters. Reentry analysis involves a suite of sophisticated computer codes; however, simple analytical models can be used to estimate reentry behavior.

Reentry safety practices have evolved since the inception of space nuclear missions. Initially, RTGs were designed to undergo complete breakup and dispersal of the radioisotope fuel at high altitude. The practice subsequently changed to intact reentry of the RTG. Sophisticated computer analysis and extensive testing are used to assure safe response of the RTG during a postulated reentry accident. Both intact and dispersed reentry approaches have been considered for space reactors.

Symbols

A	surface area	F_m	gravitational force
A_{ref}	reference area	F_D	drag force
a	semi-major axis	F	heating factor
a_t	$(r_1 + r_2)/2$	G	gravitational constant
b	semi minor axis	g	acceleration due to gravity
c	half-distance between foci	$\tilde{H}$	specific enthalpy
C_d	drag coefficient	h	heat transfer coefficient
C	1.10×10^8 W/m$^{3/2}$	I_{sp}	specific impulse
c_p	specific heat, constant pressure	K	defined by Eq. 6.66
D	diameter	Kn	Knudsen number
d_c	characteristic dimension	k	thermal conductivity
E	energy	**L**	angular momentum vector
$\tilde{E}$	specific energy	$\tilde{L}$	specific angular momentum
E_k	kinetic energy	L	length
e	eccentricity	m_e	Earth mass
f, f'	focal points	m	mass
$f(\theta)$	heat flux distribution function	Nu	Nusselt number

$\mathbf{n}$	normal to surface	t	time
$\mathcal{P}$	semi-latus rectum	U	potential energy
q	heat transfer rate	$\mathbf{v}$	velocity
q''	heat flux	v	speed
R	sphere or cylinder radius	v_0	v at atmosphere edge
R_e	Earth radius	x	horizontal direction
r	scalar distance	y	vertical direction
$\mathbf{r}$	position vector	y_0	edge of atmosphere (100 km)
St	Stanton number	Y_0	scale height
T	temperature		

α	accommodation coefficient	κ	constant = 1.67
α_m	angle of misalignment	λ	mean free path
β	ballistic coefficient (mass)	μ	Gm_e product
β_F	ballistic coefficient (force)	σ	Stefan-Boltzmann constant
ΔT	temperature difference	τ	period of orbit
$\Delta \mathrm{v}$	velocity change	θ	angle
Δr	orbital radius change	υ	polar angle
δ_s	skin thickness	ω	flight path angle
ε	emissivity	ς	density

Subscripts

0	original, reference value, or staging point	rad	radiative heat
a	apogee	ref	reference
$aero$	aerothermal	sl	stagnation line on cylinder
bl	boundary layer	sph	sphere
c	continuum flow	ss	stagnation point on a sphere
cl	continuum laminar flow	t	transfer ellipse
cs	circular orbit speed	tot	total
cyl	cylinder	w	wall
fm	free molecular flow	x	x-direction
∞	infinity, or free stream	y	y-direction
p	perigee	$\perp$	normal incidence
RP	rocket + payload		

References

1. Bate, R. R., D. D. Mueller, and J. E. White, *Fundamentals of Astrodynamics.* New York: Dover Publications, 1971.
2. Thomson, W. T., *Introduction to Space Dynamics.* New York: John Wiley & Sons, 1963.
3. Tsien, H., "Superaerodynamics, Mechanics of Rarefied Gases." *Journal of the Aeronautical Sciences*, Dec. 1946.
4. Klett, R. D., *Drag Coefficients and Heating Ratios for Right Circular Cylinders in Free Molecular and Continuum Flow from Mach 10 to 30.* Albuquerque, NM: Sandia National Laboratories Report SC-RR-64-2141, Dec. 1964.
5. Connell, L. W., unpublished, Mar. 2001.
6. Hipp, J. R., *Preliminary Solar Resonance Accident Consequence Assessment.* Briefing to Ballistic Missile Defense Organization, July 27, 1993.
7. Frisbee, R. H., S. D. Leifer, and S. V. Shah, "Nuclear Safe Orbit Basing Considerations." AIAA/NASA/OAI Conference on Advanced SEI Technologies, Sept. 1991.
8. Allen, H. J. and A. J. Eggers, Jr., *A Study of the Motion and Aerodynamic Heating of Missiles Entering the Earth's Atmosphere at High Supersonic Speeds.* National Advisory Committee for Aeronautics, NACA-TN-4047, Oct. 1957.
9. Detra, R. W., N. H. Kemp, and F. R. Riddell, "Addendum to Heat Transfer to Satellite Vehicles Reentering the Atmosphere," Jet Propulsion, Dec. 1957, p. 1256–1257.
10. Lees, L., "Laminar Heat Transfer over Blunt Nosed Bodies at Hypersonic Flight Speeds." *Jet Propulsion*, 26 (4), Apr. 1956.
11. Connell, L. W. and L. C. Trost, *Reentry Safety Issues and Analysis for the TOPAZ II Space Nuclear System.* Albuquerque, NM: Sandia National Laboratories, SAND-0484, Feb. 1994.

12. Outka, D. E., *Users Manual for TSAP*. Albuquerque, NM: Sandia National Laboratories Report, SAND88-3158, July 1985.

13. Sutton, G. W., "The Initial Development of Ablation Heat Protection, An Historical Perspective." *J. Spacecraft*, **19**, No. 1, p. 3–11, Jan.-Feb. 1982.

14. Potter, D. L., *Approximate Heating Analysis Methods for Appended Bodies*. Albuquerque, NM: Sandia National Laboratories, Internal Memorandum, Dec. 1989.

15. Moyer, C. B. and R. A. Rindal, *Finite Difference Solution for the In-Depth Response of Charring Materials Considering Surface Chemical and Energy Balances*. Report No. 66–7, Aerotherm Corporation, Mountain View, CA, Mar. 1967.

16. Moyer, C. B., B. Blackwell, and P. Kaestner, *A User's Manual for the Two-Dimensional Axi-Symmetric Transient Heat Conduction Material Ablation Computer Program (ASTHMA)*. Albuquerque, NM: Sandia National Laboratories Report, SC-DR-70-510, Dec. 1970.

17. Von Karman, T., "Aerodynamic Heating: The Temperature Barrier in Aeronautics." Proceedings of the High Temperature Symposium, Stanford Research Institute, Berkeley, CA, June 1956.

18. McAlees, S., Manager Aerospace Nuclear Safety Program, Sandia National Laboratories, Personal Communication, May 1992.

19. Elliot, R. D., *Aerospace Safety Reentry Analytical and Experimental Program, SNAP 2 and 10A (Interim Report)*. NAA-SR-8303, Sept. 30, 1963.

20. Bost, D. S., *FY 1985 Surety Program Final Report*. Rockwell International, Rocketdyne Division, RI/RD85-236, Sept. 1985.

Student Exercises

1. A spacecraft at a distance of $r = 6800$ km from the center of the Earth has an instantaneous speed of 8.0 km/s with a flight path angle $\omega = 10$ degrees. (a) Prove that the orbit about the Earth is an ellipse and calculate the basic orbital parameters, L, E, p, a, e, r_p, r_a, and t. Use $R_e = 6378$ km and $\mu_e = 3.99 \times 10^5$ km^3/s^2. (b) Calculate the Δv needed to circularize the orbit with circular radius equal to the apogee radius r_a of the elliptic orbit.

2. A space nuclear power system with a total mass of 5 metric tons is initially deployed in a 500 km altitude circular orbit around Mars ($\mu_{mars} = 4.3 \times 10^4$ km^3/s^2 and $R_{mars} = 3380$ km). At its end-of-life the system is to be boosted to a 2500 km altitude circular parking orbit. (a) Calculate the Δvs needed to accomplish this using the basic Hohmann transfer method. (b) Using the basic rocket equation (Eq. 6.22) calculate the mass of the propellant needed to achieve this transfer. Ignore the mass of the rocket stage structure (i.e., $M_{RP} = 5$ metric tons). Also assume that the propellant has a gI_{sp} of 4.5 km/s.

3. Assume that a nuclear rocket engines (NRE) and its payload are in a 400 km circular. Astronauts then bring the NRE up to power, delivering a total Δv of 4.1 km/s. Due to a software error, the thrust vector of the NRE is misaligned relative to the circular orbit velocity vector by an angle $\delta = 100$ degrees. (See Figure 6-E1.) Using $R_e = 6378$ km and $\mu_e = 3.99 \times 10^5$ km^3/s^2, compute the reentry state vector (i.e., the reentry speed and flight path angle).

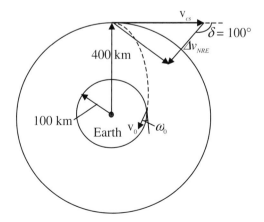

Figure 6-E1 Orbital geometry for Exercise 3.

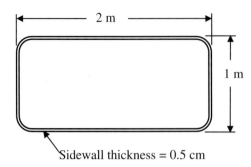

Figure 6-E2 Schematic of pressure vessel for Exercise 4.

4. A space nuclear reactor steel pressure vessel (see Figure 6-E2) has a density $\varsigma = 7.87$ g/cm³, a heat capacity $c_p = 0.45$ J/g·C, and a melt temperature of 1500 C. Using the thin skin thermal model, compute the altitude at which the pressure vessel reaches the melt temperature. Assume a constant flight path angle and use the following reentry initial conditions and properties: reentry speed $v_o = 5$ km/s, altitude $y_o = 60$ km, flight path angle $\omega_o = 20$ degrees, and air heat capacity $c_{p\text{-}air} = 1.05$ J/g·C. (a) Assume the vessel reenters end-on stable (b) Assume the vessel reenters with random tumble and spin motion.

5. A reentry test is to be performed on a mockup of a Pu-238 fueled RTG to ensure that its thermal protection system will provide for an intact reentry. The RTG mockup is a 10 cm diameter cylinder, 20 cm long, with a ballistic coefficient = 4000 N/m². The reentry speed, altitude, and flight path angle are $v_o = 8.0$ km/s, $y_o = 50$ km, and $\omega_o = 15$ degrees, respectively. Assuming a constant flight path angle and end-on stable reentry, use the simple trajectory and heating models provided in Section 3 to compute the time to reach an altitude of 20 km, the speed at this point, and the stagnation point heat flux. Compute and plot the speed and stagnation point heat flux versus altitude. At what altitude does the heat flux reach a maximum? Explain your results.

6. Assuming the same body from Exercise 5, a constant velocity from 100 km down to 50 km, and free molecule flow, calculate the stagnation point heat flux versus altitude over this altitude range. How does the free molecule flow heat flux at 50 km compare with the laminar flow value at 50 km from problem 5? Using the diameter of the cylinder as the characteristic dimension, calculate the Knudsen number and determine the correct flow regime at 50 km.

Chapter 7

Impact Accidents

James R. Coleman

The objectives of this chapter are to provide the reader with an introduction to postulated impact accidents; to present a relatively simple caculational approach for analyzing impact accidents; to provide a discussion of the importance of impact modeling for safety test planning; and to illustrate the approach using test data.

1.0 Scenarios and Issues

The evaluation of barrier failure and radioactive material release (the source term) is central to most space nuclear safety analyses; consequently, impact accidents are an important aspect of space nuclear safety analyses. Preceding steps in many safety analysis sequences are directed at developing the scenarios, probabilities, and environments for those events that could result in an impact, a breach of protective barriers, and release of radioactive materials. Subsequent steps in the analysis are directed at predicting the transport, distribution, and human interaction with released radioactive materials or the damaged system. Each space nuclear mission requires a comprehensive safety analysis program that identifies potential impact accident events and provides an analysis of impact accidents and consequences.

1.1 Pre-deployment Accident Issues

The most serious impact events occur in mission-terminating, pre-deployment accidents, particularly launch pad and early launch phase accidents. The United States has relied primarily on solid rocket boosters (SRBs) in the initial lift phase for nuclear systems. The probability of failure of these systems is generally taken to be on the order of 0.001 to 0.01 per unit launch on the basis of past experience. Generally the failure of an SRB prompts intentional launch destruct with explosive initiated high-pressure rupture of the SRBs and the subsequent ignition of any liquid fuels on board. Russia has used mostly liquid fuel type launch vehicles, but failure rates are comparable to U.S. rates. Russian rocket failure without explosion, due to an engine malfunction, is about six times more probable than a liquid fuel explosion event. Nonetheless, all rocket failures result in the destruction of the launch vehicle and the exposure of the nuclear payload to impact from a variety of fragments and shrapnel. In addition, all of these events conclude with Earth impact of the nuclear payload.

The primary safety issue with radioisotope sources is the possibility of breaching containment and dispersing a portion of the radioactive fuel into the local environment. Given that reactors are launched radiologically cold, the principal issue for pre-deployment impact accidents is the possibility of altering the reactor geometry or environment in such a way as to initiate inadvertent criticality. Although the radioactive content of essentially fresh fuel is very low relative to radioisotope sources and rarely presents a serious radiological issue, the probability and consequences of an impact causing the release of unirradiated fuel must be assessed.

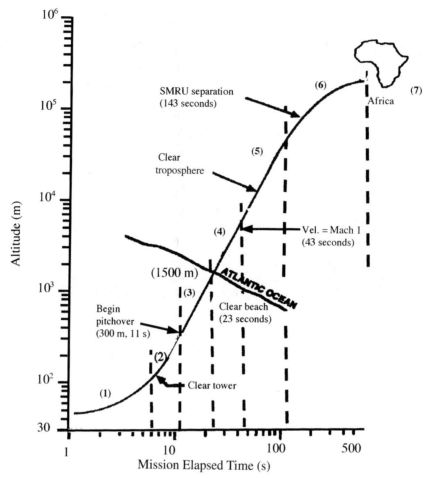

Figure 7.1 Cassini mission ascent profile [1]. *Courtesy of U.S. Interagency Nuclear Safety Review Panel.*

1.2 Pre-deployment Scenarios

The ascent profile can be divided into a number of stages. Figure 7.1 identifies seven stages of the ascent profile for the Cassini Mission [1]. Potential accident scenarios for each stage are presented in the following correspondingly numbered paragraphs:

(1) For the first 6 seconds after liftoff, potential impact accident scenarios include propellant explosions with explosion driven fragments. Subsequent surface impact of the space nuclear system will be on concrete or steel launch pad structure (e.g., Figure 7.2) and may be at velocities slightly greater than free fall (generally on the order of 35 m/s). Full or partial impact of the launch vehicle components on top of the nuclear system is a possibility.

(2) For an accident initiated at this stage, surface impact will occur at velocities up to terminal velocities (generally about 60 m/s). The debris impact point is centered on the concrete or steel pad elements until pitchover, which begins at approximately 11 seconds at an altitude of about 300 m (about 1000 feet). Flight termination may expose the space nuclear system to an explosive blast and explosion-driven fragments.

(3) For accidents initiated after pitchover begins, the debris impact point moves toward the Atlantic Ocean and the impact surface changes from concrete or steel to wet sand and then shallow water. The altitude increases to about 1500 m (about 5000 feet) at 23 seconds. Flight termination may ex-

Figure 7.2 Launch facility at Cape Kennedy launch complex. *Courtesy of U.S. National Aeronautics and Space Administration.*

pose the space nuclear system to energetic fragment impacts as well as surface impact, which would occur at about terminal velocities.

(4) For an accident initiated during stage 4, the drag corrected debris impact point, reaches the coastline after about 33 seconds (Figure 7.1 shows that land is cleared at 23 seconds, but that time is for the vacuum impact point and does not include atmospheric drag). Between about 33 and 680 seconds all impacts will be in water at or slightly above sea level terminal velocity.

(5) The vehicle reaches Mach 1 at about 43 seconds—a velocity which will cause aerodynamic breakup for any off-normal deflection (more than about 0.5 degree) of the solid rocket motor nozzles. Thus, breakup and reentry considerations become important. Ocean impact will be at terminal velocity.

(5) A failure during this stage will result in ocean impact at terminal velocity until about 680 seconds. For a brief period (between 680 and 690 seconds), a postulated failure could result in land impact in Africa.

(5) Orbit is achieved at about 690 seconds.

It is clear from this discussion that the range of possible impact events for the pre-deployment phase is very broad.

1.3 Post-deployment Scenarios

Once in orbit, the nuclear system becomes exposed to the energetic impacts from space debris and meteoroids. The possibility of space debris and meteoroid impact must be considered during and following operation. Space debris or meteoroid impact could result in premature reentry of debris or a damaged system (for LEO operation). Furthermore, for a *low Earth orbit* (LEO) mission, an internally initiated reactor accident could result in system disruption leading to reentry and Earth surface impact.

2.0 Impact Environments

The safety analysis for a postulated impact accident begins by identifying a set of potential accident environments. Of primary importance are the distributions of fragment masses and velocities from postulated catastrophic accidents. Three broad impact categories must be addressed: (1) impacts of large vehicle fragments, (2) intact impact of the system (or subsystems) on the Earth's surface, and (3) impacts of small penetrating fragments (shrapnel, space debris, or meteoroids). While the fundamental conceptual approach to the analysis is the same for all three cases, it is convenient to use this distinction to discuss accident environments.

2.1 Large Fragments

Large fragments can pose a significant challenge to space nuclear systems. The two potential sources of large fragments include propellant explosion-driven fragments and space debris.

Explosion-Driven Fragments

Explosion induced fragments are discussed in Chapter 5. The size and velocity of the fragments depends on both the strength of the blast and the materials involved. An example of the predicted distribution of fragment masses and velocities generated by a random failure of the upgraded Titan solid rocket motors is shown in Table 7.1 [2] as a function of mission elapsed time (MET). The fragment size data are devised using historic data on solid rocket explosions and the assumption that the fragments are lognormally distributed by mass. A fragment size of 0.027 kg (a cube 2.54 cm on a side) is assumed and taken as the 1st percentile value. The velocity is also stated as a lognormal stochastic variable with an uncertainty of roughly a factor of two. The fragment velocities were obtained using a computer program that models the complex interactions due to drag from the expanding combustion gasses, the separation of fragments, and venting flow. The Cassini Data Book [2] provides fragment data for several other accident events. A similar listing of fragment sizes, velocities, and impact probabilities must be prepared for any space nuclear mission.

Fragment velocity at impact and impact probability also depend on the geometry of the total system (particularly the direction and distance of the nuclear system from the center of explosion), in-

Table 7.1   Fragment Sizes and Average Speeds for Titan Solid Rocket Motor Upgrade Random Failures (Nominal Pressure) [2]

Mission Elapsed Time (s)	Percent Less Than Given Size	Aft Segment Mass (kg)	Aft Segment Speed (m/s)	Center Segment Mass (kg)	Center Segment Speed (m/s)
0	99	36,300	31	36,300	28
	50	31	61	31	56
	1	0.027	119	0.027	112
10	99	29,100	36	29,100	33
	50	28	68	28	64
	1	0.027	131	0.027	123
20	99	23,600	42	23,600	38
	50	25	77	25	71
	1	0.027	142	0.027	135
30	99	18,200	46	18,200	42
	50	22	82	22	77
	1	0.027	148	0.027	141
40	99	12,700	51	12,700	47
	50	19	88	19	83
	1	0.027	153	0.027	147
50	99	9,500	55	9,500	49
	50	16	91	16	85
	1	0.027	151	0.027	146

tervening structure, and spatial orientation (a face-on fragment may strike the system while an edge-on may miss). All of these considerations introduce additional stochastic variables into the analysis. A critical process that must precede impact analysis is devising the analytic approaches for evaluating hit probabilities, the effect of intervening structure, and the treatment of fragment orientation relative to the system.

Space Debris

The problem of relatively large items of orbiting space debris may become increasingly important. Currently there are less than 100 such large objects present in LEO in the altitude range of 700 km to 1300 km. With increasing space activities, however, the number of objects in LEO may increase significantly. As the spatial density of this debris increases, the probability for collision with space nuclear systems operating in orbit also increases. Collision with these relatively large items of space debris can cause severe impact damage to a space nuclear system, and the collision will alter the orbital characteristics of the system due to momentum considerations. For the worst case, large object impact will result in early reentry of debris or the system.

2.2 Intact Impact

Intact impact is defined as energetic Earth impact of an item, which is in an *unfailed* condition (not pristine but unfailed). Thus, intact impact includes the impact of a full radioisotope-generator, a reactor, or the intact impact of any specific subsystems in an unfailed condition. For example, consider the standard GPHS-RTG system which consists of an outer housing (the thermoelectric converter assembly), 18 GPHS modules (fine-weave-pierced-fabric graphite reentry modules), and 72 *clads* (iridium clad fuel pellets). Possible intact components include the GPHS-RTG itself, a free GPHS module, or a free fuel clad. Figures 2.14 and 2.15, in Chapter 2, provide a description of the GPHS-RTG system.

Free-fall Impact Velocity

In addition to the description of the system or subsystem being considered, the important impact parameters are the velocity of impact and the material on which impact occurs. The velocity of impact of a free falling object depends on its shape, areal mass density, the density of the atmosphere, and the height of release. The free-fall velocity is related to these variables by the difference between the forces of gravity and drag; i.e.,

$$\frac{dv}{dt} = g \frac{\varsigma_i - \varsigma_{air}}{\varsigma_i} - \frac{C_d A_i \varsigma_{air}}{2m} v^2, \tag{7.1}$$

where
g = acceleration due to gravity,
ς_i = density of article,
ς_{air} = density of air,
v = velocity of article,
C_d = dimensionless drag coefficient,
A_i = projected area of article, and
m = mass of article.

Assuming $\varsigma_i >> \varsigma_{air}$ and $\varsigma_{air} \approx$ constant, we can obtain the equation for the time dependent speed of an object in free fall by solving Eq. 7.1, yielding

$$v(t) = \sqrt{\frac{2mg}{\varsigma_{air} A_i C_d}} \tanh \left(t \sqrt{\frac{\varsigma_{air} A_i C_d g}{2m}} \right). \tag{7.2}$$

As t increases, the hyperbolic tangent goes to 1 and the speed of the article approaches the terminal velocity v_t,

$$v_t = \sqrt{\frac{2mg}{\varsigma_{air} A_i C_d}}.$$ (7.3)

Impact Materials

After orbit is achieved, the problem is more complicated. For a postulated reentry accident, impact can occur on water, a soft surface, or a hard surface. The areas of hard surface (exposed hard rock and other hard materials) vary with elevation; impact velocity also varies with elevation due to air density effects. Assuming random reentry, one must be able to predict the probability of impacting on a hard surface in a given population density at a given impact velocity. NUS Corporation has collected and documented the worldwide occurrence of surface materials between 65°35′ north and south latitudes as a function of elevation (in correlation with population density) for use in space nuclear safety analyses [3]. These data permit a calculation of the joint probability of impact on a hard surface, within a specific elevation range, associated with a specific population density. The hard surface impact probabilities presented in Table 7.2 were adapted from Reference [3].

2.3 Small Fragment or Shrapnel

Shrapnel generated in launch abort accidents is one of the two broad classifications of small fragment impacts. For example, the small shrapnel generated from the nose cone and forward closure of the Cassini SRBs includes several hundred screws, nuts, and bolts with areal mass densities between 3.5 and 11.7 g/cm^2 [2]. For a failure at about 10 s MET, the velocity of these items is estimated to range from about 180 to 280 m/s.

The other broad classifications of small fragment impacts include small debris or meteoroid impact in space. For altitudes of less than a few thousand kilometers from Earth's surface, the principal concern is space debris generated by human activities. A recent estimate is that the Earth has an envelope of about 2 million kilograms of man-made space debris orbiting below 2000 kilometers. This debris is nearly uniformly distributed in latitude and longitude and has orbital lifetimes up to thousands of years [4]. Much of the material is aluminum with a collision velocity in the range of about 9-13 km/s. The debris flux (average for altitudes between 600 km and 1100 km) as a function of projectile diameter is graphed in Figure 7.3 [4]. Several years ago Kessler [5] speculated that the impact of small fragments of space debris may well drive the design of even average-size spacecraft intended to operate in low Earth orbit. Above a few thousand kilometers meteoroids pose the major penetration risk. The average meteoroid flux [4] is also shown in Figure 7.3. The mean velocity of these meteoroids is about 22 km/s.

Table 7.2   Impact Probability for Hard Surface Correlated with Altitude and Population Density for Random Reentry Between 65° 35′ N and S Latitude [3]

Population Density (people/km^2)

Elevation (meters)	0.4 to 10	10 to 40	40 to 100	>100	Cities > 1,000,000	Cities > 5,000,000
0 to 200	0.00288	0.00243	0.00084	0.00068	0.00008	0.00008
200 to 500	0.00769	0.00415	0.00115	0.00169	0.00015	0.00008
500 to 1000	0.00931	0.00406	0.00194	0.00175	0.00013	0.00002
1000 to 2000	0.00852	0.00439	0.00090	0.00084	0.00005	0.00002
2000 to 4000	0.00198	0.00059	0.00015	0.00011	0.00002	0.00000
> 4000	0.00017	0.00007	0.00001	0.00001	0.00000	0.00000

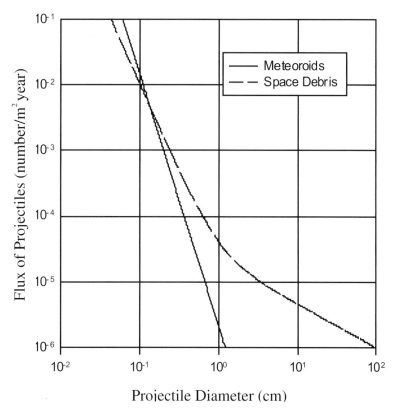

Figure 7.3 Average meteoroid and space debris flux for altitudes between 600 and 1100 km [4].

3.0 Energy Interaction Model

The distributions of fragment masses and velocities, discussed in the preceding section, are used to predict impact interactions, potential failure of safety barriers, and the release of radioactive material. Because of the wide variety of conditions that must be considered and the stochastic nature of the end states, the results in terms of both events and outcomes are probabilistic in nature. Thus, the deterministic approaches described in this section are just the starting point.

The following discussion presents a relatively simple deterministic approach to impact analysis using an energy interaction model. This approach offers a logical basis for assessing impact consequences as part of a comprehensive safety analysis for designing safety tests and for analyzing subsequent test data. Although the general approach presented here uses a radioisotope thermoelectric generator as the example case, the basic principles are applicable to reactor systems as well.

3.1 Fundamental Relationships

The fundamental assumption of this model is that the impact damage to an article is measured by the disruptive work done on that article as a result of an impact. The model starts with the simple concept of a two-body impact problem assuming complete inelasticity (no rebound; i.e., a restitution coefficient of zero) with conservation of both energy and momentum. For the purpose of this description, the two interacting bodies will be called the *item* and the *object*. The *item* is the system component of interest, and the *object* is the object with which the item interacts (e.g., fragment). Consider the case of the two centers of mass impacting each other; an object of mass m_{ob} and velocity $\mathbf{v}_{ob}$ and an item with mass m_i and velocity $\mathbf{v}_i$, move together as one with velocity $\mathbf{v}_2$ after impact (see Figure 7.4). We will use a ref-

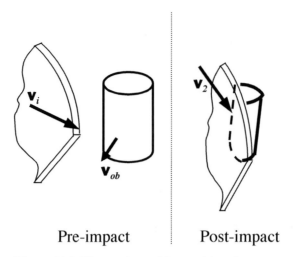

Pre-impact | Post-impact

Figure 7.4 Illustration of item-object impact.

erence frame centered at the point of impact of the two centers of mass. The energy and momentum balances for the collision are described by the following two equations:

$$\frac{1}{2}(m_i + m_{ob})\mathrm{v}_2^2 + W = \frac{1}{2}m_i\mathrm{v}_i^2 + \frac{1}{2}m_{ob}\mathrm{v}_{ob}^2,$$ (7.4)

and

$$(m_i + m_{ob})\mathbf{v}_2 = m_i\mathbf{v}_i + m_{ob}\mathbf{v}_{ob}.$$ (7.5)

Here, W is that portion of the initial kinetic energy that is internalized as work in disordering the system as a result of the collision. This work, the internalized loss of kinetic energy, is the quantity taken to be the gauge of insult for predicting impact consequences (i.e., barrier failure, penetration, reconfiguration, and fuel release). Solving Eq. 7.5 for $\mathbf{v}_2$ and taking the dot product $\mathbf{v}_2 \cdot \mathbf{v}_2$, we obtain an expression for the scalar v_2^2. Substituting this expression into Eq. 7.4 provides an expression for the work expended disordering the system in terms of $\mathbf{v}_i$ and $\mathbf{v}_{ob}$; i.e.,

$$W = \frac{1}{2}\frac{m_i m_{ob}}{(m_i + m_{ob})}(\mathbf{v}_i - \mathbf{v}_{ob})^2.$$ (7.6)

Here, $(\mathbf{v}_i - \mathbf{v}_{ob})^2$ is the dot product $(\mathbf{v}_i - \mathbf{v}_{ob}) \cdot (\mathbf{v}_i - \mathbf{v}_{ob})$.

For simplicity we will designate $(m_i m_{ob})/(m_i + m_{ob})$ as the reduced mass m_{red} of the system. The dot product $(\mathbf{v}_i - \mathbf{v}_{ob}) \cdot (\mathbf{v}_i - \mathbf{v}_{ob})$ is a scalar equal to the square of the relative speed of closure v_{rel}^2 between the two centers of mass as they approach the impact point. Thus, W can be expressed as a function of the relative closure speed and the reduced mass; i.e.,

$$W = \frac{1}{2}m_{red}\mathrm{v}_{rel}^2.$$ (7.7)

The next step is to apportion this work (internal energy) between the two interacting bodies. To do this we will define a damage coefficient fw to represent the fraction of the work expended in disordering the item and $(1 - fw)$ as the fraction that is expended in disordering the object. Basically fw is a measure of the relative stiffness of the two bodies. As our interest is related only to the work done on the item, the model is stated by the following equation:

$$W_i = \frac{1}{2} m_{red} \mathrm{v}_{rel}^2 \, fw. \qquad (7.8)$$

Equation 7.8 represents the basic energy model for impact.

3.2 Imbedded Safety Barriers and Multiple Impacts

The basic energy model can be extended to systems of imbedded safety barriers and to multiple sequential impacts. In order to illustrate the approach, we will use a GPHS-RTG as an example. The GPHS-RTG's first level of protection is provided by the RTG structure itself (the casing, the thermo-electric devices, and the supporting structure); a second level of protection is provided by the GPHS modules (the 18 graphite heat sources); and a third level of protection is provided by the fuel cladding (iridium cladding for the PuO$_2$ fuel). The disruptive work expended as a result of impact is distributed within the system by this protective construction.

The analysis begins at a time immediately following the impact event pictured in Figure 7.4. Following the initial impact, the RTG and its components will be observed in either of three major damage states: (1) an intact but possibly damaged RTG, (2) a fractured RTG case with damaged free modules, or (3) a fractured RTG case with free damaged modules and also free damaged fueled clads released from fractured modules. These damage states are illustrated in Figure 7.5. This initial event will be followed by at least one sequential impact (i.e., an initial explosion or fragment event followed by Earth impact). Thus, our analysis model needs to address the question of accumulating the damage from such sequential events and to consider the sequential damage to any of the various component states which may face sequential impact, i.e., the intact RTG, bare modules, or bare fueled clads.

We start with the basic assumption that cumulative damage relates to the cumulative work done on the various components of the system. The disruptive work is assumed to be partitioned among the nested protective levels by damage coefficients (similar to fw above), and each level is assumed to exhibit a threshold of damage W_T that must be exceeded to yield damage to the next level. Finally, we assume that if a threshold is not exceeded, its protective capability is reduced by the work already partitioned to it.

To facilitate the discussion, see Figures 7.6 and 7.7 for a schematic guide. Two subscripts, j and k, will be used to identify the important impact descriptors.

j = the protective level for which work is being calculated, and
k = the damage state at which impact occurs

(1 = RTG case, 2 = free GPHS module, 3 = free fuel capsule).

Not all of the disruptive work in excess of the threshold is partitioned to the next level. We define fw_{jk} as the fraction of work from the previous protective level that exceeds the protective threshold of that level which is partitioned to protection level j from an impact in the kth damage state. As the damage coefficient fw in Eq. 7.8 represents only the apportionment of work between the item of interest and

Damage State 1: RTG intact, possibly damaged.

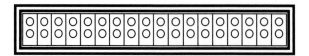

Damage State 2: RTG case ruptured, free damaged modules.

Damage State 3: Free damaged modules and free damaged clads.

Figure 7.5 Schematic illustration of three major damage states following impact.

the object impacted in an initial event, it is designated as fw_1. More generally, the damage coefficient fw_k will be used for an item in state k impacting an object. Also, W_{T11} is defined as the threshold of work that needs to be added to breach the first level of protection (the RTG casing fracture threshold). We can define similar damage coefficients and thresholds for the second and third levels of protection.

The fraction of the work added to any level is distributed within that level as a function of its construction and the type of impact. We will define two additional factors fm and fc to distribute the work among the modules in the RTG and the four clads in each module respectively. If we consider each module and each clad to be equally affected by the impact, the set fm is a single value (1/18 for an 18 module RTG) and the set fc is the single value (1/4). Other assumptions may be used for the distribution of work.

With this construction of the problem the disruptive work apportioned to each of the three levels of protection from an initial impact on an integral RTG can be computed. The impact work apportioned to the impacted item (the first level of protection) from an initial impact W_{11} is

$$W_{11} = \frac{1}{2} m_{red} v_{rel}^2 fw_1. \tag{7.9}$$

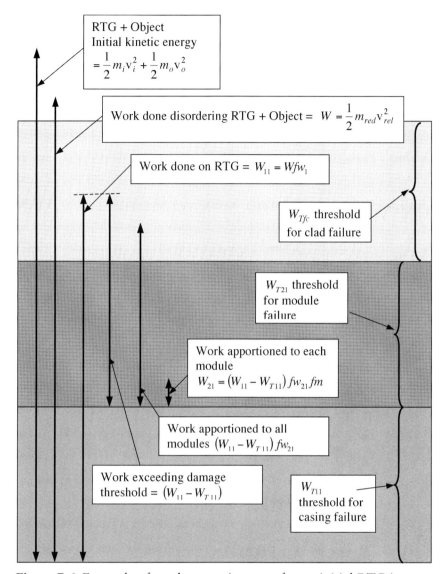

Figure 7.6 Example of work apportionment for an initial RTG impact.

The impact work apportioned to each module (the second level of protection) from the initial impact on an integral RTG, including the RTG casing failure threshold W_{T11}, is

$$W_{21} = (W_{11} - W_{T11})\, fw_{21} fm. \qquad (7.10)$$

The impact work apportioned to each fueled clad (the third level of protection) from the initial impact on an integral RTG including the GPHS module protection failure threshold is

$$W_{31} = (W_{21} - W_{T21})\, fw_{31} fc. \qquad (7.11)$$

Equations 7.9 through 7.11 describe the work done on the three components of the RTG system by an initial impact.

For sequential impacts we need to consider the state in which impact occurs. For initial impacts with W_{11} greater than the RTG case threshold but not severe enough to free fuel capsules from any

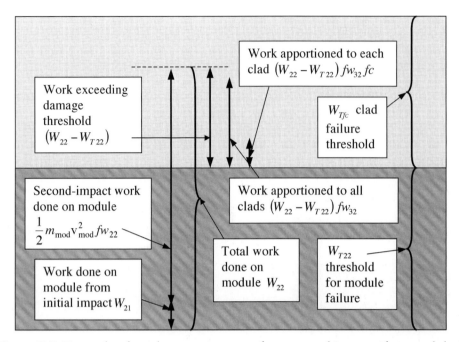

Figure 7.7 Example of work apportionment for a second impact (free modules).

modules (State 2), the modules are released from the RTG and are free to act independently to final ground impact. The work done by both impacts must be calculated. The initial disruptive energy partitioned to each module W_{21} is calculated from Eq. 7.10. The work added to a module by the ground impact is given by the apportioned kinetic energy of the free module. Thus, using m_{mod} and v_{mod} to denote the module mass and impact velocity respectively, and fw_2 as the apportioning coefficient between the impacting module and the impacted surface, the total work added to each free module is given by

$$W_{22} = \tfrac{1}{2} m_{mod}\, v_{mod}^2\, fw_2 + W_{21}. \tag{7.12}$$

Likewise the work added by the second impact to a clad within the module is given by the kinetic energy apportioned by the appropriate coefficients, with fc denoting the distribution of work among the four fueled clads and fw_{32} the apportioning coefficient between the fueled clads and the module. The total work added to each clad after the second impact is given by

$$W_{32} = (W_{22} - W_{T22})\, fw_{32} fc + W_{31}. \tag{7.13}$$

For initial impacts above the RTG case threshold that are severe enough to free fuel capsules from some modules (State 3) the fueled clads are free to interact independently to final ground impact. The initial disruptive energy partitioned to each clad W_{31} is calculated from Eq. 7.11. The total work added to each clad is given by

$$W_{33} = \tfrac{1}{2} m_{clad}\, v_{clad}^2\, fw_3 + W_{31}. \tag{7.14}$$

Equations 7.9 through 7.14, summarized in Table 7.3, are the mathematical relationships that describe the impact model for initial and subsequent impacts. This approach provides a useful tool for planning tests and understanding test results. It is not a simple task to obtain suitable values for the ap-

Table 7.3 Equations for Disruptive Work by Impacts for GPHS-RTG

Work to	Initial Impact of RTG	Subsequent Impact of Module*	Subsequent Impact of Clad*
RTG	$W_{11} = \frac{1}{2} m_{red} v_{rel}^2 fw_1$		
Modules	$W_{21} = (W_{11} - W_{T11})fw_{21}fm$	$W_{22} = \frac{1}{2} m_{mod} v_{mod}^2 fw_2 + W_{21}$	
Clads	$W_{31} = (W_{21} - W_{T21})fw_{31}fc$	$W_{32} = (W_{22} - W_{T22})fw_{32}fc + W_{31}$	$W_{33} = \frac{1}{2} m_{clad} v_{clad}^2 fw_3 + W_{31}$

*Following initial RTG impact

portioning coefficients, thresholds, and proportionality factors for any given system. These parameters are dependent on material, structure, target, orientation, and angle of impact. They can only be obtained empirically through a well-planned test program. Note that for safety tests using pristine modules and clads, W_{21} in Eq. 7.12 and W_{31} in Eqs. 7.13 and 7.14 are set to zero.

The model discussed in this section is based on a series of critical assumptions: that the gauge of damage is the disruptive work apportioned as a result of impact; that the disruptive work can be partitioned within the imbedded system by relative damage coefficients; that imbedded systems have unique failure thresholds; and that if a threshold is not exceeded, the system survivability (its threshold of failure) is reduced only by the work added. These are rather bold assumptions, and only detailed engineering analysis and testing can evaluate their significance to a specific system.

Example 7.1

Consider a postulated launch abort for a mission with an RTG payload at an altitude of about 300 m (approximately 10 s MET). Include an in-air, face-on impact by a median-sized fragment that precedes the Earth impact on the concrete pad. The mass of the RTG is 56.2 kg. For the fragment impact calculation, treat the RTG as stationary. Use $fw_1 = 0.33$ for the first impact and $fw_1 = 1$ for the second impact. Also use $W_{T11} = 45,000$ J, $fw_{21} = 0.252$, and $fm = 1/18$. Use a ground impact velocity of 42.6 m/s. Determine if an RTG failure is predicted for the first or second impact. Calculate the work done on each module resulting from both impacts.

Solution:

Using the fragment mass and velocity from Table 7.1, the reduced mass is

$$m_{red} = \frac{28 \text{ kg} \times 56.2 \text{ kg}}{28 \text{ kg} + 56.2 \text{ kg}} = 18.7 \text{ kg},$$

and the work done by the fragment on the RTG for the first impact is (from Eq. 7.9)

$$W_{11}(1^{st} \text{ impact}) = 1/2 \times 18.7 \text{ kg} \times (68)^2 \text{ (m/s)}^2 \times 0.33 = 1.427 \times 10^4 \text{ J}.$$

The RTG failure threshold has not been exceeded, so the RTG falls intact but damaged. The work done on the RTG from ground impact (second impact) is

$$W_{11}(2^{nd} \text{ impact}) = 1/2 \times 56.2 \text{ kg} \times (42.6)^2 \text{ (m/s)}^2 (1) = 5.1 \times 10^4 \text{ J}.$$

The total accumulated work on the RTG is

$$W_{11}(\text{total}) = [50,995 + 14,267] = 6.5 \times 10^4 \text{ J}.$$

The RTG failure threshold is exceeded for the second impact. The work apportioned to each module is given by Eq. 7.10; i.e.,

$$W_{21} = (6.5 \times 10^4 - 4.5 \times 10^4)(0.252)(1/18) = 284 \text{ J}.$$

4.0 Penetration by Small Fragments or Shrapnel

Two types of penetration by small fragments need to be considered. The first type includes fragments impacting with velocities up to several hundred m/s, such as the fragments produced from launch abort accidents. The second type includes hypersonic impacts (relative to sound speed in the fragment or target materials). Hypersonic fragment impacts can occur for collisions with space debris or meteors. The mechanics of these two types of impacts are quite different, and separate penetration models are given for the two cases. However, our interest in either case is to describe the penetration potential of the fragments as functions of their physical characteristics (size and velocity).

4.1 Subsonic Velocity Impact

For subsonic velocities, the theory of ballistic perforation is well studied [6–9] and derives from the same initial energy and momentum balances given in Eqs. 7.4 and 7.5. Consider the case of a small projectile (the object) striking a stationary plate (the item). The projectile penetrates the plate and ejects a plug. The projectile and plug move together after penetration. The loss of kinetic energy of the projectile is expended as work in compressing both the projectile and target, and as shear work on the target.

The energy balance given by Eq. 7.4 can be rewritten to express the kinetic energy lost as disruptive work in two parts; i.e., W_c the work utilized in inelastic compression and W_s the work expended in shearing the plug free. Thus the energy balance for the system is

$$\frac{1}{2} m_{ob} v_{ob}^2 = W_c + W_s + \frac{1}{2}(m_{ob} + m_i)v_r^2, \tag{7.15}$$

where v_r is the residual velocity of the projectile and plug after penetration. As the only velocity component is in the direction of impact, the velocity does not need to be treated as a vector. The momentum equation for this system is

$$(m_i + m_{ob})v_r = m_{ob}v_{ob}. \tag{7.16}$$

These two equations allow us to devise analytical relationships for both W_c and W_s. Consider the case where there is no shear; i.e., the plug is simply free of the surrounding target. In that case, the two preceding equations can be solved for W_c (with $W_s = 0$). Thus,

$$W_c = \frac{1}{2} \frac{m_{ob} m_i}{(m_{ob} + m_i)} v_{ob}^2. \tag{7.17}$$

At this point it is convenient to introduce the ballistic limit velocity v_{lim}, defined as that projectile velocity which provides just enough energy for complete penetration. Consider a case where the impact velocity is just v_{lim}. For this case the residual velocity v_r in Eq. 7.15 is zero and inserting W_c from Eq. 7.17, we can solve for W_s in terms of v_{lim}; i.e.,

$$W_s = \frac{1}{2} \frac{m_{ob}^2}{(m_{ob} + m_i)} v_{lim}^2. \tag{7.18}$$

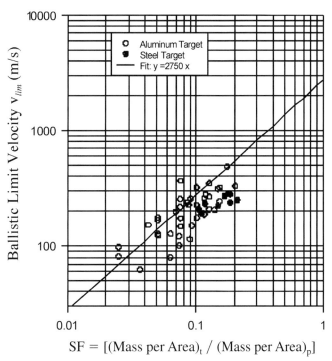

Figure 7.8 Ballistic limit velocity vs. scale factor [8].

The ballistic limit velocity is a function of the projectile and target materials and geometry as well as of the impact velocity. Unfortunately, no general theoretical relationship is available for calculating the ballistic limit velocity. The limits for particular materials and geometries need to be obtained from an appropriate testing program using Eqs. 7.15–7.18. The impact and exit velocities, the projectile description, the target material and thickness are used together with Eqs. 7.15–7.18 to calculate a value for v_{lim} in each case. Testing included both regular and armor piercing lead bullets and both aluminum and steel targets. The individual test results for v_{lim} can be normalized, to some extent, for target and projectile materials using the *scale factor* equal to the dimensionless ratio of areal densities of the target t to the projectile p; i.e., $[(\text{mass/area})_t/(\text{mass/area})_p]$. While not perfect, this ratio significantly reduces the scatter in the data. An example of this approach, calculated from the data of Awerbach and Bodner [7] and Marom and Bodner [8], is shown in Figure 7.8. The scatter in the data in Figure 7.8 provides a measure of the uncertainty for safety analysis purposes. The scale factor *SF* is given by

$$SF = \frac{\varsigma_t \delta_t}{m_p} A_p. \tag{7.19}$$

A best-fit line approximating the data is provided in Figure 7.8. The equation for the fitted line (where v_{lim} is in m/s) is

$$v_{lim} = 2750(SF). \tag{7.20}$$

The parameters ς_t, δ_t, m_p, and A_p are the density of the target, the thickness of the target, the mass of the projectile, and the cross-sectional area of the projectile, respectively. The ballistic limit velocities range from a few tens of meters per second to about a thousand meters per second and include aluminum and steel targets with plain lead and armor piercing lead bullets. We may go beyond these limits with care, but applicable test data are extremely important in the safety analysis of a specific system.

4.2 Hypersonic Velocity Impact

As for lower velocities, no general theoretical model is available for calculating the ballistic limit velocities for high-speed impacts in which the velocity of impact exceeds the sonic velocity in one of the materials (above several km/s). Nonetheless, a number of empirical relationships have been proposed to estimate the ballistic limit velocity. The following relationship, from Fraas [10] referenced to Micro-meteoroid Hazard Workshop [11], is an appropriate example:

$$\delta = Km_p^{0.352}\varsigma_p^{0.167}v_{rel}^{0.875}. \tag{7.21}$$

Here, δ = penetration thickness (cm),
               K = a constant depending on the physical properties of the target ,
               m_p = mass of projectile (g),
               ς_p = density of projectile (g/cm^3), and
               v_{rel} = relative velocity of impact (km/s).

For space debris the important constituents are steel and aluminum. The values of K for aluminum and steel are 0.57 and 0.32, respectively. The penetration thickness as a function of impact velocity, calculated using Eq. 7.21, is graphed in Figure 7.9 for a projectile having a density of 0.5 g/cm^3 (characteristic of meteoroid density) and masses of 0.001, 0.01, 0.1, and 1.0 grams. Table 7.4 provides the ratio of the threshold hypersonic penetration thicknesses for various targets and projectile materials to the penetration thickness for an aluminum target and a projectile density of 0.5 g/cm^3. Table 7.4 can be used with Figure 7.9 to estimate the penetration thickness for several other projectile materials and three different target materials. The uncertainty in calculated values of v_{lim} using Eq. 7.21 is not given but may be assumed to be about a factor of two, similar to the uncertainty in low speed impacts.

Equations 7.15–7.21 provide a generalized approach to analyzing the penetration by small fragments and shrapnel. However, these relationships are applicable only to impacts normal to the surface impacted; they are limited to the materials and ranges investigated and they include a significant range of uncertainty. For any system where event probabilities suggest that small fragment impact is important, these equations need to be augmented and verified with a well-planned test program.

5.0 Impact Consequence Evaluation

Estimating the physical damage that results from an impact is only the first step in satisfying the safety analysis needs. The next step is to estimate the physical consequences of that damage. The three end-point consequences identified in Section 1.0 are altered reactor geometry, breached containment, and release of radioactive fuel.

5.1 Reconfiguration and Criticality

Serious impact accidents, either the impact of large fragments or ground impact, will result in the reconfiguration of a reactor system with the potential for an accidental criticality. A breach of boundaries followed by water flooding can also result in an unplanned criticality event. Criticality is a complex function of reactor geometry and materials (see Chapter 8). The analyst must relate potential impacts, par-

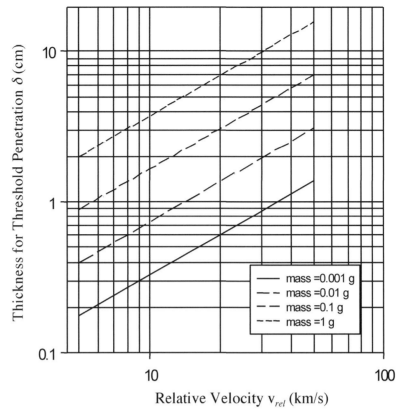

Figure 7.9 Meteoroid ($\varsigma = 0.5$ g/cm^2) penetration of 2024 aluminum vs. relative speed.

ticularly sequential impacts, to a description of reactor reconfiguration. In this regard we need to recognize that impacts, including intact ground impacts, are highly variable in terms of intensity (velocity and target material) and impact orientation (essentially random).

The typical approach is to devise relatively simple limiting cases for analysis (see Chapter 8 for several examples). Although this general approach is useful if limiting cases demonstrate that reconfiguration will not produce an inadvertent criticality accident, the approach is often overly conservative. A series of hydrocode analyses may allow the analyst to hypothesize a relationship between the disrup-

Table 7.4 Ratio of Threshold Hypersonic Penetration Thickness for Various Target and Projectile Materials to an Aluminum Target and a Projectile of Density 0.5 g/cm^3 [11]

	Penetration Thickness Ratio		
	Target Material		
Projectile Material	Aluminum (K = 0.57)	Steel (K = 0.32)	Niobium (K = 0.34)
Meteoroid (0.5 g/cm^3)	1.00	0.56	0.60
Plastic (1 g/cm^3)	1.12	0.63	0.67
Aluminum (2.77 g/cm^3)	1.33	0.75	0.79
Stainless Steel (8.0 g/cm^3)	1.59	0.89	0.95
Niobium (8.58 g/cm^3)	1.61	0.90	0.96
Tungsten (19.2 g/cm^3)	1.84	1.03	1.10

tive work done on the system calculated from Eq. 7.8 and some measure of reconfiguration, e.g., relative compaction (see Chapter 8.). The relative compaction might then be related to dimensional changes to establish more realistic boundaries for limiting calculations. A detailed Monte Carlo neutronic analysis of predicted reactor core geometries, for several representative impact scenarios, are also valuable to determine the degree of conservatism inherent in the limiting analysis.

5.2 Deformation Induced Cladding Failure

Deformation induced failure, with the possible release of radioisotope fuel, is an extremely important issue for a radioisotope system. Impact tests for the GPHS-RTG resulted in observed fuel clad failures at more-or-less random locations (not related to penetration or impact face). Although the probability of failure and the probable number of cracks appear to increase with impact energy, both failure and survival are observed at most impact energies. This same probabilistic failure outcome for fueled clads is also observed in impacts involving bare module and generator impacts. These observations demonstrate that the outcome of an impact at any given energy is best treated as a stochastic event.

Several simplifying assumptions are used to model this type of failure: (1) non-specific impact failures are assumed to result from the failure of one or more random flaws (or weak spots) distributed in the cladding; (2) given an insult (impact work) above some threshold value, each flaw has a probability pf_f for failing; and (3) the probability of failure is assumed to be proportional to the disruptive work partitioned to the individual fueled clad that exceeds the failure threshold. Thus,

$$pf_f = k(W_{33} - W_{Tfc}),$$
(7.22)

where k is a proportionality constant, W_{33} is the work apportioned to an individual clad for a bare clad impacting after initially being freed by a prior event, and W_{Tfc} is an empirical failure threshold for the fuel cladding (the impact work that must be exceeded for clad failure to occur).

For n flaws, the expected number of failures per clad is just $n \cdot pf_f$; and the number of failures per clad is Poisson distributed. For a Poisson distribution of failures, the probability of survival (the probability that no failures occur) is given by $\exp(-n \cdot pf_f)$. (See Chapter 11 for a discussion of probability distributions.) The failure probability (at least one failed flaw) for an individual clad pf_c as a function of work added is just,

$$pf_c = 1 - \exp[-\alpha_f W_{eff}].$$
(7.23)

The parameters, $\alpha_f \equiv nk$ and $W_{eff} \equiv (W_{33} - W_{Tfc})$ represents the proportionality of the expected number of failed clads and the internal work partitioned to the clad, respectively. Figure 7.10 illustrates the failure probability as a function of energy for three different values of α_f and with $W_{Tfc} = 0$.

Equations 7.9 through 7.14 together with Eqs. 7.22 and 7.23 provide a complete failure model for the fueled clads of the GPHS-RTG. Although this model has been devised for the GPHS-RTG, a similar approach to cladding failure could be applied to any system. In this regard, the first step would be to construct a system-specific impact model and from that a failure model. The combined impact-failure model would provide the analytical foundation on which to develop a test plan directed at validating the model and obtaining the data necessary for the safety analysis.

5.3 Fuel Release

The next step for calculating the consequences of impacts is estimating the quantity and character of released radioactive materials. Accidental releases are generally described by the composition, mass

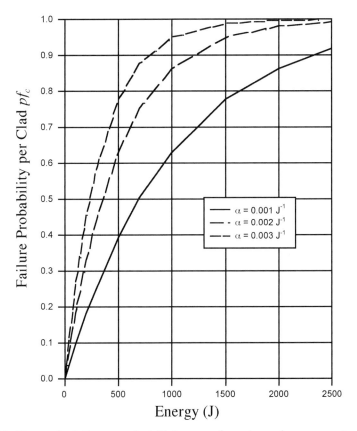

Figure 7.10 Example failure probabilities as a function of energy using Eq. 7.23.

quantity, and particle size distribution of radioactive materials. This information is commonly referred to as the source term. For radioisotope fuels and fresh reactor fuel, the composition is fixed. For postulated reentry of a reactor following operation, the composition of the radioactive materials can be predicted using established *burnup* computer codes. A generally acceptable theory or methodology for making mass and particle size estimates, however, has not been established for impact accidents; and we are limited to empirical data. For the purpose of illustrating the methodology, the GPHS-RTG will be used as an example.

Fuel release data for the GPHS-RTG appear to be comprised of two individual distributions (when considered on a fraction per fuel capsule basis) that have no clear relationship to any of the key parameters of insult (e.g., impact speed, calculated distortion, crack size, number of cracks). The smaller fractional releases (median 2.6×10^{-4} of the fuel) constitute about 80% of all releases, and the larger fractional releases (median 0.13) constitute the other 20% of all releases. The data have not been resolvable into a deterministic relationship relating source terms to impact insult. In fact, it is not clear that a deterministic solution is possible.

The particle size distribution of the released material is needed for environmental transport calculations and for calculations predicting the effects on humans. The impact model described here relates to the work added to the system by impact. A linear approximation is used to estimate the proportion of small material following impact fracture in various nuclear applications. The approximation arises from the observation of a general linear relationship between the fraction of material in the less than 10 μm-diameter range and the impact work added (in J/cm³) for a range of brittle materials [12, 13]. With this approach, the fraction of material in the less than 10 μm size range F_{10} can be approximated from the work apportioned to the fuel. Thus,

$$F_{10} = a_p \frac{\varsigma_f}{m_f} (W_{33} - W_{Tfc}) = a_p(w_\rho), \tag{7.24}$$

where a_p is the fitted proportionality constant for a particle size in the < 10 μm range, and the work terms are in Joules. The parameters m_f, ς_f, and w_ρ are the mass of the fuel in grams, the density of the fuel in g/cm³, and the work density added in J/cm³, respectively. Considering all of the release data for which particle sizing is available for the GPHS-RTG, a median value of a_p was found to be 1.25×10^{-4} cm³/J.

In addition, it has been observed that many of the particle size distributions can be approximated by a first order Weibull distribution; i.e., the particle size distributions can be approximated using

$$CMF = 1 - \exp\left(-\frac{D}{B}\right). \tag{7.25}$$

Here, CMF is the cumulative mass fraction; i.e., the fraction of particles with diameter greater than D, where D is the particle size in μm and B is the Weibull scale (or location) parameter in μm. With this construction of the problem, we see that B can be estimated using Eqs. 7.24 and 7.25 as

$$B = \frac{-10}{\ln(1 - a_p w_p)}. \tag{7.26}$$

Thus, the calculation of impact work provides the information necessary to estimate particle size distributions as well as cladding failure. As a first approximation, one can assume that the released particle size distribution is given by Eq. 7.25.

5.4 Consequence of Large Debris Impact in Orbit

Consider the collision of an operating space nuclear system in near Earth orbit with a large piece of space debris. The probability of impact and the physical consequences of the impact event must be determined. First, consider the probability of impact.

In general, the probability of impact can only be determined through knowledge of the debris inventory and the orbital characteristics of those items. Unfortunately, that information is not available. In its absence, however, one can make an order of magnitude approximation by assuming that the impact probability is a product of the effective volume swept out by an orbiting system and the number density of large debris items within that volume. The effective volume swept out by an orbiting system can be calculated as $V_{eff} = A_{eff} v_{ob} \Delta t$, where V_{eff} is the effective volume of interest, A_{eff} is the effective cross-sectional area of the orbiting system, v_{ob} the orbital speed of the system, and Δt the time period considered for calculation. For a 5-year (1.58×10^8 s) lifetime of a 10-m² system in a circular orbit at 1000 km (orbital velocity ~ 7.35 km/s) this effective volume is 1.16×10^{13} m³.

The number density of large items of space debris is a function of spatial position. However, for this simple example, the 100 or so large objects distributed between 700 and 1300 km can be taken to be randomly distributed in regard to altitude and angular position within that band. With this simplifying assumption the effective number density of large debris ς_{eff} is given by N_{tot}/V_{tot}, where N_{tot} is the total number of large debris items in V_{tot}, the total volume of space being considered. For the volume between an altitude of 700 and 1300 km, $\varsigma_{eff} = 100/(4.3 \times 10^{20}) = 2.44 \times 10^{-19}$ items/m³. Thus, the collision probability for our example can be approximated as $p_{col} = V_{eff} \varsigma_{eff} = (1.16 \times 10^{13})(2.44 \times 10^{-19}) = 2.83 \times 10^{-6}$. The calculated probability appears to be only a marginal risk, but this calculation is only a rough estimate. Furthermore, larger systems, longer mission lifetimes, increasing debris population, and other factors will increase the calculated risk.

One important consequence of an impact accident is the disruption of the space nuclear system itself. The other important consequence is the possibility of altering orbital characteristics which could lead to premature reentry of the nuclear system or radioactive debris. Both of these postulated consequences may be addressed with the impact model presented in Sections 3.1 and 3.2 of this chapter. Depending on the elasticity of the constituents, kinetic energy will be lost to disruptive work. The altered orbital velocity of the center of mass of the two impacting items after impact is given by $\mathbf{v}_2$ in Eq. 7.5 (assuming complete inelasticity). The energy lost in impact will necessarily reduce the constituent velocities of the components and as such will result in an early reentry (unless corrective actions are taken to prevent reentry). The specific consequence of this change in velocity in terms of orbital lifetime can only be determined by a detailed analysis of the specific systems.

6.0 Impact Testing

An accident response model is a mathematical description of how the nuclear system is projected to respond to potential threats. Careful planning is needed to assure that appropriate test data is obtained to validate the model.

6.1 General Approach to Test Planning

The fundamental goal of the test plan is to verify (or revise) an accident response model. The development of the test plan includes delineation of potential accident environments and use of this information to devise a credible accident response model. The early test plans are not inviolate; the process must be dynamic. Each new set of data, whether providing new insights about accident environments or providing results from initial testing, can lead to a reevaluation of the model and the plan itself. Thus, a formal written test plan, including the accident event basis and the accident response model, is necessary at each step to document the evolution of the plan and the changing technical rationale for testing.

The first step, when developing a test plan, is to review the mission and hardware and document the initial knowledge of events, environments, and responses. The following information should be tabulated and summarized from previous experience with similar vehicles and systems.

- **Event schedule.** A list of those potential accident events that appear to be significant for each mission phase (e.g., main engine failures, solid rocket booster failures, explosions in upper stages).
- **Potential blast environments.** Approximations of the static and dynamic overpressure and impulse at the location of the nuclear system, together with the occurrence probability distributions of those conditions.
- **Potential fragment environments.** Initial distributions characterizing the important fragment parameters (e.g., number, material, mass, velocity, ballistic coefficient, intervening structures).
- **Potential thermal environments.** Potential thermal environments including time-dependent temperature profiles that may be experienced by the system as a result of the accidents considered.
- **Reentry response.** An early estimate of the reentry response of the system and its components. This information is needed to assure a complete evaluation of impact testing requirements.
- **Impact media.** A summary of potential impact media by mission phase with first-cut approximations of occurrence probabilities.

- **Existing test data.** If the system has been flown before, a sizable body of information will already be available. If so it should be identified and evaluated in light of the new mission. For new systems, a body of system engineering test data should exist and experience and data from other systems should be reviewed for analysis insight. These data will provide the basis for an initial accident response model and the associated testing needs.

The second step is to document the progression of important postulated accidents from the viewpoint of the nuclear system. For each such *insult progression* an accident event timeline must be developed specifying the general location and possible configurations of the nuclear system together with the associated accident environments. This effort should be directed at identifying the impact environments and system states that must be investigated as part of the test plan.

The third step is to prepare an initial engineering damage model for the physical response of the system (and its components) to the impact environments. The last step is to design a test program that provides the information necessary for verifying and applying the damage model.

A critical part of designing the test program is to identify the most appropriate (primary) test and outcome variables. The engineering damage model is the connection between the primary test variable and the primary outcome variable. In general, for impact testing, the primary test variable will be impact velocity with consideration of mass, stiffness, orientation, temperature, etc., as modifying parameters. The most relevant outcome variable is not as easy to determine and may differ for different systems. For radioisotope-fueled systems, the most appropriate outcome variable is barrier failure; but release quantity could also be considered as the primary outcome variable. In either case, the impact damage model must describe a clear and predictable (mathematical) relationship between the primary test variable and the primary outcome variable. Without this clear and predictable relationship, the impact damage model cannot be tested.

For reactor systems, failure and release may not be the most significant issue. Typically, an unplanned criticality is the principal issue for launch accidents involving a cold reactor. In general, inadvertent criticality accidents can involve an essentially limitless number of possible post-impact configurations and environments. Planning to evaluate impact events without showing a clear and calculable relationship to criticality can result in a wasted effort. A more efficient approach may be to assure that the design of the system is such that inadvertent criticality cannot occur for any credible worst-case accident (e.g., full compaction, optimum water flooding). If this approach is used, impact testing may not be needed.

Once the test and outcome variables have been selected, a testing program can be designed. Simulated systems and surrogate materials are an important option; their efficacy should be evaluated by considering the uncertainty introduced by their use in comparison to the overall cost and safety of using actual systems and materials. More tests with well chosen simulated systems (or surrogate materials) may result in less uncertainty and lower cost than fewer tests with actual systems.

Documenting uncertainty is a key component of safety analyses. It is important that the test plan includes a careful discussion of the data and analyses necessary to document uncertainty. The uncertainty in the calculated outcomes of impact events can arise from three sources: (1) modeling uncertainty (reflecting our inability to completely describe physical processes); (2) model parameter uncertainty; and (3) measurement uncertainty. Each of these sources needs to be considered as a part of the test plan, and the plan should clearly describe how uncertainty is to be treated in the safety analysis.

6.2 U.S. Impact Tests

For the purpose of constructing an impact model, the most useful data are those obtained from the simulated generator tests; i.e., the radioisotope thermoelectric generator (RTG) tests and the large fragment tests (LFT). The module tests, including the cold-process verification (CPV) tests and the Safety verification tests (SVT), are also useful. The results of these tests provide a sufficient basis for illustrating the energy interaction model [14].

Radioisotope Thermoelectric Generator (RTG) Tests

The RTG tests, RTG-1 and RTG-2, were used to provide information on the response of a Cassini RTG to end-on impact against a concrete target. For the RTG-1 test an impact velocity of 57 m/s was used, and for the RTG-2 test the impact velocity was 77 m/s. These two tests incorporated a half-stack RTG housing loaded with urania fueled GPHS test modules (five in RTG-1 and six in RTG-2), molybdenum weighted dummy modules (three in RTG-1 and two in RTG-2), and one POCO (polycrystalline) graphite module. The urania loaded modules were at the impact end of the half-stack RTG, and the POCO module was at the trailing end. The test stack was heated for each test (RTG-1 to 1071 °C and RTG-2 to 1090 °C).

Test results include data on the distortion, failure, and release of fuel from the individual test clads. No clad failures were experienced in RTG-1 at 57 m/s. Three clad failures were experienced in RTG-2 at 77 m/s. Of additional interest is the fact that the RTG-2 test included four clads that were apparently used in the previous RTG-1 test. These four clads were not only impacted previously, but withstood additional heating cycles.

Large SRB Fragment Tests (LFT)

The purpose of these tests was to acquire information on the response of an RTG and its components to a simulated collision of an SRB fragment. In these tests an SRB plate fragment (142 cm square and 1.27 cm thick with a mass of approximately 193 kg) was propelled into a simulated half-section (one-half axial length) of an RTG by a rocket sled. The simulated RTG section contained two urania fueled GPHS test modules and six dummy modules loaded with molybdenum slugs. The test modules were heated to 1090 °C for each test. Three tests were run: LFT-1, a 114.9 m/s face-on fragment impact; LFT-2, a 212 m/s face-on fragment impact; and LFT-3, a 95.4 m/s vertical edge-on fragment impact. Clad failure and releases were reported for the 212 m/s face-on fragment (two breached clads) and the 95.4 m/s edge-on tests (two breached clads).

Safety and Cold-Process Verification Tests (SVT and CPV)

The purpose of the safety verification test series was to provide information on the response of the GPHS modules to Earth impact following reentry. The tests were carried out in the 178 mm (diameter) pneumatic gas gun at Los Alamos National Laboratory. In the individual tests, modules were impacted at various impact angles, temperatures, and velocities on hard targets (12 on steel; 1 on concrete). All of these tests used plutonia fuel. The test included 14 modules containing 56 clads; all but one of the tests were carried out at impact velocity of about 54 m/s. A total of 13 breached clads were observed.

The cold-process verification (CPV) test series, run to determine the significance of "hot" or "cold" pressing of the urania (UO_2) pellet, included one bare module test. The test (CPV-12), impact at 54.4 m/s on steel, was similar to the side-on SVT impacts. No clad failures were observed in this test.

Example 7.2

Using RTG-1 and RTG-2 test results (discussed in this section) together with the module impact tests, obtain the damage thresholds, damage coefficients, and failure functions for the damage model presented in this chapter. Check the model by making clad failure predictions for comparison with the results of the large fragment tests.

Use the following information:

- For bare module tests CPV and SVT, using the known parameters and curve fitting to the test data yielded $\alpha_f f w_{32} = 0.0005$ J^{-1} and both W_{T22} and $W_{Tfc} \approx 0$. (The zero thresholds are an artifact of the simple model and curve fitting.)
- The outcome of the RTG-1 test was that the RTG casing was fractured and several modules fell free. One fuel clad was free of its module, but no clad failures were observed.

Solution:

Fit to Bare Module Tests

The failure function can be obtained using results from the CPV and SVT bare module tests with Eq. 7.23. For bare modules impacting on a hard surface fw_2 is 1.0. The parameter W_{eff} in Eq. 7.23 is the effective work added to each clad which for the case of bare module impact is ($W_{32} - W_{Tfc}$). Because these tests were for bare modules W_{21} and $W_{31} = 0$. Using Eqs. 7.12 and 7.13 in Eq. 7.23 gives

$$pf_c = 1 - \exp\left\{-a_f\left[\left(\tfrac{1}{2}m_{\text{mod}}v_{\text{mod}}^2 - W_{T22}\right) fw_{32}\, fc - W_{Tfc}\right]\right\}.$$

Using the curve fit data, Eq. 7.23 for an RTG impact becomes

$$pf_c = 1 - \exp\left[-0.0005\left(\tfrac{1}{2}m_{\text{mod}}v_{\text{mod}}^2\right) fc\right], \text{ or from Eqs. 7.12 and 7.13,}$$

$$pf_c = 1 - \exp\left(-0.0005\,\frac{W_{32}}{fw_{32}}\right).$$

Fit to RTG-1 and RTG-2 Tests

The impact velocity RTG-1 test was 57.6 m/s and fw_1 is taken as 1.0 for intact impact on hard surfaces [14]. A half-generator (nine modules with a generator mass of 28 kg) was used for the impact tests. Using Eq. 7.9 gives

$$W_{11} = \frac{1}{2}(28)(57.6)^2(1) = 46{,}448 \text{ J}.$$

From the RTG-1 and 2 test results (second bullet in problem statement), the failure threshold W_{T11} is judged to be moderately exceeded; thus, a value of 45,000 J is assigned to W_{T11}. Equation 7.10 can be used to calculate W_{21} (the work done on the module stack) in the RTG-2 impact at 77.1 m/s. Equation 7.10 gives the work done on the module stack as $W_{21} = (W_{11} - W_{T11})fw_{21}fm$. Assuming $fm = 1/9$ (even distribution among 9 modules), the work added to each module for RTG-2 is

$$W_{21} = [(1/2)(28)(77.1)^2(1) - 45{,}000]\, fw_{21}(1/9) = (4247)\, fw_{21}.$$

Equation 7.11 is applied in order to relate W_{21} to W_{31}; i.e.,

$$W_{31} = (W_{21} - W_{T21})fw_{31}fc.$$

In the preceding it was shown that W_{T22} and $W_{Tfc} \approx 0$. We assume that the same work threshold holds true for RTG impacts, therefore, $W_{T21} \approx 0$ and $W_{31} = (1/4)(W_{21})fw_{31}$. The work partitioned from the modules to the clads is assumed to be the same whether the work is added to the modules from an impact of the RTG or an impact of the free module. Thus, the expression for pf_c for the bare module tests can be used for the RTG tests with W_{31}/fw_{31} substituted in place of W_{32}/fw_{32},

$$W_{31} = (1/4)\,W_{21}fw_{31}, \qquad \text{or} \qquad \frac{W_{31}}{fw_{31}} = \frac{1}{4}\,W_{21} = (1061)fw_{21},$$

and
$$pf_c = 1 - \exp[-(0.53)fw_{21}].$$

With this result, and using the observed failures in RTG-2 (3 out of 24, $pf_c = 3/24$) the foregoing equation can be solved to obtain $fw_{21} = 0.252$.

Model Applied to LFT Tests

The parameter values now obtained include:

$$W_{T11} = 45{,}000 \text{ J}, \qquad W_{T21} = W_{T22} = W_{Tfc} \approx 0, \qquad fc = 1/4, \qquad fm = 1/9,$$

$$fw_{21} = 0.252, \qquad fw_{31} = fw_{32}, \qquad \text{and} \qquad \alpha_f\, fw_{32} = 0.0005 \text{ J}^{-1}.$$

In addition, $m_i = 28$ kg, $m_{ob} = 193$ kg, and $fw_1 = 0.33$ (LFT-1 and LFT-2) and $fw_1 = 1.0$ for LFT-3. Given the above parameter values, the probable clad failures in the three tests can be estimated. Using Eqs. 7.9 and 7.10 in Eq. 7.23 gives

$$pf_c = 1 - \exp\!\left(-a_f\!\left\{\left[\left(\tfrac{1}{2}m_{red}v_{rel}^2 fw_{11} - W_{T11}\right)fw_{21}fm - W_{T21}\right]fw_{31}fc - W_{Tfc}\right\}\right),$$

and by substitution,

$$pf_c = 1 - \exp\!\left\{-3.5 \times 10^{-6}\!\left[24.5\,(v_{rel})^2\,fw_1 - 45{,}000\right]\right\} \qquad .$$

LFT-1: For a face-on impact of large SRB fragments, the Cassini data book gives a value of $fw_1 = 0.3$. The above relationship with $v_{rel} = 114.9$ m/s and $fw_1 = 0.3$ yields a clad failure probability $pf_c = 0.012$. For 8 clads at risk, this relates to an expectation of 0.1 failed clads. The standard deviation of the Poisson distribution of failed clads is 0.3. No failures were observed.

LFT-2: The above relationship with $v_{rel} = 212$ m/s and $fw_1 = 0.3$ yields a clad failure probability $pf_c = 0.343$. For 8 clads at risk, this relates to an expectation of 2.71 failed clads. The standard deviation of the Poisson distribution of failed clads is 1.6. Two failed clads were observed.

LFT-3: The above relationship with $v_{rel} = 95.4$ m/s and $fw_1 = 1.0$, yields a clad failure probability $pf_c = 0.207$. For 8 clads at risk, this relates to an expectation of 1.66 failed clads. The standard deviation of the Poisson distribution of failed clads is 1.29. Two failed clads were observed.

These results are quite encouraging and provide not only expected values for failure but also a measure of uncertainty through the Poisson distributed results.

6.3 The U.S. GPHS-RTG Test Program in Retrospect

Having completed a number of test programs, one can trace the evolution and development of one U.S. program (the GPHS-RTG) and critique its technical aspects. The design of the GPHS-RTG was fixed by mid-1981, and safety testing was initiated in the 1982–83 time frame. Safety tests proceeded in support of the Galileo and Ulysses missions (launched October 18, 1989, and October 6, 1990, respectively) and continued through the Cassini review (launched October 15, 1997). The Cassini FSAR [14] and the Cassini INSRP-PSSP report [1] identify three test phases.

1. Pre-1986 Tests
   - Shock Tube Tests (BMT and CST)
   - Fragment/Projectile Tests
   - Solid Propellant Tests
   - Bare Clad Impact Tests (BCI)
   - Module Impact Tests (DIT and SVT)
2. Post-1986 Tests
   - Bare Clad Impact Tests (BCI)
   - SRB Fragment Tests in Gas Gun (FGT)
   - Large Fragment Tests (LFT)
   - Fragment/Fuselage Tests (FFT)
3. Cassini Tests
   - End-on Converter Tests (RTG)
   - Aluminum Fragment Tests
   - Cold-Process Verification Tests (CPV)

Although the Galileo and Ulysses missions were launched aboard space shuttles in 1989 and 1990, the safety test program started prior to the *Challenger* accident in 1986. In essence, the *Challenger* accident was the catalyst for changes that differentiate the earlier studies from those that followed. Prior to the *Challenger* accident, the accepted management position was that launch accidents involving the shuttle were incredible (although a number of engineers opposed this view), and the accepted accident probability was expressed as being in the range of 10^{-6} to 10^{-5} per launch. In addition it was generally accepted, at that time, that if an accident did occur, it would be catastrophic with essentially the entire liquid propellant inventory (1.6 million pounds) being involved. Pre-1986 tests reflect this thinking; they were focused primarily on relatively large explosive shocks and reentry survival.

In the early shock tests, bare GPHS modules were exposed to a series of explosive shocks with static overpressures ranging from about 1.4 MPa to 7 MPa (BMT series). Subsequent to these bare module tests, a full scale simulated RTG was tested in conjunction with the DIRECT COURSE event [15]. DIRECT COURSE involved the detonation of an 11 m diameter sphere of ammonium nitrate and fuel oil (ANFO). A test RTG was located 11.6 m from the surface of the sphere. Explosive static overpressures in the range of 10 MPa to 14 MPa (1500 psi to 2000 psi) were projected. At the time it was anticipated that, even if the RTG itself did not survive, survival and recovery of damaged but intact modules and fueled clads would reduce the need for further testing of high static overpressures. Unfortunately, the test item and its components were completely destroyed, with the largest parts recovered being a few centimeters on a side. The quick simple solution did not work.

Subsequent analysis concluded that static overpressure was not the appropriate measure of potential damage for the near field insults from large explosions. The principal causes of damage in DIRECT COURSE were fragments of unburned explosive, explosive by-products, and the fiberglass debris from the casing of the ANFO charge. The dynamic overpressure ($\varsigma v^2/2$) of the explosion debris impinging on the test article, rather than static overpressure, was concluded to be the significant parameter. Although it was concluded that DIRECT COURSE conditions were inapplicable to launch abort accidents, the test was quite valuable in focusing attention on the important role that intervening structure and dynamic pressure played in damage. This led to a significant change in subsequent shock tests (CST series). In these tests a simulated converter segment was placed between the shock source and the test article to simulate an RGT configuration and thus more closely represent reality.

The *Challenger* accident on January 28, 1986, brought about another change in the view of launch pad accidents. Solid rocket motor (SRB) destruction events, which had been assigned accident probabilities in the range of 10^{-6} to 10^{-5} per launch prior to *Challenger,* assumed much greater significance, with accident probabilities now taken to be in the range of 0.001 to 0.01 per launch. Furthermore, it shifted the attention from liquid fuel explosions, which were the focus of earlier tests, to impacts from large fragments, particularly SRB casing fragments. Large fragments remained the major focus through Cassini. In addition, the Cassini analysis raised the important issue of final ground impact; i.e., the RTG impacting on hard launch pad surfaces with or without debris impacting on top of the RTG.

Clearly the GPHS-RTG test program was flexible and dynamic, changing as new information became available. However, the focus on initial events obscured both the need for an impact damage model and the importance of sequential impacts. Finally, but less obvious, the lack of a clearly defined damage model may have contributed to the inattention to uncertainty considerations in the test program. Thus, three issues are raised by a brief review of the GPHS-RTG test program: the absence of a specific damage model, the impact of undamaged components, and uncertainty considerations.

The Absence of a Specific Damage Model

Rather than a mathematical damage model, the test program was directed at measuring distortion [14]. The use of distortion in the absence of a predictable connection between the primary test and outcome variables led to practical as well as theoretical difficulties. Single valued two-dimensional distortion is reasonable only for single impacts normal to the cylindrical axis of the clad where more-or-less uniform deformation (along the cylindrical axis) is experienced. However, such uniform distortions are not generally expected in an accident. For example, Figure 7.11 illustrates the type of damage observed in an end-on converter impact tests [14]. Simple distortion is not meaningful for such damage. In the absence of a method to relate distortion to more realistic physical changes and those physical changes to failure, distortion is useless as a gauge of impact damage. In any event, distortion is a relatively uncertain and inconsistent index of damage; and its use was confounded by the lack of any analysis that described its relationship to fundamental physical or material properties.

The lack of an applicable mathematical model relating velocity to damage and to failure made much of the impact test data unusable. Consider the early BMT series consisting of bare modules lightly clamped between two graphite blocks and exposed to various levels of explosion-generated shock. Three points need be made about the sequence of test events: first, since the modules are contained within the RTG, they are shielded from direct shock in an initial event; second, if the RTG casing is removed by a severe initial explosion, we need to postulate a second serial explosion for bare modules to be exposed to a shock; and third, if such a second explosion can be postulated, the modules and the fueled clads within them are not pristine (they have already been damaged). The CST tests were similar to the BMT. In the absence of a detailed understanding of the physical parameters of the test structure and a model relating to the RTG structure, the test results cannot be related to postulated RTG accident conditions.

Impact of Undamaged Components

A review of accident progressions shows that released modules or fueled clads must lead to a secondary impact of the damaged components. Figure 7.12 shows the remains of the RTG case after one of the large fragment tests (LFT-2). In this test, 2 of the 8 simulated-fuel clads in this test were breached. In a 77 m/s end-on converter test (RTG-2), 3 out of 24 simulated-fuel clads were breached. In both of these tests the RTG casing was breached, and the components were freed for subsequent impact.

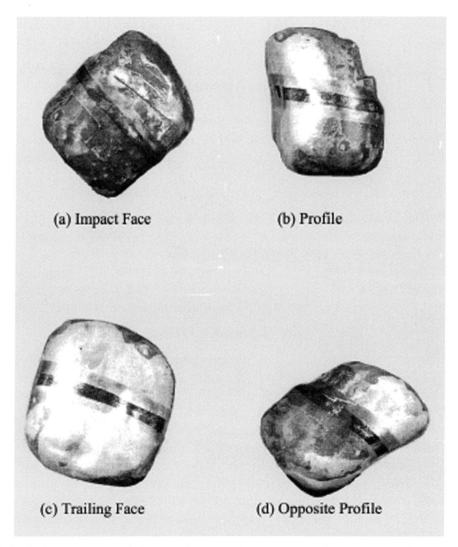

Figure 7.11 Fueled Clad SC0096 (Module #2) after 77 m/s impact test. *Courtesy of U.S. Department of Energy.*

Figure 7.12 RTG case after LFT-2 test. *Courtesy of U.S. Department of Energy.*

ORBIT SERIES • AVAILABLE TITLES

ORBIT
A FOUNDATION SERIES

SPACE
NUCLEAR
POWER

Joseph A. Angelo, Jr., Ph.D.
and David Buden

ORBIT
A FOUNDATION SERIES

CHEMICAL
PRINCIPLES APPLIED
TO SPACECRAFT
OPERATIONS

Rex E. Dalton
Daniel S. McKnight

NEW

ORBIT
A FOUNDATION SERIES

Space
Nuclear
Safety

Albert C. Marshall

ORBIT
A FOUNDATION SERIES

FUNDAMENTALS
OF SPACE LIFE
SCIENCES
Volumes 1 & 2

Susanne E. Churchill

ORBIT
A FOUNDATION SERIES

ARTIFICIAL
SPACE
DEBRIS
Updated Edition

Nicholas L. Johnson
and Darren S. McKnight

ORBIT
A FOUNDATION SERIES

Changing Patterns
of International
Cooperation in Space

Joan Johnson-Freese

ORBIT
A FOUNDATION SERIES

SPACE
STATIONS AND
PLATFORMS

Gordon R. Woodcock

ORBIT
A FOUNDATION SERIES

ORBITAL
MECHANICS

Richard G. Madonna

ORBIT
A FOUNDATION SERIES

Updated!

INTRODUCTION
TO THE SPACE
ENVIRONMENT

Thomas F. Tascione

ORBIT
A FOUNDATION SERIES

THE
CHINESE
SPACE
PROGRAM
A Mystery Within a Maze

Joan Johnson-Freese

ORBIT
A FOUNDATION SERIES

ADVANCED
AIRBREATHING
PROPULSION

T. H. Tinoco

Although free modules or fueled clads must have experienced some previous damage, all of the test impacts using modules (DIT, SVT, and CPV) and all but four of the tests using bare clad (BCI and CPV) were performed on pristine test items. It would be of immense value to have impact tested (at terminal velocity) the breached clads and, at least, an equal number of unbreached clads recovered following the LFT and RTG series.

Uncertainty and the Test Plan

In contrast to the previous FSARs, the Cassini FSAR [14] made an effort to document uncertainty. Because the test program and the associated tests did not arise from a mathematical impact response model, the first two areas of uncertainty (modeling and parameter uncertainties) are beyond determination. In addition, measurement errors were not reported (most important would be the measurement variation in the dimensions used for distortion calculations). As a result, uncertainty cannot be directly calculated. INSRP [1] made an estimate of the measurement error that might be inherent in reported distortion for bare clad impacts.

7.0 Mitigation Methods

The design of a system generally starts with a basic concept that evolves throughout development to meet mission requirements. Mitigation methods for postulated accidents are accomplished during design, analysis, and testing. Tests that fail to demonstrate adequate nuclear system integrity or impact response to a postulated impact accident must be redesigned. In some situations, a different basic approach may be required. The redesign may include alternative materials, changes in the thickness of encapsulating materials, altered geometry, or other features. Redesigns and new designs may be required to respond to changing policies. For example, contemporary radioisotope sources are designed to survive reentry and impact rather than burn up in the atmosphere. The design for intact reentry and impact reflected a revision in the basic approach to radioisotope source reentry safety.

8.0 Computer Methods

Computer models have not been developed for the specific purpose of performing the impact analysis discussed in this chapter. However, Lockheed Martin [14] has developed a Launch Accident Scenario Evaluation Program—Titan IV Version (LASEP-T) that includes an evaluation of impact damage. The impact model assumes a number of linear functions relating impact velocity to distortion (for various damage pathways), and two failure functions (one for bare clads and one for bare modules) similar to Eq. 7.23, except that the exponent is a linear function of distortion. LASEP-T simulates the probabilistic nature of launch accidents using the Monte Carlo method. Probability distributions are constructed for the variable elements of the accident by drawing a random number distributed according to the probability distribution of that specific variable. The specific scenario calculations are repeated many times (on the order of 100,000 times) so that the output provides an approximation of the distribution of calculated launch accident consequence. This code and its predecessors have provided the heart of the safety analyses for the Galileo, Ulysses, and Cassini space missions.

In addition to this specific analysis code, a number of hydrodynamic codes are available for evaluating the physical response of impacts. One example, the PISCES-2D-ELK (and predecessors) has been used extensively to analyze GPHS impact events. PISCES is a two-dimensional Lagrange-Euler hydrocode used extensively for GPHS-RTG analyses [15, 16, 17]. The code was developed by Physics International [18] and was specifically modified for calculations made as part of the Cassini Safety Analysis Report. [14]. An example output is illustrated in Figure 7.13. Two difficulties arise in relating the reported output from this code to a comprehensive safety analyses. First, the results are not related to breaching the cladding; therefore, they offer little help in exposing the relationship of impact to failure

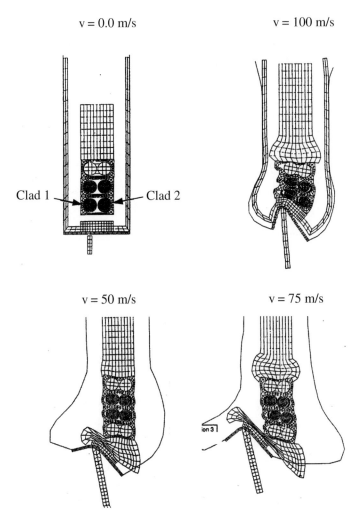

v = 0.0 m/s

Clad 1 ———→ ←——— Clad 2

v = 100 m/s

v = 50 m/s

v = 75 m/s

Figure 7.13 Output of PISCES-2D-ELK. Effect of fragment velocity on RTG response of a 142×142 cm fragment along the fueled clad axis [14]. *Courtesy of Lockheed Martin Co.*

(the outcome of major importance). Second, the results reflect a focus on the "distortion parameter" discussed earlier. Because of this focus on distortion, the PISCES investigations have all been of impacts normal to the central axis of the cylindrical fueled clads. These results add little to an understanding of the impact-failure-release chain of events for the GPHS-RTG.

9.0 Summary

The analysis of impact events is a major element of space nuclear safety analyses. Every space nuclear mission must include a comprehensive program directed at identifying potential impact events, cataloguing the impact descriptors, and describing a general analysis structure for assessing impact consequences. The potential for impact accidents exists throughout the entire space mission. Three broad types of impact events need to be evaluated: (1) the impact of large vehicle fragments on the system, (2) the intact impact of the system (or subsystems) on the Earth's surface, and (3) the impact of small penetrating fragments (shrapnel, space debris, or meteoroids) on the system. A launch phase accident may result in all of these types of impacts, either nearly simultaneously or sequentially. Any serious accident subsequent to liftoff will conclude by energetic Earth impact of the nuclear system.

A relatively simple energy interaction model was presented to estimate the consequences of sequential impacts on a systems of nested safety protection. The analysis model, including clad failure, is summarized by the equations in Table 7.3, together with Eq. 7.23. The model relates damage to impact velocity and mass using damage coefficients. Penetration by small fragments were categorized as either subsonic or hypersonic impacts. Simple formulations were provided for both categories.

The usual practice for reactors is to launch the reactor shut down and radiologically cold. The principal impact consideration for postulated impact accidents for reactor systems is the possibility of an inadvertent criticality. For radioisotope sources, the principal concern is the release of the radioisotope fuel. The probability of cladding failure is expressed by a Poisson distribution in terms of a proportionality coefficient and the work partitioned to the clad. The cumulative mass fraction for released fuel is expressed as a function of particle diameter and a Weibull scale parameter.

Three requirements for designing and carrying out an effective safety analysis program include: (1) documentation of the progression of postulated major accidents; (2) a clear mathematical damage model describing how the system (and its components) are expected to respond to the impact environments identified in the accident event-timelines; and (3) a test program that will obtain the information necessary for verifying and applying the damage model as a part of the actual safety analysis. It is also necessary to include the explicit consideration of uncertainty as an analysis component at each step in the safety analysis process. In the event that safety tests or safety analysis indicate that the design of the nuclear system does not meet safety standards, a redesign may be needed.

Symbols

A_{eff}	orbiting system effective area	m_{clad}	mass of clad
A_i	projected area of item of interest	n	number of flaws in clad
A_p	cross-sectional area of a projectile	pf_f	flaw failure probability
a_f	coefficient of work to clad failure	pf_c	clad failure probability
a_p	proportionality coefficient of work done on fuel to <10 m diameter size range	p_{col}	collision probability
		$\mathbf{v}$	velocity vector
B	scale parameter for Weibull distribution	v	speed of item
C_d	dimensionless drag coefficient	v_i	speed of impacting item
CMF	cumulative mass fraction as a function	v_{imp}	speed at impact
D	particle diameter	v_{lim}	projectile speed for target penetration
F_{10}	fraction of particles in range <10 m	v_{ob}	speed of impacting object
fc	damage coefficient for individual clads	v_r	residual speed
fw_k	damage coefficient for an item in state k impacting an object	v_{rel}	closure speed between two objects
		v_t	terminal velocity
fw_{jk}	damage coefficient for j^{th} protective level from impact in the kth impact state	SF	scale factor for ballistic penetration
		w_ρ	work density in J/cm^3
fm	damage coefficient for modules	W	work done on impacting articles
g	acceleration due to gravity	W_c	inelastic compression work for projectile penetration
k	failure probability proportionality constant		
K	constant for hypersonic impact	W_{eff}	effective work partitioned to clad
m_{clad}	mass of a impacting bare clad	W_i	work done on the impacted item
m_i	mass of item of interest	W_s	impact work used in shear for penetration
m_f	mass of fuel	W_{jk}	impact work done on j^{th} protective level from impact in the kth state
m_{ob}	mass of object impacting item		
m_p	mass of a projectile	W_{Tfc}	empirical failure threshold for cladding
m_{red}	reduced mass for impact purposes	W_{Tjk}	work threshold for the jth protective barrier from impact in k^{th} impact state
m_{mod}	module mass		

α_f	clad failure proportionality constant	ς_i	density of the item falling free-fall
Δt	time duration	ς_f	density of fuel
δ	penetration thickness	ς_p	density of the projectile for impact
δ_t	target thickness	ς_t	density of the target material
ς_{air}	density of air		

Special Subscripts

j	Subscript for protective level (see fw_{jk})	p	projectile
k	Subscript for state at which impact occurs	t	target

References

1. Interagency Nuclear Safety Review Panel, "Supporting Technical Studies Prepared for the Cassini Mission." INSRP Safety Evaluation Report, Oct. 1997.
2. Lockheed Martin Company, *Cassini Titan IV/Centaur RTG Safety Databook.* Aug. 1995.
3. NUS Corporation, and KRDAS Incorporated, *Worldwide Hardrock Distribution Study.* Nov. 1983.
4. Kessler, D. J., "The Orbital Debris Environment." *Physics and Society,* July 3, 1991.
5. Kessler, D. J., "Decision Time on Orbital Debris—Predicting Debris." *Aerospace America,* June 1988.
6. Recht, R. F. and T. W. Ipson, "Ballistic Perforation Dynamics." *J. Appl. Mech.,* Sept. 1963.
7. Awerbuch, J. and S. R. Bodner, "Analysis of the Mechanics of Performation of Projectiles in Metallic Plates." *Int. Jr. Solids Structures,* 10, 1974.
8. Marom, I. and S. R. Bodner, "Projectile Perforation of Multi-layered Beams." *Int. Jr. of Mech. Science,* 21, 1979.
9. Bodner, S. R., *Modeling Ballistic Perforation, Structural Impact and Crash Worthiness.* 1, Keynote Lecture, G. A. O. Davies (Ed.), Elsevier Applied Science Publishers, 1984.
10. Fraas, A. P., *Protection of Spacecraft from Meteoroids and Orbital Debris.* ORNL/TM-9904, Oak Ridge National Laboratory, Feb. 1986.
11. Wagner, M. H. and K. H. Kreyenhagen, "Review of the Hydro-Elastic-Plastic Code Analyses as Related to the Hypervelocity Particle Impact Hazard." Proceedings of the Comet Halley Micrometeoroid Hazard Workshop, Noordwijk, Netherlands, Apr. 1979.
12. Mecham, Q. J. L. J., Jardine, R. H. Peto, G. T. Reedy, and M.J. Steindler, ANL-81-27, Argonne National Laboratory, 1981.
13. Jardine, L. J., W. J. Mecham, G. T. Reedy, and M. J. Stendler, *Final Report of Experimental Laboratory Scale Brittle Fracture Studies of Glasses and Ceramics.* Argonne National Laboratory, ANL-82-39, Oct. 1982.
14. Lockheed Martin Company, *Final Safety Analysis Report.* Volume II, Book 2 of 2, In Support of the Cassini Mission, CDRL C.3, Nov. 1996.
15. General Electric Co., *GPHS-RTG System Explosion Test, DIRECT COURSE EXPERIMENT 5000.* Advanced Energy Programs Dept., General Electric Co., GESP-7181, Mar. 1, 1985.
16. Eck, M. and M. Mukunda, *On the Response of the GPHS Fueled Clad to Various Impact Environments.* Fairchield Space Co., FSC-ESD-217/88/427, July 1989.
17. Mukunda, M., *Hydrocode Analyses for the Response of the RTG to Accident Environments.* Orbital Sciences Corp., FSC-ESD-217-96-550.
18. Hancock, S. L., *PISCES-2DELK Theoretical Manual.* Physics International, Gouda, Netherlands, 1985.

Student Exercises

1. Intact impact: Consider the launch of a space nuclear system with an early mission profile similar to that shown in Figure 7.1. The nuclear payload consists of a single GPHS-RTG. The RTG weighs 56.2 kg. and has a projected area of about 0.481 m² with the radiator fins intact and a projected area of about 0.289 m² with the radiator fins removed. Estimate the number of fuel clads that would be expected to fail due to Earth impact if the mission were aborted at approximately 8 seconds at an altitude of 150 m: (a) with the radiator fins intact; (b) with the radiator fins removed.

2. Face-on fragment followed by intact impact: Reconsider the launch described in Exercise 1, but this time with an abort at an altitude of about 300 m (approximately 10 s MET), and including an in-air impact by a large frag-

ment that precedes the Earth impact. Assume the fragment impacts face-on. As the abort occurs at 10 seconds, assume the median sized fragment from Table 7.1 for the 10 s MET, center segment listing—a fragment of 28 kg and a velocity of 68 m/s. Consider the two cases (with and without fins).

3. Very large face-on fragment followed by intact impact: Reconsider Exercise 2, but this time consider the in-air impact by the maximum (99.9 percentile) fragment for the 10 s MET case in Table 7.1—a fragment of 29,100 kg and a velocity of 36 m/s.

4. Large edge-on fragment followed by intact impact: Reconsider the 10 s MET case in Table 7.1, (the previous exercises) but this time consider the in-air edge-on impact by the fragment that produces the maximum work on the RTG system—a fragment of 250 kg and a velocity of 56.3 m/s. For this case, the fraction of the fragment work that is partitioned to the RTG (fw_1) is taken as 1.0 rather than 0.33.

5. Penetration of the RTG by shrapnel: In addition to large fragments a failed SRMU can produce relatively small fragments consisting of nuts, bolts, screws, etc. The approximate upper limit of projectile areal mass density (AMD_p) for Titan IV shrapnel is a bolt with an AMD_p of 12.2 g/cm^2. For a 10 s MET failure (nominal pressure) this bolt will have a velocity of approximately 165 m/s. Can this bolt penetrate the RTG casing?

6. Penetration by space debris: Small fragments from launch accidents are only one type of shrapnel to which space systems are exposed. As mentioned in Section 2.0, the Earth has an envelope of about 2 million kilograms of man-made space debris orbiting below about 2000 kilometers. Much of this material is aluminum with collision velocities in the range of 9–18 km/s. Assume that we are planning a low-orbit space project intended to remain in operation for 7 years. We are asked two questions: (a) What thickness of stainless steel and aluminum can be penetrated by spherical aluminum projectiles having diameters (D) of 0.3, 1.0, 3.0 cm and traveling at 10 km/s? (b) Would any actions be required to assure that the probability of penetration by a 1 cm (or greater) spherical projectiles would be less that 0.001 over the lifetime of the mission?

7. Penetration by meteoroids: On space missions beyond the space debris belt, the space system will be exposed to impacts from meteoroids. Consider a manned, round trip mission to Mars taking roughly 2 years. If the exposed critical area is approximately 40 m^2, how many meteoroid projectiles of 0.1 cm or greater would be expected to impact the system? What thickness would they be expected to penetrate? Meteoroids have a density of about 0.5 g/cm^3 and a mean velocity of about 22 km/s.

Chapter 8

Reactor Criticality Safety

Albert C. Marshall

The objective of this chapter is to provide the reader with an understanding of: (1) inadvertent criticality safety issues for space reactor missions, (2) the effect of postulated accidents on the neutronic state of reactors, and (3) the methods used to prevent inadvertent criticality. A brief introduction to criticality fundamentals and reactor physics is provided. Approximate computational methods are presented to enhance understanding of the effects of postulated accidents on reactor criticality. Computer methods used in criticality safety analysis are briefly described.

1.0 Scenarios and Issues

The typical approach for deploying nuclear reactors into space is to launch the reactor radiologically cold and in a subcritical configuration. Because radiological inventories are low for a reactor that has never been operated at high power, the projected consequences of postulated launch accidents will be very small if the reactor remains subcritical. Although an inadvertent criticality does not necessarily imply a significant radiological hazard, the assertion of low consequences is far more difficult to demonstrate if criticality is predicted for a postulated accident sequence. As a consequence, space reactors are typically designed to remain safely subcritical for all credible accident environments. Launch phase reactor safety assessments focus on assuring that the probability of an inadvertent criticality is extremely small.

Some radioisotope fuels are also capable of achieving criticality if sufficient quantities are properly configured. However, the quantity, composition, and configuration of the radioisotope fuel make inadvertent criticality for radioisotope power sources a virtual impossibility. Thus, this discussion of inadvertent criticality is limited to space reactor systems. The focus of this chapter is on assessments of the effect on the neutronic state of the core due to changes in reactor configuration or the reactor environment during postulated accident sequences.

1.1 Criticality Issues

As stated in Chapter 2, a reactor is said to be critical when the neutron production rate from fission equals the neutron loss rate due to neutron absorption and neutron leakage. A reactor is placed in a critical, subcritical, or supercritical state by adjusting the position of control rods or reflector elements. For a critical reactor, the neutron flux is sustained at a constant level, and for a supercritical reactor, the neutron flux level will increase with time. Reactors must be critical to produce useful steady power, and brief supercritical transients are required to increase power level. Reactors are designed to function safely in these normal operational modes for planned operating conditions and environments. Accidents postulated for space reactor systems, however, have the potential to produce an unplanned criticality. An inadvertent criticality can create conditions and environments with safety implications. For example, a critical core emits neutron and gamma radiation and generates fission products and activation products. For normal terrestrial operations, on-site personnel, the public, and the environment are protected from the harmful effects of emitted radiation by shielding and an exclusion area. For space reactors, vast distances between the operating reactor and Earth's surface protect the public and environment. Partial radiation shielding can protect astronauts in the vicinity of an operating reactor (see Chapter 4). If a space reactor accidentally becomes critical on Earth, however, protection from emitted radiation may not be adequate. The issue of direct exposure to emitted radiation is important only if humans are in the vicinity of the critical reactor.

Some inadvertent criticality scenarios can present safety and environmental issues by processes other than direct exposure to neutron and gamma radiation emitted from a critical reactor. During normal power operation the reactor core is heated by nuclear fission and during post operation the subcritical core is heated by the radioactive decay of fission products. Heat transport and rejection systems used during normal operation and post operation are designed to maintain core materials within design temperature limits. These heat rejection systems may not be available during postulated inadvertent criticality accidents. As a consequence, core heating during an inadvertent criticality may result in melting or failure of reactor barriers and the release of radioactive material into the environment. Radioactive material released into the atmosphere or water bodies can be transported beyond the accident site by wind, water currents, and other mechanisms discussed in Chapter 11.

Planned supercritical excursions are controlled, brief, and limited to modest increases in the neutron multiplication factor above the just-critical state. An accidental prompt supercritical excursion, however, is uncontrolled and may for some conditions result in explosive disassembly of the core (discussed in Chapter 9). Explosive disassembly makes prolonged criticality impossible, but the potential for creation and atmospheric dispersion of radioactive particulate presents a radiological risk.

Possible criticality accidents must be carefully studied to assure that the potential for an inadvertent criticality is extremely small. If the probability of inadvertent criticality accidents is not sufficiently small, design changes, additional safety features, or procedural changes will be required. Under some conditions, brief supercritical excursions may be acceptable if they do not present a significant radiological risk.

1.2 Criticality Scenarios

Previous chapters discussed fire, explosion, launch abort, reentry, and impact accidents. Scenarios leading to inadvertent criticality can be postulated for each of these accident categories. For example, a reentry-impact accident scenario can be postulated that results in compaction of the reactor core. The reduced surface area of the compacted core reduces neutron leakage losses, possibly resulting in an inadvertent criticality. Compaction accidents can be postulated for other impact scenarios such as a launch failure or dropping the reactor system during mating with the launch vehicle. Other categories of reconfiguration accidents include reshaping the core and fuel ejection into a critical configuration outside the reactor pressure vessel.

Inadvertent criticality due to a core flooding accident is one of the most commonly discussed potential space reactor accidents. The typical sequence of events includes core impact resulting in a breach of core barriers followed by submersion in a moderating fluid. The moderating fluid flows into the core through the disruptions in the barriers resulting in enhanced neutron moderation and a possible inadvertent criticality. Typical candidates for moderating fluids include ocean water, bodies of fresh water, and liquid rocket propellants. Core immersion without subsequent core flooding may result in criticality for some reactor designs if immersion significantly enhances neutron reflection at the core boundary. Common reflecting materials include water, rocket propellants, sand, and soil. The ability of these materials to significantly increase neutron reflection will depend on the specific reactor design and on the reactor configuration during immersion. Accident scenarios that result in ejection of neutron poisons (required to maintain a subcritical core) and scenarios involving credible combinations of core reconfiguration, flooding, reflection, and poison ejection are also evaluated.

Inadvertent criticality scenarios and issues must be addressed for the prelaunch stage as well as the launch and postlaunch phases. Potential inadvertent criticality scenarios are studied for all prelaunch activities including fuel manufacturing, transportation of nuclear fuel, fuel loading of the core, testing, mating with the launch vehicle, and on-pad operations. Criticality scenarios for the prelaunch phase are similar to scenarios for the fabrication, testing, and operation of many terrestrial reactors; consequently, less emphasis is given in this text to these scenarios. Several unique aspects of the prelaunch phase are notable. For example, unlike commercial reactors, transporting of the entire core may be required for space reactors. The possibility of inadvertent criticality due to impact, fire, and explosive environments must be addressed for postulated accidents during the transportation of space reactor fuel. These scenarios differ from launch and postlaunch scenarios in the possible sequence of events and the characteristics of the accident environments. Furthermore, for ground transportation, reactor fuel is transported over roadways or railways in specially designed rugged shipping casks.

2.0 Criticality Fundamentals

A basic understanding of reactor physics is needed for this chapter; consequently, a condensed discussion of criticality fundamentals is presented here. The concepts and equations developed in this section will be used in subsequent discussions to obtain approximate or upper limit predictions of the effect of accidents on the critical state of the core. For a more complete discussion of reactor criticality see any basic nuclear engineering text, such as Lamarsh's *Introduction to Nuclear Reactor Theory* [1] or Weinberg and Wigner's *The Physical Theory of Neutron Chain Reactors* [2].

The discussion that follows is a departure from the more common approach to introduce reactor physics methods. Space reactors are relatively small, use highly enriched fuel, and are often fast reactors; hence, the traditional focus on large thermal reactors using low enrichment fuel will not be given. The discussion begins with the continuity equation for neutrons of all energies. The neutron flux is assumed to be separable into energy and space dependent components, and the diffusion equation is then presented in terms of cross sections that have been averaged over the entire energy spectrum. A multigroup form of the diffusion equation is derived to illustrate the effects of neutron moderation.

2.1 The Diffusion Equation for a Bare Homogenous Reactor

Monte Carlo, transport theory, and neutron diffusion theory methods can be used to carry out a criticality safety analysis. Computer codes using these methods are briefly described in this chapter. Although Monte Carlo methods and transport theory methods are more accurate than diffusion theory, Monte Carlo methods are not well suited for explaining functional dependencies, and transport theory methods are beyond the scope of this text. For the purpose of explaining basic principles and providing student exercises, this chapter will focus on the more approximate diffusion theory analysis approach. The reader should regard the equations and methods developed in this chapter as approximations useful for a qualitative understanding of the factors affecting criticality.

The Neutron Continuity Equation

For any given volume in a reactor, the time rate of change in neutron density must equal the neutron production rate minus the loss rate per unit volume. If no additional neutron sources are assumed, the rate of change in the total neutron density $n(\mathbf{r},t)$ is

$$\frac{\partial n(\mathbf{r},t)}{\partial t} = \mathcal{P}_f(\mathbf{r},t) - \mathcal{L}_a(\mathbf{r},t) - \mathcal{L}_L(\mathbf{r},t). \qquad (8.1)$$

Here, $\mathcal{P}_f$, $\mathcal{L}_a$, and $\mathcal{L}_L$ are the densities of the neutron production rate by fission, loss rate by absorption, and loss rate by leakage integrated over all neutron energies. The parameters $\mathcal{L}_a$, $\mathcal{L}_L$, and $\mathcal{P}_f$ have units of n/cm^{3}s. Equation 8.1 is the neutron continuity equation. For an exactly critical core, the neutron density is constant with time, hence

$$\mathcal{P}_f(\mathbf{r},t) = \mathcal{L}_a(\mathbf{r},t) + \mathcal{L}_L(\mathbf{r},t). \tag{8.2}$$

The neutron production rate by fission and the neutron loss rate by absorption can be written in terms of the reaction rates for fission and absorption. As discussed in Chapter 1, the reaction rates per unit volume are obtained by integrating the product of the macroscopic cross sections and the neutron flux over the full energy range of the neutron energy spectrum. When a region of the core contains a homogenous mixture of materials, reaction rates for the region can be expressed in terms of a total macroscopic cross section for the homogenous mixture. For a core region at location $\mathbf{r}$ containing a homogenous mixture of i nuclides, the total macroscopic cross section for reaction j is

$$\Sigma_j(\mathbf{r},E) = \sum_{all\ i} \Sigma_{i,j}(\mathbf{r},E). \tag{8.3}$$

The total absorption cross section $\Sigma_a(\mathbf{r},E)$ for all N types of nuclear absorption reactions (e.g., fission, capture) is

$$\Sigma_a(\mathbf{r},E) = \sum_{j=1}^{N} \Sigma_j(\mathbf{r},E). \tag{8.4}$$

The total neutron loss rate by absorption, per unit volume, is

$$\mathcal{L}_a(\mathbf{r}) = \int_0^{\infty} \Sigma_a(\mathbf{r},E)\Phi(\mathbf{r},E)dE, \tag{8.5}$$

where $\Phi(\mathbf{r},E)$ is the energy-dependent flux at location $\mathbf{r}$. The total neutron production rate by fission, per unit volume, is

$$\mathcal{P}_f(\mathbf{r}) = \int_0^{\infty}\int_0^{\infty} \upsilon(\mathbf{r},E)\chi(E')\Sigma_f(\mathbf{r},E)\Phi(\mathbf{r},E)dEdE', \tag{8.6}$$

where $\Sigma_f(\mathbf{r},E)$ is the energy dependent macroscopic fission cross section at location $\mathbf{r}$, $\upsilon(\mathbf{r},E)$ is the yield (n/fiss) for neutron fission at energy E, and $\chi(E')$ is the fraction of neutrons emitted at energy E'. Other neutron production processes [e.g., (n,2n) reactions] make only a small contribution and are ignored here.

Monoenergetic Neutrons

An approximation can be used to predict the neutron leakage rate by assuming that neutron transport through a medium is somewhat analogous to molecular diffusion from regions of high concentration to regions of low concentration. Diffusion theory can be applied to the transport of monoenergetic neutrons in an idealized homogenous medium. The idealized medium is assumed to have a relatively small cross section for neutron absorption, and scattering collisions are assumed to occur without neutron energy loss [1]. The diffusion equation can provide reasonably accurate predictions of the neutron flux in regions that are more than a few neutron mean free paths from boundaries, strong neutron absorbers, or additional neutron sources. (A neutron mean free path is the average distance between neutron-nuclei interactions).

For a homogenous medium, the neutron leakage rate per unit volume predicted by diffusion theory is given by

$$\mathcal{L}_L(\mathbf{r}) = -D\nabla^2\phi(\mathbf{r}) \tag{8.7}$$

where ∇^2 is the Laplacian operator and D is the neutron diffusion coefficient. In rectangular (x,y,z) co-ordinates the Laplacian of the neutron flux ϕ is

$$\nabla^2\phi = \frac{\partial^2\phi}{\partial x^2} + \frac{\partial^2\phi}{\partial y^2} + \frac{\partial^2\phi}{\partial z^2}. \tag{8.8}$$

The diffusion coefficient is approximately equal to

$$D = \frac{1}{3[\Sigma_T - \Sigma_s(2/3A_r)]}, \tag{8.9.}$$

where Σ_T is the total macroscopic cross section (i.e., the cross section for all neutron-nucleus interactions). The parameter Σ_s is the macroscopic cross section for neutron scattering, and A_r is the relative atomic mass of the scattering nucleus. For $\Sigma_s >> \Sigma_a$, the diffusion coefficient is approximately equal to

$$D = \frac{1}{3\Sigma_s[1 - (2/3A_r)]}. \tag{8.10}$$

E-r Separable Approximation for a Bare Reactor

A spectrum of neutron energies is present in an operating reactor and neutrons can lose an appreciable fraction of their kinetic energy E during collisions with nuclei. These conditions imply that the assumption of diffusion of monoenergetic neutrons given in the preceding discussion is not directly applicable to reactor systems. Nonetheless, several methods have been developed using the basic diffusion theory approach that provide good predictions of neutron leakage for many reactor designs. For example, the diffusion approximation can be used to treat neutron leakage over segments of the energy spectrum called energy groups. This method yields coupled neutron balance equations for each energy group. Another approach is the so-called age-diffusion approximation, used for thermal reactors, that combines the diffusion equation with a continuous slowing down model for fast neutrons.

Here, diffusion theory will be applied to a homogenous bare (unreflected) reactor to illustrate the relationship of neutron leakage to other reactor parameters. Neither the energy group method nor the age-diffusion approximation will be used at this stage of our discussion. Here, we make the assumption that the neutron flux is separable in space and energy over the entire energy range; i.e.,

$$\Phi(\mathbf{r},E) = \varphi(E)\phi(\mathbf{r}). \tag{8.11}$$

Although this assumption is valid only for a bare homogenous reactor, the approximation is useful for understanding criticality principles. This approach also assumes that the energy spectrum of the flux $\varphi(E)$ is known for the particular reactor. Note that the symbol ϕ is used to denote the flux without energy dependence. Thus, ϕ is used for the position dependent flux, the core average flux. The symbol ϕ is also used for the one-group flux and the multi-group flux.

Taking the Laplacian of the expression for the flux in Eq. 8.11 and multiplying by the diffusion constant gives the leakage rate per unit volume of neutrons with energies between E and $E + dE$,

$$D(E)\nabla^2\Phi(\mathbf{r},E)dE = D(E)\varphi(E)\nabla^2\phi(\mathbf{r})dE. \tag{8.12}$$

The spatial dependence $\mathbf{r}$ for the diffusion coefficient and cross sections has been dropped because we have made the assumption of a homogenous core. If $\varphi(E)$ is normalized such that the integral of $\varphi(E)$

over all energies equals unity, the integral of Eq. 8.12 over all energies gives the neutron leakage rate per unit volume in terms of the energy averaged diffusion coefficient $\bar{D}$,

$$\mathcal{L}_L(\mathbf{r}) = -\bar{D}\nabla^2\phi(\mathbf{r}). \tag{8.13}$$

Energy averaged cross sections $\bar{\Sigma}_a$ and $\overline{\nu\Sigma}_f$ are also obtained and Eq. 8.2 can now be written

$$\overline{\nu\Sigma}_f\phi(\mathbf{r}) = \bar{\Sigma}_a\phi(\mathbf{r}) - \bar{D}\nabla^2\phi(\mathbf{r}). \tag{8.14}$$

2.2 Solutions of the Diffusion Equation

The spatial dependence of the neutron flux for a bare homogenous core can be obtained by solving Eq. 8.14. The equation can also be written as

$$\nabla^2\phi + B^2\phi = 0. \tag{8.15}$$

Combining Eqs. 8.14 and 8.15, we obtain

$$B^2 \equiv \frac{\overline{\nu\Sigma}_f - \bar{\Sigma}_a}{\bar{D}}, \tag{8.16}$$

where B^2, defined by Eq. 8.16, is called the material buckling and the units for B^2 are cm^{-2}.

Equation 8.15 can be solved for a number of core geometries. For a homogenous bare spherical core of radius r, the general solution for the flux that satisfies the boundary condition of a finite flux at the center of the sphere is

$$\phi(r) = C\frac{\sin(Br)}{r}, \tag{8.17}$$

where C is a constant. In order to solve for B^2 we make the approximation that the neutron flux goes to zero at some radial distance R_e greater than the bare-core radius R_{bc}. The permitted values for B^2 that satisfy the zero flux boundary condition at $r = R_e$ for a spherical core are $B^2 = (m\pi/R_e)^2$, where m is an integer. It can be shown that only the solution for $m = 1$ gives an expression for the steady-state neutron flux [3]. Thus,

$$B^2 = \left(\frac{\pi}{r}\right)^2. \tag{8.18}$$

The buckling, as given by Eq. 8.18, is referred to as the geometric buckling. For a critical reactor, the geometric buckling and the material buckling are identical. Substituting Eq. 8.18 into Eq. 8.17, the spatially dependent neutron flux for a bare sphere is obtained; i.e.,

$$\phi(r) = \phi_{max}R_e\frac{\sin(\pi r/R_e)}{\pi r}, \tag{8.19}$$

where ϕ_{max} is the maximum value of the neutron flux.

Table 8.1 provides the expressions for the neutron flux and buckling for a bare homogenous core with spherical, rectangular, and cylindrical geometries. The neutron flux shapes for a sphere, semi-infinite slab, and semi-infinite cylinder are plotted as a function of r or x in Figure 8.1. The extrapolated radius used in this development is slightly larger than the actual bare core critical radius R_{bc} and is approximated by

$$R_e = R_{bc} + 2\bar{D}. \tag{8.20}$$

Equation 8.20 is the standard formula used to obtain approximate agreement of neutronic predictions by diffusion theory with predictions using the more accurate (but more difficult) transport theory method [3].

Equation 8.16 can also be written

$$B^2 = \frac{\bar{\Sigma}_a}{\bar{D}}(k_\infty - 1), \tag{8.21}$$

where

$$k_\infty \equiv \frac{\overline{\nu\Sigma_f}}{\bar{\Sigma}_a} = \frac{\mathscr{P}_f}{\mathscr{L}_a}. \tag{8.22}$$

k_∞ is called the infinite medium neutron multiplication factor.

Combining Eqs. 8.18, 8.20, and 8.21, the expression for the critical radius of a bare spherical core is obtained in terms of the energy-averaged neutronic parameters

$$R_{bc} = \pi \sqrt{\frac{\bar{D}}{\bar{\Sigma}_a(k_\infty - 1)}} - 2\bar{D}. \tag{8.23}$$

The fuel mass contained within this critical dimension is called the critical mass m_{cr} equal to

$$m_{cr} = \frac{4}{3}\pi R_{bc}^3 \varsigma_F(VF_F), \tag{8.24}$$

where ς_F and VF_F are the fuel density (g/cm³) and the volume fraction of fuel in the core, respectively.

Table 8.1   Expressions for Buckling and Flux Distributions for Bare-Core Geometries

Geometry	Buckling	Flux Distribution
(a) Sphere	$\left(\dfrac{\pi}{R_e}\right)^2$	$\dfrac{R_e}{\pi r}\sin\left(\dfrac{\pi r}{R_e}\right)$
(b) Rectangular Parallel piped	$\left(\dfrac{\pi}{a_e}\right)^2 + \left(\dfrac{\pi}{b_e}\right)^2 + \left(\dfrac{\pi}{c_e}\right)^2$	$\cos\left(\dfrac{\pi x}{a_e}\right)\cos\left(\dfrac{\pi y}{b_e}\right)\cos\left(\dfrac{\pi z}{c_e}\right)$
(c) Finite Cylinder	$\left(\dfrac{2.405}{R_e}\right)^2 + \left(\dfrac{\pi}{H_e}\right)^2$	$J_0\left[\dfrac{2.405(r)}{R_e}\right]\cos\left(\dfrac{\pi z}{H_e}\right)$

Here, J_0 is the Bessel function of the first kind of zero order, R_e is the extrapolated radius of a critical spherical core or cylindrical core, a_e, b_e, and c_e are the extrapolated dimensions of a critical rectangular parallel piped core, and H_e is the extrapolated height of a critical cylindrical core. The variable r is the radial coordinate for a spherical or cylindrical coordinate system, and x, y, and z are the coordinates for a rectangular coordinate system. For cylindrical geometry, z is the dimension parallel to the axis.

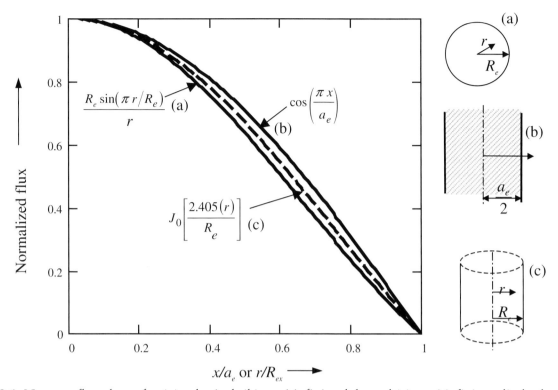

Figure 8.1 Neutron flux shape for (a) spherical, (b) semi-infinite slab, and (c) semi-infinite cylinder bare cores.

2.3 Critical Mass Minimization Methods

Critical mass requirements are reduced by surrounding the reactor core by neutron reflectors to minimize neutron leakage. The reduction in critical dimensions, referred to as reflector savings δ, can be approximated by several different formulations. The critical radius of a reflected core R_c is related to the critical radius of a bare core R_{bc} by

$$R_c = R_{bc} - \delta. \tag{8.25}$$

A reflector often provides additional moderation of the neutron flux in its vicinity, resulting in a distortion of the flux shape, as illustrated in Figure 8.2. The distortion of the thermal flux implies that the assumption of simple separation of space and energy variables (using a single group for the entire energy spectrum) is not justified when a reflector is present.

Because space reactors must fit within the dimensions and mass constraints of a launch vehicle, special measures are used to minimize space reactor mass and size. For example, the fuel for space reactor designs is usually enriched to over 90% in ^{235}U. The fuel for terrestrial power reactors is typically enriched to only a few percent in ^{235}U (the natural abundance of ^{235}U in uranium is about 0.7%). Increasing the fuel enrichment increases the core-averaged macroscopic fission cross section and decreases the cross section for parasitic capture in ^{238}U resulting in an increase in k_∞. Enriching in uranium-235 will also increase the value of $\bar{\Sigma}_a$ relative to the value of $\bar{D}$, reducing neutron leakage. Equation 8.23 indicates that the net effect of the increase in k_∞ and $\bar{\Sigma}_a/\bar{D}$ is a decrease in the critical dimensions of the core.

Neutron moderation can also be used to reduce critical mass requirements. Neutron moderation will usually have a relatively small effect on k_∞; however, the microscopic cross section for thermal neutron

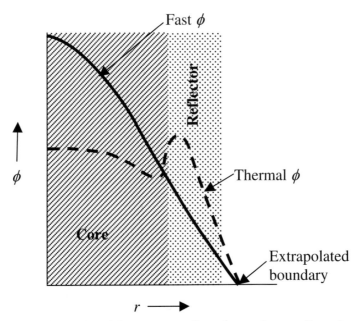

Figure 8.2 Thermal and fast neutron flux shapes for a reflected reactor.

absorption in ^{235}U is much larger than the cross section in the fast range (see Figure 1.17). The larger microscopic cross section implies an increase in the $\overline{\Sigma}_a/\overline{D}$ ratio, and a decrease in neutron leakage. As a result, the quantity of fuel required for criticality in a thermal reactor is usually much less than the critical mass for a fast reactor. However, for some designs and applications, the core volume required for the moderator can be very large, resulting in a net increase in system size and mass rather than a decrease. Temperature limitations for some moderators may place additional constraints on their use. Given these considerations, many reactor designs proposed for use in space are fast reactors. Nonetheless, for some system options and mission requirements a moderator can reduce system mass and can provide other advantages.

2.4 The Effective Neutron Multiplication Factor

In the preceding section we discussed the analysis of an exactly critical reactor. An infinite medium neutron multiplication factor k_∞ was defined as the ratio of the neutron production rate by fission to the neutron loss rate by absorption. Neutron leakage is not included in the ratio; consequently, k_∞ can be regarded as an indicator of the neutronic state of an infinitely large reactor. An effective neutron multiplication factor, k_{eff}, can be defined that includes neutron loss rate by leakage as well as absorption; i.e.,

$$k_{eff} \equiv \frac{\mathcal{P}_f}{\mathcal{L}_a + \mathcal{L}_L}. \tag{8.26}$$

The value of k_{eff} indicates the neutronic state of actual (finite) reactors. When $k_{eff} = 1.0$, the reactor will be exactly critical, and Eq. 8.26 reduces to Eq. 8.2. If $k_{eff} < 1.0$ the reactor is subcritical, and if $k_{eff} > 1.0$ the reactor is supercritical. The effective neutron multiplication factor is an important parameter in the design and safety analysis of reactors.

If a reactor deviates from exact criticality, the time dependent form of the continuity equation applies. However, Eq. 8.26 can be rearranged to give

$$\mathcal{P}_f - k_{eff}(\mathcal{L}_a + \mathcal{L}_L) = 0, \tag{8.27}$$

where k_{eff} can be regarded as an adjustment to the loss term required to achieve criticality, e.g., an adjustment of control drum positions or internal neutron poisons. Although less intuitive, the adjustment can also be associated with the fission source term as $\mathcal{P}_f/k_{eff}$; either association is mathematically equivalent. With this approach, a time independent equation can be used to obtain the solutions for k_{eff}. Inserting the expressions for the terms into Eq. 8.27 gives

$$\overline{\nu \Sigma_f}\,\phi(\mathbf{r}) = k_{eff}[\overline{\Sigma}_a\,\phi(\mathbf{r}) - \overline{D}\nabla^2\phi(\mathbf{r})]. \tag{8.28}$$

Equation 8.28 can be rewritten as

$$\overline{D}\nabla^2\phi(\mathbf{r}) + \left(\frac{\overline{\nu \Sigma_f}}{k_{eff}} - \overline{\Sigma}_a\right)\phi(\mathbf{r}) = 0. \tag{8.29}$$

The buckling is now defined as

$$B^2 \equiv \frac{1}{\overline{D}}\left(\frac{\overline{\nu \Sigma_f}}{k_{eff}} - \overline{\Sigma}_a\right). \tag{8.30}$$

The lowest eigenvalue solution of the resulting equation provides the expression for the self-sustaining neutron flux for this pseudo-static reactor configuration. Using Eqs. 8.22 and 8.30, we obtain

$$k_{eff} = \frac{k_\infty}{1 + (\overline{D}/\overline{\Sigma}_a)B^2}. \tag{8.31}$$

Eq. 8.31 can also be written

$$k_{eff} = k_\infty P_{NL}, \tag{8.32}$$

where
$$P_{NL} = \frac{1}{1 + (\overline{D}/\overline{\Sigma}_a)B^2}. \tag{8.33}$$

The parameter P_{NL} is defined as the non-leakage probability; i.e., the probability that neutrons will be absorbed rather than leak from the core.

2.5 Neutron Moderation

Neutron moderation is fundamental to thermal reactor operation and is an important criticality safety consideration for all types of reactors and nuclear fuels. Methods have been developed to compute neutron spectra for moderated systems. The development of these methods is beyond the scope of this text; however, a brief summary of the basic principles of neutron moderation and its effect on the spectrum is presented in this section. For a detailed discussion of neutron moderation see Refs. [1] or [2].

Effect on Spectrum

Most neutrons released during fission are fast neutrons, with energies well above the thermal energy range. Fast neutrons can give up some of their kinetic energy during elastic or inelastic scattering collisions with the nuclei of core materials. As discussed in Chapter 1, the maximum possible energy loss per elastic scattering collision is given by the approximate relationship $\Delta E_{max} \approx (1 - \alpha_s)E$, where $\alpha_s = (A + 1)^2/(A - 1)^2$ and A is the mass number of the scattering nucleus. The parameter α_s increases with increasing mass number, as shown in Table 8.2. For uranium the maximum neutron energy loss in a col-

Table 8.2 Moderating Properties of Nuclides

Element	Atomic Mass Number	α_s	ξ	Average Number of Collisions to Thermalize
Hydrogen	1	0	1.0	18
Deuterium	2	0.111	0.725	25
Helium	4	0.360	0.425	43
Lithium	7	0.563	0.268	67
Beryllium	9	0.640	0.209	86
Carbon	12	0.716	0.158	114
Oxygen	16	0.779	0.120	150
Uranium	238	0.983	0.00838	2172

lision is only 1.7%. For a thermal reactor almost all neutron moderation is due to elastic scattering collisions with a low atomic mass moderator in the core.

Below the fission energy range, the fast flux spectrum in a moderated reactor can be shown [1] to have the relationship,

$$\varphi(E) \sim \frac{1}{E\xi\Sigma_s(E)}. \tag{8.34}$$

The parameter ξ is defined as the average logarithmic energy decrement per collision,

$$\xi \equiv \overline{\ln(E/E')}, \tag{8.35}$$

where E and E' are the neutron energies before and after a collision, respectively. The average logarithmic energy decrement can also be shown [3] to have the relationship

$$\xi = 1 + \frac{\alpha_s}{1 - \alpha_s}\ln\alpha_s, \tag{8.36}$$

and for $A > 10$, $\xi \approx 2/(A + 2/3)$. Table 8.2 also provides the values of ξ for several nuclides, as well as the average number of collisions to reach thermal energies.

In the low energy range, the neutron flux departs from the approximate $1/E$ dependence of the fast flux, given in Eq. 8.34, and takes on an approximate Maxwellian energy spectrum $f(E)$ given by Eq. 1.16 in Chapter 1.

Figures of Merit

A number of figures of merit have been used to judge the ability of moderators to increase a reactor's k_{eff} (or decrease critical mass requirements). The parameters A, α_s, and ξ can be viewed as moderator figures of merit, because a small A or α, and a large ξ are indicative of a large neutron energy loss per collision. However, these figures of merit do not include the probability that a scattering collision will occur and do not account for parasitic absorptions by the moderator. In order to account for both the probability of a scattering collision and the energy loss per collision, a *slowing down power* is defined by $\xi\Sigma_s$. Parasitic neutron capture by the moderator can reduce the effectiveness of a moderator. This consideration can be included by defining a *moderating ratio* given by $\xi\Sigma_s/\Sigma_a$, where Σ_a is the moderator absorption cross section. Both figures of merit are useful for comparing moderator effectiveness. The slowing down power and moderating ratio for several moderators are presented in Table 8.3.

Table 8.3   Figures of Merit for Important Neutron Moderator Materials

Moderator	High Figure-of-Merit for Enhanced Moderation		Low Figure-of-Merit for Enhanced Moderation	
	$\xi\Sigma_s$ (cm^{-1})	$\xi\Sigma_s/\Sigma_a$	τ (cm^2)	L^2 (cm^2)
Zirconium Hydride	1.47	49	27	9
Water	1.28	58	31	7.62
Beryllium	0.16	130	85	441
Beryllium Oxide	0.12	163	100	900
Graphite (Reactor grade)	0.065	200	350	2938

Although no longer used in the design of nuclear reactors, the so-called *age-diffusion method* is useful for understanding neutron moderation and for comparing moderator effectiveness. Using the age-diffusion method, it can be shown [2] that the effective neutron multiplication factor for a thermal reactor can be approximated by

$$k_{eff} \approx \frac{k_\infty e^{-B^2\tau}}{1 + L^2B^2} \tag{8.37}$$

where τ is defined as the Fermi age, expressed in units of cm^2. The Fermi age is equal to one-sixth the mean square distance a neutron travels from the point of origin (fission emission) to the point at which the neutron reaches the thermal energy range. The quantity L^2 is equal to one-sixth the mean square distance a neutron travels from birth as a thermal neutron (reaching the thermal energy range) to absorption by a nucleus. L is defined as the thermal diffusion length (cm) and is equal to the quantity $\sqrt{D/\overline{\Sigma}_a}$ averaged over the thermal energy range. The age-diffusion approximation, used to obtain Eq. 8.37, is not applicable to very light nuclei, such as hydrogen.

The exponential in Eq. 8.37 can be expanded, and for larger cores the relationship for the effective neutron multiplication factor is approximately

$$k_{eff} \approx \frac{k_\infty}{(1 + \tau B^2)(1 + L^2B^2)}. \tag{8.38}$$

The terms $e^{-B^2\tau}$ and $(1 + \tau B^2)^{-1}$ represent the non-leakage probability for fast neutrons, and $(1 + L^2B^2)^{-1}$ is the non-leakage probability for thermal neutrons. The smaller the values for τ and L^2, the lower the probability of neutron leakage for the fast and thermal energy ranges, respectively. Thus, the Fermi age and the diffusion length for moderators (presented in Table 8.3) can also be used as a figure of merit for moderator options.

Moderated Core Critical Mass

Equation 8.37, along with Eqs. 8.18 and 8.22, can be used to obtain the relationship of the moderator-to-fuel molecular density ratio N_M/N_F to the critical radius for a bare spherical reactor; i.e.,

$$\frac{N_M}{N_F} \approx \frac{[(v\sigma_{fF}/\sigma_{aF})P_{NL1} - 1]}{[1 + L_M^2(\pi/R_{bc})^2]}\left(\frac{\sigma_{aF}}{\sigma_{aM}}\right). \tag{8.39}$$

Here, P_{NL1} is the fast neutron non-leakage probability, and the subscripts F and M refer to the fuel and the moderator, respectively. As noted previously, $P_{NL1} = e^{-\tau B^2}$ is not suitable for hydrogenous modera-

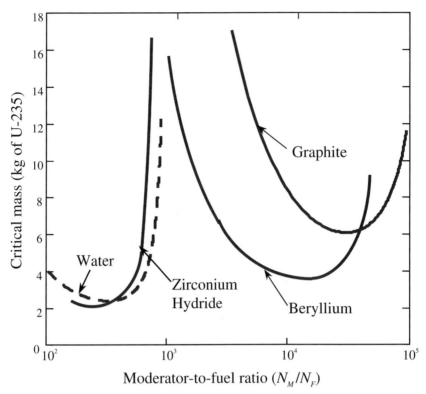

Figure 8.3 Critical mass vs. moderator-to-fuel ratio (using Eqs. 8.39 and 8.40) for fully enriched homogenous bare spherical cores.

tors. An equation having the form of Eq. 8.38 (i.e., Eq. 8.45) can be used in place of Eq. 8.37 for hydrogenous moderators. For hydrogenous moderators the non-leakage probability is better approximated using $P_{NL1} = (1 + \tau'B^2)^{-1}$, where τ' is an *effective age* (discussed in Section 2.6). The critical mass m_{cr} for a bare spherical core is

$$m_{cr} = \left(\frac{N_M}{N_F}\right)^{-1} \frac{N_M 4\pi R_{bc}^3}{3} \frac{M_{rF}}{N_A},$$

(8.40)

where M_{rF} is the relative molecular mass of the fuel compound and N_A is Avogadro's number. Using Eqs. 8.39 and 8.40, the critical mass of bare, homogenous, fully enriched cores were estimated for a variety of possible moderators. The critical masses from the computation are presented as a function of moderator-to-fuel ratio in Figure 8.3. From this figure, the hydrogenous moderators are observed to be far more effective moderators than the non-hydrogenous moderators. Figure 8.3 also shows that at some point increasing the moderator-to-fuel ratio increases the required critical mass due to the effect of parasitic neutron capture in the moderator. Furthermore, as shown in the following example, a minimum critical mass does not imply a minimum critical radius or a minimum system mass.

Example 8.1

Calculate the required moderator-to-fuel ratio, the critical fuel mass, and the combined mass of fuel and moderator for a beryllium moderated bare spherical core with a radius of 35 cm. Repeat these calculations for a critical core radius of 55 cm. The fuel is in the form of fine UO_2 particles homogeneously dispersed in a Be matrix with a density of 1.84 g/cm³. Assume that the ratio of the average cross sections is approximately equal to the ratio of the 2200 m/s cross sections; i.e., use $\sigma_{aF} = 683$ b and σ_{aM}

= 0.01 b. Also assume fully enriched fuel with $v\sigma_{fF}/\sigma_{aF} = 2.06$. Ignore the small neutronic effect of the oxygen in the UO_2 fuel. For simplicity assume no other core materials and no internal cooling channels or voids. What do the results suggest about the use of a fully moderated core for space reactors?

Solution:

From Table 8.3, the τ and L^2 for beryllium are 85 cm^2 and 441 cm^2, respectively. Using Eq. 8.39 we obtain

$$\frac{N_M}{N_F} = \frac{2.06\left[e^{-85\text{ cm}^2(\pi/35\text{ cm})^2}\right] - 1}{1 + 441\text{ cm}^2(\pi/35\text{ cm})^2}\left(\frac{683\text{ b}}{0.01\text{b}}\right) = 5.8 \times 10^2.$$

Ignoring the small fraction of the core volume due to the UO_2 fuel, the core average atom density for Be is

$$N_{Be} = \frac{1.84\text{ g/cm}^3\,(6.02 \times 10^{23}\text{ atoms/g·mole})}{(9\text{ g/g·mole})} = 1.231 \times 10^{23}\text{ atoms/cm}^3.$$

Using Eq. 8.40 we obtain

$$m_{cr} = \frac{4\pi(35\text{ cm})^3(1.231 \times 10^{23}\text{ atoms/cm}^3)(267\text{ g/g·mole})}{3(5.8 \times 10^2)(6.02 \times 10^{23}\text{ atoms/g·mole})} = 17 \times 10^3\text{ g} = 17\text{ kg.}$$

Note that the mass computed here is for the UO_2 molecule and differs slightly from the critical mass of uranium metal given in Figure 8.3. The mass of Be in the core is

$$m_{Be} = \frac{4\pi(35\text{ cm})^3}{3}1.84\text{ g/cm}^3 = 3.3 \times 10^5\text{ g} = 330\text{ kg,}$$

and the combined mass is $m_{tot} = 17 + 330 = 347$ kg.

For a critical core radius of 55 cm, we compute a moderating ratio of 1.57×10^4, $m_{cr} = 2.4$ kg, and $m_{Be} = 1282$ kg; consequently, $m_{tot} = 1284$ kg.

From Figure 8.3 observe that the moderating ratio for the 55-cm radius core corresponds to an optimally moderated (lowest critical U-235 mass) core. The optimally moderated core has a critical mass much smaller than the undermoderated 35-cm radius core. However, the larger optimally moderated core requires a significant moderator mass and the combined fuel and moderator mass is three times that of the undermoderated core. The large combined mass and core size of the optimally moderated core illustrate that using optimal moderation to reduce critical mass can be a poor choice for space reactor designs.

2.6 Two-Group Method

In the forgoing discussion the energy dependence of the neutron flux was implicitly included by defining and using reactor parameters averaged over the entire energy spectrum. Although this one-group approach was useful for explaining some types of functional dependencies, the neutron spectrum must be known for the particular reactor of interest, and single-group cross sections can be very sensitive to the reactor spectrum. The spectrum of each reactor type will depend on the spatially dependent core composition, reflector composition and configuration, and operating temperatures. Furthermore, the composition and configuration can change during operation. Consequently, the spectrum can be spatially dependent, temperature dependent, and time dependent.

Given these considerations, single-group methods are never used in the design and analysis of reactors. Another shortcoming of the single-group approach is that it provides little insight into the effect of neutron moderation. The age-diffusion approximation addresses, in part, the latter shortcoming. The usefulness of the age-diffusion method is limited, however, and the approach does not demonstrate the multigroup method widely used in neutronic design and safety analysis. A two-energy group approximation is presented in this section to better understand the effects of neutron moderation on reactor criticality. The two-group approach will be derived using two different approaches; i.e., using the definition of the neutron multiplication factor and using coupled diffusion equations.

Using k_{eff} Definition

The two-group approximation divides the spectrum into a high-energy group and a low-energy group. The low-energy group, referred to as the thermal group, includes the entire thermal energy range. The high-energy group, referred to as the fast group, includes all neutrons with energies greater than the upper bound of the thermal energy range. Equation 8.31 can be expressed with fast and thermal reactor parameters explicitly included in the formulation; i.e.,

$$k_{eff} = \frac{k_\infty}{1 + \frac{(D_1\phi_1 + D_2\phi_2)}{(\Sigma_{a1}\phi_1 + \Sigma_{a2}\phi_2)} B^2}. \tag{8.41}$$

The subscripts 1 and 2 for D and Σ_a refer to flux weighted averages over the fast and thermal energy ranges, respectively. The fluxes ϕ_1 and ϕ_2 are the energy-dependent neutron flux integrated over the fast and thermal energy ranges, respectively. In a thermal reactor with $\Sigma_{a2}\phi_2 >> \Sigma_{a1}\phi_1$ Eq. 8.41 reduces to

$$k_{eff} \approx \frac{k_\infty}{1 + \frac{D_1}{\Sigma_{a2}} \frac{\phi_1}{\phi_2} B^2 + \frac{D_2}{\Sigma_{a2}} B^2}. \tag{8.42}$$

Most neutrons released during fission are fast neutrons; consequently, the majority of thermal neutrons in a thermal reactor are produced by moderation. The neutron balance equation for thermal neutrons can be approximated by

$$\Sigma_{q1}\phi_1 = \Sigma_{a2}\phi_2 + D_2 B^2 \phi_2, \tag{8.43}$$

where Σ_{q1} is the cross section for neutron energy loss from the fast to thermal energy range. From Eq. 8.43 we obtain the ratio of the fast flux to the thermal neutron flux

$$\frac{\phi_1}{\phi_2} = \frac{\Sigma_{a2} + D_2 B^2}{\Sigma_{q1}}. \tag{8.44}$$

Substituting Eq. 8.44 into Eq. 8.42 gives

$$k_{eff} = \frac{k_\infty}{\left(1 + \frac{D_1}{\Sigma_{q1}} B^2\right)\left(1 + \frac{D_2}{\Sigma_{a2}} B^2\right)}. \tag{8.45}$$

The ratio D_2/Σ_{a2} is the definition for L^2 and it can be shown [4] that D_1/Σ_{q1} is roughly equivalent to the Fermi age (although the numerical value may differ from those in Table 8.3). Thus Eq. 8.45 is consistent with Eq. 8.38.

Using Coupled Diffusion Equations

Equation 8.45 was derived from Eq. 8.43 and the definition of the effective neutron multiplication factor. Equation 8.45 can also be derived using coupled diffusion equations for both the fast and thermal energy groups. Using the approximation that essentially all neutrons produced by fission are fast neutrons, the neutron diffusion equation for the fast group can be written

$$\frac{1}{k_{eff}}(v_1\Sigma_{f1}\phi_1 + v_2\Sigma_{f2}\phi_2) = (\Sigma_{a1} + \Sigma_{q1} + D_1 B^2)\phi_1. \tag{8.46}$$

The diffusion equation for the thermal group is given by Eq. 8.43, which was rearranged to give Eq. 8.44. Using Eq. 8.44 in Eq. 8.46 gives

$$k_{eff} = \left(\frac{v_2\Sigma_{f2}\Sigma_{q1}}{\Sigma_{a2} + D_2 B^2} + v_1\Sigma_{f1}\right)\left(\frac{1}{\Sigma_{a1} + \Sigma_{q1} + D_1 B^2}\right). \tag{8.47}$$

For thermal reactors, $v_1\Sigma_{f1}$ and Σ_{a1} are much smaller than the other terms within the parenthesis in Eq. 8.47; hence, $v_1\Sigma_{f1}$ and Σ_{a1} can be dropped to obtain an approximate expression for the neutron multiplication factor for a thermal reactor; i.e.,

$$k_{eff} = \left(\frac{v_2\Sigma_{f2}}{\Sigma_{a2} + D_2 B^2}\right)\left(\frac{\Sigma_{q1}}{\Sigma_{q1} + D_1 B^2}\right). \tag{8.48}$$

Dividing the numerator and denominator by Σ_{a2} for the left parenthetical expression and by Σ_{q1} for the right parenthetical expression, and noting that $k_\infty \approx v_2\Sigma_{f2}/\Sigma_{a2}$, yields Eq. 8.45. Thus, the same basic expression is obtained using the two-group diffusion equation, the age diffusion method, and a derivation using the definition of the effective multiplication factor. For a fast reactor $\Sigma_{q1} \approx 0$ and $\phi_2 \approx 0$ in Eq. 8.46; consequently, the neutron multiplication factor for a fast reactor is just

$$k_{eff} = \frac{v_1\Sigma_{f1}/\Sigma_{a1}}{1 + (D_1/\Sigma_{a1})B^2}. \tag{8.49}$$

3.0 Core Reconfiguration

As stated in Section 1.0, core reconfiguration due to impact, explosions, or fire has the potential for causing an inadvertent criticality accident. Core reconfiguration, as defined here, includes core compaction, reshaping, redistribution, and ejection. The effect of a net gain or loss of moderating materials in the core and the effect of changes in neutron reflection are treated separately in subsequent sections of this chapter. The various types of core reconfigurations are illustrated schematically in Figure 8.4.

3.1 Core Compaction

To illustrate the potential effect of core compaction, we will narrowly define core compaction as a uniform compression of the core, such that the core shape and core inventory before and after compaction do not change. It should be clear, however, that uniform compaction of this type is usually an extremely unlikely event. Core reconfiguration due to compaction is illustrated in Figure 8.4 (a).

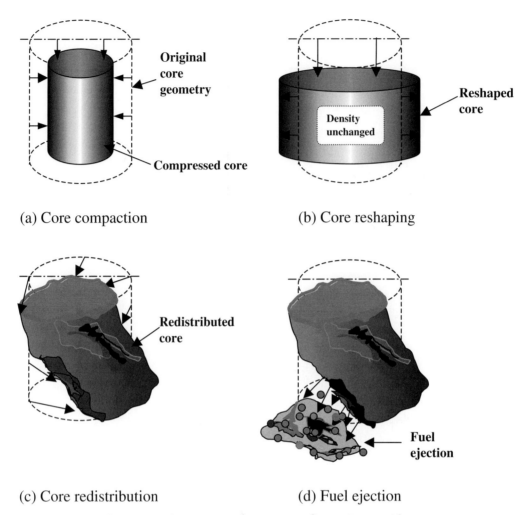

(a) Core compaction   (b) Core reshaping

(c) Core redistribution   (d) Fuel ejection

Figure 8.4 Categories of core reconfiguration accidents.

The effect of core compaction on the neutron multiplication factor can be estimated using Eq. 8.31. The diffusion coefficient in Eq. 8.31 can be expressed in terms of a transport cross section $\bar{\Sigma}_{tr}$ defined as $\bar{\Sigma}_{tr} \equiv 1/3\bar{D}$. We can express $\bar{D}/\bar{\Sigma}_a$ as

$$\frac{\bar{D}}{\bar{\Sigma}_a} = \frac{1}{3\Sigma_{tr}\Sigma_a} = \left[3 \sum_{all\,i} N_i \sigma_{tri} \sum_{all\,i} N_i \sigma_{ai} \right]^{-1} = C_D V_c^2, \tag{8.50}$$

where C_D is a constant for any particular choice of core materials. This result is obtained from the expression for the atom density; i.e.,

$$N_i = \frac{m_i N_A}{V_c A_{ri}} = \left(\frac{1}{V_c}\right) \frac{m_i N_A}{A_{ri}}, \tag{8.51}$$

where V_c is the core volume, N_A is Avogadro's number, and m_i and A_{ri} are the total mass and the relative atomic mass of nuclide i. Observe that the core volume factors out and all other terms are unchanged during compaction.

From Table 8.1, for spherical geometry, the buckling is proportional to $1/R_e^2$. Furthermore, the buckling for a cylindrical core can be expressed as

$$B^2 = \left(\frac{2.405}{R_e}\right)^2 + \left(\frac{\pi}{H_e}\right)^2 = \left(\frac{1}{R_e^2}\right)\left[(2.405)^2 + \left(\frac{\pi}{2a}\right)^2\right],$$

(8.52)

where a is the aspect ratio (H/D). Likewise, the buckling for a parallelepiped can be expressed as a function of a single dimension with proportionality constants for two of the dimensions. For uniform compaction only R_e changes. In a similar fashion, the core volume can be expressed in terms of a single dimension and aspect ratios; e.g., for a cylinder $V_c = 2\pi R_E^3 a$. Consequently, we can express the buckling as

$$B^2 = \frac{C_B}{V_c^{2/3}},$$

where C_B is a constant for any particular geometry. Using this and Eq. 8.50 with Eq. 8.31, the neutron multiplication factor can be written

$$k_{eff} = \frac{k_\infty}{1 + C_D(V_c^2)(C_B/V_c^{2/3})} = \frac{k_\infty}{1 + C(V_c)^{4/3}},$$

(8.53)

with $C \equiv C_D C_B$. For a compacted core, the relative compaction F_c is defined as

$$F_c \equiv \frac{V_{cc}}{V_c},$$

(8.54)

where V_{cc} is the volume of the compacted core. Hence, the neutron multiplication factor for the uniformly compressed core $k_{c,eff}$ can be written

$$k_{c,eff} = \frac{k_\infty}{1 + (\overline{D}/\overline{\Sigma}_a)B^2 F_c^{4/3}},$$

(8.55)

or using Eq. 8.31, the multiplication factor for the compressed core can be expressed in terms of the neutron multiplication factor of the uncompressed core $k_{c,eff}$,

$$k_{c,eff} = k_{eff} \frac{1 + (\overline{D}/\overline{\Sigma}_a)B^2}{1 + (\overline{D}/\overline{\Sigma}_a)B^2 F_c^{4/3}}.$$

(8.56)

The minimum compaction for criticality F_{cmin} is obtained by setting $k_{c,eff}$ equal to one in Eq. 8.56 and solving for F_{cmin}

$$F_{cmin} = \left\{\frac{k_{eff}[1 + (\overline{D}/\overline{\Sigma}_a)B^2] - 1}{(\overline{D}/\overline{\Sigma}_a)B^2}\right\}^{3/4}.$$

(8.57)

The sensitivity to compaction is illustrated in the following example.

Example 8.2

For a postulated reentry accident, estimate the neutron multiplication factor due to compaction. The cylindrical core has a radius of 16 cm, a length of 35 cm, and consists of assemblies of stainless steel clad fully enriched UN fuel rods. The fuel has a solid density of 14 g/cm³ and fuel makes up 80% of the core volume. No axial reflectors are used and poison control elements are located at the core radial periphery to maintain subcriticality. For this configuration, assume no neutron reflection before or during impact. If k_{eff} before impact is 0.92, plot the neutron multiplication factor as a function of relative compaction and find the minimum compaction required to cause an inadvertent criticality. Ignore the neutronic effect of the steel cladding in the calculation and use $\bar{\sigma}_a = 1.85$ b and $\bar{\sigma}_{tr} = 6.4$ b for fully enriched uranium.

Solution:

The uranium atom density is obtained from Eq. 8.51; i.e.,

$$N_{UN} = \frac{\varsigma_{UN} N_A}{M_r}, \quad \text{or}$$

$$N_{UN} = \frac{14 \text{ g/cm}^3 (0.80)(6.022 \times 10^{23} \text{ atoms/g·mole})}{(235+14) \text{ g/g·mole } (10^{24} \text{ b/cm}^2)} = 0.027 \text{ atom/b·cm},$$

$$\Sigma_{tr} = (0.027 \text{ atom/b·cm})(6.4 \text{ b/atom}) = 0.173 \text{ cm}^{-1},$$

$$\Sigma_a = (0.027 \text{ atom/b·cm})(1.85 \text{ b/atom}) = 0.05 \text{ cm}^{-1}, \quad \text{and}$$

$$\frac{\bar{D}}{\bar{\Sigma}_a} = \frac{1}{3\Sigma_{tr}\Sigma_a} = \frac{1}{3(0.173 \text{ cm}^{-1})(0.05 \text{ cm}^{-1})} = 38.5 \text{ cm}^2.$$

From Table 8.1

$$B^2 = \left(\frac{2.405}{R_e}\right)^2 + \left(\frac{\pi}{H}\right)^2 = \left(\frac{2.405}{16}\right)^2 + \left(\frac{\pi}{35}\right)^2 = 0.0307 \text{ cm}^{-2}, \quad \text{and}$$

$$\frac{\bar{D}}{\bar{\Sigma}_a} B^2 = (38.5 \text{ cm}^2)(0.0307 \text{ cm}^{-2}) = 1.18.$$

Using these parameters in Eq. 8.57 we obtain

$$F_{c\min} = \left\{\frac{0.92[1 + 1.18] - 1}{1.18}\right\}^{3/4} = 0.886.$$

Thus, criticality occurs when compaction is 11.4% or greater. From Eq. 8.57 we obtain the relationship of $k_{c,eff}$ to F_c given in Figure 8.5.

Although ~11% compaction is required to cause an inadvertent criticality for this example, the likelihood of uniform compaction is very small. Equation 8.56 shows that uniform compaction will

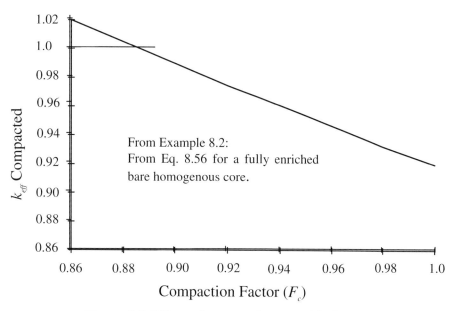

From Example 8.2:
From Eq. 8.56 for a fully enriched
bare homogenous core.

Figure 8.5 Effect of compaction on critical mass.

always result in an increase k_{eff} regardless of the core type, as long as these limiting (and unrealistic) assumptions are used; i.e., the basic geometric shape is unchanged and the quantity and relative position of all core materials remain unchanged.

3.2 Core Reshaping

For core compaction accidents the core average density of the fuel is assumed to increase, but the basic geometric shape and core integrity are assumed to be preserved during compaction. For core reshaping (as defined here) the core deviates from its original shape, but the fuel density of the reconfigured core is assumed to remain unchanged, as shown in Figure 8.4 (b). The definition of reshaping also assumes that any core heterogeneities (e.g., arrangement of fuel and moderator regions) are unchanged. This separation of reshaping and compaction is used simply to clarify the different effects on the neutronic state of the core. Compaction without reshaping and reshaping without some density change are highly unlikely.

From the definition of core reshaping, the multiplication factor for the reshaped core $k_{rs,eff}$ can be expressed by

$$k_{rs,eff} = k_{eff} \frac{1 + (\bar{D}/\bar{\Sigma}_a)B^2}{1 + (\bar{D}/\bar{\Sigma}_a)B_{rs}^2},$$
(8.58)

where B_{rs}^2 is the buckling for the reshaped core. If the core density and core inventory are unchanged, then the volumes of the original core and the reshaped core are the same. If the reshaped core is also assumed to maintain a cylindrical configuration, the reshaped core radius is $R_{rs} = R_0(a_0/a_{rs})^{1/3}$, where R_{rs}, a_{rs}, R_0, and a_0 are the radius and the aspect ratio of the reshaped core and the radius and aspect ratio of the original core, respectively. From Eq. 8.52, the buckling for the reshaped core is

$$B_{rs}^2 = \frac{1}{R_0^2} \left(\frac{a_{rs}}{a_0}\right)^{2/3} \left[(2.405)^2 + \left(\frac{\pi}{2a_{rs}}\right)^2\right].$$
(8.59)

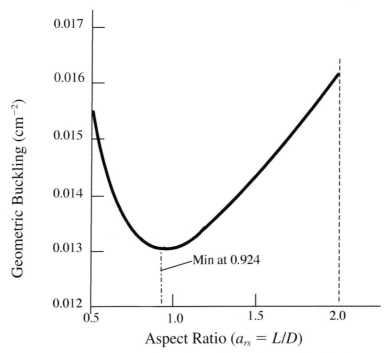

Figure 8.6 Geometric buckling vs. aspect ratio of a reshaped core for a cylindrical core with an initial radius of 20 cm.

Using Eq. 8.59, the buckling was computed as a function of aspect ratio for an initial radius $R_0 = 20$ cm and is plotted in Figure 8.6. As expected the minimum buckling occurs close to $a_{rs} = 1.0$, i.e., at $a_{rs} = 0.924$. Equations 8.58 and 8.59 are used in the following example to estimate the effect of reshaping on a the multiplication factor for a thermal reactor.

Example 8.3

For a postulated launch abort, an onboard space reactor is assumed to impact a concrete surface. The pre-impact core geometry is cylindrical with a radius of 28 cm and a length of 112 cm. As for Example 8.2, assume no neutron reflection before or during impact. Also assume that an end-on impact reduces the aspect ratio while maintaining the basic cylindrical shape and density of core materials. The core is a gas-cooled design with close packed prismatic fuel elements incorporating coolant channels that make up 25% of the core volume. Each fuel element is composed of fully enriched UO_2 powder dispersed in a BeO matrix with a moderator-to-fuel ratio of 380. If the k_{eff} before impact is 0.94, compute and plot the neutron multiplication factor as a function of the core aspect ratio of the reshaped core. Use Table 8.4 for the average microscopic cross sections and use a BeO density of 3.0 g/cm^3.

Solution:

Ignoring the small contribution of UO_2 to the core volume and accounting for 25% void for the cooling channels, we obtain from Eq. 8.51

$$N_{BeO} = \frac{3 \text{ g/cm}^3(0.75)(6.022 \times 10^{23} \text{ atoms/g·mole})}{(25) \text{ g/g·mole } 10^{24} \text{ b/cm}^2} = 0.054 \text{ atom/b·cm},$$

Table 8.4 One-group U-235 and BeO Microscopic Cross Sections
for Moderator/Fuel Ratio = 380

	U-235	BeO
$\bar{\sigma}_{tr}$	11.4	10.1
$\bar{\sigma}_a$	47.9	0.027
$\nu\bar{\sigma}_f$	87.6	0

and $N_u = \dfrac{0.054}{380} = 1.42 \times 10^{-4}$ atom/b·cm. Thus,

$$\Sigma_{tr} = (0.054 \text{ atom/b·cm})(10.1 \text{ b/atom}) + (1.42 \times 10^{-4} \text{ atom/b·cm})(11.4 \text{ b/atom}) = 0.547 \text{ cm}^{-1},$$

$$\Sigma_a = (0.054 \text{ atom/b·cm})(0.027 \text{ b/atom}) + (1.42 \times 10^{-4})(47.9 \text{ b/atom}) = 8.25 \times 10^{-3} \text{ cm}^{-1},$$

$$\frac{\bar{D}}{\bar{\Sigma}_a} = \left(\frac{1}{3 \times 0.547 \text{ cm}^{-1}}\right)\left(\frac{1}{0.00825 \text{ cm}^{-1}}\right) = 73.87 \text{ cm}^2,$$

$$B^2 = (2.405/28 \text{ cm})^2 + (\pi/112 \text{ cm})^2 = 8.16 \times 10^{-3} \text{ cm}^{-2}, \quad \text{and}$$

$$\frac{\bar{D}}{\bar{\Sigma}_a} B^2 = (73.87 \text{ cm}^2)(8.16 \times 10^{-3} \text{ cm}^{-2}) = 0.603.$$

The neutron multiplication factor for reshaped core can be computed as a function of aspect ratio using Eqs. 8.58 and 8.59. For an aspect ratio of 0.924, Eq. 8.59 gives $B_{rs}^2 = (0.924/28)^2(1/2)^{2/3}[(2.405)^2 + (\pi/2 \cdot 0.924)^2] = 6.61 \times 10^{-3}$. Inserting the above parameters and this value for B_{rs}^2 into Eq. 8.57 we obtain the maximum $k_{rs,eff}$

$$k_{rs,eff} = (0.94)(1 + 0.603)/[1 + (73.87)(6.61 \times 10^{-3})] = 1.012.$$

Thus, for this particular case, criticality due to reshaping is predicted to occur. Repeating this procedure for aspect ratios from 2.0 to 0.5 we obtain the plot presented in Figure 8.7. As expected k_{eff} increases to a maximum for an aspect ratio of 0.924 then decreases again as the aspect ratio drops below 0.924.

The foregoing simple equations and Examples 8.2 and 8.3 illustrate that for core compaction and reshaping, the general behavior of neutron multiplication factor is independent of the reactor type. This observation may not be generally valid, however, if the relative position of the core constituents (e.g., moderator and fuel) change as a result of impact.

3.3 Core Redistribution and Fuel Ejection

As mentioned in the previous section, the artificial distinction between compaction and reshaping scenarios was used to clarify the different effects. The more likely scenarios involve a combination of non-uniform core reshaping and regional density changes, as illustrated in Figure 8.4 (c). In addition, fuel may be ejected from the core and collect in a critical configuration outside the core. Fuel ejection is schematically illustrated in Figure 8.4 (d). Monte Carlo computer code analysis, discussed at the end

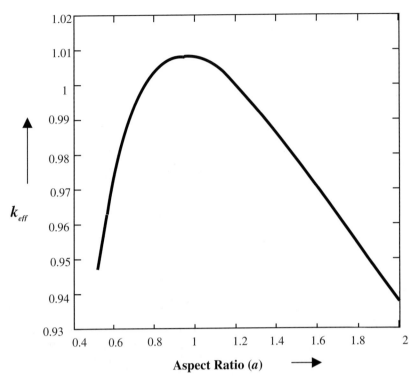

Figure 8.7 Neutron multiplication factor vs. aspect ratio of a reshaped core from Example 8.3.

of this chapter, is typically used to more accurately predict the neutronic effect of these potentially complex geometric changes.

3.4 Mitigation Techniques

Although criticality due to reconfiguration is possible for some reactor designs and accident scenarios, more realistic analyses often predict a loss in k_{eff} due to reconfiguration into a non-optimum geometry. If analysis or testing can convincingly demonstrate that inadvertent criticality by core reconfiguration is virtually impossible, mitigation techniques are unnecessary. If mitigation methods are required, the core may be redesigned or additional structure may be used to minimize core reconfiguration during postulated accidents. An alternative approach is to insert a removable neutron poison in the core or on the core periphery. The worth of the poison is chosen to assure core subcriticality during all credible core reconfiguration accident scenarios. After the reactor is safely deployed in space the poison is removed to permit reactor startup. Removable poisons differ from control rods in that they are designed to be removed just prior to operation and remain out of the core throughout operation. Removable poisons are not intended for reactor control or other uses during reactor operation. Removing some of the fuel from the core prior to launch has been proposed as an alternative to the removable poison approach. In this case, the fuel is inserted in the core after safe deployment to permit reactor startup. Each of these alternatives has potential advantages and disadvantages; the best choice will depend to a large extent on the reactor type and mission.

4.0 Core Flooding

Core flooding scenarios usually postulate a sequence of accident conditions, such as: (1) a launch abort, (2) impact, (3) breach of the core pressure vessel, (4) immersion in a moderating fluid, (5) flooding of core cavities by the moderating fluid flowing through the pressure vessel breach, resulting in (6) inadvertent criticality. The equations developed in Section 2.5 can be used to estimate the effect of core flooding on criticality. Although Eqs. 8.39 and 8.40 were useful for estimating the effect of the moderator-to-fuel ratio on critical mass requirements, Eq. 8.39 applies only to moderated reactors with an appreciable moderator-to-fuel ratio. Space reactors are often fast reactors; consequently, the more broadly applicable equation derived in Section 2.6 (i.e., Eq. 8.47) will be used to determine the effect of core flooding on criticality.

4.1 Approximate Analysis Method

Equation 8.47 can be rewritten in another form that more clearly illustrates the effect of core flooding on the neutronic state of the core. We begin by multiplying the terms in the left bracket by the right bracket. The numerator and denominator for the first term of the resulting expression are then divided by $\Sigma_{a2}\Sigma_{q1}$ and the numerator and denominator for the second term are divided by Σ_{a1} to obtain

$$k_{eff} = \frac{v_2\Sigma_{f2}/\Sigma_{a2}}{(1 + (D_2/\Sigma_{a2})B^2)(\Sigma_{a1}/\Sigma_{q1} + 1 + (D^1/\Sigma_{q1})B^2)} + \frac{v_1\Sigma_{f1}/\Sigma_{a1}}{\Sigma_{q1}/\Sigma_{a1} + 1 + (D_1/\Sigma_{a1})B^2}. \quad (8.60)$$

For a fast reactor with no flooding, Σ_{q1} is very small. In the limit where Σ_{q1} is assumed to equal zero, the first term vanishes and the second term equals the one-group expression for a fast reactor. For significant flooding, Σ_{q1} is large and the second term is small relative to the first term.

A more intuitive and compact form of Eq. 8.60 can be obtained by defining

$$\Pi \equiv \Sigma_{q1}/\Sigma_{a1}, \quad \tau' = D_1/\Sigma_{q1}, \quad \text{and} \quad L_1^2 \equiv D_1/\Sigma_{a1}. \quad (8.61)$$

The prime on the age symbol is used to indicate that this definition of age may yield numerical values somewhat different from published values for the Fermi age. Here, Π is the probability of slowing down to thermal energy relative to the probability of absorption of a fast neutron. The parameter L_1^2 is defined here as the diffusion length for fast neutrons. With these substitutions and replacing D_2/Σ_{a2} with L^2, Eq. 8.60 becomes

$$k_{eff} = \frac{v_2\Sigma_{f2}/\Sigma_{a2}}{(1 + L^2B^2)(1 + 1/\Pi + \tau'B^2)} + \frac{v_1\Sigma_{f1}/\Sigma_{a1}}{1 + \Pi + L_1^2B^2}. \quad (8.62)$$

Further simplification can be obtained by using the following definitions:

$$k_{\infty 1} \equiv \frac{v\Sigma_{f1}}{\Sigma_{a1}} \equiv \text{fast-only } k_\infty, \qquad k_{\infty 2} \equiv \frac{v\Sigma_{f2}}{\Sigma_{a2}} \equiv \text{thermal-only } k_\infty,$$

$$P_1 \equiv \frac{1}{1 + \Pi + L_1^2B^2}, \qquad P_2 \equiv \frac{1}{1 + L^2B^2}, \qquad \text{and} \quad P_S \equiv \frac{1}{1 + (1/\Pi) + \tau'B^2}. \quad (8.63)$$

Substituting these parameters into Eq. 8.62 we obtain

$$k_{eff} = k_{\infty 1}P_1 + k_{\infty 2}P_2P_S. \quad (8.64)$$

The parameter P_1 is the probability that neutrons in the fast energy group will be absorbed rather than lost by leakage or by slowing down to the thermal energy range. The probability that thermal neutrons will be absorbed, rather than lost by leakage, is given by P_2. The parameter P_S is the probability that fast neutrons will reach thermal energies rather than being absorbed or lost by leakage while slowing down. These definitions and formulations are a departure from traditional formulations. This approach, however, provides insights into the effect of core flooding on k_{eff} that cannot be obtained using traditional approximate methods.

The effect of moderation on core parameters is approximated as follows. We first define a moderator-flooding fraction F_M

$$F_M = f_c f_{ff}, \tag{8.65}$$

where f_c is the volume fraction of core that consists of empty cavities in the core and f_{ff} is the fraction of the cavity volume assumed to be filled by the moderating fluid. The macroscopic cross sections with flooding are given by

$$\Sigma_q = \Sigma_{qc} + F_M\Sigma_{qM}, \quad \Sigma_a = \Sigma_{ac} + F_M\Sigma_{aM}, \quad \text{and} \quad D = 1/3(\Sigma_{trc} + F_M\Sigma_{trM} \tag{8.66}$$

for each energy group. The subscripts c and M refer to the unflooded core and the moderating fluid-filled core, respectively. For simplicity, the moderating fluid is assumed to be homogeneously distributed throughout the core. This simplification is not always valid because space reactor core geometries are typically heterogeneous with distinct void or coolant regions. A more thermalized flux, due to moderation, can be attenuated when passing through thick fuel and structural regions. If a homogenized core model is used and attenuation is significant, the effective cross sections of core materials must be flux weighted to account for self-shielding of the neutron flux.

Although the two group method is generally more useful than using a single group for the entire neutron energy spectrum, a single fast energy group is still very sensitive to changes in the fast spectrum due to the addition of a moderating fluid. This sensitivity can be accommodated in the two-group equation (Eq. 8.64) by computing a flux-weighted average fast group cross section. In order to obtain the approximate fast flux spectrum, we begin by writing the neutron diffusion equation for three fast and one thermal energy group;

$$(D_{(1)}B^2 + \Sigma_{q(1)} + \Sigma_{a(1)})\phi_{(1)} = \chi_{(1)}(S_f), \tag{8.67}$$

$$(D_{(2)}B^2 + \Sigma_{q(2)} + \Sigma_{a(2)})\phi_{(2)} = \chi_{(2)}(S_f) + \Sigma_{q(1)}\phi_{(1)}, \tag{8.68}$$

$$(D_{(3)}B^2 + \Sigma_{q(3)} + \Sigma_{a(3)})\phi_{(3)} = \Sigma_{q(2)}\phi_{(2)}, \tag{8.69}$$

$$(D_{(4)}B^2 + \Sigma_{a(4)})\phi_{(4)} = \Sigma_{q(3)}\phi_{(3)}, \tag{8.70}$$

where
$$S_f = \sum_{(g)=1}^{(4)} v\Sigma_{f(g)}\phi_{(g)}. \tag{8.71}$$

Here, the numbers in parentheses in the subscripts indicates four-group values and $\chi_{(1)}$ and $\chi_{(2)}$ are the fission neutron fractional yields to groups (1) and (2) respectively. Fission yields to groups (3) and (4) are assumed to be negligible. Furthermore, these four-group equations contain the assumption that neutrons slow down only to the next lower energy group (this assumption may not be valid if substantial quantities of hydrogenous media are present). Solving simultaneous Eqs. 8.67–8.70, the four-group fluxes per source neutron (i.e., for $S_f = 1$) are

$$\phi_{(1)} = \frac{\chi_{(1)}}{(D_{(1)}B^2 + \Sigma_{q(1)} + \Sigma_{a(1)})}, \tag{8.72}$$

$$\phi_{(2)} = \frac{\chi_{(2)} + \Sigma_{q(1)}\phi_{(1)}}{(D_{(2)}B^2 + \Sigma_{q(2)} + \Sigma_{a(2)})}, \tag{8.73}$$

$$\phi_{(3)} = \frac{\Sigma_{q(2)}\phi_{(2)}}{(D_{(3)}B^2 + \Sigma_{q(3)} + \Sigma_{a(3)})}, \tag{8.74}$$

and $$\phi_{(4)} = \frac{\Sigma_{q(3)}\phi_{(3)}}{(D_{(4)}B^2 + \Sigma_{a(4)})}, \tag{8.75}$$

The flux-weighted one-group fast cross sections are then obtained from

$$\nu\Sigma_{f1} = \frac{\phi_{(1)}\nu\Sigma_{f(1)} + \phi_{(2)}\nu\Sigma_{f(2)} + \phi_{(3)}\nu\Sigma_{f(3)}}{\phi_{(1)} + \phi_{(2)} + \phi_{(3)}}, \tag{8.76}$$

$$\Sigma_{a1} = \frac{\phi_{(1)}\Sigma_{a(1)} + \phi_{(2)}\Sigma_{a(2)} + \phi_{(3)}\Sigma_{a(3)}}{\phi_{(1)} + \phi_{(2)} + \phi_{(3)}}, \tag{8.77}$$

$$D_1 = \frac{\phi_{(1)}D_{(1)} + \phi_{(2)}D_{(2)} + \phi_{(3)}D_{(3)}}{\phi_{(1)} + \phi_{(2)} + \phi_{(3)}}, \tag{8.78}$$

and $$\Sigma_{q1} = \frac{\phi_{(3)}\Sigma_{q(3)}}{\phi_{(1)} + \phi_{(2)} + \phi_{(3)}}, \tag{8.79}$$

The coarse fast spectrum provided by Eqs. 8.72 through 8.74 does not account for changes in the detailed energy spectrum in the resonance energy region. Although approximate methods can be used to account for resonance region spectrum effects and core heterogeneities [5], the additional complexity resulting from their inclusion may be more confusing then enlightening. Consequently, these potential effects will be ignored.

4.2 Flooding Effect on k_{eff}

The equations developed in the previous subsection can be used to estimate the effect of moderating fluid flooding on the neutron multiplication factor for either a fast or thermal reactor. The effect of a postulated flooding accident for a fast space reactor is illustrated in the following example.

Example 8.4

For the core described in Example 8.2 with a 10% void fraction, compute the neutron multiplication factor for full water flooding ($f_{ff} = 1.0$) and compare to the unflooded case. Use the four-group microscopic cross sections in Tables 8.5 and 8.6.

Solution:

Using the UN atom density from Example 8.2 and the above four group cross sections, we obtain the macroscopic cross sections in Table 8.7 for the unflooded case. The macroscopic cross sections in Table

Table 8.5 H_2O Microscopic Cross Sections (barns)

Group	Lower Energy (eV)	σ_a	σ_{tr}	σ_q
(1)	1.353×10^6	0	3.08	2.81
(2)	9.12×10^3	0	10.52	4.04
(3)	0.4	0.035	16.55	4.14
(4)	0.0	0.57	68.6	0.0

Table 8.6 ^{235}UN Microscopic Cross Sections (barns)

Group	χ	$\nu\sigma_f$	σ_a	σ_{tr}	σ_q
(1)	0.575	3.45	1.4	4.7	1
(2)	0.425	3.57	1.7	7	0.5
(3)	0.0	57.5	41.0	51	0.5
(4)	0.0	1230	587	597	0.0

Table 8.7 ^{235}UN Macroscopic Cross Sections (cm^{-1}) for Example 8.5

Group	$\nu\Sigma_f$	Σ_a	Σ_{tr}	Σ_q	D (cm)
(1)	0.0934	0.038	0.127	0.0406	2.62
(2)	0.0967	0.046	0.190	0.0135	1.76
(3)	1.558	1.111	1.381	0.0135	0.241
(4)	33.32	15.9	16.17	0.0	0.021

8.7 are next used with the buckling from Example 2 and Eqs. 8.72–8.75 to obtain the normalized fluxes for groups (1), (2), and (3).

$$\phi_{(1)} = \frac{\chi_{(1)}}{D_{(1)}B^2 + \Sigma_{q(1)} + \Sigma_{a(1)}} = \frac{0.575}{2.62 \times 0.0307 + 0.0406 + 0.038} = 3.616 \text{ n/cm}^2\text{s/S}_f,$$

$$\phi_{(2)} = \frac{0.425 + 0.0406 \times 3.616}{1.76 \times 0.0307 + 0.0135 + 0.046} = 5.037 \text{ n/cm}^2\text{s/S}_f,$$

$$\phi_{(3)} = \frac{0.0135 \times 5.037}{0.241 \times 0.0307 + 0.0135 + 1.111} = 0.0601 \text{ n/cm}^2\text{s/S}_f,$$

and $\phi_{(1)} + \phi_{(2)} + \phi_{(3)} = 3.616 + 5.037 + 0.0601 = 8.713 \text{ n/cm}^2\text{s/S}_f.$

Flux weighting the fast cross sections, we obtain the one-fast-group cross sections

$$\nu\Sigma_{f1} = \frac{3.616 \times 0.093 + 5.037 \times 0.097 + 0.0601 \times 1.558}{8.713} = 0.1054 \text{ cm}^{-1},$$

$$\Sigma_{a1} = \frac{3.616 \times 0.038 + 5.037 \times 0.046 + 0.0601 \times 1.1106}{8.713} = 0.050 \text{ cm}^{-1},$$

$$D_1 = \frac{2.62 \times 3.616 + 1.76 \times 5.037 + 0.241 \times 0.0601}{8.713} = 2.107 \text{ cm},$$

and $\qquad \Sigma_{q1} = \dfrac{0.0601 \times 0.0135}{8.713} = 9.30 \times 10^{-5} \text{ cm}^{-1}.$

Next we calculate the parameters using Eqs. 8.61 and 8.62

$$k_{\infty 1} = \frac{v\Sigma_{f1}}{\Sigma_{a1}} = \frac{0.1054}{0.050} = 2.108, \quad k_{\infty 2} = \frac{v\Sigma_{f2}}{\Sigma_{a2}} = \frac{33.32}{15.9} = 2.096,$$

$$L_1^2 = \frac{D_1}{\Sigma_{a1}} = \frac{2.107}{0.050} = 42.13, \qquad L^2 = \frac{D_2}{\Sigma_{a2}} = \frac{0.021}{15.9} = 0.0013,$$

$$\Pi = \frac{\Sigma_{q1}}{\Sigma_{a1}} = \frac{9.30 \times 10^{-5}}{0.050} = 0.0018, \qquad \tau' = \frac{D_1}{\Sigma_{q1}} = \frac{2.107}{9.30 \times 10^{-5}} = 2.266 \times 10^4,$$

$$P_1 = \frac{1}{1 + \Pi + L_1^2 B^2} = \frac{1}{1 + 0.0018 + (42.13 \times 0.0307)} = 0.436,$$

$$P_2 = \frac{1}{1 + L^2 B^2} = \frac{1}{1 + (0.0013 \times 0.0307)} = 1$$

$$P_S = \frac{1}{1 + (1/\Pi) + \tau' B^2} = \frac{1}{1 + (1/0.0018) + (2.266 \times 10^4 \times 0.0307)} = 8.10 \times 10^{-4}.$$

Using the computed parameter values and Eq. 8.64 we obtain

$$k_{eff} = k_{\infty 1} P_1 + k_{\infty 2} P_2 P_s = (2.108 \times 0.436) + (2.096 \times 1 \times 8.10 \times 10^{-4}) = 0.9199.$$

In the absence of flooding, the probability Π of slowing down to thermal energies, relative to fast absorptions, is very small and τ' is enormous. Consequently, P_1 is fairly large resulting in a significant fast contribution to k_{eff}. The small Π and large τ' also result in a small P_S and a very small thermal contribution to k_{eff}.

Using Eq. 8.51, the core-averaged atom density for water is

$$N_M = \frac{1 \text{ g/cm}^3 (0.1)(1)(6.022 \times 10^{23} \text{ atoms/g·mole})}{18 \text{ g/g·mole } (10^{24} \text{ b/cm}^2)} = 0.0033 \text{ atom/b·cm}.$$

Using N_M and the microscopic cross sections from Table 8.6, the macroscopic cross sections for water are computed and added to the values in Table 8.7 to obtain the core flooded cross sections in Table 8.8.

Table 8.8   Flooded Core Macroscopic Cross Sections (cm^{-1}) for Example 8.5

Group	$v\Sigma_f$	Σ_a	Σ_{tr}	Σ_q	D(cm)
(1)	0.0934	0.038	0.137	0.050	2.43
(2)	0.0967	0.046	0.225	0.027	1.48
(3)	1.558	1.111	1.436	0.027	0.232
(4)	33.32	15.9	16.40	0.0	0.021

The normalized fluxes for groups (1), (2), and (3) for the flooded core are

$$\phi_{(1)} = \frac{0.575}{2.43 \times 0.0307 + 0.05 + 0.038} = 3.56 \text{ n/cm}^2\text{s/S}_f.$$

Also, $\phi_{(2)} = 5.09 \text{ n/cm}^2\text{s/S}_f$ and $\phi_{(3)} = 0.12 \text{ n/cm}^2\text{s/S}_f$,   and

$$\phi_{(1)} + \phi_{(2)} + \phi_{(3)} = 3.56 + 5.09 + 0.12 = 8.77 \text{ n/cm}^2\text{s/S}_f.$$

Flux weighting the fast cross sections we obtain the one-group fast cross sections

$$\nu\Sigma_{f1} = 0.1155 \text{ cm}^{-1}, \qquad \Sigma_{a1} = 0.0573 \text{ cm}^{-1},$$

$$D_1 = 1.85 \text{ cm}, \qquad \Sigma_{q1} = 3.70 \times 10^{-4} \text{ cm}^{-1}.$$

We then obtain the parameters

$$k_{\infty 1} = \frac{0.1155}{0.0573} = 2.015, \qquad k_{\infty 2} = \frac{33.32}{15.9} = 2.096, \qquad \Pi = \frac{3.70 \times 10^{-4}}{0.0574} = 0.0065$$

$$L_1^2 = \frac{1.85}{0.0573} = 32.2, \qquad L^2 = \frac{0.021}{15.9} = 1.32 \times 10^{-3}, \qquad \tau' = \frac{1.85}{3.70 \times 10^{-4}} = 5.00 \times 10^3,$$

$$P_1 = \frac{1}{1 + 0.0065 + (32.2 \times 0.0307)} = 0.501, \qquad P_2 = \frac{1}{1 + (0.00132 \times 0.0307)} = 1, \quad \text{and}$$

$$P_s = \frac{1}{1 + (1/0.0065) + (5.00 \times 10^3 \times 0.0307)} = 3.23 \times 10^{-3}.$$

$$k_{eff} = (2.015 \times 0.501) + (2.096 \times 1 \times 3.23 \times 10^{-3}) = 1.010 + 6.77 \times 10^{-3} = 1.0163.$$

Thus, our approximate method predicts that the fully flooded core will achieve criticality.

Note that for full flooding, the fast diffusion length decreases resulting in a significant increase in the fast component of k_{eff}. Although the thermal component of k_{eff} increases somewhat, the thermal contribution to the increase in k_{eff} is not significant. This prediction may seem surprising; however, the moderator to fuel ratio for full flooding is only $(0.0033)/(0.027) = 0.12$. For this low ratio the age is still very large; consequently, fission from thermal neutron absorption remains very small.

4.3 Flooding Effect with Parasitic Resonance Capture

From Example 8.5, one might conclude that flooding of a fast reactor core will always result in an increase in the core multiplication factor. One might also suspect that flooding of a thermal reactor would not have a significant effect. However, if a thermal reactor is undermoderated, core flooding can cause a significant increase in neutron multiplication factor. On the other hand, many fast reactor concepts use cladding and structural materials that possess significant neutron absorption resonances in the epithermal region. A shift in the spectrum toward the epithermal region, as a result of core flooding, can increase parasitic neutron capture and may significantly reduce any tendency of core flooding to increase k_{eff}. Fast reactors that incorporate structure exhibiting strong resonance absorption may even experience a decrease in the neutron multiplication factor as a result of core flooding. The effect of resonance absorption is illustrated in the following example.

Example 8.5

Repeat the calculations in Example 8.4, for a slightly larger core in which the stainless steel cladding is replaced by Ta cladding. Use a core radius of 17 cm, a core length of 35 cm, and assume 10% of the core volume is Ta with a density of 16.6 g/cm³. Use the following approximate four-group microscopic absorption cross sections for Ta. Ignore the effect of Ta on Σ_{tr} and Σ_q.

Ta microscopic absorption cross sections (barns)

Group	(1)	(2)	(3)	(4)
σ_a	0.08	0.7	70	16

Solution:

From Table 8.1, the buckling for the core is

$$B^2 = (2.405/17 \text{ cm})^2 + (\pi/35 \text{ cm})^2 = 0.0281 \text{ cm}^{-2},$$

and the Ta atom density is

$$N_{Ta} = \frac{16.6 \text{ g/cm}^3 (0.1)(6.022 \times 10^{23} \text{ atoms/g·mole})}{(181 \text{ g/g·mole}) (10^{24} \text{ b/cm}^2)} = 0.0055 \text{ atom/b·cm}.$$

Using these atom densities and the cross sections in the above table, the Ta macroscopic absorption cross sections are computed. The Ta macroscopic absorption cross sections are then added to the values given in Table 8.7 to obtain the core macroscopic absorption cross sections presented in the following table.

Core macroscopic absorption cross sections (cm⁻¹) with 10% Ta

Group	(1)	(2)	(3)	(4)
Σ_a	0.0384	0.0499	1.496	16.00

For the unflooded case the three fast group fluxes are computed using Eqs. 8.72–8.74, yielding

$$\phi_{(1)} = \frac{0.575}{2.62 \times 0.0281 + 0.0406 + 0.0384} = 3.766 \text{ n/cm}^2\text{s/S}_f,$$

$$\phi_{(2)} = 5.123 \text{ n/cm}^2\text{s/S}_f \qquad \text{and} \qquad \phi_{(3)} = 0.0456 \text{ n/cm}^2\text{s/S}_f.$$

Then, $\phi_{(1)} + \phi_{(2)} + \phi_{(3)} = 3.776 + 5.123 + 0.0456 = 8.935 \text{ n/cm}^2\text{s/S}_f.$

Flux weighting the fast cross sections we obtain the one-fast-group cross sections

$$\nu\Sigma_{f1} = 0.1028 \text{ cm}^{-1}, \qquad \Sigma_{a1} = 0.0525 \text{ cm}^{-1}, \qquad D_1 = 2.115 \text{ cm} \qquad \Sigma_{q1} = 6.9 \times 10^{-5} \text{ cm}^{-1},$$

$$k_{\infty 1} = \frac{0.1028}{0.0525} = 1.9578, \qquad\qquad k_{\infty 2} = \frac{33.32}{16.00} = 2.0825,$$

$$L_1^2 = \frac{2.115}{0.0525} = 40.29, \qquad\qquad L^2 = \frac{0.021}{16.00} = 0.0013,$$

$$\Pi = \frac{6.9 \times 10^{-5}}{0.0525} = 0.0013, \qquad\qquad \tau' = \frac{2.115}{6.9 \times 10^{-5}} = 3.07 \times 10^4,$$

$$P_1 = 0.469, \qquad\qquad P_2 = 1, \qquad P_S = 6.2 \times 10^{-4}, \quad \text{and}$$

$$k_{eff} = (1.9578 \times 0.469) + (2.0825 \times 1 \times 6.2 \times 10^{-4}) = 0.9177 + 0.0013 = 0.9190.$$

For the flooded core we obtain

$$\phi_{(1)} = 3.669 \text{ n/cm}^2\text{s/S}_f, \qquad \phi_{(2)} = 5.1372 \text{ n/cm}^2\text{s/S}_f, \qquad \phi_{(3)} = 0.0907 \text{ n/cm}^2\text{s/S}_f.$$

$\phi_{(1)} + \phi_{(2)} + \phi_{(3)} = 8.897$ n/cm²s/S$_f$. Also,

$$\nu\Sigma_{f1} = 0.1104 \text{ cm}^{-1}, \qquad \Sigma_{a1} = 0.0600 \text{ cm}^{-1}, \qquad D_1 = 1.859 \text{ cm} \qquad \Sigma_{q1} = 2.75 \times 10^{-4} \text{ cm}^{-1},$$

$$k_{\infty 1} = \frac{0.1104}{0.0600} = 1.8415, \qquad\qquad k_{\infty 2} = \frac{33.32}{16.00} = 2.0825,$$

$$L_1^2 = \frac{1.859}{0.0600} = 30.99, \qquad\qquad L^2 = \frac{0.0021}{16.00} = 1.31 \times 10^{-3},$$

$$\Pi = \frac{2.75 \times 10^{-4}}{0.0600} = 0.0046, \qquad\qquad \tau' = \frac{1.859}{2.75 \times 10^{-4}} = 6.75 \times 10^3,$$

$$P_1 = 0.5332, \qquad\qquad P_2 = 1, \qquad P_S = 2.45 \times 10^{-3}, \quad \text{and}$$

$$k_{eff} = (1.8415 \times 0.5332) + (2.0825 \times 1 \times 2.45 \times 10^{-3}) = 0.9819 + 5.10 \times 10^{-3} = 0.9870.$$

The fast flux spectrum is observed to shift toward the epithermal region, as a result of flooding, in both Examples 8.4 and 8.5. For Example 8.5, however, the shift in spectrum results in more parasitic absorptions in the epithermal resonances when a Ta cladding is used. The increased absorption results in an appreciable reduction in k_∞, relative to the steel cladding case. The method illustrated in Examples 8.4 and 8.5 was used to obtain the plot of k_{eff} as a function of f_f as presented in Figure 8.8.

4.4 Mitigation Techniques

In Section 4.3 it was mentioned that mitigation methods may not be needed to prevent inadvertent criticality during a postulated water flooding accident. The neutron multiplication factor for a fully moderated thermal reactor will decrease if flooded with a moderating fluid. If a fast reactor contains structure with significant capture resonances, the shift in the flux spectrum due to flooding can increase parasitic capture in the resonance energy region. The increase in parasitic capture may prevent inadvertent criticality for a flooding accident, and in some cases it may cause a decrease in the neutron multiplication factor. If analysis or testing can demonstrate that core flooding will not induce inadvertent criticality, mitigation techniques are unnecessary. For many space reactor designs, however, flooding will result in an inadvertent criticality unless specific mitigative measures are incorporated. For example, removable neutron poison or removable fuel elements, described in the previous section for re-

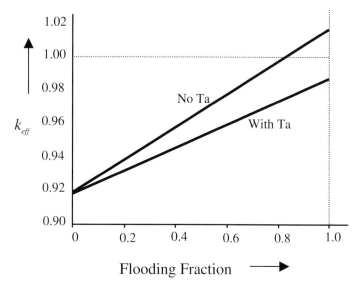

Figure 8.8 Critical mass vs. core flooding fraction, (obtained from Examples 8.4 and 8.5).

configuration accidents, can also be used to prevent flooding criticality. For fast reactors, materials with large parasitic capture cross sections may be added or selected as structural materials to enhance resonance capture during a postulated flooding accident.

5.0 Reflection Accidents

Many space reactor designs employ neutron poisons or adjustable reflectors at the core boundaries in order to assure shutdown or for reactor control. Prior to deployment, the poison or reflector elements can be configured to minimize neutron reflection; consequently, the assumption of no reflection prior to a postulated accident is a reasonable approximation. Accidents can be postulated, however, that result in a change in neutron reflection conditions. For example, an impact accident can be postulated for which neutron poison elements at the core periphery are displaced or ejected. If the reactor is also fully or partially submerged or buried, the surrounding medium will reflect neutrons at the core boundary and could lead to an inadvertent criticality. Water, liquid propellants, sand, and soil are typically proposed as reflecting materials. Although the atmosphere will also reflect some neutrons, the density of the atmosphere is too low to have a significant reflection effect.

The methods developed to this point assume that the neutron flux extrapolates to zero at a distance $2\bar{D}$ beyond the core radius. The flux was also assumed to be separable into energy dependent and spatially dependent components, as given by Eq. 8.11. These assumptions are valid for a bare core homogenous reactor, or a homogenous core surrounded by a highly absorbing neutron poison. For the purpose of understanding the effect on k_{eff} due to core compaction or flooding, these assumptions are reasonable approximations. However, the effect on k_{eff} due to a postulated reflection accident cannot be based on the assumption of zero flux at $R_{bc} + 2\bar{D}$. Furthermore, if the reflecting material is also a good moderator, fast neutrons entering the reflector region will slow down causing a peak in the thermal flux near the core reflector boundary, as illustrated in Figure 8.2. Methods have been developed that account for spectral differences in the core and reflector regions [4]. These methods, however, are tedious and are not presented here. For this elementary introduction to the effects of neutron reflection, we will use a simple one-group model approximation applicable to non-hydrogenous reflectors. Although this simple model cannot account for spectral effects, the principal considerations and characteristics of postulated reflection accidents can be illustrated using a one-group approach.

From Eq. 8.29 we can write the diffusion equation for the core region as

$$\bar{D}_c \nabla^2 \phi_c(\mathbf{r}) + \left[(v\bar{\Sigma}_f/k_{eff}) - \bar{\Sigma}_{ac}\right]\phi_c(\mathbf{r}) = 0, \tag{8.80}$$

where c refers to the core region. Equation 8.80 can also be written as

$$\bar{D}_c \nabla^2 \phi_c(\mathbf{r}) + B^2 \phi_c(\mathbf{r}) = 0, \tag{8.81}$$

where B^2 is the core buckling. For the reflector region, the one-group diffusion equation is

$$\bar{D}_R \nabla^2 \phi_R(\mathbf{r}) - \bar{\Sigma}_{ar} \phi_R(\mathbf{r}) = 0, \tag{8.82}$$

or

$$\nabla^2 \phi_R(\mathbf{r}) - \frac{1}{L_r^2} \phi_R(\mathbf{r}) = 0, \tag{8.83}$$

where R refers to the reflector region and L_R is the diffusion length for the reflector.

For a spherical core, the solution to Eq. 8.81 is

$$\phi_c(r) = A_c \frac{sinBr}{r}, \tag{8.84}$$

where A_c is an arbitrary constant determined by the boundary conditions. It can also be shown [3] that the solution to Eq. 8.83 is

$$\phi_R(r) = A_R \frac{\sinh[(1/L_R)(R_c + \delta t_R - r)]}{r/L_R}, \tag{8.85}$$

where A_R is an arbitrary constant, R_c is the core radius, and δt_R is the reflector thickness. The neutron flux and neutron current must be continuous at the core-reflector interface; i.e.,

$$\phi_c(R_c) = \phi_R(R_c), \tag{8.86}$$

and

$$D_c \frac{d\phi_c}{dr} = D_R \frac{d\phi_R}{dr} \qquad \text{at } r = R_c. \tag{8.87}$$

Thus,

$$A_c \frac{\sin(BR_c)}{R_c} = A_R \frac{\sinh(\delta t_R/L_R)}{R_c/L_R}, \tag{8.88}$$

and

$$D_c A_c \left(\frac{B\cos(BR_c)}{R_c} - \frac{\sin(BR_c)}{R_c^2}\right) = -D_R A_R \left[\frac{\cosh(\delta t_R/L_R)}{R_c} + \frac{\sinh(\delta t_R/L_R)}{R_c^2/L_R}\right]. \tag{8.89}$$

Dividing Eq. 8.89 by Eq. 8.88 we obtain

$$D_c \left(B\cot(BR_c) - \frac{1}{R_c}\right) = -D_R \left[\frac{1}{L_R}\coth(\delta t_R/L_R) + \frac{1}{R_c}\right]. \tag{8.90}$$

For most space reactor safety studies, the bounding assumption is made that for a reflection accident, the surrounding reflector material (e.g., ocean water) is infinitely thick. This assumption is quite reasonable since the reflector effect on k_{eff} saturates within less than a meter for most reflecting materials. With the assumption of an infinitely thick reflector, Eq. 8.90 becomes the transcendental equation,

$$\cot(BR_c) = \frac{1}{B}\left[\frac{1}{R_c}\left(1 - \frac{D_R}{D_c}\right) - \frac{D_R}{D_c L_R}\right]. \tag{8.91}$$

If R_c, D_c, D_R, and L_R are known, Eq. 8.91 can be solved graphically to determine the buckling. The neutron multiplication factor for the reflected core is then computed from

$$k_{eff} = \frac{v\Sigma_f/\Sigma_a}{1 + (\overline{D}/\overline{\Sigma}_a)B^2}. \tag{8.92}$$

The effect of reflection on k_{eff} is shown, from Eqs. 8.91 and 8.92, to depend on core size and diffusion parameters for the core and the reflector. The use of these equations is illustrated in the following example.

Example 8.6

A cylindrical UO_2/BeO thermal space reactor uses radial reflector drums with poison segments for control. No axial reflector is used. Prior to space deployment, poison segments in the radial reflector are positioned to prevent significant neutron reflection. For a postulated launch abort accident scenario, the launch vehicle's liquid oxygen (LOX) tank ruptures and LOX fills a crater produced by the impact. During impact, the reactor loses its radial reflector and subsequently rolls into the pool of LOX. Assume that no core flooding occurs, but submersion results in neutron reflection at the radial and axial core boundaries. The core is 62 cm in length with a 31-cm radius, and the core composition is the same as the composition given in Example 8.3. Use the equations developed in this section to estimate the k_{eff} of the submerged reactor core. Also use $v\Sigma_f = 0.01244$, $\Sigma_a = 0.00825$, $D_C = 0.6094$, $L_C = 8.595$, $L_R = 997.4$, and $D_R = 4.629$.

Solution:

From Table 8.1, the buckling for the unreflected (bare) core is

$$B_{bc}^2 = (2.405/31 \text{ cm})^2 + (\pi/62 \text{ cm})^2 = 0.0086 \text{ cm}^{-2},$$

and from Eq. 8.92, the multiplication factor for the unreflected core is

$$k_{eff} = \frac{v\Sigma_f/\Sigma_a}{1 + L_c^2 B_{bc}^2} = \frac{0.01244/0.00825}{1 + (8.595)^2(0.0086)} = 0.9221.$$

The radius R_{es} of an equivalent spherical reactor with the same leakage is

$$R_{es} = \frac{\pi}{(0.0086)^{1/2}} = 33.9 \text{ cm.}$$

Using Eq. 8.91, we define:

$$X(B) \equiv \cot(BR_c) = \cot(B \times 33.9) \quad \text{and}$$

$$Y(B) \equiv \frac{1}{B}\left[\frac{1}{R_c}\left(1 - \frac{D_R}{D_c}\right) - \frac{D_R}{D_c L_R}\right] = \frac{1}{B}\left[\frac{1}{33.9}\left(1 - \frac{4.269}{0.6094}\right) - \frac{4.269}{(0.6094)(997.4)}\right].$$

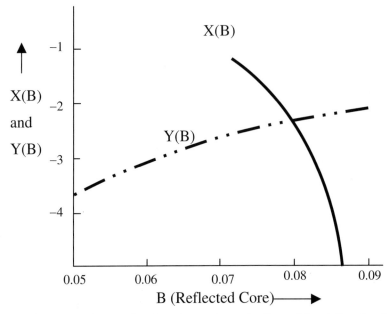

Figure 8.9 Plot of X(B) and Y(B) for determination of reflected buckling for Example 8.6.

Solving for $X(B)$ and $Y(B)$ as a function of B, and plotting (Figure 8.9), the intersection of $X(B)$ and $Y(B)$ is at $B = 0.0805$. Thus,

$$k_{eff}(reflected) = \frac{\upsilon\Sigma_f/\Sigma_a}{1 + L_c^2 B^2} = \frac{0.01244/0.00825}{1 + (8.595)^2(0.0805)^2} = 0.9456.$$

consequently, criticality is not predicted for this hypothetical LOX reflection accident.

Mitigation techniques described in previous sections (e.g., removable poisons) can also be used to prevent inadvertent criticality due to postulated reflection accidents.

6.0 Poison Displacement Accidents

Neutron poisons, such as boron, are typically used to assure safe shutdown and to control the operation of reactors. For space reactors, neutron poisons are often contained in segments of radial reflector drums, as described in Chapter 2. Prior to deployment, the poison segments of the control drums are turned toward the core. A launch accident can be postulated in which an impact either causes the control drums to turn or causes the drums to be ejected from the periphery of the core. If the core is subsequently surrounded by a reflecting medium, neutron leakage will decrease and inadvertent criticality is possible. The latter case was discussed in the previous subsection. If control drums turn during a postulated accident and the reflecting portion of the drums face the core, neutron reflection will increase, possibly inducing an inadvertent criticality. The possibility of an inadvertent criticality by this scenario will depend on the design of the control features, the angle of drum rotation, the number of drums rotated, other shutdown features, and the probability that an accident of this type can occur.

Large space reactors may require internal control rods. Furthermore, space reactor designs may include additional shutdown features, such as neutron poisons in the form of rods inserted in channels within the reactor core. These poison elements, sometimes called safety rods, are used to assure reactor shutdown during postulated accidents. After safe reactor deployment in space, the safety rods are removed from the core to permit reactor startup. Safety rods and control rods are often in the form of cylindrical rods of B_4C contained within a metal cladding. Other rod shapes and neutron poison materials are possible. Neutron poisons in the form of removable wires and beads have also been studied. The effect of safety or control rod ejection on k_{eff} must be computed to determine the potential neutronic effect.

Poison rods are typically highly absorbing and are placed in only a few locations of the core. These conditions result in a highly heterogeneous core configuration that is not amenable to the approximate methods developed in the previous sections of this chapter. Detailed Monte Carlo or transport theory computer calculations are used to compute the effect of poison rod ejection. For the purpose of gaining a feel for the effect of poison rod ejection on the k_{eff} for the core, an approximate equation, developed by Glasstone and Edlund [3], is used here. Their equation,

$$\Delta k_{eff} = \frac{7.5L^2}{R_c^2}\left[0.116 + \ln\left(\frac{R_c}{2.4\ r_{eff}}\right)\right]^{-1},\tag{8.93}$$

gives the approximate change in reactivity Δk_{eff} due to the loss of a central poison rod from a cylindrical core of radius R_c. Here, r_{eff} is the effective radius of the poison rod. The effective poison rod radius is defined as $r_{eff} \equiv r_{pr} - d_{pr}$, where r_{pr} is the actual poison rod radius and d_{pr} is the linear extrapolation distance into the rod. (From a plot of the neutron flux external to the poison rod, the flux at the rod surface can be extrapolated linearly to zero at some point within the rod. The linear extrapolation distance is defined as the distance from the surface to the location at which the extrapolated flux equals zero.) It has been assumed here that the poison rod is black for thermal neutrons; i.e., all thermal neutrons incident on the rod are absorbed. For a small neutronically black cylindrical rod, $d_{pr} = 2D$. A simple example follows.

Example 8.7

For the cylindrical reactor in Example 8.6, compute the change in k_{eff} due to the ejection of a central safety rod. The safety rod consists of 2 cm radius cylindrical steel tube containing B_4C. Use a diffusion length for B_4C of 0.6083 cm.

Solution:

From Table 8.1, the buckling for the unreflected (bare) core is

$$B_{bc}^2 = (2.405/31\ cm)^2 + (\pi/62\ cm)^2 = 0.0086\ cm^{-2},$$

and the multiplication factor for the unreflected core is obtained from Eq. 8.93:

$$\Delta k_{eff} = \frac{7.5(8.62)^2}{(31)^2}\left(0.116 + \ln\left\{\frac{31}{2.4[2 - 2(0.6083)]}\right\}\right)^{-1} = 0.199.$$

Carefully designed and tested latching mechanisms can be used to prevent poison rod ejection. The possibility of poison rod withdrawal or control drum operation due to a spurious signal will be discussed in Chapter 9.

7.0 Combined Effects

Many of the postulated inadvertent criticality accident scenarios described in the previous subsections may occur together or in sequence. Combined and sequential accident scenarios may increase or decrease the neutron multiplication factor. In general, the combined effects of two or more types of postulated accidents cannot be determined by simply adding the change in k_{eff} due to each type of accident. For example, an impact accident that causes spreading of the core will reduce the neutron multiplication factor. Core spreading, however, can increase the fraction of empty core space. If the expanded core is unmoderated or undermoderated and is subsequently flooded, the resulting increase in moderator fraction can cause an increase in k_{eff} far in excess of the effect of flooding the core in its original geometry. Some types of scenarios, such as core flooding and reflection accidents are logically considered together in the neutronic analysis. Some scenarios will be possible for a particular combination of reactor design and mission profile, and impossible for a different set of design and operational considerations. One of the safety analyst's responsibilities is to assess the set of possible scenarios for the particular mission and reactor design features under consideration.

8.0 Computer Methods

The calculational methods described in the preceding sections are too approximate for reliable criticality safety assessment. Neutronic safety analysis is usually performed using either Monte Carlo or transport theory computer codes.

8.1 Monte Carlo Methods

Monte Carlo methods are especially important for determining the neutronic state of unsymmetrical core configurations. The predicted core geometries resulting from core reconfiguration accidents are often highly asymmetric. Standard transport theory and diffusion theory methods for predicting neutron multiplication factors are not well suited for these potentially unusual geometries. Monte Carlo neutronic analysis tools, such as the MCNP computer code [6], are more appropriate for neutronic analysis of non-symmetric core geometries. The Monte Carlo method predicts the neutron flux and the core neutron multiplication factor by tracing the individual histories of many neutrons through successive neutron-nucleus collisions. Until fairly recently, tracking many neutron histories required substantial computer computational time. As a consequence, the number of core geometries analyzed using Monte Carlo methods was fairly limited. With the rapid growth in computer power, however, the computational time required for a Monte Carlo analysis is no longer a major constraint.

Required Monte Carlo code input includes a detailed description of the core geometry, the atom densities of all materials in each core region, and energy/angle dependent microscopic neutron cross sections for each nuclide in the core. Using a random number generator and assuming an isotropic fission source, the initial angle is selected for the trajectory of each neutron. Sets of random numbers are then used, along with the macroscopic neutron cross sections for each core region, to obtain the location and the consequences of the first collision for each neutron. An individual neutron may be captured parasitically, scatter, cause a fission event, or escape from the core. If the neutron is scattered, random numbers and neutronic data are used to determine the energy and angle of the scattered neutron. If a fission event occurs, the trajectory of the liberated fission neutrons is determined. The histories of subsequent collision events are determined by using this approach for all scattered neutrons and all neutrons produced by fission. The individual histories are then combined to determine the spatial and energy distribution of the neutron flux. The ratio of the neutron production rate due to fission to the neutron loss rate due to absorption or leakage provides the neutron multiplication factor for the core.

8.2 Transport Theory Methods

Transport theory methods, such as TWODANT [7], use a deterministic approach rather than the probabilistic approach used in a Monte Carlo analysis. The transport equation is an adaptation of the Boltzmann equation to neutron transport in fission reactors. For steady state conditions, with no external source, the transport equation can be expressed as

$$\mathbf{\Omega} \cdot \nabla \psi(\mathbf{r}, E, \mathbf{\Omega}) + \Sigma_T(\mathbf{r}, E)\psi(\mathbf{r}, E, \mathbf{\Omega}) = \int\int \Sigma_S(\mathbf{r}, E', \mathbf{\Omega}' \to E, \mathbf{\Omega})\psi(\mathbf{r}, E, \mathbf{\Omega})dE'\,d\mathbf{\Omega}'$$

$$+ \frac{\chi}{4\pi} \int\int \nu\Sigma_f(\mathbf{r}, E)\psi(\mathbf{r}, E, \mathbf{\Omega})dE'\,d\mathbf{\Omega}' \qquad (8.94)$$

In this equation, ψ is the angular dependent neutron flux, $\mathbf{r}$ is the position vector, $\mathbf{\Omega}$ is the unit vector in the direction of motion, Σ_T is the total cross section for all neutron-nucleus collisions, and $\Sigma_S(\mathbf{r}, E', \mathbf{\Omega}' \to E, \mathbf{\Omega})$ is the macroscopic scattering cross section for scattering from energy E' with direction $\mathbf{\Omega}'$ in solid angle $d\mathbf{\Omega}'$ to energy E and solid angle $d\mathbf{\Omega}$ about direction $\mathbf{\Omega}$. The first and second terms on the left side of Eq. 8.94 give the neutron loss from element $d\mathbf{r}dEd\mathbf{\Omega}$ due to leakage and collisions, respectively. The first and second terms on the right side of the equation are the neutron gain due to scattering and direct fission yield. Eq. 8.94 is intractable to direct solution for most reactor physics problems. An approximate solution to the transport equation can be obtained by representing the angle dependent terms as expansions of spherical harmonics polynomials [8]. For plane geometry, the angular dependent terms are expanded in a series of Legendre polynomials, $P_n(\mu)$, where μ is the direction cosine and n is the index of the nth term in the series. This approach is frequently referred to as the P_n approximation. Only the first few terms of the expansions are needed to obtain accurate solutions for most problems. It can be shown that the P_1 approximation (i.e., truncating the series at $n = 1$) reduces to the neutron diffusion equation [8].

Energy dependence is included in P_n codes by using discrete energy groups with cross sections averaged over the energy range for each group. The detailed energy spectrum for averaging cross sections is typically obtained for each distinct region of the reactor using a separate "spectrum" computer code; e.g., [9]. The spectrum codes employ many energy groups to obtain the detailed energy spectrum. A number of calculational schemes have been developed to properly account for the very fine structure of resonance cross sections. Simple approximations are often used to account for the effect on the spectrum due to neutron leakage. Energy-averaged cross sections for each energy group are obtained for each region using the computed spectra and a flux-weighting method similar to the approach used in the previous subsection for diffusion theory calculations.

Major reactor regions must be identified and appropriate cross section sets and atom densities are assigned to each region. The positions of region boundaries are specified and a geometric mesh is defined within each region in order to carry out finite difference calculations. The spatially dependent neutron flux and the neutron multiplication factor are then obtained using iterative finite difference methods. Often the specified reactor regions contain heterogeneous structure that is not explicitly included in the transport calculation; i.e., the materials within these regions are homogenized to simplify the calculations. If the flux is locally perturbed by the individual components within a region, position-dependent flux weighting of cross sections may be necessary. For example, a core region containing identical clad fuel rods surrounded by a moderator and coolant channels may show significant thermal flux peaking in the moderator and flux attenuation in the fuel. One method for obtaining flux-weighting factors is to perform a separate calculation for a unit cell of the region components. Typically, a simple one-dimensional calculation is used to obtain the average fluxes in each component and energy group. This method assumes that a core region can be considered as an array of identical cells. Each cell, for example, may consist of cylindrically symmetric zones containing a central fuel pellet, cladding, and annular regions for the moderator and coolant. The average fluxes from the one-dimensional calculation are then used to obtain weighting factors ω_{cg} given by

$$\omega_{cg} = \frac{\overline{\phi}_{cg}}{\overline{\phi}_{Rg}}, \qquad (8.95)$$

where ϕ_{cg} is the average flux for group g in component c, and ϕ_{Rg} is the group g flux averaged over the cell for region R. Thus, the homogenized macroscopic absorption cross section for group g used in region R in the transport calculation is given by

$$\Sigma_{a,Rg} = \sum_{all\ i} N_{i,c}\omega_{c,g}\sigma_{a,i,g}. \tag{8.96}$$

Here, $N_{i,c}$ is the atom density of nuclide i in component c homogenized over region R.

An alternative method for approximating Eq. 8.94 is called the discrete ordinates method. For the discrete ordinates method the flux is divided into components representing discrete directions for the neutron flux, rather than representing the angular flux in terms of a P_n expansion. In other words the flux is divided into components for angular directions between μ_k and μ_{k+1}, with $k = 1, 2, \ldots n$. This approach is also referred to as the S_n method. The TWODANT computer code is one of the more commonly used two-dimensional discrete ordinates transport theory code.

9.0 Criticality Testing

Often the neutron cross sections required for criticality safety analysis are not sufficiently certain to assure accurate criticality safety predictions. In order to verify criticality predictions, it may be necessary to perform criticality safety tests. Nuclear criticality tests typically involve the construction of a *critical mockup* of the reactor using the actual nuclear materials and the most important structural materials that will be used to construct the space reactor. Furthermore, the geometry of the mockup is close to that of the actual reactor. Criticality mockups are only tested at very low power levels (often called zero-power testing). As a consequence, many details of the actual reactor design (coolant pumping system, pressure boundaries, etc.) are not required for the criticality mockup. The simplified design of a mockup permits relatively inexpensive, early verification of cross sections and analysis methods. If criticality predictions do not accurately predict the critical configuration of the mockup, alternative cross section sets may be required, or the deviation of predictions from measurement (Δk_{eff}) may be used as an adjustment to criticality predictions for the space reactor.

The use of a criticality mockup also permits rapid change-out of materials and components. Thus, a variety of materials and configurations can be studied. A critical mockup was used at the Kurchotov Institute to perform a variety of design and safety tests for the Russian Enisey space reactor. In addition to the normal configuration and reactor environment, tests were performed for simulated accident configurations. Figure 8.10 shows the critical assembly used to mock up the Enisey reactor for a postulated accident in which the reactor is flooded with water and reflected by water. This configuration simulates the conditions for a postulated impact accident that results in a breach of the reactor vessel, loss of radial control drums, and subsequent immersion in water. This mockup was subsequently modified to perform criticality tests simulating immersion in wet sand.

In addition to mockup criticality testing, criticality testing before launch is desirable to assure that criticality safety margins of the actual space reactor agree with predictions. Ground criticality testing of the actual space reactor must be at essentially zero power to minimize the buildup of fission and activation products. A post-test cooling period can be used to further reduce the radioactive content of the space reactor. Minimizing radioactivity due to prelaunch testing is necessary to facilitate safe handling during transportation and mating with the launch vehicle, and to minimize the potential source term in the event of a launch accident.

10.0 Summary

For a critical reactor, the neutron production rate by fission is equal to the neutron loss rate by absorption and neutron leakage, and the critical mass was defined as the quantity of nuclear fuel required

Figure 8.10 Critical assembly mockup for water-flooded, water-reflected Enisey reactor.

to achieve criticality. In this chapter, neutron leakage was approximated using diffusion theory. For this approximation, the neutron leakage rate is given as the product of a diffusion constant and the Laplacian of the neutron flux. The neutron diffusion equation can be written terms of energy averaged reactor parameters (cross sections) and the neutron flux (Eq. 8.14). The parameters of the diffusion equation can be rearranged to define a material buckling and a geometric buckling. For a just critical core, the material buckling and the geometric buckling are equal. In order to solve the diffusion equation for a bare reactor, the neutron flux is assumed to extrapolate to zero at points approximately two diffusion lengths greater than the core dimensions. The critical mass of a reactor can be reduced by surrounding the core with neutron reflector materials.

An infinite media neutron multiplication factor is defined as the ratio of the neutron production rate to the absorption rate for an infinite quantity of core material, and an effective neutron multiplication factor is defined as the ratio of the neutron production rate to the neutron loss rate by absorption and by neutron leakage from the core. The effective neutron multiplication factor is, consequently, the product of the infinite medium neutron multiplication factor and the non-leakage probability. Materials with low relative atomic mass, called moderators, can be used in the reactor core to slow down neutrons emitted during fission. As neutrons slow down to thermal energies, the probability of absorption by the fuel increases; hence, moderation generally increases the neutron multiplication factor. Multigroup techniques are commonly used to properly include energy spectrum effects. Multigroup methods divide the energy spectrum into several energy groups and a separate equation is used for each energy group.

Impact accident scenarios can be postulated that lead to core reconfiguration. Core reconfiguration can include core compaction and core reshaping, resulting in reduced neutron leakage and inadvertent criticality. Fuel ejection from the core may also result in a critical configuration outside the core. For one class of accident scenarios, an impact accident results in a breach in reactor barriers followed by submersion in a moderating fluid, such as ocean water. Enhanced moderation or reflection by the moderating fluid may result in an inadvertent criticality accident. Rugged design and the use of poison safety rods can be used to prevent inadvertent criticality accidents.

Symbols

a	aspect ratio	M_r	relative molecular mass
A	mass number	N	atom density
A_r	relative atomic mass	N_A	Avogadro's number
b	barn	n	neutron density
B^2	buckling	P_1	$(1 + \Pi + L_1^2 B^2)^{-1}$
C, C_B, C_D	core compaction constants	P_2	$(1 + L^2 B^2)^{-1}$
d_{pr}	poison rod extrapolated distance	$\mathcal{P}_f$	neutron fission production rate
D	neutron diffusion coefficient	P_s	$(1 + /\Pi + \tau' B^2)^{-1}$
E	energy	P_{NL}	non-leakage probability
F_C	relative compaction	P_{NL1}	fast non-leakage probability
F_M	moderator flooding fraction	r	radius
f_C	cavity core volume fraction	$\mathbf{r}$	position vector
f_{ff}	cavity fraction fluid-filled	r_{eff}	poison rod effective radius
k_{eff}	effective n-multiplication factor	r_{pr}	poison rod radius
$k_{c,eff}$	k_{eff} for compacted core	R_{bc}	bare core radius
k_∞	infinite medium n-mult. factor	R_e	extrapolated core radius
L	thermal-n diffusion length	R_{es}	equivalent sphere core radius
L_1	fast-n diffusion length	t	time
$\mathcal{L}_a$	neutron absorption loss rate/vol.	V	volume
$\mathcal{L}_L$	neutron leakage loss rate/vol.	VF	volume fraction
m_{CR}	critical mass	x,y,z	Cartesian coordinates

α_s	$(A + 1)^2/(A - 1)^2$	σ_γ	for capture
χ	n fission fraction yield	Σ	macroscopic cross section
δ	reflector savings	τ	Fermi age
δ_{tR}	reflector thickness	υ	neutrons/fission
Φ	space/energy dependent n-flux	ξ	logarithmic energy decrement
ϕ	space-dependent or group n-flux	$\xi\Sigma_s$	slowing down power
φ	energy-dependent neutron flux	$\xi\Sigma_s/\Sigma_a$	moderating ratio
κ^D	energy yield/nuclear decay	Ω	unit vector in direction of motion
λ	decay constant	ψ	angle-dependent n flux
Π	Σ_{q1}/Σ_{a1}	ς	density
σ	microscopic cross section		

Special Subscripts/Superscripts

0	original	M	moderator or moderated
a	neutron absorption	pr	poison rod
bc	bare core	q	slowing down (cross section)
c	core	R	reflector
cc	compact core	rs	reshaped core
f	nuclear fission	s	scattering (cross section)
F	fuel	tr	transport (cross section)
i	radioisotope identifier	1,2	two energy group indices
j	reaction type	(1)-(4)	four energy group indices

References

1. Lamarsh, J. R., *Introduction to Nuclear Reactor Theory*. Addison-Wesley, Reading MA, 1965.
2. Weinberg, A. M. and E. P. Wigner, *The Physical Theory of Neutron Chain Reactors*. Univiversity of Chicago Press, Chicago, IL, 1978.
3. Glasstone, S. and M. C. Edlund, *Nuclear Reactor Theory*. Van Nostrand Reinhold Co., Princeton, NJ, 1952.

4. Glasstone, S. and A. Sesonske, *Nuclear Reactor Engineering*. Van Nostrand Reinhold Co., New York, 1967.
5. Argonne National Laboratory, *Reactor Physics Constants*. ANL-5800. USAEC, 1965.
6. Briesmeister, J. F. (Ed.), "MCNP- A General Monte Carlo Code for Neutron and Photon Transport, Version 3A." Los Alamos National Laboratory LA-7396-M Rev. 2, Los Alamos, NM, 1986.
7. Alcouffe, R. E., F. W. Brinkley, D. R. Marr, R. D. O'Dell, "User's Guide for TWODANT: A Code Package for Two-Dimensional, Diffusion Accelerated, Neutral-particle Transport." LA-10049-M Los Alamos National Laboratory Report Rev. 1, Los Alamos, NM, Feb. 1990.
8. Bell, G. I. and S. Glasstone, *Nuclear Reactor Theory*. Van Nostrand Reinhold Company, New York, 1970.
9. Greene, N. M., J. L. Lucius, L. M. Petrie, W. E. Ford, J. E. White, and R. Q. Wright, "AMPX: A Modular Code System for Generating Coupled Multigroup Neutron Gamma Libraries from ENDF/B." Oak Ridge National Laboratory ORNL TM –3706, Oak Ridge, TN, 1976.

Student Exercises

1. (a) A particle bed space reactor design is proposed for nuclear thermal propulsion. The reactor consists of porous cylinders containing a bed of lightly coated fully enriched UC fuel particles, with a UC density of 13.0 g/cm^3. The volume fraction of uranium in the bed is 55%. The fraction of the core volume for the fuel bed, coolant channels, and core structure is 0.75 0.24, and 0.01, respectively. The core is a cylinder with a diameter and length of 44 cm. No axial reflectors or moderator are used and poison elements are located at the radial periphery of the core. For this configuration, assume no neutron reflection. A scenario is postulated for which an impact accident ruptures the fuel cylinders and the fuel fills all of the available coolant channel space in the bottom of the core (assuming an upright cylinder). Determine the new height of the fuel in the core following reconfiguration, and calculate the buckling for the pre-impact and post-impact core. (b) Develop the equation required to estimate the change in the neutron multiplication factor. (c) If the k_{eff} prior to impact was 0.90, compute the new value for the k_{eff} following the postulated core reconfiguration accident. Ignore self-shielding effects and the effects of other materials, and use the fuel cross sections $\bar{\sigma}_a = 1.85$ b and $\bar{\sigma}_{tr} = 6.4$ b.

2. For the space reactor described in Exercise 1, assume that an impact ejects all of the fuel particles into a deep cylindrical trench. Compute the neutron multiplication factor for the ejected fuel as a function of trench diameter. Assume the particle packing fraction used in Exercise 1, and assume that the height of the bed in the trench is uniform.

3. For the reactor discussed in Examples 8.2 and 8.5, assume that an impact accident causes core spreading (in the radial direction) and doubles the void fraction in the core. Also assume reactor submersion with full core flooding. Ignoring self-shielding effects, compute the neutron multiplication factor following flooding. Compare results to the neutron multiplication factor from Example 8.5 and discuss findings.

4. For the reactor discussed in Examples 8.2 and 8.5, assume that an impact accident causes half of the fuel to be crushed to a powder and ejected into a water-filled cylindrical hole 60 cm deep and 60 cm in diameter. Assume that the fuel forms a uniform suspension in the water-filled hole and ignore reflection effects. Compute the neutron multiplication factor, compare to results from Exercise 3, and discuss findings.

5. Repeat the calculation in Example 8.6 assuming submersion and reflection by dry sand. Assume that the sand is pure SiO$_2$ with a density of 1.6g/cm^3, and use the following thermal cross sections.

	σ_a	σ_{tr}
Si	0.14	2.34
O	0.0002	3.64

6. Using Eq. 8.93, develop a plot of Δk_{eff} vs. core radius between 25 and 100 cm. Present results for effective rod radii of 1 cm and 5 cm, for reactor systems exhibiting a diffusion length of 2 cm and a diffusion length of 25 cm.

Chapter 9

Reactor Transient Analysis

Albert C. Marshall and Edward T. Dugan

The objectives of this chapter are to identify safety issues associated with postulated reactor transient accidents and to present simple analysis approaches for understanding the characteristics of reactor transients. The focus of the transient analysis discussion in this chapter will be on potential transient overpower accidents and cooling failure accidents. An introduction to reactor kinetics is given to provide the background required for understanding the development of transient overpower analysis models.

1.0 Scenarios and Issues

In Chapter 8, the potential for accidental reactor criticality and supercriticality was explored. In this chapter, the dynamic behavior of a reactor system is studied for supercritical accident conditions. In addition, this chapter will examine reactor dynamic response during postulated cooling failure accidents. Some accident scenarios relating to space reactor dynamics are postulated for the prelaunch phase. Other reactor dynamic accidents may result from the consequences of an unplanned reentry. Missions involving astronauts in the vicinity of an operating space reactor must consider the potential safety risk to astronauts resulting from postulated transient accidents. For some missions, space reactor transient accidents during the operational phase may present public safety and environmental issues. For example, reactor disruption could result from an operational transient accident. If the system were operating in LEO, a severely damaged system may be incapable of boost to high orbit for disposal. As a consequence, the system may reenter Earth's biosphere before fission product activity decays to a safe level. Another possibility is that debris generated by an operational transient accident might have a small ballistic coefficient. The low ballistic coefficient could result in premature reentry of the radioactive debris. Space reactor accidents during the operational phase do not necessarily present safety or environmental issues. If astronauts are not in the vicinity of an operating reactor and the reactor is in a sufficiently high orbit or on the surface of another planet or moon, reactor accidents may have no safety consequence for humans or Earth's biosphere.

1.1 Transient Overpower Accidents

In Chapter 8, scenarios and issues were postulated that could lead to an inadvertent criticality accident. In general, however, postulated accidental criticality accidents will result in a supercritical condition. Thus, accidental criticality scenarios proposed in Chapter 8 apply to transient overpower accidents. Prelaunch transient overpower accidents include potential accidental startup and supercriticality prior to launch. Such postulated accidents may result from a spurious signal to the startup and control system or from a prelaunch impact, fire, or explosion accident that causes movement of control or shutdown elements. The principal issues include potential exposure of the ground crew to radiation from a supercritical reactor and the possibility of reactor disruption and dispersion of radioactive material. Disruption can be postulated to result from explosive disassembly. Disruption may also result from prolonged high power operation without adequate heat removal, leading to melting of the fuel and containing structures (e.g., cladding and pressure vessel). If radioactive materials are released and dispersed, radiological exposure of the public must be considered.

Similar scenarios can be postulated for reentry accidents; i.e., reentry heating or impact may result in movement of control elements causing an inadvertent supercriticality. Again, the issues relate to the possibility of direct exposure to an operating reactor and the release of radioactive material due to reactor disruption. For reentry or launch abort accidents, however, the possibility of a flooding accident must be considered. As discussed in Chapter 8, reactor immersion and core flooding by water and other fluids can be postulated to lead to an inadvertent criticality or supercriticality accident. Flooding leading to prompt-supercritical explosive disassembly is one of the classic accidents proposed for space reactor safety analysis. A more likely accident, however, is the so-called chug-criticality accident. For this scenario, the reactor core becomes flooded with water resulting in supercriticality, heating of the fuel and water, and expulsion of water due to boiling. Expulsion of water from the core results in a subcritical condition, shutting down the reactor; as the core cools, water refloods the core, taking the reactor supercritical again. The process then repeats, resulting in cyclical supercritical/subcritical oscillations. Potential exposure to radiation from the reactor during chug-critical operation is the principal issue; however, chug-criticality accident scenarios can be postulated that lead to the release of fission products.

Some postulated supercritical transient accidents for reactors operating in space are similar to those analyzed for terrestrial reactors. However, missions employing space reactors may present accidental supercriticality scenarios that have no terrestrial reactor counterpart. For example, space reactors can be subject to meteoroid or space debris impact that could alter the position of reflector elements, perturb the core geometry, or damage the control system or instrumentation. Space reactors operate in remote, harsh, and unfamiliar environments; consequently, periodic maintenance may be impossible or impractical. Some space missions may require years of unattended operation. Nuclear thermal propulsion reactors may require rapid startup, and the sudden introduction of a hydrogen propellant typically results in an increase in the core neutron multiplication factor. These unique designs, environments and operational conditions for space reactors present unique accidental supercriticality scenarios.

1.2 Cooling Failure Accidents

Postulated operational accidents also include cooling failure accidents due to flow blockages, loss of flow, and loss of coolant. Flow blockages may result from the buildup of particulate carried by the reactor coolant, or blockages may result from reactor material that has broken free during operation. Loss of flow is usually postulated to result from a failure of a coolant pump, and a loss of coolant may result from leaks due to corrosion or from meteoroid or space debris impact with the coolant system. Typically, coolant failure accidents are assumed to result in automatic core shutdown by the reactor protection system. Nonetheless, decay heating from fission products may overheat the reactor fuel if cooling is insufficient.

The principal concern associated with cooling failure accidents is the possibility of overheating and failure of reactor components. Fuel disruption from overheating can result in the release of radioactive fission products into the primary system. In addition, fuel disruption can lead to blocking of coolant channels and possible progression to core melt, pressure vessel failure, and release of fission products into the terrestrial or space environment. Movement of core materials as a result of a cooling failure may affect the core neutron multiplication factor, possibly progressing to a supercriticality accident.

2.0 Fundamentals of Reactor Kinetics

An understanding of the fundamentals of reactor kinetics is needed to study postulated supercriticality accidents. Basic reactor kinetics concepts and governing equations are presented in this section. Approximate methods are developed to permit first order analyses of space reactor transient accidents.

For steady state reactor analysis, no distinctions are made between prompt and delayed neutrons. The prompt and delayed neutrons are lumped together by assuming a fission source term of the form $v\Sigma_f(\mathbf{r})\phi(\mathbf{r})$, where v is the total number of neutrons (prompt and delayed) emitted per fission. Although this approach is acceptable for steady state analysis, for transient analysis, the very large differences in the time required to produce prompt and delayed neutrons must be explicitly included in the analysis. Despite the fact that delayed neutrons constitute a very small fraction of fission neutrons (less than 1%), their relatively long time scales can have a dominant effect on the time behavior of a reactor transient. The influence of delayed neutrons on reactor kinetics is especially important when reactors are only slightly supercritical or subcritical.

2.1 Point Reactor Kinetics Equations

From Chapter 8, the neutron continuity equation for a reactor system without an external source was given as

$$\frac{\partial n(\mathbf{r},t)}{\partial t} = \mathcal{P}_f(\mathbf{r},t) - \mathcal{L}_a(\mathbf{r},t) - \mathcal{L}_L(\mathbf{r},t), \tag{9.1}$$

where, $\mathcal{P}_f$, $\mathcal{L}_a$, and $\mathcal{L}_L$ are the densities of the neutron production rate by fission, loss rate by absorption, and loss rate by leakage integrated over all neutron energies. As in Chapter 8, we make the simplifying assumption that the reactor is homogenous and the cross sections are position independent; however, for transient overpower accidents, we cannot eliminate the time derivative term from Eq. 9.1. In order to explicitly account for the differences in time scales for prompt and delayed neutrons, separate differential equations are required for the time dependent neutron flux and the time dependent delayed neutron precursor concentrations. Recall that the precursors are fission products that emit neutrons during decay.

We have previously shown that $\mathcal{L}_a(\mathbf{r},t) = \bar{\Sigma}_a\Phi(\mathbf{r},t)$, and for the diffusion approximation $\mathcal{L}_L(\mathbf{r},t) = -\bar{D}\nabla^2\Phi(\mathbf{r},t)$. As pointed out in Chapter 1, a fraction β of neutrons produced by fission are delayed because they are released by the decay of fission products rather than emitted during fission. In order to accurately account for the time dependence of delayed neutrons, the precursor concentrations C_i are represented by a number of delayed neutron groups. Table 9.1 presents the delayed neutron fractions β_i and decay constants λ_i for a typical six-precursor group scheme. The prompt neutron production rate can be written as $k_\infty\bar{\Sigma}_a\Phi(\mathbf{r},t)(1 - \beta)$, and the delayed neutron production rate is $\sum_i\lambda_iC_i(\mathbf{r},t)$. Using these expressions and $\Phi(\mathbf{r},t) = n(\mathbf{r},t)\bar{v}$ in Eq. 9.1, where $\bar{v}$ is the average neutron speed (all other parameters were defined in Chapter 8) we obtain

$$\frac{1}{\bar{v}}\frac{\partial\Phi(\mathbf{r},t)}{\partial t} = \bar{D}\nabla^2\Phi(\mathbf{r},t) - \bar{\Sigma}_a\Phi(\mathbf{r},t) + (1 - \beta)k_\infty\bar{\Sigma}_a\Phi(\mathbf{r},t) + \sum_i\lambda_iC_i(\mathbf{r},t). \tag{9.2}$$

Table 9.1 Six-Group Delayed Neutron Precursor Properties
for Thermal Fission of U-235

Group	Fraction β_i	Decay Constant λ_i (s^{-1})
1	0.00021	0.0124
2	0.00142	0.0305
3	0.00127	0.111
4	0.00257	0.301
5	0.00075	1.14
6	0.00027	3.01
Total	0.00649	

If $C_i(\mathbf{r},t)$ is the delayed neutron precursor density at location $\mathbf{r}$ and at time t for the ith delayed neutron group, then the precursor removal rate is $\lambda_i C_i(\mathbf{r},t)$ and the precursor production rate is $\beta_i \bar{v}\bar{\Sigma}_f \Phi(\mathbf{r},t)$. Thus, the differential equation describing the behavior of the precursor density for the ith-delayed group is,

$$\frac{\partial C_i(\mathbf{r},t)}{\partial t} = -\lambda_i C_i(\mathbf{r},t) + \beta_i \bar{v}\bar{\Sigma}_f \Phi(\mathbf{r},t). \tag{9.3}$$

Equations 9.2 and 9.3 are the coupled dynamic equations for the reactor neutron density and the delayed neutron precursor concentrations. Noting that for the one-energy-group model, $k_\infty = \bar{v}\bar{\Sigma}_f/\bar{\Sigma}_a$, Eq. 9.3 can be written

$$\frac{\partial C_i(\mathbf{r},t)}{\partial t} = -\lambda_i C_i(\mathbf{r},t) + \beta_i k_\infty \bar{\Sigma}_a \Phi(\mathbf{r},t), \qquad i = 1,2,\ldots m, \tag{9.4}$$

where the number of delayed groups m is usually taken to be 6.

Next we make the assumption that the neutron density and the precursor concentration are separable in space and time; i.e.,

$$\Phi(\mathbf{r},t) = [\bar{v} n(t)]\,\phi(\mathbf{r}), \tag{9.5}$$

and

$$C_i(\mathbf{r},t) = c_i(t)\,g_i(\mathbf{r}). \tag{9.6}$$

Here, $n(t)$ and $c_i(t)$ are the amplitude functions and $\phi(\mathbf{r})$ and $g_i(\mathbf{r})$ are the shape functions. If we assume that $\phi(\mathbf{r})$ and $g_i(\mathbf{r})$ have the same shape (usually a good assumption) and for convenience assume $[\phi(\mathbf{r})/g_i(\mathbf{r})] = 1$, then the precursor equation becomes

$$\frac{\partial c_i(t)}{\partial t} = -\lambda_i C_i(t) + \beta_i k_\infty \bar{\Sigma}_a \bar{v} n(t) \qquad i = 1,2,\ldots m, \tag{9.7}$$

and the neutron flux equation becomes

$$\frac{dn(t)}{dt} = \bar{D}\bar{v}\frac{\nabla^2\phi(\mathbf{r})}{\phi(\mathbf{r})} n(t) - \bar{v}\bar{\Sigma}_a n(t) + (1-\beta)k_\infty \bar{v}\bar{\Sigma}_a n(t) + \sum_i \lambda_i C_i(t). \tag{9.8}$$

The spatial dependence in the above equation can be removed by employing the bare, one-energy-group reactor equation

$$\nabla^2\phi(\mathbf{r}) + B^2\phi(\mathbf{r}) = 0, \tag{9.9}$$

where B^2 is the reactor buckling. Using Eq. 9.9 allows the neutron density equation to be written as

$$\frac{dn(t)}{dt} = -\bar{D}B^2\bar{v}n(t) - \bar{v}\bar{\Sigma}_a n(t) + (1-\beta)k_\infty \bar{v}\bar{\Sigma}_a n(t) + \sum_i \lambda_i c_i(t). \tag{9.10}$$

The neutron lifetime in an infinite medium l_∞ is equal to the mean free path between neutron absorptions $\bar{\Sigma}_a^{-1}$ divided by the mean neutron speed $\bar{v}$; thus, $l_\infty = 1/(\bar{v}\bar{\Sigma}_a)$, and using $L^2 = \bar{D}/\bar{\Sigma}_a$, we can write

$$l = \frac{l_\infty}{1 + B^2 L^2}. \tag{9.11}$$

We can also use
$$k_{eff} = \frac{k_\infty}{1 + B^2 L^2}. \tag{9.12}$$

The parameter l is the average neutron lifetime in a finite reactor. With these substitutions, the neutron density and precursor equations can be written as

$$\frac{dn(t)}{dt} = \left[\frac{(1 - \beta)k_{eff} - 1}{l}\right] n(t) + \sum_i \lambda_i c_i(t), \tag{9.13}$$

and
$$\frac{\partial c_i(t)}{\partial t} = \beta_i \left(\frac{k_{eff}}{l}\right) n(t) - \lambda_i c_i(t) \qquad i = 1, 2, \ldots m. \tag{9.14}$$

The time between successive neutron generations, called the *neutron generation time Λ*, is given by

$$\Lambda = \frac{l}{k_{eff}}. \tag{9.15}$$

We also define the *reactivity ρ* as

$$\rho \equiv \frac{(k_{eff} - 1)}{k_{eff}}. \tag{9.16}$$

Using Eqs. 9.15 and 9.16 in Eqs. 9.13 and 9.14, we obtain the *point reactor kinetics* (PRK) equations:

$$\frac{dn(t)}{dt} = \left(\frac{\rho(t) - \beta}{\Lambda}\right) n(t) + \sum_i \lambda_i c_i(t), \tag{9.17}$$

and
$$\frac{dc_i(t)}{dt} = \left(\frac{\beta_i}{\Lambda}\right) n(t) - \lambda_i c_i(t), \qquad i = 1, 2, \ldots m. \tag{9.18}$$

The first term on the right side of Eq. 9.17 gives the rate of change in the neutron density due to prompt neutron production, and the second term gives the rate of change due to delayed neutrons. The first term on the right side of Eq. 9.18 gives the rate of increase in the *i*th precursor concentration due to the yield from fission, and the second term gives the rate of decrease due to decay of the *i*th precursor. If diffusion theory is invalid for the reactor system being analyzed, the identical PRK equations can be derived by starting with the time-dependent Boltzmann transport equation. The preceding development, which used separation of variables and time-dependent diffusion theory, is a simpler and more restrictive approach.

The PRK equations are a set of coupled ordinary first order differential equations that describe the time dependence of the neutron density (or neutron flux) and the delayed neutron precursor concentrations in a reactor. The reactor power can be expressed as

$$P(t) = [\kappa \bar{\Sigma}_f n(t) \bar{v}] V_c, \tag{9.19}$$

where κ is the energy per fission and V_c is the core volume. Thus, the PRK equations can be used to describe the time dependent behavior of the reactor power.

2.2 Solution of the PRK Equations

This section discusses the general solution of the PRK equations, introduces the concept of the asymptotic period, and discusses common units for reactivity. The general solution of the PRK equations provides a method for predicting reactor dynamic behavior for a full range of possible reactivity insertions.

General Solution

In Reference [1] it is shown that the solutions to the PRK equations are of the form,

$$n(t) = \sum_{j=0}^{m} N_j e^{\omega_j t}, \qquad (9.20)$$

and

$$c_i(t) = \sum_{j=0}^{m} C_{ij} e^{\omega_j t}, \qquad (9.21)$$

where each of the ω_j must satisfy the equation

$$\rho = \omega \Lambda + \sum_i \frac{\omega \beta_i}{\omega + \lambda_i}. \qquad (9.22)$$

The constants N_j and C_{ij} are determined from the initial conditions. From the definition of ρ, only the real values of ρ between plus and minus one have physical significance. Note that another form of Eq. 9.22, more common in older texts, can be obtained by using l/k_{eff} instead of Λ in Eq. 9.22. With this substitution we obtain

$$\rho = \frac{\omega l}{1 + \omega l} + \frac{\omega}{1 + \omega l} \sum_i \frac{\beta_i}{\omega + \lambda_i}. \qquad (9.23)$$

Equations 9.22 and 9.23 show that the characteristic roots depend on the parameters ρ, β, and Λ (or l). The precursor decay constants λ_i are relatively independent of the reactor fuel and other reactor properties. The solution of Eq. 9.22 is presented in Figures 9.1 (top and bottom figures for narrow and wider ω ranges, respectively). The figure was generated for a ^{235}U-fueled reactor with $\Lambda = 10^{-6}$ seconds. Note that the poles shown in the figure are the negatives of the decay constants for the precursor groups. The roots ω_j are shown in the figure for reactivity insertions of $\rho = +0.02, + 0.003,$ and -0.02. Because of the limited range used for ρ and ω in the plot, not all roots appear in the figure. The β_i λ_i and given in Table 9.1 were used to generate Figure 9.1.

Asymptotic Period

For positive reactivities, one root of Eq. 9.22 is positive and six roots are negative. The six terms with negative roots in the preceding equations make only transient contributions to the solutions. Each N_j coefficient in Eq. 9.20 has the same sign as the corresponding ω_j. Following a small sudden change in reactivity, the negative root terms decrease rapidly to zero (within a few hundredths of a second). Hence, the solution of Eq. 9.20 quickly approaches,

$$n(t) \sim e^{\omega_0 t} \sim e^{t/T}, \qquad (9.24)$$

where $T \equiv 1/\omega_0$ is defined as the *asymptotic period*.

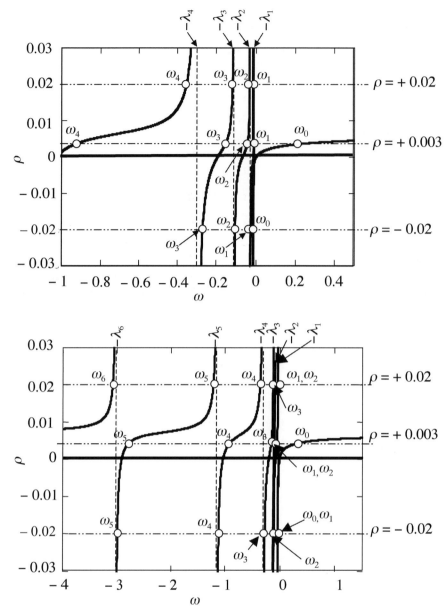

Figure 9.1 ρ as a function of ω for ^{235}U, $\Lambda = 10^{-6}$ second.

From Figure 9.1 we observe that for negative reactivity insertions, all of the ω roots are negative. Furthermore, all of the N_j coefficients are positive. Very quickly, all of the contributions from the larger $|\omega_j|$ roots decay away, and only the smallest contributor $|\omega_j| = |\omega_0|$ persists. In this case, both ω and $\mathcal{T}$ are negative.

Units

Reactivity, as defined by Eq. 9.16, is unitless. For convenience, however, a number of units for ρ have been established. For example, reactivity is often expressed in percent by multiplying ρ by 100. Perhaps the most common units for reactivity are dollars ($) and cents (¢). To express reactivity in dollars, the value of ρ, as defined in Eq. 9.16, is divided by β; thus, a reactivity of $\rho = \beta$ is equal to $1.00. For reactivity values $\geq$$1.00, reactor dynamics is not affected by delayed neutrons and the reactor power level increases very rapidly at a rate dependent on the prompt neutron lifetime. The reactor condition

$\rho \geq \$1.00$ is called superprompt criticality. Using units of dollars simplifies many expressions that possess ρ/β terms. Furthermore, units of dollars reduce the sensitivity of predictions to differences and changes in β. Reactivity is expressed in cents simply by multiplying the reactivity in dollars by 100. Both dollars and cents are referred to as relative reactivity units.

2.3 Approximate Methods

The general solution of the PRK equations represents a simplification of a complete treatment of the coupling of space, energy, and time dependence of the neutron flux. Nonetheless, the general PRK solution is too cumbersome to use for simple student exercises. Consequently, this section develops several approximate methods.

One Delayed Group: Small Positive ρ Insertion

Figure 9.2 presents a plot of positive reactivity as a function of the asymptotic period for a number of neutron lifetimes. From Figure 9.2 it is apparent that the asymptotic period for small reactivity insertions is independent of the prompt neutron lifetime, and the delayed neutrons are controlling reactor dynamics. For small reactivity insertions the change in the neutron flux is sufficiently slow that we can approximate the delayed neutron contribution by one delayed group. For a single delayed group, the total delayed neutron fraction is

$$\beta = \sum_{i=1}^{6} \beta_i, \tag{9.25}$$

and the one delayed group decay constant is

$$\bar{\lambda} = \left[\frac{1}{\beta} \sum_{i=1}^{6} \left(\frac{\beta_i}{\lambda_i} \right) \right]^{-1} \approx 0.08 \text{ s}^{-1}. \tag{9.26}$$

With this approximation, the PRK equations simplify to

$$\frac{dn(t)}{dt} = \left(\frac{\rho(t) - \beta}{\Lambda} \right) n(t) + \bar{\lambda} c(t), \tag{9.27}$$

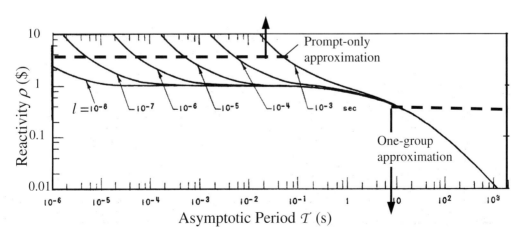

Figure 9.2 Positive reactivity vs. asymptotic period for various neutron lifetimes l for ^{235}U. Adapted from [2] *courtesy of Argonne National Laboratory.*

and
$$\frac{dc(t)}{dt} = \left(\frac{\beta}{\Lambda}\right) n(t) - \bar{\lambda}\, c(t),$$ (9.28)

For one delayed group, we have only two coupled first order differential equations and only two characteristic roots. The one-delayed-group characteristic equation is

$$\rho = \omega \Lambda + \frac{\omega \beta}{\omega + \bar{\lambda}},$$ (9.29)

or

$$\rho = \frac{\omega l}{1 + \omega l} + \left(\frac{\omega}{1 + \omega l}\right)\frac{\beta}{\omega + \bar{\lambda}}.$$ (9.30)

For reasonably small reactivity insertions ($\rho < \$0.25$), it can be shown that the two roots are given approximately by

$$\omega_0 = \frac{\rho \bar{\lambda}}{\beta - \rho} \qquad \text{or} \qquad T = \frac{\beta - \rho}{\rho \bar{\lambda}},$$ (9.31)

and

$$\omega_1 = -\frac{(\beta - \rho)}{\Lambda}.$$ (9.32)

The only time parameter appearing in the expression for the asymptotic period is $\bar{\lambda}$, the one-delayed-group decay constant. In other words, the asymptotic time behavior of the reactor is independent of the neutron generation time and is entirely controlled by the delayed neutrons.

For a step reactivity insertion (instantaneous reactivity insertion to a constant value of ρ) we can use the preceding expressions and initial conditions to obtain

$$n(t) = n_0 \left[\frac{\beta}{\beta - \rho}\exp\left(\frac{\bar{\lambda}\rho}{\beta - \rho}t\right)\right] - n_0 \left[\frac{\rho}{\beta - \rho}\exp\left(-\frac{\beta - \rho}{\Lambda}t\right)\right].$$ (9.33)

The first term is the asymptotic term, and the second term is the *fast transient* term that very rapidly decays away. Subsequent to the decay of the transient term, the neutron density is given by

$$n(t) \approx n_0 \frac{\beta}{\beta - \rho}\exp\left[\left(\frac{\bar{\lambda}\rho}{\beta - \rho}\right)t\right].$$ (9.34)

The multiplier $\beta/(\beta - \rho)$ in Eq. 9.34 is called the prompt jump (or prompt drop for negative reactivity insertions). Figure 9.3 illustrates the prompt jump and the time behavior of the reactor flux following a 0.0010 positive step reactivity insertion. The prompt jump, the initial rise in neutron density due only to prompt neutrons, occurs almost instantaneously (note the short time scale in Figure 9.3). The subsequent slower rise in neutron density (or flux or power) is due to the "holdback" effect of the delayed neutrons.

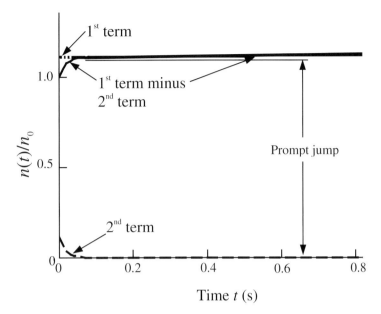

Figure 9.3 Relative neutron density as a function of time for 0.0020 ρ insertion, $\Lambda = 10^{-4}$ s.

The above one-delayed-group model gives reasonably good agreement with the six-group model for positive reactivity insertions $\rho_{in} < \$0.25$. Figure 9.4 compares the time response of the neutron flux ($\phi = n\bar{v}$) for a \$0.10 and \$0.25 step reactivity insertion.

Prompt-Only: Large Positive ρ Insertions

For large positive reactivity insertions worth several dollars, the period depends on the lifetime l and not on the delayed neutrons. For the short neutron lifetimes typical for space reactors, the ω values in the six-delayed-group characteristic equation are all much greater than the values for λ_i, for positive reactivity insertions $> \$1$. Thus, for $\rho_{in} > \$1$, Eq. 9.22 can be written as

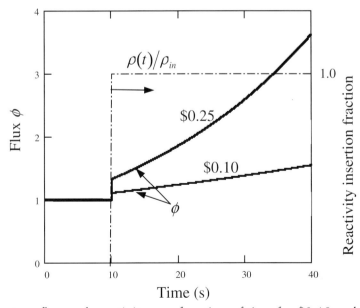

Figure 9.4 Relative neutron flux and reactivity as a function of time for \$0.10 and \$0.25 ρ step insertions.

$$\rho = \omega \Lambda + \beta. \tag{9.35}$$

Hence, for large positive reactivity insertions,

$$T = \frac{\Lambda}{\rho - \beta}. \tag{9.36}$$

The only time parameter in Eq. 9.36 is the prompt neutron generation time Λ. The precursor decay constants do not appear, and the delayed neutrons do not control the time behavior. Thus, for Eq. 9.17 we drop the second term and solve to obtain

$$n = n_0 \exp\left(\frac{\rho - \beta}{\Lambda} t\right). \tag{9.37}$$

The power and neutron flux increase from prompt neutron generation alone; i.e., the prompt jump continues until some feedback mechanism reduces the reactivity insertion. This condition, where $\rho > \$1$, is referred to as prompt supercriticality. For $\rho > \$10$, the expression for the asymptotic period is approximately $T \approx \Lambda/\rho$. Figure 9.5 compares the time-dependent neutron flux for positive step reactivity insertions of \$3.00 and \$4.00 for $\Lambda = 10^{-5}$ s and for a \$3.00 step insertion assuming $\Lambda = 5 \times 10^{-6}$ s. The time dependence of the flux for prompt critical excursions is strongly dependent on the neutron generation time.

As indicated in Figure 9.2, the one-group approach can provide a good approximation for $\rho < \$0.25$, and the prompt-only approximation can be used for ρ of several dollars or more. For insertions that fall between \$0.50 and a few dollars, the full six-delayed-group approach must be used.

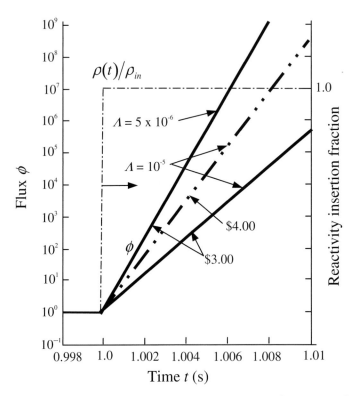

Figure 9.5 Relative neutron flux and reactivity as a function of time for \$3.00 and \$4.00 ρ step insertions with $\Lambda = 5 \times 10^{-6}$ and $\Lambda = 10^{-5}$ s.

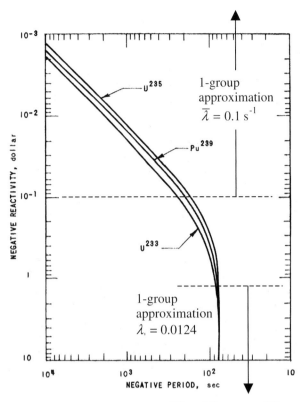

Figure 9.6 Negative reactivity vs. asymptotic period for ^{235}U, ^{233}U, and ^{239}Pu. Adapted from [2] *courtesy of Argonne National Laboratory.*

Negative Reactivity Insertions

Figure 9.6 shows the asymptotic period as a function of negative reactivity for ^{235}U, ^{233}U, and ^{239}Pu for all neutron lifetimes. The asymptotic period is independent of the prompt neutron lifetime for all negative reactivities. For small negative reactivity insertions ($|\rho| < \$0.10$), the one-group approximation can be used with an *effective mean decay constant* $\bar{\lambda} = 0.1$ s^{-1} for a ^{235}U-fueled reactor. For small positive reactivity insertions, the effective mean decay constant also gives better agreement with the six-delayed-group method than does the true effective constant $\bar{\lambda} = 0.08$ s^{-1}.

The holdback effect of delayed neutrons is more significant for negative reactivities than for positive reactivities. For large negative reactivity insertions ($|\rho_{in}| \geq \$1$), the accumulation of delayed neutron emitter fission products limits how rapidly the reactor power can be reduced following the initial prompt drop. The limit is governed by the longest-lived precursor decay constant λ_1. Thus, for large negative reactivities the one group equation is

$$n(t) = n_0 \left[\frac{\beta}{\beta - \rho} \exp\left(\frac{t}{\mathcal{T}}\right) \right], \tag{9.38}$$

where

$$\mathcal{T} \approx -\frac{1}{\lambda_1}. \tag{9.39}$$

The parameter λ_1 is the decay constant for the longest-lived delayed neutron precursor group. For ^{235}U and a large negative reactivity insertion, $\mathcal{T} \approx -80.65$ s. Whether \$1 or \$20 of negative reactivity is inserted, the asymptotic period is about -80 seconds. The larger negative insertion, however, will lead

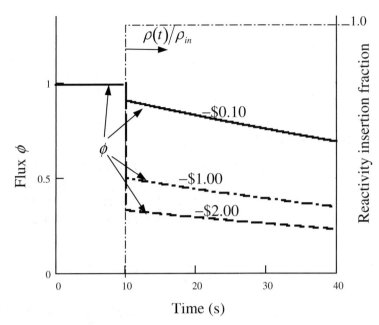

Figure 9.7 Relative neutron flux and reactivity as a function of time for $-\$0.10$, $-\$1.00$ and $-\$2.00$ ρ step insertions.

to a larger initial prompt drop. These observations are illustrated in Figure 9.7 for negative reactivity insertions of $\$0.10$, $\$1.00$, and $\$2.00$.

3.0 Reactivity Feedback

The change in a reactor's power level depends on the reactivity inserted by control or safety mechanisms, such as reflector drum movement. However, reactivity is also affected by the reactor power level and component temperatures. For example, a positive reactivity insertion will increase reactor power, resulting in an increase in the fuel temperature; and a change in fuel temperature typically changes the reactivity level of the core. If the reactivity feedback due to the increase in fuel temperature is negative, then the rate of power level growth will decline. If the feedback is positive, then the reactor power will increase more rapidly. In order to prevent a runaway reactivity excursion accident and to enhance reactor stability, reactor materials and designs are generally selected such that feedback is negative,

Reactivity feedback typically results from changes in temperatures of core components. Feedback effects can also result from changes in coolant pressure, flow rate, and other operational parameters. Feedback is generally characterized by a reactivity coefficient as

$$\alpha_x = \frac{\Delta\rho}{\Delta x},\qquad(9.40)$$

where x is the parameter affecting feedback (e.g., fuel temperature). Thus, for a change in fuel temperature T_F, the change in reactivity will be $\Delta\rho_{T_F} = \alpha_{T_F}\Delta T_F$. Often feedback effects can be represented by an overall power coefficient such that the feedback reactivity is given by $\Delta\rho_P = \alpha_P\Delta P$.

3.1 Temperature Feedback Mechanisms

Reactivity feedback due to a change in core component temperature can arise from several possible mechanisms. Depending on the component, the feedback effect may be prompt or delayed.

Prompt Feedback

Because fuel reactivity feedback effects are usually prompt, fuel feedback can provide a rapid inherent mechanism for preventing runaway power excursions in the event of an unplanned reactivity insertion. Doppler broadening and fuel expansion are generally the most common mechanisms for fuel temperature reactivity feedback. Doppler broadening results from the increase in random motion of the fuel nuclei as temperature increases. The increased random motion affects the relative velocity between core neutrons and the target nuclei. The increase in random motion is equivalent to lowering the maximum value of cross section resonances, while broadening the energy-width of the resonance (i.e., the width of the cross section resonance from a plot of cross section as a function of energy). The net effect of Doppler broadening is to increase the probability of neutron absorption within the resonance cross section energy range. For low-enrichment reactors, the principal effect of Doppler broadening is an increase in parasitic capture of neutrons in the resonance energy range by ^{238}U. Thus, Doppler broadening in reactors employing slightly enriched uranium fuel will generally have a negative reactivity feedback effect. Space reactors are almost always highly enriched in ^{235}U; consequently, the Doppler broadening effect is typically small compared to reactors using low-enriched uranium fuel. Furthermore, the Doppler feedback effect for space reactors is not necessarily negative. On the other hand, fuel expansion from increased fuel temperature almost always results in a net negative prompt feedback coefficient. As discussed in Chapter 8, an increase in the core region surface-to-mass ratio (as a result of fuel expansion) increases neutron leakage and decreases core reactivity.

Delayed Feedback Effects

Because of the time delay between changes in fuel temperature and the subsequent change in component temperature, reactivity feedback effects due to other components are commonly delayed effects. Feedback from the reactor coolant or moderator (when present) usually results from changes in coolant or moderator density. A decrease in coolant or moderator density due to temperature increase can increase neutron leakage and alter the neutron spectrum. Although feedback from moderating materials is typically delayed, the zirconium hydride moderator for the SNAP-10A space reactor is integral with the fuel and reactivity feedback is prompt. Thermal feedback from structural materials usually results from Doppler broadening. For fast reactors, the net effect of neutron leakage and spectrum changes is strongly dependent on the details of the design and materials choice.

Reactivity Temperature Coefficients

Reactivity temperature coefficients can be determined for various core materials and core regions. Temperature coefficients are often reported as *prompt coefficients* and *delayed coefficients* corresponding to prompt and delayed effects, respectively. A so-called *isothermal temperature coefficient* is often used that gives the overall reactor temperature coefficient, assuming that all regions of the reactor core are at the same temperature and that temperature changes are the same for all core regions. When a *total temperature coefficient* is provided, it includes both prompt and delayed feedback effects, but the temperature of each core region is at predicted operating temperatures rather than assuming isothermal conditions. Prompt, isothermal, and total temperature coefficients are given in Table 9.2 for various space reactors and terrestrial reactors. Reactivity temperature coefficients are generally temperature dependent. The temperature coefficients provided in Table 9.2 are for the operating temperature range.

3.2 Supercritical Excursions with Feedback

Reactor kinetic behavior with feedback is generally more complex than the analysis with the assumption of no feedback. Nevertheless, some useful approximate methods have been developed for several possible reactivity insertion scenarios. Approximate methods are presented here for step and ramp reactivity insertions for super prompt critical and delayed prompt critical excursions.

Table 9.2   Typical Space Reactor and Terrestrial Reactor
Temperature Coefficients of Reactivity

Usage	Reactor		Coefficient (K^{-1})
Space	SNAP-10A	prompt	-2.0×10^{-5}
	Enisey	isothermal	$+1.5 \times 10^{-5}$
		prompt	-1.0×10^{-6}
Terrestrial	Shippingport (PWR)	total	-5.5×10^{-4}
	EBR-1 (Fast reactor)	total	-3.5×10^{-5}
	HTGR	isothermal	-2.0×10^{-5}
		prompt	-4.0×10^{-5}

Superprompt Critical Excursions: Step ρ Insertion

For reactivity excursions in which $\rho > \beta$, the reactor period is very small; and the reactor power level increases too rapidly for the reactor safety system to terminate the transient. If the prompt temperature coefficient of reactivity is sufficiently large and negative, inherent feedback due to Doppler or expansion effects may shut the reactor down. Here we examine superprompt critical excursions during reactor startup. For a step reactivity insertion ρ_{in} the time-dependent reactivity of the reactor can be written as

$$\rho(t) = \rho_{in} - |\alpha_{T_F}|\overline{T}_F(t), \tag{9.41}$$

where α_{T_F} is the fuel temperature coefficient of reactivity (a negative quantity for this scenario) and $\overline{T}_F(t)$ is the time-dependent change in the average fuel temperature. For rapid heat-up during the excursion, the core can be assumed to behave adiabatically; consequently, the fuel temperature is given by

$$\overline{T}_F(t) = \int_0^t \frac{P(t)}{m_F c_F}\, dt + \overline{T}_F(0), \tag{9.42}$$

where m_F is the mass of the fuel, and c_F is the specific heat of the fuel. Equation 9.41 can now be written

$$\rho(t) = \rho_{in} - |\alpha_E| \int_0^t P(t)dt, \tag{9.43}$$

where $\alpha_E \equiv \alpha_{T_F}/m_F c_F$ is an energy coefficient of reactivity with units J^{-1}.

For a prompt excursion, delayed neutrons can be ignored; consequently, the differential equation for the reactor power during the excursion is obtained from Eq. 9.17 by dropping the delayed neutron terms and multiplying through by $\kappa \overline{v} \Sigma_f$. Then replacing $\rho(t)$ by the expression given in Eq. 9.43 we obtain

$$\frac{dP(t)}{dT} = \left[\rho_{in} - |\alpha_E| \int_0^t P(t)dt - \beta \right] \frac{P(t)}{\Lambda} \tag{9.44}$$

Lewis has shown [3] that the solution of Eq. 9.44 is

$$P(t) = \frac{2\Lambda U^2}{|\alpha_E|} \frac{Y \exp(-Ut)}{[1 + Y \exp(-Ut)]^2}, \tag{9.45}$$

where
$$U = \sqrt{\frac{(\rho_{in} - \beta)^2}{\Lambda^2} + \frac{2|\alpha_E|P(0)}{\Lambda}}, \tag{9.46}$$

and
$$Y = \frac{\Lambda U + \rho_{in} - \beta}{\Lambda U - \rho_{in} + \beta}. \tag{9.47}$$

This approximation is called the Fuchs-Nordheim model. The relationships for maximum power and total energy released during the excursion are

$$P_{max} = \frac{\Lambda U^2}{2|\alpha_E|}, \tag{9.48}$$

and
$$E = \frac{\Lambda}{|\alpha_E|}\left(U + \frac{\rho_{in} - \beta}{\Lambda}\right), \tag{9.49}$$

respectively. For low initial power, $U \approx (\rho_{in} - \beta)/\Lambda$. With this assumption, the preceding expressions are used to obtain the following approximations for the low initial power case:

$$P_{max} \approx \frac{(\rho_{in} - \beta)^2}{2|\alpha_E|\Lambda}, \tag{9.50}$$

$$P(t) \approx P_{max}\,\text{sech}^2\left(\frac{\omega \hat{t}}{2}\right), \tag{9.51}$$

$$E = 2\frac{(\rho_{in} - \beta)}{|\alpha_E|}, \tag{9.52}$$

$$\Gamma = \frac{3.52\Lambda}{(\rho_{in} - \beta)}, \tag{9.53}$$

$$T_{max} \approx 2\frac{(\rho_{in} - \beta)}{|\alpha_E|}, \tag{9.54}$$

and
$$T_{Pmax} \approx \frac{T_{max}}{2}. \tag{9.55}$$

Here, $\hat{t} = 0$ is the time referenced to P_{max}, Γ is the burst full width at half maximum, and T_{Pmax} is the fuel temperature at $\hat{t} = 0$. Note that the total energy release does not depend on $P(0)$ or Λ.

The symmetric power shape given by Eq. 9.50 does not include the effect of delayed neutrons produced during the pulse. Hetrick has shown [1] that the power produced by delayed neutrons that immediately follows the burst is approximately

$$P_{dn} \approx \frac{2}{|\alpha_E|}\sum_i \beta_i \lambda_i \approx \frac{2\lambda'\beta}{|\alpha_E|}, \tag{9.56}$$

where $\lambda' \equiv (\beta^{-1})\sum_i \beta_i \lambda_i \approx 0.4\ \text{s}^{-1}$ for ^{235}U.

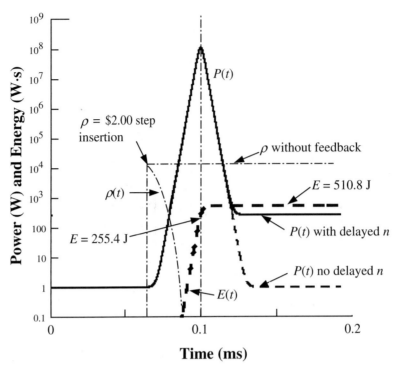

Figure 9.8 Predicted power and energy as a function of time for a $2.00 step insertion with $|\alpha_E| = 2 \times 10^{-5}$ joules^{-1} and $\Lambda = 10^{-8}$ s (with and without delayed neutrons).

Using Eqs. 9.51 and 9.56, the power for a $2.00 step insertion was computed as a function of time, assuming $|\alpha_E| = 2 \times 10^{-5}$ joules^{-1} and $\Lambda = 10^{-8}$ s. The power profile for this case is presented in Figure 9.8, with and without the delayed neutron tail. The figure also includes the energy produced as a function of time and a schematic illustration of the core reactivity. Note that the total energy released at the at the end of the pulse is twice the energy at the peak of the pulse (the increase in energy is difficult to discern from the logarithmic plot). Although the maximum power level for accidents of this type can be very high, the pulse width is very narrow. Thus, the deposited energy may not be sufficient to cause fuel damage. For scenarios of this type, the reactor system may be designed to assure that the maximum credible reactivity insertion does not result in an energy release sufficient to cause fuel damage.

Subprompt Critical Excursions: Step ρ Insertions

For reactivity step insertions in which $\rho_{in} = \beta$, the maximum power achieved as a result of feedback may be estimated for a range of conditions. These conditions require that ρ_{in} is sufficiently small such that the prompt-jump approximation can be used, and sufficiently large to use the adiabatic heating approximation. For the prompt-jump approximation, the assumption is made that $\Lambda \to 0$. For these conditions, Hetrick shows [1] that the maximum power is given by

$$P_{max} = \frac{\bar{\lambda}\beta}{|\alpha_E|}\left[1 - \frac{\bar{\lambda}}{\omega + \bar{\lambda}}\sqrt{1 + \frac{2\omega}{\bar{\lambda}}}\right]. \tag{9.57}$$

A plot of the maximum power as a function of inverse period is presented in Figure 9.9 using Eq. 9.57 for the prompt-jump model and for the Fuchs-Nordheim model. A coefficient of reactivity of $\alpha_E = 2 \times 10^{-3}$ J^{-1} was used. Hetrick has shown that using $\bar{\lambda} = 0.08$ for small ω and using $\lambda' = 0.40$ for large ω, gives the best agreement with detailed (six-group) model predictions. Note that the point of departure from the prompt-jump model to the Fuchs-Nordheim model depends on the neutron generation time. For small ω (large periods), the adiabatic approximation is invalid.

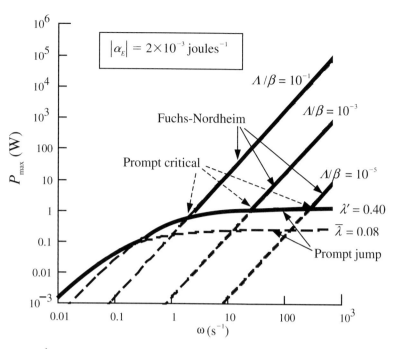

Figure 9.9 Maximum power for a step ρ insertion with feedback as a function of inverse period using the prompt jump and the Fuchs-Nordheim models.

Superprompt Critical Excursions: Ramp ρ Insertion

For a ramp reactivity insertion (reactivity insertion at constant rate) with feedback, the reactor's kinetic behavior differs from the behavior following a step reactivity insertion. The power increases rapidly upon achieving superprompt criticality, reaches a maximum power level, and then drops rapidly as a result of the negative temperature coefficient taking the core subprompt critical. However, the reactivity insertion continues to increase and the reactor again achieves superprompt criticality causing the power to increase. The temperature feedback again takes the reactor subprompt. This process continues until delayed neutron effects cause the reactor power to stabilize at some maximum value. The stable power level P_∞ is shown in Reference [3] to be given by

$$P_\infty = \frac{\dot{\rho}}{|\alpha_E|}, \tag{9.58}$$

where $\dot{\rho}$ is the reactivity insertion rate. Typically, the first burst is the most damaging. Lewis has shown [3] that the maximum power can be estimated using

$$P_{\max} = \frac{\dot{\rho}}{|\alpha_E|}\left[\ln\left(\frac{P_{\max}}{P(0)}\right)\right]. \tag{9.59}$$

Equation 9.59 must be solved iteratively or graphically. Assuming that the time-dependent power trace is approximately symmetric about the maximum power, Lewis provides the following equation to estimate the energy released during the first burst:

$$E = \frac{2}{|\alpha_E|}\sqrt{2\dot{\rho}\Lambda\left[\ln\left(\frac{\dot{\rho}}{|\alpha_E|}\right) - 1\right]}. \tag{9.60}$$

Example 9.1

Compute the stable power level, the maximum power level, and the energy produced during the first burst for a space reactor undergoing an unplanned reactivity insertion of $30/s. Assume an initial power of 0.01 W, $|\alpha_E| = 3 \times 10^{-6}$ joules^{-1}, $\beta = 0.0073$, and $\Lambda = 10^{-7}$ s.

Solution:

The stable power level is $P_\infty = \dfrac{(\$30/s)(0.0073 \ \$^{-1})}{3 \times 10^{-6}} = 7.3 \times 10^4$ W.

Using Eq. 9.59 we define

$$u = \frac{\dot{\rho}}{|\alpha_E|}\left[\ln\left(\frac{P_{trial}}{P(0)}\right)\right] - P_{trial}$$

where P_{trial} are the trial values for P_{max} and $u = 0$ when $P_{trial} = P_{max}$. The solution is plotted in Figure 9.10 and shows that $P_{max} = 1.37 \times 10^6$ W.

The energy produced during the first burst is

$$E = \frac{2}{3 \times 10^{-6}} \sqrt{2(30 \times 0.0073)(10^{-7})\left[\ln\left(\frac{30 \times 0.0073}{3 \times 10^{-6}}\right) - 1\right]} = 446 \text{ joules}$$

3.3 Reactor Stability

Operational stability and dynamic performance considerations usually require a negative overall delayed feedback for a reactor. Both the magnitude of the feedback coefficient and the associated time constants are important. Delayed feedback that is too small, too slow, too large, or too fast can lead to undesir-

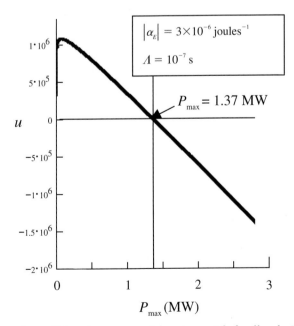

Figure 9.10 Residual u as a function of P_{max} for a ramp insertion with feedback. Used to find P_{max} in Example 9.1.

able dynamic performance. If the PRK equations can be linearized, they can be readily solved by means of Laplace transforms. The well-known classical method of stability analysis and control can then be used to analyze the reactor system dynamics. A discussion of operational stability is beyond the scope of this text. A thorough discussion of this topic can be found in *Dynamics of Nuclear Reactors* by Hetrick [1].

4.0 Reactivity Excursion Thermal Response

Direct exposure to a high neutron and gamma flux is one of the possible safety considerations associated with unplanned reactivity excursions. More commonly, however, safety considerations focus on destruction of reactor barriers and the release of fission products as a result of overheating of reactor components. Thus, we need to examine the thermal effects on reactor components in response to unplanned reactivity excursions. For subprompt critical experiments, the principal concern is overheating and possible melting of the reactor fuel and subsequent penetration of the pressure vessel or other containment structure. The full analysis of the reactor thermal response to reactivity excursions requires the use of coupled kinetic and thermal analysis computer models. For our purposes, however, we can examine the case of very slow transients by quasi-steady-state solutions, and we can use a simple lumped parameter model for faster transients.

4.1 Slow Overpower Transients

For overpower transients without significant feedback or compensation, power production as a function of time can be estimated using Eqs. 9.19 and 9.34 to obtain

$$P(t) = P(0) \frac{\beta}{\beta - \rho} \exp\left[\left(\frac{\bar{\lambda}\rho}{\beta - \rho}\right) t\right]. \tag{9.61}$$

The linear heat flux $q'(z,t)$ (in W/cm) for the hottest fuel rod at axial position z is

$$q'(z,t) = \frac{P(t)f(z)F_r}{n_e}, \tag{9.62}$$

where F_r is the core radial peaking factor, n_e is the number of fuel elements, and $f(z)$ is the axial power shape with units of cm^{-1}, normalized such that the integral of $f(z)$ over the length of the fuel element equals 1.0. The power peaking factor is the ratio of the power of the hottest fuel element to the average fuel element power. The coolant temperature in the hot channel is obtained from

$$T_{co}(z,t) = T_{coi} + \frac{1}{(\dot{m}/n_e)c_{co}} \int_{-H/2}^{z} q'(z',t)dz'. \tag{9.63}$$

Here, $\dot{m}$ is the coolant mass flow rate (g/s), T_{coi} is the coolant inlet temperature (K), c_{co} is the coolant specific heat, and H is the height of the active (fueled) zone of the fuel pin (cm), with axial position $z = 0$ at the axial midpoint of the active zone.

For slow overpower transients we can make the approximation that temperatures across the fuel element, cladding, and coolant achieve steady state conditions at each power level as the fuel element power gradually increases. For this assumption, the fuel element centerline temperature T_{Fc} at location z and time t can be determined using Eq. 9.63 and computing the temperature difference between the fuel centerline and the coolant; i.e., $\Delta T(z,t) = R_T q'(z,t)$. Here, R_T is the total thermal resistance (in cm·K/W) across the fuel element to the bulk coolant. The use of a total thermal resistance for heat

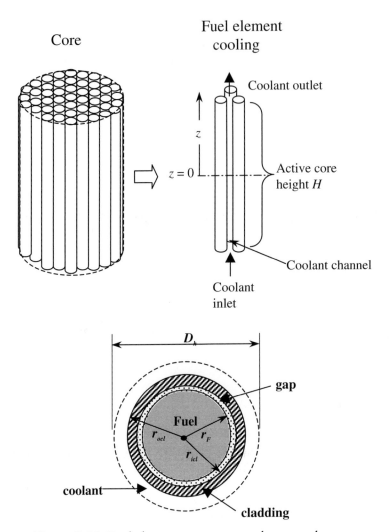

Figure 9.11 Fuel element geometry and nomenclature.

transfer calculations is analogous to the net electrical resistance of components in an electrical circuit. For typical fuel element geometries, the total thermal resistance is given as the sum of the thermal resistances of the fuel element components and coolant. For the reactor design schematically illustrated in Figure 9.11, the core consists of clad fuel pins with a coolant flowing in the interstitial spaces between the pins. It is easily shown that the component resistances are given by

$$R_F = \frac{1}{4\pi k_F}, \tag{9.64}$$

$$R_g = \frac{1}{2\pi r_{icl} h_g}, \tag{9.65}$$

$$R_{cl} = \frac{\ln(r_{ocl}/r_{icl})}{2\pi k_{cl}}, \tag{9.66}$$

and
$$R_{co} = \frac{1}{2\pi r_{ocl} h_{co}}. \tag{9.67}$$

Here, R_F, R_g, R_{cl}, and R_{co} are the thermal resistances of the fuel, fuel-clad gap, cladding, and coolant, respectively. The parameters, k_F and k_{cl} are the thermal conductivities (W/cm·K), and h_g and h_{co} are the heat transfer coefficients (W/cm²·K) for the gap and coolant, respectively. The fuel radius and the inner and outer radii (cm) of the cladding are r_F, r_{icl}, and r_{ocl}, respectively.

If the gap heat transfer is dominated by thermal conduction through the gap gas, the gap heat transfer coefficient can be approximated by

$$h_g = \frac{k_g}{r_{icl} \ln(r_{icl}/r_F)}, \tag{9.68}$$

where k_g is the gap gas conductivity. The coolant convective heat transfer coefficient can be obtained using

$$h_{co} = \frac{k_{co}}{D_h} (0.0023 \, \text{Re}^{0.8} \, \text{Pr}^{0.4}). \tag{9.69}$$

Here, k_{co} is the coolant thermal conductivity, D_h is the hydraulic diameter of the coolant channel, Re is the Reynolds number, and Pr is the Prandtl number for the coolant. The hydraulic diameter is an effective diameter defined as

$$D_h = 4 \frac{A}{C_p}, \tag{9.70}$$

where A is the flow area and C_p is the wetted perimeter. The Reynolds and Prandtl numbers are dimensionless parameters used in heat transfer analysis given by

$$\text{Re} = \frac{D_h \bar{v}_{co} \varsigma_{co}}{\mu_{co}} \tag{9.71}$$

and

$$\text{Pr} = \frac{c_{co} \mu_{co}}{k_{co}}, \tag{9.72}$$

where $\bar{v}_{co}$, c_{co}, μ_{co}, and ς_{co} are the mean coolant speed (cm/s), specific heat (W·s/g·K), viscosity (g/cm·s), and density (g/cm³) of the coolant. Using these equations, the individual thermal resistances can be computed and the total thermal resistance is

$$R_T = R_F + R_g + R_{cl} + R_{co}. \tag{9.73}$$

The fuel centerline temperature can now be obtained using

$$T_{Fc}(z,t) = T_{co}(z,t) + R_T q'(z,t), \tag{9.74}$$

and the cladding temperature can be computed from

$$T_{cl}(z,t) = T_{Fc}(z,t) - R_{Fc} q'(z,t), \tag{9.75}$$

where $R_{Fc} = R_F + R_g + R_{cl}$.

4.2 Rapid Overpower Transients

For more rapid subprompt critical overpower transients, the quasi-steady-state approach cannot be used. For this case, the thermal analysis can use the approximation given by the following equation:

$$\frac{m_{FE}\, c_{FE}}{n_e H}\frac{d\overline{T}_{FE}(z,t)}{dt} = q'(z,t) - \frac{1}{\overline{R}_{FE}}[\overline{T}_{FE}(z,t) - \overline{T}_{co}(z,t)]. \tag{9.76}$$

Here, $\overline{T}_{FE}$ and c_{FE} are the average fuel element temperature and specific heat. The parameter m_{FE} is the total mass of the fuel elements in the core (fuel + cladding). The parameter $\overline{R}_{FE}$ is the thermal resistance of the fuel element with respect to the average fuel temperature [3], given by

$$\overline{R}_{FE} = \frac{m_F c_F \overline{R}_T + m_{cl} c_{cl} \overline{R}_C}{m_{FE}\, c_{FE}}, \tag{9.77}$$

where

$$\overline{R}_T = \frac{R_F}{2} + R_g + R_{cl} + R_{co}, \tag{9.78}$$

$$\overline{R}_C = \frac{1}{4\pi k_{cl}} - \left(\frac{r_{icl}^2}{r_{ocl}^2 - r_{icl}^2}\right)R_{cl} + R_{co}, \tag{9.79}$$

and

$$c_{FE} = \frac{m_F c_F + m_{cl} c_{cl}}{m_F + m_{cl}}. \tag{9.80}$$

Here m_{cl} and c_{cl} are the mass and specific heat of the cladding.

Rearranging Eq. 9.76, we can write

$$\frac{d\overline{T}_{FE}(z,t)}{dt} = \frac{q'(z,t)}{(m_{FE}/Hn_e)c_{FE}} - \frac{1}{\tau}[\overline{T}_{FE}(z,t) - \overline{T}_{co}(z,t)]. \tag{9.81}$$

Here, τ is the time constant defined by

$$\tau \equiv \frac{m_{FE}\, c_{FE}\, \overline{R}_{FE}}{n_e H}. \tag{9.82}$$

For a modest reactivity insertion ($< 25\cent$), Eq. 9.62 can be written in the form

$$q'(z,t) = P(0)\frac{\beta}{\beta - \rho}\exp\left[\left(\frac{\overline{\lambda}\rho}{\beta - \rho}\right)t\right]\frac{f(z)F_r}{n_e}. \tag{9.83}$$

If we assume that the bulk coolant temperature remains essentially constant ($\overline{T}_{co}(z,t) = T_{co}$) and use the substitution $\Theta(z,t) = \overline{T}_{FE}(z,t) + T_{co}$, Eq. 9.81 becomes

$$\frac{d\Theta(z,t)}{dt} = A_{FE}(z)\exp\left(\frac{t}{T}\right) - \frac{\Theta(z,t)}{\tau}, \tag{9.84}$$

where

$$A_{FE}(z) = \frac{P(0)\beta f(z)F_r H}{(\beta - \rho)\, m_{FE}\, c_{FE}}.$$

Multiplying through by the integrating factor $\exp(t/\tau)$, solving, and substituting the expression for Θ back into the resulting equation yields

$$\overline{T}_{FE}(z,t) = T_{co} + \frac{A(z)}{\left(\dfrac{1}{\mathcal{T}} + \dfrac{1}{\tau}\right)}[\exp(t/\mathcal{T}) - \exp(-t/\tau)] + [\overline{T}_{FE}(z,0) - T_{co}]\exp(-t/\tau). \qquad (9.85)$$

For $t \gg \tau$, then $\exp(-t/\mathcal{T}) \gg \exp(-t/\tau)$ and the last term in Eq. 9.85 is $\ll T_{co}$, and Eq. 9.85 simplifies to

$$\overline{T}_{FE}(z,t) = T_{co} + \frac{P(0)f(z)F_r H}{m_{FE}\, c_{FE}(\mathcal{T}^{-1} + \tau^{-1})}\frac{\beta}{(\beta - \rho)}\exp(t/\mathcal{T}). \qquad (9.86)$$

Note that Eq. 9.86 provides the average fuel element temperature for the hottest fuel element, rather than the fuel centerline temperature in the hottest fuel element. If we use the axial power peaking factor F_z divided by H in place of $f(z)$ in Eq. 9.86, we obtain the fuel element radially averaged temperature at the peak axial power location $\hat{T}_{FE}$ for the hottest fuel element. Thus,

$$\hat{T}_{FE}(t) = T_{co} + \frac{\hat{P}_0}{m_{FE}\, c_{FE}(\mathcal{T}^{-1} + \tau^{-1})}\frac{\beta}{(\beta - \rho)}\exp(t/\mathcal{T}), \qquad (9.87)$$

where $\hat{P}_0 = P(0)F_r F_z$.

Example 9.2

A space reactor operating at 2 MW$_{th}$ for 2 years suffers a malfunction that causes a 20¢ positive reactivity insertion. Assuming that the reactivity change is instantaneous, compute the fuel element average temperature as a function of time in the hottest fuel element at the core centerline. The values for the applicable reactor parameters are as follows: $m_F = 183$ kg, $m_{cl} = 70$ kg, $r_F = 0.325$ cm, $r_{icl} = 0.338$ cm, $r_{ocl} = 0.387$ cm, $k_F = 0.26$ W/cm·K, $k_{cl} = 0.52$ W/cm·K, $k_{gap} = 0.0046$ W/cm·K, $h_{co} = 4$ W/cm²K, $c_F = 0.26$ J/g·K, $c_{cl} = 0.50$ J/g·K, $n_e = 1000$ pins, $F_r = 1.25$, $T_{co} = 800$ K, $H = 40$ cm, and axial power peaking factor $F_z = 1.15$. The fuel element is predicted to fail when the local average fuel temperature reaches 1800 K. Assuming no reactivity feedback, how long will it take for the fuel element to fail?

Solution:

The thermal parameters are:

$$R_F = \frac{1}{4\pi(0.26)} = 0.306 \text{ cm·K/W}, \qquad R_g = \frac{\ln(0.338/0.325)}{2\pi(0.0046)} = 1.36 \text{ cm·K/W},$$

$$R_{cl} = \frac{1}{2\pi(0.52)}\ln\left(\frac{0.387}{0.338}\right) = 0.041 \text{ cm·K/W}, \qquad R_{co} = \frac{1}{2\pi(0.387)(4)} = 0.103 \text{ cm·K/W}.$$

$$c_{FE} = \frac{183(0.26) + 70(0.50)}{253} = 0.326 \text{ W·s/g·K},$$

$$\bar{R}_T = \frac{0.306}{2} + 1.36 + 0.041 + 0.103 = 1.65 \text{ cm·K/W}$$

$$\bar{R}_C = \frac{1}{4\pi(0.52)} - \left[\frac{(0.338)^2}{(0.387)^2 - (0.338)^2}(0.041)\right] + 0.103 = 0.124 \text{ cm·K/W},$$

$$\bar{R}_{FE} = \frac{183(0.26)(1.65) + 70(0.50)(0.124)}{253(0.326)} = 1.00 \text{ cm·K/W},$$

and $\tau = \dfrac{253 \times 10^3 \text{ g}}{(40 \text{ cm})(1000 \text{ pins})}(0.326 \text{ W·s/g·K})(1.00 \text{ cm·K/W}) = 2.07 \text{ s}.$

Using Eq. 9.31 with $\bar{\lambda} = 0.08 \text{ s}^{-1}$, the asymptotic period is

$$\mathcal{T} = \frac{1 - (0.20)}{(0.08)(0.20)} = 50 \text{ s, and using Eq. 9.87 the hot-spot fuel element temperature is}$$

$$\hat{T}_{FE} = 800 + \frac{(2 \times 10^6 \text{ W})(1.25)(1.15)}{(253 \times 1000 \text{ g})(0.326 \text{ W·s/g·K})} \cdot \frac{1}{1 - 0.20} \cdot \frac{\exp(t/50)}{[(1/2.07) + (1/50)]}, \text{ or}$$

$$\hat{T}_{FE} = 800 + 86.61 \exp(t/50) \text{ K}$$

A plot of the fuel temperature increase is presented in Figure 9.12, and the time required to reach 1800 K is

$$t = (50)\ln\left[\frac{1800 - 800}{86.61}\right] = 122.3 \text{ s}.$$

Observe that for this analysis $t \gg \tau$; consequently, for this case, the approximations used to obtain Eq. 9.86 are justified. However, the prediction of a fairly rapid temperature increase for this example is unrealistic because no feedback was assumed; and space reactors are invariably designed with negative prompt reactivity coefficients.

5.0 Reactivity Excursion Disassembly Accidents

Accidents have been postulated for reactors in which rapid reactivity excursions are sufficiently severe to cause fuel vaporization resulting in destructive disassembly of the reactor. For terrestrial reactors, reactivity-induced disassembly accidents are generally a consideration for fast reactors and not for thermal reactors. The focus on fast reactors is associated with the several key characteristics of terrestrial fast and thermal reactors. One major consideration is that, although the probability is low, large sudden reactivity insertions can be postulated for some fast reactor accident scenarios. These insertions may result from sudden movement of fuel or from the accidental addition of a moderating liquid into the reactor core. For most terrestrial thermal reactors, the presence of a moderator makes large sudden reactivity insertions highly improbable. In addition, as shown in Table 9.3, the prompt neutron lifetime

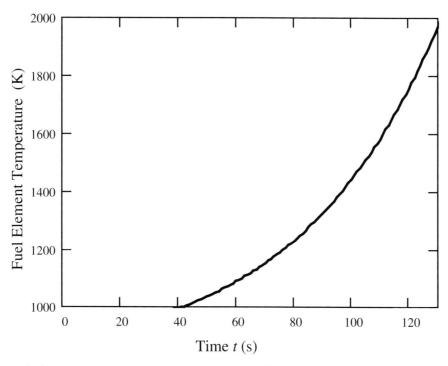

Figure 9.12 Fuel element temperature vs. time for $\rho_{in} = \$\ 0.20$ with no feedback, from Example 9.2.

for fast reactors is typically much shorter than for terrestrial thermal reactors. Finally, the negative Doppler coefficient for low-enrichment thermal reactors is usually much greater than for higher en-richment fast reactors.

In the early days of space reactor safety analysis, the unique characteristics of space reactor designs and potential accident scenarios raised the possibility of reactivity induced disassembly accidents for both fast and moderated reactors. Scenarios can be postulated (such as a reentry-impact-reflection ac-cident) that have the potential for large sudden reactivity insertions. Furthermore, as shown in Table 9.3, even moderated space reactors have relatively short prompt neutron generation times. Contrary to commercial water-cooled reactors, both fast and moderated space reactors are highly enriched and may

Table 9.3   Typical Space Reactor and Terrestrial Reactor Prompt Neutron Lifetimes

Neutron Lifetime (s)	Space Reactors	Terrestrial Reactors	Spectrum
10^{-3}			
10^{-4}		Heavy Water-Cooled Graphite-Moderated Light Water-Cooled	Thermal
10^{-5}	Enisey		
10^{-6}	SNAP-10A		Epithermal
10^{-7}	Liquid Metal-Cooled and Gas-Cooled	Liquid Metal-Cooled and Gas-Cooled	Fast
10^{-8}			
10^{-9}			Pure ^{235}U

not exhibit large Doppler feedback coefficients. Some investigators saw a potential advantage of accident scenarios that led to self-destruction for space reactors. Accidents of this type were generally postulated to occur prior to reactor operation; consequently, core disruption by low-yield explosive disassembly provided an inherent mechanism for permanent shutdown in the event of a criticality accident. The SNAP-10A design did not include provisions for subcriticality during a core-reflection accident. Reactivity induced disassembly was considered to be the principal reactor shutdown mechanism for this scenario.

As a result of these considerations, much effort was directed toward the study of reactivity induced destructive disassembly accidents for space reactors. Full-scale destructive disassembly tests were carried out for both SNAP-10A and for NERVA. A reactivity induced explosive shutdown for a SNAP-10A reflection accident was, however, the result of the peculiarities of the core design. The uranium zirconium hydride fuel rapidly dissociates hydrogen at only slightly elevated temperatures and SNAP-10A reflector ejection followed by submersion resulted in high reactivity insertion rates. Few other space reactor designs exhibit the necessary features to undergo an explosive disassembly of this type. Although developers must at least examine space reactors for the potential of reactivity induced explosive disassembly, the emphasis once placed on accidents of this type is generally unwarranted for current space reactor designs and safety approaches.

5.1 Simple Disassembly Analysis Methods

A full analysis of reactivity induced disassembly accidents is complex and requires a detailed computer model. Behte and Tait [1, 3] developed a relatively simple method to place an upper bound on the explosive release that could result from a reactivity induced disassembly accident for a fast terrestrial reactor. The Behte-Tait approach can be used for space reactors with the understanding that the predicted consequences may be grossly overestimated. This postulated accident is divided into a pre-disassembly phase and a disassembly phase. During the pre-disassembly phase it is assumed that the reactor achieves prompt criticality with a ramp rate $\dot{\rho}$ at a power level too small for any reactivity feedback. When sufficient energy density has been generated, fuel vaporizes resulting in pressure driven material motion. The disruption of the core during the disassembly phase provides the inherent shutdown mechanism.

Pre-disassembly Phase

During the pre-disassembly phase, the kinetic behavior for the reactor power is approximated by dropping the delayed contribution (second term) in Eq. 9.17 and substituting $P(t)$ for $n(t)$ to obtain

$$\frac{dP(t)}{dt} = \frac{[\rho(t) - \beta]}{\Lambda} P(t).$$ (9.88)

For a very rapid reactivity insertion rate, we can ignore the small effect prior to prompt supercriticality and approximate the time-dependent reactivity as

$$\rho(t) = \beta + \dot{\rho} t.$$ (9.89)

Here $t = 0$ at the point the reactor first achieves prompt supercriticality; thus, we can use $\rho(t) = \dot{\rho} t$ in Eq. 9.88

$$\frac{dP(t)}{dt} = \frac{\dot{\rho} t}{\Lambda} P(t).$$ (9.90)

Integrating Eq. 9.90 yields

$$P(t) = P(0) \exp\left(\frac{\dot{\rho}}{2\Lambda} t^2\right).$$ (9.91)

If we define $\tilde{E}_{th}$ as the specific energy threshold for pressure development, then $E_{th} = \tilde{E}_{th} m_c$ is the energy produced by the core at pressurization initiation and m_c is the core mass. The threshold energy can be expressed by the integral

$$E_{th} = \int_0^{t_{th}} P(t)\, dt,$$ (9.92)

where t_{th} is the time of initiation of the disassembly phase. For $\dot{\rho} t^2/(2\Lambda) >> 1$, the solution to Eq. 9.92 is approximated by

$$E_{th} \approx \frac{\Lambda P(0)}{\dot{\rho} t_{th}} \left[\exp\left(\frac{\dot{\rho}\cdot t_{th}^2}{2\Lambda}\right)\right].$$ (9.93)

Equation 9.93 can be rewritten as

$$\frac{\dot{\rho}\cdot t_{th}^2}{\Lambda} - \ln\left(\frac{\dot{\rho}\cdot t_{th}^2}{\Lambda}\right) = \ln\left\{\frac{\dot{\rho}}{\Lambda}\left[\frac{E_{th}}{P(0)}\right]^2\right\}.$$ (9.94)

The second term on the left is small compared to the first term and can be neglected; thus, the time to pressure initiation is approximately

$$t_{th} \approx \sqrt{\frac{\Lambda}{\dot{\rho}} \ln\left[\frac{\dot{\rho}}{\Lambda}\left(\frac{E_{th}}{P(0)}\right)^2\right]}.$$ (9.95)

The reactivity inserted at the end of the pre-disassembly phase is $\Delta\rho_{th} = \dot{\rho} t_{th}$; thus, from Eq. 9.95 we obtain

$$\Delta\rho_{th} = \sqrt{\dot{\rho}\Lambda \ln\left[\frac{\dot{\rho}}{\Lambda}\left(\frac{E_{th}}{P(0)}\right)^2\right]}.$$ (9.96)

Feedback Reactivity

The derivation of the reactivity feedback equation for the disassembly phase is somewhat involved and is discussed in other texts (e.g., [1] and [3]). As a consequence, we will review only the basic steps for the derivation approach by Lewis [3]. Lewis derives an equation for $(d\rho_d/dt)$ expressed in terms of ς_c, **u**, $\phi(\mathbf{r})$ and other parameters. The parameter ρ_d is the disassembly feedback reactivity, **u** is the velocity of a core volume element dV driven by fuel vapor pressure, ς_c is the density of the core, and $\phi(\mathbf{r})$ is the spatially dependent neutron flux. The velocity **u** is then related to the local pressure $p(\mathbf{r},t)$ with the equation of motion. Next, the pressure relationship is expressed as $p(\mathbf{r},t) \sim (\gamma - 1)(E(\mathbf{r},t) - E_{th})$ for $E > E_{th}$, $(p(\mathbf{r},t) = 0$ for $E < E_{th})$. Here γ is the ratio of the specific heat at constant pressure to the specific heat at constant volume for the fuel. Assuming separation of variables for $E(\mathbf{r},t)$, and assuming that at disassembly $E >> E_{th}$, Lewis uses these relationships to obtain an expression of the form

$$\frac{\partial^2 \rho_d}{\partial t^2} = -\frac{1}{C m_c r_c^4} E(t),$$ (9.97)

where C is a constant for a particular core composition.

Energy Release

For the disassembly stage, the approximation is made that the inserted reactivity and the rate of power increase do not change from their values at the end of the pre-disassembly phase until the reactor becomes subprompt critical due to disassembly. These approximations are justified by the fact that disassembly occurs over an extremely brief time span. Thus, the time-dependent power increase during the disassembly phase is

$$P(t) = P(t_{th}) \exp\left(\frac{\Delta\rho_{th}}{\Lambda}(t - t_{th})\right). \tag{9.98}$$

Integrating Eq. 9.98 between t_{th} and t yields

$$E(t) = E_{th} \exp\left(\frac{\Delta\rho_{th}}{\Lambda}(t - t_{th})\right). \tag{9.99}$$

Substituting Eq. 9.99 into Eq. 9.97 and integrating twice, the approximate solution is

$$\rho_d(t) \approx -\frac{1}{C m_c r_c^4}\left(\frac{\Lambda}{\Delta\rho_{th}}\right)^2 E(t). \tag{9.100}$$

When $\rho_d = -\Delta\rho_{th}$, the reactor becomes subprompt critical again; consequently, from Eq. 9.100 the total energy release is approximately

$$E \approx C m_c r_c^4 \frac{(\Delta\rho_{th})^3}{\Lambda^2}. \tag{9.101}$$

Substituting $\Delta\rho_{th}$ from Eq. 9.96, the energy released from a disassembly accident is

$$E \approx C m_c r_c^4 \frac{\dot{\rho}^{3/2}}{\Lambda^{1/2}}\left\{\ln\left[\frac{\dot{\rho}}{\Lambda}\left(\frac{E_{th}}{P(0)}\right)^2\right]\right\}^{\frac{3}{2}}. \tag{9.102}$$

Equation 9.102 shows that the energy released is a strong function of reactor size, the reactivity insertion rate, and the prompt neutron generation time. Using a somewhat different approach, Hetrick derives an equation of the form

$$E \approx C_{pd} \frac{\dot{\rho}^{3/2}}{\Lambda^{1/2}}\left\{\ln\left[\frac{\dot{\rho}}{\Lambda}\left(\frac{E_{th}}{P(0)}\right)^2\right]\right\}^{\frac{3}{2}}. \tag{9.103}$$

Equations 9.102 and 9.103 are identical, except that the latter does not explicitly express the reactor mass and size dependence. Note that Eq. 9.102 is used to obtain the total energy production by fission and does not represent the explosive energy released during a disassembly accident. The explosive energy release is typically a small fraction of the fission energy. Stratton et al. have calculated the fission and explosive energy release for a 611 kg spherical UO_2-fueled reflected reactor using the PAD code [4]. They also performed calculations for a 960 kg spherical UO_2-fueled unreflected reactor. Using their data, the ratio of explosive-to-fission energy released was obtained for a postulated disassembly accident; the ratio is plotted as a function of reactivity insertion rate in Figure 9.13.

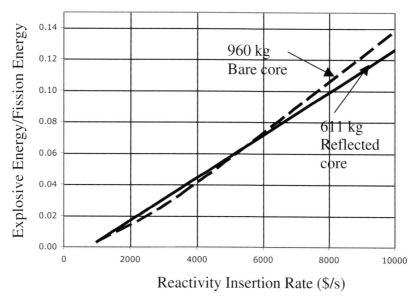

Figure 9.13 Disassembly accident calculated explosive-to-fission energy ratio for a spherical UO_2-fueled-reactor.

5.2 Reactor Disassembly Experiments

Reactivity excursion disassembly experiments have been carried out for both the NERVA and SNAP-10A space reactor designs. A single test, Kiwi-TNT, was executed for an early design of the NERVA reactor, referred to as the Kiwi. The reactor fuel was highly enriched uranium in a graphite matrix. In order to assure an explosive disassembly, the reactor control servo system was upgraded to provide a very rapid reactivity insertion. Following an intentional rapid reactivity insertion, a reactor period of 6 × 10^{-4} seconds was achieved; and in a few milliseconds the reactor was destroyed by graphite vapor pressure generated by high-energy deposition in the fuel matrix. The explosive energy released was about 345 MJ.

Three test series (Snaptran-1, 2, and 3) were performed for the SNAP-10A program. The reactor fuel for these tests consisted of the standard SNAP-10A uranium fuel in a ZrH_2 matrix. For Snaptran-1, a series of reactivity excursions were executed with reactor periods insufficient to cause explosive disassembly (> 0.001 second). The Snaptran-2 test was carried out in air, without its NaK coolant. Reactivity was inserted by very rapid rotation of control drums, resulting in a reactor period of about 2 × 10^{-4} second. The energy released was estimated to be 0.76 MJ, equivalent to about 0.17 kg of TNT. The Snaptran-3 test was carried out underwater. For this test the NaK coolant was retained in the core. In order to provide rapid reactivity insertion for the underwater test, the SNAP reactor reflectors were replaced by a cylinder of boral (boron carbide and aluminum) surrounding the core. Very rapid reactivity insertion for this test was achieved by rapid removal of the boral cylinder, resulting in a reactor period of about 6 × 10^{-4} second. The explosive energy release was estimated to be about 1.8 MJ; however, a significant fraction of the explosion energy was attributed to the chemical reaction of NaK with water [5].

Stratton has pointed out, "Small reactor nuclear explosions are more difficult to create than is commonly believed [6]." The message from Stratton's observations is that space reactor designs and accident scenarios should be carefully examined to determine if explosive disassembly accidents are credible before undertaking a costly and possibly unnecessary disassembly accident experiment. The need for these tests is unlikely for most space reactor concepts and missions, and the conduct of such a test, in itself, is a safety consideration. The other message is that to rely on explosive disassembly as a reactor shutdown approach for accident scenarios is probably unwise. The conditions required for prompt disassembly are difficult to achieve for most designs, and the varieties of possible reactor and environmental conditions are essentially unlimited. A more likely scenario for a water immersion induced reactor criticality is the induction of chug criticality, discussed in the following section. The prevention of accidental criticality, as discussed in Chapter 8, provides a more defendable approach for criticality safety issues.

6.0 Chug Criticality

Although early space reactor studies put much emphasis on prompt disassembly accidents following water flooding, chug criticality is generally a more likely scenario. For this scenario, as described in Chapter 8, an impact accident is postulated that causes a rupture in the reactor boundaries. Water floods the open spaces in the core, and the increased moderation and reflection cause the reactor to achieve supercriticality. The core power level increases, heating up the water that has filled the empty spaces. The heated water either boils or rapidly vaporizes, resulting in a negative reactivity insertion causing the reactor to become subcritical. The core fills with water again and the process is repeated. In the following we explore the approximate reactor behavior for a water-flooding scenario that induces boiling chug criticality.

6.1 Initial Excursion

We can perform a simple analysis for this scenario by beginning with the method described in Section 4.1 to analyze the initial power rise following water flooding. For this example, we assume that the core fills rapidly but that criticality is not achieved until the core has nearly filled. For the scenario in which water flooding causes a step reactivity insertion of <\$0.25, we can use Eq. 9.87. Defining t_b as the time for initiation of boiling, assuming that boiling occurs when $\hat{T}_{FE}(t_b) = 400$ K, and rearranging Eq. 9.87 gives

$$t_b = \mathcal{T} \ln\left[(400 - T_{co})\left(\frac{\beta - \rho}{\beta}\right)\frac{m_{FE}\, c_{FE}\, (\mathcal{T}^{-1} + \tau^{-1})}{\hat{P}_0} \right]. \qquad (9.104)$$

Using Eq. 9.104, and assuming that after a time interval Δt_b voiding will provide a sufficient reduction in water volume to take the core subcritical, the power increase is terminated at $t_1 = t_b + \Delta t_b$. The reactor power level at this time is

$$\hat{P}_1 = \hat{P}_0\, \frac{\beta}{\beta - \rho}\, \exp(t_1/\mathcal{T}), \qquad (9.105)$$

and

$$\hat{T}_{FE1} = T_{co} + \frac{\hat{P}_0}{m_{FE}\, c_{FE}\, (\mathcal{T}^{-1} + \tau^{-1})}\, \frac{\beta}{(\beta - \rho)}\, \exp(t_1/\mathcal{T}). \qquad (9.106)$$

6.2 Boiling Shutdown

Boiling typically results in a rapid and significant negative reactivity insertion. For large negative reactivity insertions, the heating during the rapid power drop is usually insignificant. Furthermore, for the small change in surface temperature during boiling oscillations, the temperature difference $\hat{T}_{FE}(t) - T_{co}$ in Eq. 9.81 changes very little. Thus dropping the first term on the right-hand side of Eq. 9.81 and approximating the temperature difference between the fuel and water as $\hat{T}_{FE1} - T_{co}$, the equation is solved to obtain

$$\hat{T}_{FE}(t) = \hat{T}_{FE1} - \frac{(\hat{T}_{Fe1} - T_c)}{\tau_{co}} t. \tag{9.107}$$

Here, τ_{co} is the time constant accounting for boiling heat transfer. Note that the time t in Eq. 9.107 is indexed to zero at the start of the negative reactivity insertion. From Eq. 9.107, the time t_{co} to cool down to saturation temperature (assumed to be 400 K) is

$$t_{co} = \tau_{co} \frac{(\hat{T}_{Fe1} - 400)}{(\hat{T}_{Fe1} - T_{co})}. \tag{9.108}$$

The bulk coolant temperature is assumed to remain at the temperature of the water source (e.g., 300 K). At a subsequent time $t_2 = t_{co} + \Delta t_{co}$, bubbles remaining in the bulk water collapse. For the assumed large negative reactivity insertion ρ_{co} during boiling, the period is $\mathcal{T}_{co} = -\lambda_1^{-1}$. Thus, using Eq. 9.38, the power at time t_2 is

$$\hat{P}_2 = \hat{P}_1 \left[\frac{\beta}{\beta - \rho_{co}} \exp\left(\frac{t_2}{\mathcal{T}_{co}}\right) \right]. \tag{9.109}$$

Also, from Eqs. 9.107 we obtain $\hat{T}_{FE2} = \hat{T}_{FE}(t_2)$.

6.3 Second Excursion

Assuming that at time t_2 the net reactivity returns to the original inserted value, the time-dependent fuel element temperature can be obtained by modifying Eq. 9.81. Using Eq. 9.83, the assumption of an approximately constant temperature difference between the fuel and cladding, and other substitutions in Eq. 9.81, we obtain the solution

$$\hat{T}_{FE}(t) = T_{FE2} + \frac{\hat{P}_2}{m_{FE} c_{FE}} \frac{\beta \mathcal{T}}{(\beta - \rho)} [\exp(t/\mathcal{T}) - 1] - \frac{(T_{FE2} - T_{co})}{\tau} t. \tag{9.110}$$

The time is again indexed to zero at the time of reactivity insertion. Equation 9.110 can be written as a series expansion and for the case $t \ll \mathcal{T}$, and approximated by the first two terms of the expansion as $1 + t/\mathcal{T}$. Thus rearranging Eq. 9.110 and using this truncated expansion, the time $t_3 = t_b + \Delta t_b$ is

$$t_3 = \Delta t_b + \frac{(400 - T_{FE2})}{\left[\frac{\hat{P}_2}{m_{FE} c_{FE}} \frac{\beta}{(\beta - \rho)} - \frac{(T_{FE2} - T_{co})}{\tau} \right]}. \tag{9.111}$$

This process is repeated, yielding a prediction of an oscillating power about a quasi equilibrium value.

The preceding method was used to provide the predicted power history in Figure 9.14 for a postulated water flooding accident. For this example it was assumed that water flooding, at a core filling rate of 0.5 cm/s, produces a step reactivity insertion of $0.25 and the voiding reactivity insertion was worth $-$1.05. It is also assumed that $\Delta t_b = 1.0$ s and $\Delta t_{co} = 0.01$ s. Given the approximations and assumptions used to generate Figure 9.14, the predicted power history should be regarded as a qualitative estimate.

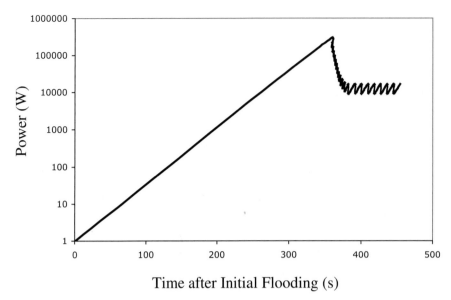

Figure 9.14 Power vs. time for a postulated chug criticality accident. Assumed $0.25 flooding reactivity, void worth $-$1.05, and fill rate $= 0.5$ cm/s.

7.0 Cooling Failure Accidents

Terrestrial reactor accident analysis typically places much emphasis on cooling failure accidents. For reactors operating in high orbit or on the surface of a planet or moon, cooling failure accidents may not have significant safety implications. Nonetheless, as for other operational accidents postulated for space reactors, cooling failure accidents may have safety consequences for some classes of missions. In this section we examine some of the more common postulated cooling failure accidents. These include flow blockages, loss-of-flow (LOF), and loss-of-coolant accident (LOCA) scenarios.

7.1 Flow Blockages

Space reactors using a coolant to transport reactor heat to a power conversion system are designed with adequate coolant flow to assure that the reactor fuel elements do not exceed safe temperature limits. However, a blockage of a flow channel could result if a piece of hardware in the reactor system breaks free and is swept into the core region by the flowing coolant. Blockages can also result from fuel swelling or bowing or from the buildup of particulate entrained in the coolant. The blockage could prevent adequate cooling, resulting in excessive fuel element temperatures and possible melting of fuel and cladding boundaries. Fuel melting and cladding failure can release fission products into the coolant steam. Fuel melting and relocation may also lead to large-scale flow blockages and more extensive fuel failure, possibly progressing to pressure vessel failure and the release of radioactive materials into the environment.

For a normal flow channel, the pressure drop across the core can be determined from

$$\Delta p = \frac{\mathcal{K}_0}{2\varsigma_{co}} \left(\frac{\dot{m}_{ch}}{A_{ch}}\right)^2. \tag{9.112}$$

Here, $\mathcal{K}_0$ is the dimensionless hydraulic resistance, ς_{co} is the coolant density, $\dot{m}_{ch}$ is the channel mass flow rate, and A_{ch} is the cross sectional flow area of the channel. The hydraulic resistance for the unobstructed channel $\mathcal{K}_0$ is given by

$$\mathcal{K}_0 = f_r \frac{H}{D_h}. \tag{9.113}$$

Here, f_r is the friction factor, H is the channel length, and D_h is the hydraulic diameter. If we assume that the flow restriction is in the form of a sharp edge orifice, the hydraulic resistance for the restriction $\mathcal{K}_R$ can be approximated by

$$\mathcal{K}_R = 2.7 \, \chi^2 \frac{(2 - \chi)}{(1 - \chi)^2}. \tag{9.114}$$

Here, $\chi \equiv A_R/n_{ch}A_{ch}$, A_R is the cross sectional area of the flow restriction, and n_{ch} is the number of communicating coolant channels [3, 7].

For a fairly localized flow blockage, the pressure drop across the core is not significantly affected, and the flow resistance due to the blockage represents a series resistance in the flow channel. Thus, equating the pressure drops before and after the flow restrictions gives

$$\mathcal{K}_0 \dot{m}_{ch}^2 = (\mathcal{K}_0 + \mathcal{K}_R)\dot{m}_R^2, \tag{9.115}$$

where $\dot{m}_R$ is the mass flow rate in the restricted channel. Combining the above equations yields

$$\frac{\dot{m}_R}{\dot{m}_{ch}} = (1 - \chi)\left[(1 - \chi)^2 + \frac{2.7}{\mathcal{K}_0}\chi^2(2 - \chi)\right]^{-\frac{1}{2}}. \tag{9.116}$$

Using Eq. 9.116, the fractional reduction in flow rate in the obstructed channel is plotted as a function of the number of communicating channels in Figure 9.15 for several flow area restriction ratios, assuming $\mathcal{K}_0 = 2$. If the flow restriction is sufficient to block one channel ($A_R/A_{ch} = 1$), about ten communicating coolant channels are required to maintain the original flow rate. If all cooling channels are isolated (e.g., the NERVA design), Figure 9.15 shows that a 50% blockage will result in about a 40% reduction in the coolant mass flow rate in the obstructed channel.

From this example, it is seen that a number of communicating cooling channels can significantly reduce the consequences of flow blockages. Good design and careful quality assurance can prevent or reduce the consequences of flow blockages. Protective grid plates are typically used upstream of the core to prevent any debris from being swept into the core where it could create a blockage.

7.2 Loss-of-Flow Accidents

If a coolant pump or compressor fails, inadequate cooling will result, possibly leading to fuel failure and core damage. For liquid metal-cooled space reactor systems using an electromagnetic (EM) coolant pump, the analysis is simplified somewhat because pump inertia is absent. Here we consider two loss-of-flow scenarios for liquid metal-cooled space reactors; i.e., (1) an orbiting power system and (2) a planet-surface-based reactor.

Orbiting Reactor

For an orbiting liquid metal-cooled reactor with an EM pump, both pump inertia and hydrostatic pressure are absent. For a flow transient with a single-phase coolant, the pump pressure Δp_p must equal the sum of the fluid inertia and the viscous pressure drops; i.e.,

$$\Delta p_p = (L/A)_T \frac{d\dot{m}}{dt} + \frac{\mathcal{K}_T}{A_T 2 \varsigma_c}\dot{m}^2. \tag{9.117}$$

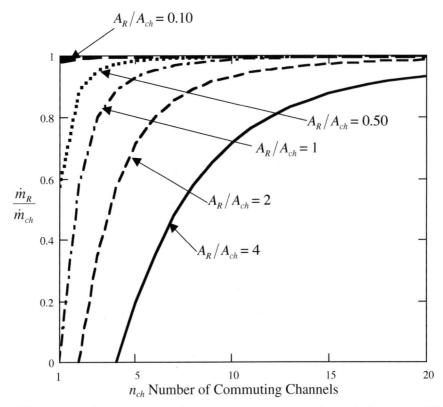

Figure 9.15 Channel flow rate reduction vs. number of communicating channels for several flow area restriction ratios, assuming $\mathcal{K}_0 = 2$.

Here, $(L/A)_T$ and $\mathcal{K}_T$ are the total (effective) ratio of the loop length to flow area and the hydraulic resistances. Lewis [3] shows that

$$(L/A)_T = (L/A)_c + \frac{1}{N_L}(L/A)_L \tag{9.118}$$

and

$$\frac{\mathcal{K}_T}{A_T^2} = \frac{\mathcal{K}_c}{A_c^2} + \frac{1}{N_L^2}\frac{\mathcal{K}_L}{A_L^2}. \tag{9.119}$$

The subscripts c and L refer to parameters for the core and external (to the core) loops, and N_L is the number of identical external loops. For a space reactor, the external loops may refer to coolant passages in a heat exchanger.

Following a loss of an electromagnetic pump Δp_p in Eq. 9.117 instantaneously goes to zero resulting in a rapid coolant flow coast down. Solving Eq. 9.117 with $\Delta p_p = 0$ yields

$$\dot{m}(t) = \frac{\dot{m}(0)}{1 + t/\tau_T}, \tag{9.120}$$

where τ_T is the total loop time constant. For a single coolant loop, the total loop time constant is

$$\tau_T = \frac{2\varsigma_c}{\dot{m}(0)(\mathcal{K}_T/A_T^2)(A/L)_T}. \tag{9.121}$$

For the simple cases of either (1) constant power or (2) rapid drop to zero power (no decay heating) following pump loss, the method given by Lewis [3] can be used to obtain an approximate prediction of the coolant temperature rise ΔT_{co}; i.e.,

$$\Delta T_{co}(t) = \Delta T_{co}(0) \frac{\dot{m}(0)}{\dot{m}(t)} f_\tau(t), \qquad (9.122)$$

where

$$
\begin{aligned}
f_\tau(t) &= 1 &&\text{for constant power} \\
f_\tau(t) &= \exp\left(-\frac{t}{\tau}\right) &&\text{for power} \to 0
\end{aligned}
\qquad (9.123)
$$

and τ is defined by Eq. 9.82. Substituting $\dot{m}(t)$ from Eq. 9.120 into 9.122 gives

$$\Delta T_{co}(t) = \Delta T_{co}(0)(1 + t/\tau_T)\, f_\tau(t). \qquad (9.124)$$

Example 9.3

An orbiting liquid metal-cooled space reactor has just begun operation at 1 MW$_{th}$ when the power to all EM pumps is lost. Assuming that the reactor is immediately shut down, estimate the time-dependent behavior of the coolant temperature rise across the core using the following data: $\Delta T_{co}(0) = 150$ K, $\dot{m}(0) = 4$ kg/s, $\tau = 5$ s, $\varsigma_{co} = 0.74$ g/cm^3, $H = 60$ cm, $A_c = 50$ cm^2, $\mathcal{K}_c = 1.1$. Assume that the external loop can be ignored and ignore the effect of fission product heating.

Solution:

The total loop time constant is

$$\tau_T = \frac{2(0.74 \text{ g/cm}^3)}{(4 \times 10^3 \text{ g/s})[1.1/(50 \text{ cm}^2)^2](50 \text{ cm}^2/60 \text{ cm})} = 1.01 \text{ s}.$$

Thus, the time dependence of the coolant temperature rise across the core is

$$\Delta T_{co}(t) = (150 \text{ K})(1 + t/1.01 \text{ s})\exp(-t/5 \text{ s}).$$

Using this relationship, the coolant temperature rise across the core is plotted as a function of time following pump failure in Figure 9.16.

Surface System with Decay Heating

For a reactor operating on the surface of a planet or moon, the effect of gravitation permits natural circulation in the event of pump failure. For this example, we assume that the reactor is immediately shut down, but the power level does not immediately drop to zero. If the reactor has been operating at high power, fission product decay will continue heating the core at a low power level. The time-dependent reactor power level due to decay heating for light water-cooled reactors (LWRs) is often approximated by

$$P_d(t) = 0.0061\, P(0)[(t - t_0)^{-0.2} - t^{-0.2}], \qquad (9.125)$$

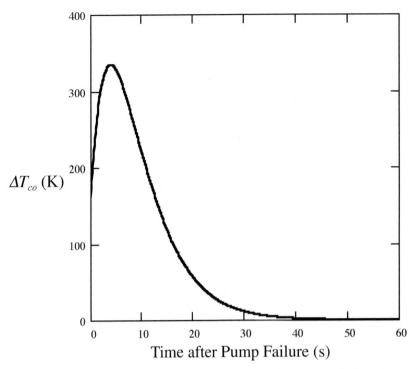

Figure 9.16 Coolant temperature rise across core vs. time after pump failure from Example 9.3.

where t_0 is the elapsed time in days of constant full power operation at power $P(0)$ at the moment of shutdown. The total time t is the time in days of full power operation plus the time following shutdown [8]. Although Eq. 9.125 was developed for LWRs, the equation provides very good predictions of the decay power level for other types of reactors, including space reactors.

The rise in the core coolant temperature can be estimated by adapting the formula for terrestrial reactor natural cooling [3]; i.e.,

$$\Delta T_{co}(t) = \left(\frac{P_d(t)}{\overline{\varsigma}_{co} c_p}\right)^{2/3} \left[\frac{(\mathcal{K}_T/A_T^2)}{2 g_p \vartheta_V (z_{hx} - z_{mc})}\right]^{1/3}. \tag{9.126}$$

Here $\overline{\varsigma}_{co}$ is the average coolant density, g_p is the gravitational acceleration for the planet or moon, z_{hx} is the elevation at the top of the heat exchanger, z_{mc} is the core midplane elevation, and ϑ_V is the coefficient for volumetric thermal expansion.

LOF Accident Mitigation

Electromagnetic pumps do not contain moving parts that can fail or seals that can develop leaks. The principal issue associated with EM pumps is the possibility of loss of electrical power to the pumps. One method for providing assured continuous pumping during an LOF accident is to use thermoelectric EM (TEM) pumps. TEM pumps use heat from the hot coolant lines and thermoelectric devices (discussed in Chapter 2) to assure uninterrupted electrical power. For gas-cooled reactors, loss-of-flow accidents can be mitigated by the use of redundant circulators. In the event of pump or circulator failure for a surface-based reactor, overheating of the core may be prevented by designs that allow for natural circulation.

7.3 Loss-of-Coolant Accidents

Actively cooled reactor systems proposed for space applications typically require a liquid metal or gas coolant. Leaks in the coolant system may result from inherent defects or from an impact with meteoroids or space debris. In the event of a loss-of-coolant accident (LOCA), reactor systems are typically designed to automatically shut down. However, after reactor shutdown, decay heating due to fission products will continue to heat the fuel. Because the coolant volume will generally be small for space reactor systems, a break in the line or a significant leak will result in a rapid depressurization and rapid loss of coolant. For very rapid coolant loss, adiabatic heating of the fuel can be assumed in order to obtain an upper bound for the fuel temperature behavior. If core coolant loss is essentially instantaneous, the fuel temperature will be primarily determined by the heat capacity of the core. Thus, the fuel temperature can be approximated by

$$T_{FE}(t) = \frac{1}{m_{FE}\, c_{FE}} \int_0^t P_d(t')dt' + T_{FE}(0). \tag{9.127}$$

The specific heat c_{FE} in Eq. 9.127 must be expressed in (W·d/g·K) rather than (W·s/g·K) to be consistent with Eq. 9.125 in which time is given in days.

In the vacuum of space, a liquid metal coolant will rapidly vaporize if a break or significant breach occurs in the coolant loop. In this event, a rapid blow-down will occur, similar to a pressurized water reactor blow-down. Some large liquid metal-cooled terrestrial reactor designs exhibit positive reactivity void coefficients and are consequently predicted to undergo a rapid power increase during a LOCA. If this were the case for space reactors, the decay power in Eq. 9.127 would need to be replaced by the expression for the power excursion. Space reactors are generally small, however, and coolant loss typically results in a sufficiently large increase in neutron leakage such that the void coefficient of reactivity is negative.

Including the time dependence of coolant loss as well as radiative and conductive heat transfer will result in lower predicted fuel temperatures. The time dependence of coolant loss can be determined as the mass flow rate of the leak due to a hole of area A_h by rearranging Eq. 9.112; i.e.,

$$\dot{m}(t) = A_h \sqrt{\frac{2\, p(t)\, \varsigma_{co}}{\mathcal{K}}}, \tag{9.128}$$

where the system pressure p is the time-dependent system pressure.

The complete analysis of a LOCA for a liquid metal-cooled reactor in a microgravity environment typically requires the use of a detailed computer model. A LOCA analysis was performed by General Electric in 1987 [9] for an orbiting SP-100 reactor. Their analysis looked at the effect of leaks in the coolant system due to flaws and meteoroid or space debris impact. The predicted maximum temperature and void time from the General Electric Co. analysis is shown in Figure 9.17.

Total loss of the coolant does not occur during a LOCA for terrestrial gas-cooled reactors because the reactor is within Earth's atmosphere and the coolant gas is constrained by atmospheric pressure. For reactors operating in the vacuum of space, however, a LOCA can result in essentially total loss of a gas coolant. Lewis shows [3] that the coolant pressure drop due to a LOCA for a gas-cooled reactor as a function of time after pressure system puncture is approximately

$$p(t) = p(0) \exp\left(-\frac{t}{\tau^*}\right), \tag{9.129}$$

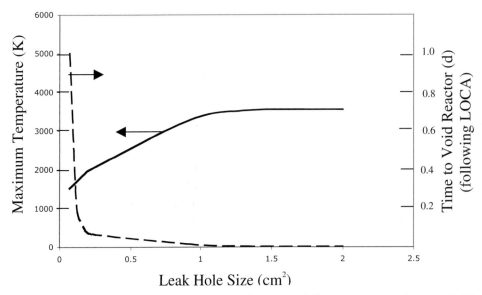

Figure 9.17 Predicted core temperature and time to void reactor following a postulated LOCA for an orbiting SP-100 reactor system.

where

$$\tau^* = \frac{V}{A_b}\left[RT_{co}\gamma\left(\frac{2}{\gamma+1}\right)^{\frac{(\gamma+1)}{(\gamma-1)}}\right]^{-1/2}.$$

(9.130)

Here, R is the universal gas constant and γ is the specific heat ratio. For helium, $\tau^* = (0.302)(V/A_b)T_{co}^{1/2}$. Using the time-dependent pressure drop and other reactor parameters, one can compute the thermal response of the core due to a LOCA. As for liquid metal-cooled reactors, the complete analysis of a LOCA typically requires the use of a detailed computer model.

Prevention of LOCAs for space reactors is addressed, as for terrestrial reactors, by good design, proper selection of materials, and quality assurance to reduce the risk of leaks resulting from flaws and corrosion during operation. In addition, loop and radiator plumbing typically incorporate meteoroid protection to prevent puncture by meteoroids and space debris (see Chapter 7). If LOCA accidents can present a credible safety hazard (e.g., for some LEO missions), redundant coolant loops may be required.

8.0 Computer Codes

A number of computer codes have been developed to predict neutron kinetic behavior for reactor excursions. These codes range from simple point kinetics, multi-precursor-group codes to multi-precursor-group codes coupled to three-dimensional spatial calculations. Large and very non-uniform reactivity insertions generally require full space-time models. For small space reactors, time-dependent S_n transport theory codes are generally required when spatial effects are important. A number of approximations can be used that are more accurate than the PRK approximation but are simpler than a full space-time treatment. For some slow transients, the time derivatives can be ignored. In this case, the delayed neutron source can be combined with the prompt neutron source to permit eigenvalue calculations of the space-dependent flux. This approach is referred to as the adiabatic approximation.

Another approximate method, called the quasistatic approximation, is used for cases where $\partial\phi(\mathbf{r})/\partial t$ can be neglected, but $dn(t)/dt$ must be considered. Delayed and prompt neutrons cannot be combined and $\phi(\mathbf{r})$ cannot be obtained from an eigenvalue calculation. The computation of the shape factor $\phi(\mathbf{r})$ requires that the prompt and delayed neutron equations are combined, and its solution is based on the conditions at time t. Delayed effects are then explicitly included for the time-dependent solution, with $\partial\phi(\mathbf{r})/\partial t$ set to zero. The term $dn(t)/dt$ can be evaluated from the prompt neutron equation for the last few sets of time intervals. The advantage of this approach is that the shape factor is computed infrequently, resulting in shorter computational times.

In addition to reactor kinetics computer codes, coupled kinetics-thermal hydraulic codes have been developed. For example, the SAS4A code [10] was designed to analyze severe accidents in terrestrial liquid metal-cooled reactors. The reactor power level is computed using point kinetics, and heat transfer within fuel pins is modeled using the two-dimensional heat conduction equation. The model also includes dimensional changes and cladding failure during accidents, fission product production and release, fuel and cladding melting and relocation, reactivity feedback, and other effects. The SAS4A code may not be suitable in its present form for the analysis of space reactor dynamic accidents; however, for some missions, detailed analyses of this type may be unnecessary.

9.0 Summary

Some accident scenarios relating to space reactor dynamics are postulated for the prelaunch phase. Furthermore, reactor dynamic accidents may be induced as a consequence of an unplanned reentry. Missions involving astronauts in the vicinity of operating space reactor must consider the potential safety risk to astronauts resulting from postulated transient accidents. For some missions, space reactor transient accidents during the operational phase may present public safety and environmental issues. For example, reactor disruption could result from an operational transient accident. If the system was operating in LEO, a severely damaged system may be incapable of boost to high orbit for disposal. As a consequence, the system may reenter Earth's biosphere before fission product activity decays to a safe level. Space reactor accidents during the operational phase do not necessarily present safety or environmental issues. If astronauts are not in the vicinity of an operating reactor and the reactor is in a sufficiently high orbit or on the surface of another planet or moon, reactor accidents may have no safety consequence for humans or Earth's biosphere.

Reactor kinetics analysis methods explicitly model the behavior of both prompt neutrons and delayed neutrons. Delayed neutron effects are included by the use of coupled equations for the reactor neutron density and delayed neutron precursor concentrations. Delayed neutrons are typically modeled as six delayed neutron precursor groups. Reactor kinetic behavior is generally expressed in terms of reactivity (deviation from criticality) and the prompt neutron generation time. For a broad class of postulated reactivity excursion accidents for space reactors, the spatial dependence during the transient can be ignored. For these types of accidents, the point reactor kinetics equation can be used to obtain reasonable predictions of reactor behavior. The general solution of the PRK equation is given as a sum of j exponential terms to the $\omega_j t$ power. Following a small sudden positive change in reactivity, all of the negative ω_j roots decrease rapidly to zero and the neutron density varies with time as $\sim \exp(\omega_0 t) = \exp(t/\mathcal{T})$, where $\mathcal{T}$ is defined as the asymptotic period. For negative reactivity insertions, the neutron time behavior also quickly approaches $\exp(t/\mathcal{T})$, where $\mathcal{T}$ is negative.

For small positive reactivity insertions and a range of negative reactivity insertions, a simple one-delayed precursor group model can be used to obtain the approximate kinetic behavior of the reactor. For large prompt reactivity insertions, delayed effects are insignificant and the neutron density $n = n_0 \exp(t/\mathcal{T})$, where the period $\mathcal{T} = \Lambda/(\rho - \beta)$. Changes in reactor fuel temperature and other parameters during a reactivity excursion usually provide reactivity feedback. These feedback effects can play an important role in reactor safety in that they slow down, speed up, or terminate a reactivity transient. Good safety practices generally incorporate designs that provide prompt negative reactivity feedback. Simple lumped parameter thermal models can be used to determine the change in fuel temperature during a reactor transient.

One of the classic accidents postulated for space reactors is explosive disassembly due to a reactivity excursion following a water flooding accident. For this scenario, a reactor is assumed to experience a very rapid and large reactivity insertion due to neutron moderation by water flooding. Rapid fuel vaporization causes the reactor to undergo explosive disassembly, shutting down the reactor. Although this scenario is usually postulated for water flooding accidents, the more likely outcome is a chug criticality. During chug criticality, moderation due to core flooding results in a reactivity insertion causing the reactor to become supercritical. The core power increases, causing boiling of water that has entered the core. The changes in neutron moderation from repeated boiling and bubble collapse causes oscillations in reactor power.

Several types of cooling failure transients are possible, such as flow blockages, loss-of-flow accidents, and loss-of-coolant accidents. Although the reactor is typically assumed to shut down rapidly, decay heating by fission products may lead to excessive fuel temperatures and subsequent damage. Armor surrounding coolant lines can reduce the risk of a loss-of-coolant accident due to meteoroid and space debris impact. As for other types of operational accidents for some space missions, coolant failure accidents may not present significant safety issues.

Symbols

(A/L)	area/length	$\mathcal{L}_a$	neutron abs. loss rate/volume
A_{FE}	function in Eq. 9.84	$\mathcal{L}_a$	neutron leakage loss rate/volume
A	flow area	m	mass
B^2	buckling	$\dot{m}$	coolant mass flow rate
c_i	t-depen. ith precursor concen.	N_L	number of loops
c_x	specific heat of material x	n	neutron density
C	constant in Eq. 9.102	n_e	number of fuel elements
C_{pd}	constant in Eq. 9.103	n_{ch}	number of coolant channels
C_i	$\mathbf{r}$, t-depen. ith precursor concen.	P	reactor power level
C_p	wetted perimeter	p	pressure
D	neutron diffusion coefficient	Pr	Prandtl number
D_h	hydraulic diameter	$\mathcal{P}_f$	n fission production rate/volume
E	energy	q'	linear heat flux
$\tilde{E}_{th}$	threshold specific energy	R	thermal resistance
f_r	friction factor	$\bar{R}$	thermal resistance for FE avg.
$f_\tau(t)$	defined by Eq. 9.123	R	universal gas constant
$f(z)$	axial shape factor	Re	Reynolds number
F_r	radial power peaking factor	r	radius
F_z	axial power peaking factor	$\mathbf{r}$	position vector
g_i	$\mathbf{r}$-dependence of ith precursor	t	time
g_p	gravitational accel. for planet	T	temperature
H	core height	$\mathcal{T}$	asymptotic period
h_{co}	coolant heat transfer coefficient	u	defined in Example 9.1
h_g	gap heat transfer coefficient	U	defined by Eq. 9.46
k	thermal conductivity	$\bar{v}$	average neutron velocity
k_{eff}	effective n-multiplication factor	$\bar{v}_c$	average coolant velocity
k_∞	infinite medium multipl. factor	V	volume
$\mathcal{K}$	hydraulic resistance	Y	defined by Eq. 9.47
l	prompt neutron lifetime	z	axial position
L	neutron diffusion length		

α	reactivity feedback coefficient		ρ_{in}	inserted reactivity
β_i	ith group delayed neutron fraction		$\dot{\rho}$	reactivity insertion rate
Δp	pressure drop		Σ	macroscopic cross section
Φ	space/energy dependent n-flux		τ	core time constant
ϕ	space-dependent neutron flux		τ_{co}	time constant with boiling
γ	specific heat ratio		τ_T	total loop time constant
Γ	full width/half-max for burst $P(t)$		τ^*	gas leak time constant
κ	energy yield per fission		υ	neutrons/fission
$\bar{\lambda}$	1-group precursor decay constant		ω	root for PRK solution
λ_i	ith precursor decay constant		χ	A_R/A_{ch}
Λ	prompt neutron generation time		ς	mass density
μ	viscosity		$\$$	reactivity unit $= \rho/\beta$
ϑ_V	thermal volume expansion coef.		$\cent$	reactivity unit $= 100$
ρ	reactivity			

Special Subscripts/Superscripts

0	initial value		i	precursor group
1,2,3...	chug critical oscil. no.		icl	inner cladding
a	neutron absorption		ocl	outer cladding
C	clad + coolant		j	root index for PRK solution
c	core		L	coolant loop
ch	coolant channel		max	maximum
cl	cladding		mc	core midplane
co	coolant or cool-down		P	perimeter
d	decay heating		p	*pump*
e	electrical		R	restricted flow
E	energy		T	total
F	fuel		T_x	temperature for material x
FE	fuel element		th	threshold
Fc	fuel centerline		∞	stable conditions
h	hole		$\wedge$	peak
hx	heat exchanger			

References

1. Hetrick, D. L., *Dynamics of Nuclear Reactors*. Chicago: University of Chicago Press, 1971.
2. Argonne National Laboratory, *Reactor Physics Constants*. ANL-5800, USAEC, 1965.
3. Lewis, E. E., *Nuclear Power Reactor Safety*. New York: John Wiley & Sons, 1977.
4. Stratton, W. R., L. B. Engle, D. M. Peterson, "Energy Release from Meltdown Accidents." Proceedings of the 1973 Winter Meeting of the American Nuclear Society, San Francisco, CA, 1973.
5. Stratton, W. R., L. B. Engle, D. M. Peterson, "Reactor Power Excursion Studies." Proceedings of the International Conference on Engineering of Fast Reactors for Safe and Reliable Operation, San Francisco, CA, 1973. Karlsruhe, Germany, Oct. 9–12, 1972.
6. Stratton, W. R., "Severe Accident Analysis, Philosophical Approach, Assumptions and Analytical Technologies." Los Alamos National Laboratory, unpublished memo to A. Walter, Mar. 28, 1990.
7. Kramers, H., *Physische Transportverschijnselen*. Hogeschool, Delft, Holland, 1958.
8. Glasstone, S. and A. Sesonske, *Nuclear Reactor Engineering*. New York: Van Nostrand Reinhold Co., 1967.
9. Magee, P. M., J. M. Berkow, D. R. Damon, B. Deb, U. N. Sinha, D. C. Wadekamper, R. Yahalom, *Assessment of Loss of Primary Coolant in Orbit*. General Electric Report, Mar. 1987.
10. Calalan, J. E. and T. Wei, "Modeling Development for the SAS4A and SASSYS Computer Codes." Proceedings of the International Fast Reactor Safety Meeting, American Nuclear Society, Snowbird, UT, Aug. 1990.

Student Exercises

1. During prelaunch testing, a space reactor is maintained at a thermal power level of 10 W. (a) Compute and plot the power level as a function of time following a $0.24 step reactivity insertion using the one-delayed group PRK approximation with no feedback. For this reactor, the prompt neutron generation time is 10^{-3} s and $\beta = 0.0067$. Obtain separate plots for times up to 1 second and up to 60 seconds. (b) Repeat Exercise (a) assuming a prompt neutron generation time is 10^{-7} s. (c) Compare plots and comment on the effect of prompt neutrons and the first and second terms in the equation.

2. A space reactor is dropped while mating the reactor system to the launch vehicle. The impact with the pad ejects the safety rods causing a positive step reactivity insertion of $3.00. (a) Use $\Lambda = 10^{-7}$ s, $\beta = 0.0065$, $m_F c_F = 2.8 \times 10^4$ J/K, and $\alpha_{T_F} = 3 \times 10^{-5}$ K^{-1} to compute the maximum power level during the excursion, the net energy produced by the pulse, the maximum fuel temperature increase, and the fuel temperature at the maximum power level. (b) Repeat Exercise (a) assuming $4.0 insertion and a prompt neutron generation time is 10^{-5} s. (c) Plot the time dependence of the power level, in MW, during the burst for exercise (b), assuming an initial power level of 10 W due to the presence of a neutron source.

3. A pin-type liquid metal-cooled space reactor operating on the Moon at 2.5 MW$_{th}$ is given a step reactivity insertion of $0.03/s. Assume the coolant inlet temperature is 500 K, and no reactivity feedback. The axial power shape is a cosine and the power is assumed to go to zero at the top and bottom of each fuel element. Also, use the following reactor parameters to calculate the initial temperatures and the temperatures after 2 minutes for the coolant, the fuel centerline, and the cladding. Plot all temperatures as a function of axial position.
$F_r = 1.3$, $k_F = 0.20$ W/cm·K, $k_{cl} = 0.45$ W/cm·K, $k_g = 0.0045$ W/cm·K, $k_{co} = 0.3$ W/cm·K, $c_{co} = 0.27$ W·s/g·K, $c_F = 0.26$ W·s/g·K, $c_{cl} = 0.5$ W·s/g·K, $\mu_{co} = 4 \times 10^{-3}$ g/cm·s, $n_e = 1000$, $H = 40$ cm, $r_F = 0.325$ cm, $r_{icl} = 0.338$ cm, $r_{ocl} = 0.387$ cm, $A_{co} = 0.093$ cm^2, Pr $= 0.004$, $\varsigma_{co} = 0.9$ g/cm^3, $\dot{m} = 20$ kg/s, $m_F = 220$ kg, and $m_{cl} = 83$ kg.

4. During a launch abort an onboard space reactor reenters and hits a rock along the shoreline at high speed, causing the pressure vessel to rupture and the coolant to escape. The reactor then falls into the ocean and fills rapidly with water. Use $C = 2 \times 10^{-11}$ cm^{-2} as a conservative estimate of the constant in the equation for the prompt disassembly fission energy yield (Eq. 9.102), and use the following parameters: $\tilde{E}_{th} = 1.8 \times 10^3$ W·s/g, $\Lambda = 3 \times 10^{-8}$ s, $\beta = 0.0070$, $P(0) = 1$ W, $m_c = 82$ kg, $\varsigma_c = 5$ g/cm^3. (a) If we assume that it is possible for water flooding to introduce reactivity at a rate in the range $1000/s to $4000/s (unrealistically high values necessary to induce explosive disassembly), calculate and plot the fission energy release as a function of reactivity insertion rate over the postulated range. Estimate the explosive energy release using Figure 9.3, and plot on the same figure. (b) Calculate the fission energy release for an insertion rate of $2000/s for each of the following deviations from the original and plot data point on same figure. (Unless stated otherwise, all other parameters are unchanged from the original specifications for each case.)

- For $\Lambda = 3 \times 10^{-9}$ s
- For an initial power of 0.001 W
- For a 150 kg core mass (the core radius must be recalculated)

5. For the same scenario given in Exercise 4, assume that the reactivity insertion due to water flooding is $0.25/s and boiling results in chug criticality. Use $P_0 = 1$ W, $T_{co} = 300$ K (water temperature), $F_z = 1.2$, $h_{co} = 1.1 \times 10^{-3}$ W/cm^{2}K, $\Delta t_b = 1.3$ s, and $\Delta t_{co} = 0.2$ s. Assume that the water temperature is 300 K, the boiling heat transfer coefficient equals 1 W/cm^{2}K, and the net reactivity due to boiling is $-$$0.90. For all other parameters, use the reactor parameters from Exercise 3. Estimate the initial power rise and the time-dependent power level during the first two power oscillations.

6. The tip of a coolant instrument probe for Mars-based reactor becomes severely corroded and breaks free. The broken tip is carried by the coolant into the reactor region where it becomes lodged at the inlet to a coolant channel. The reactor consists of a solid moderator with 400 cylindrical holes, each containing a cylindrical fuel rod and an annular space for coolant flow. The coolant channels do not interconnect. Assume that the probe blocks 30% of a single coolant channel. (a) Using $\mathcal{K}_0 = 1.7$, estimate the reduction in the mass flow rate in the obstructed coolant channel. (b) Calculate the equilibrium temperature of the coolant and fuel in the normal channel for a reactor thermal power level of 2.05 MW. Plot all temperatures as a function of axial position. The total resist-

ance is , $R_T = R_F + R_{co}$, where $R_F = 2.6$ cm·K/W, $R_{co} = 0.63$ cm·K/W, $H = 55$ cm, $T_{ci} = 550$ K, $\dot{m} = 25$ kg/s, and $c_{co} = 0.27$ W·s/g·K. Use a cosine axial power shape, in which the power goes to zero at the top and bottom of the active core regions. (c) Compute the temperatures in the blocked channel and plot on the same figure obtained from Exercise (b). The radial peaking factor for the blocked channel is $F_r = 1.22$.

7. An orbiting space reactor operating at 1 MW$_{th}$ for 1.5 years is hit by a piece of space debris. A large hole in the coolant system results in a rapid loss of coolant. The fuel element temperature at the time of impact is 1400 K. Assume that the coolant loss is instantaneous and that no heat loss occurs following the loss-of-coolant. Use a total fuel element mass of 90 kg and $c_F = 0.2$ W·s/g·K. (a) Calculate and plot (vs. time) the core power level and fuel element temperature as a function of time for 10 minutes. Assume that the reactor is shut down at the moment of coolant loss.

Chapter 10

Risk Analysis

F. Eric Haskin

Risks to the public posed by a space nuclear power source must be small when compared to other societal risks and the benefits provided by the proposed application. This chapter introduces the reader to risk, related concepts, and the process used to formally assess the risks associated with nuclear power sources. Results from terrestrial and space nuclear risk assessments are presented, and the distinction between natural variability and state-of-knowledge uncertainty is discussed.

1.0 Risk and Related Concepts

Intuitively, risk is synonymous with danger, hazard, peril, or exposure to some adverse consequence such as death, injury, or financial loss. Many different definitions of risk can be found in the literature. The colloquial definition favored here says *risk is unrealized potential for harm*. For example, consider the statements:

Being struck by a meteor is highly unlikely.

Dying in an automobile accident is comparatively likely.

The risk of dying from cancer is moderate.

This is a high-risk investment.

In each statement both *harm* and its *unrealized potential* are implied. Both of these elements are essential ingredients of risk. In the statement "the risk is high," the *harm* is unclear. In the statement "death occurred at midnight," the *harm* is evident, but there is no *unrealized potential*. When harm is actually realized, it is no longer a risk; it is an actual death, injury, loss, or other adverse consequence.

1.1 Risk Assessment

A risk assessment is simply an assessment of the unrealized potential for harm. In the simplest case, a single negative consequence is stipulated and its likelihood is estimated from available data. For example, the risk of death while driving in the Indianapolis 500 is roughly one in a hundred. In considering diseases, acts of nature, human activities, or man-made facilities and devices, a more complicated assessment is required. Likelihoods and adverse health or economic consequences need to be characterized for a spectrum of identified scenarios. In this broader context, *risk assessment* addresses three basic questions:

1. What can happen?
2. How likely is it?
3. How bad is it?

Specifically, in assessing the safety of man-made facilities and devices, a risk assessment is undertaken to systematically address these questions by

1. identifying potential accidents that could endanger the public health and safety,

2. assessing the likelihoods of such accidents, and

3. characterizing the health and economic consequences that would result from such accidents.

The characterizations of likelihoods and consequences may be categorical (e.g., high, medium, low) or quantitative. Often a risk assessment progresses from a categorical to a quantitative characterization. The complete spectrum of methods that can be applied to address the preceding questions is beyond the scope of this chapter. The emphasis here is on event trees and fault trees, which are widely used in risk and system reliability analyses for commercial and space applications of nuclear power. Additional information regarding such applications is provided in the cited references. Reference [1] explains, illustrates, and compares a number of methods including: preliminary hazard analysis, what-if analysis, what if/checklist analysis, failure modes and effects analysis, failure mode and effects criticality analysis, fault tree analysis, event tree analysis, cause-consequence analysis, and human reliability analysis.

1.2 Event Tree Analysis

Figure 10.1 is a simplified event tree for a ground test of a thermal propulsion reactor as indicated by the initiating event on the left-hand side. The probability of a test is taken to be one. The consequences associated with a test depend on the success or failure of three safety functions. The next event in the tree is success or failure of the power/flow control function. By convention, success is denoted by the upper branch and failure is denoted by the lower branch. If power/flow control succeeds, radionuclide releases are minimal, and success or failure of other safety functions is irrelevant. If power/flow control fails, fission products will be released from the reactor fuel and discharged with the hot propellant gas through the nozzle of the test assembly into the surrounding containment building. A water spray system is provided to remove heat and released radionuclides from the exhaust gases. The containment building is designed to withstand the pressure rise associated with the buildup of exhaust gases. The building is tested to demonstrate a leakage of less than 0.1% by volume at the maximum pressure consistent with an op-

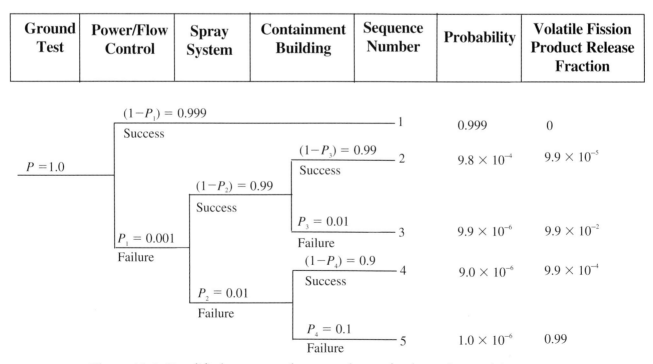

Figure 10.1 Simplified event tree for ground test of a thermal propulsion reactor.

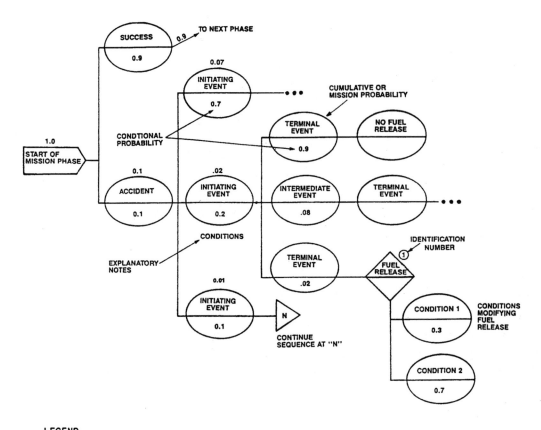

LEGEND

Pentagon	-	start of a mission phase sequence
Oval	-	indicates initiating, intermediate, and terminal events
Triangle	-	indicates where a sequence branch is continued
Diamond	-	indicates a terminal event in which fuel release occurs

Figure 10.2 Example of failure abort sequence trees for a space nuclear mission. *Courtesy of U.S. Department of Energy.*

erating spray system. Success and failure branches for the containment and the heat and fission product removal functions appear on the power/flow control failure branch in Figure 10.1. In response to the question, "What can happen?" the event tree delineates five sequences of events. Sequence 1, in which power/flow control succeeds, has negligible potential for releasing radionuclides to the atmosphere. Sequences 2, 3, 4, and 5 have different likelihoods and atmospheric releases, which will be discussed shortly.

Another event tree format is often used to represent accident sequences for space missions in which nuclear power sources are utilized. Figure 10.2 shows the starting portion of a so-called failure abort and sequence tree. The event on the left denotes the start of a mission phase. Success of the phase transfers the reader to the event tree for the next phase. Failure indicates an accident occurs during the phase, and possible initiating events and their event trees follow. In this format, any desired level of detail can be used to represent the sequence of events following a particular initiating event. Each branch ends with a terminal event and an indication of the nature of any release of radioactive material.

In Figures 10.1 and 10.2, the event probabilities are indicated on the trees. For example, $P_1 = 0.001$ is the power/flow control failure probability in Figure 10.1. A basic premise of this simple example is that each function either succeeds or fails. The success probability for each function is, therefore, one minus its failure probability. For example, the power/flow control success probability is $1 - P_1 = 0.999$. The numerical values in Figure 10.1 are for illustration purposes only. They are representative of values obtained in actual risk assessments, but they do not correspond to a specific test assembly or test facility. Methods used to model failure probabilities of safety functions accomplished by complex systems are discussed in Section 2. It is important to note here that the probability of an event on a particular branch of a tree may depend on events occurring earlier on the branch. For example, in Figure 10.1, the probability of containment building failure is an order of magnitude less when the spray system succeeds ($P_3 = 0.01$) than when it fails ($P_4 = 0.1$). This reflects the capability of the sprays to remove considerable heat from the exhaust gases and thereby limit the pressure buildup inside the containment building.

The probability of a sequence of events is the product of the probabilities of the successive events. For example, in Figure 10.1, the probability of Sequence 3, which involves failure of power/flow control, success of the water sprays, and failure of the containment building is

$$P_{test}*P_1(1 - P_2)*P_3 = (1)(0.001)(0.99)(0.01) = 9.9 \times 10^{-6}.$$

The frequencies of all five branches are calculated in this manner and indicated after the branch numbers in Figure 10.1. Note that the sum of the probabilities of all five outcomes is the probability of the initiating event (in this case $P_{test} = 1.0$).

To this point, the example depicted in Figure 10.1 has been used to illustrate how event trees can be used in answering the questions: "What can happen?" and "How likely is it?" The remaining question is, "How bad is it?" By including a quantitative consequence estimate for each accident sequence, Figure 10.1 can be used to addresses all three questions and illustrate how risk can be viewed as a set of triplets: the accident sequences, their likelihoods, and their consequences. Various consequence measures including doses to the public, public health effects, and economic costs may be appropriate for a given risk assessment. To keep the example in Figure 10.1 simple, the fraction of volatile core fission products (cesium and iodine) released to the atmosphere is used as a consequence measure, and consequence calculations are based on the following simple model. If the power/flow control function succeeds, releases of fission products from the fuel are negligible. If the power/flow control function fails, 99% of the volatile fission products are released from the fuel. The fraction of the released fission products that reach the containment building air volume is 10% if the spray function succeeds, but 100% if the spray function fails. If the containment building fails, 100% of the unremoved volatile fission products escape to the atmosphere. If the containment function succeeds, only 0.1% of the unremoved volatile fission products leaks to the atmosphere. The resulting consequence estimates for each accident sequence are included on the right-hand side of Figure 10.1. In the worst case, when all three safety functions fail, 99% of the volatile fission products are released to the atmosphere. If only the power/flow control function fails, only 0.01% of the volatile fission products is released to the atmosphere.

1.3 Risk Curves

Another way to display the information contained in Figure 10.1 is to construct a *complementary cumulative distribution* (*ccdf*) or *risk curve*. A risk curve plots the probability $P(C \geq c)$ of consequences exceeding magnitude c as a function of c. Table 10.1 develops the risk curve values for the example. For this purpose, the four event sequences are listed in order of increasing consequence. The resulting risk curve is presented as Figure 10.3. Note that the intercept of the risk curve with the vertical-axis (ordinate) is the probability of any release of volatile fission products to the atmosphere, and the intercept with the horizontal-axis (abscissa) is the largest possible release.

Actual risk assessments of man-made facilities and devices typically involve very large numbers of accident scenarios, and the resulting risk curves are usually smoother and more continuous than the one in Figure 10.3. For example, Figures 10.4 and 10.5 are classic risk curves from the pioneering Reactor Safety Study, which was published in 1975 [2]. For any point on one of the curves, the ordinate repre-

Table 10.1   Risk Curve Coordinates for Example in Figure 10.1

Release Sequence	c Consequence Level[a]	Probability	$Pr(C > c)$ Probability of Release with Consequences c
	0		1.0×10^{-3}
2	9.9×10^{-5}	9.8×10^{-4}	2.0×10^{-5}
4	9.9×10^{-2}	9.0×10^{-6}	1.1×10^{-5}
3	9.9×10^{-4}	9.9×10^{-6}	1.0×10^{-6}
5	9.9×10^{-1}	1.0×10^{-6}	0
		1.0×10^{-3}	

[a] Fraction of volatile core fission products released to atmosphere

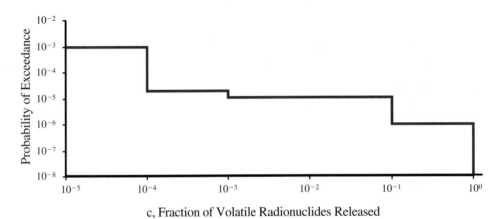

Figure 10.3 Risk curve for the example in Figure 10.1.

sents the probability that a consequence exceeding the corresponding abscissa value will occur. For example, in Figure 10.4, the probability of a nuclear power plant accident involving more than 1000 fatalities in any given year is approximately 10^{-6}.

This study was the first attempt to make a realistic estimate of the potential effects of accidents at commercial nuclear power plants on public health and safety. One boiling water reactor, Peach Bottom, and one pressurized water reactor, Surry, were analyzed in detail. In the figures, it is assumed that there are 100 power reactors and that each has risks identical to those estimated for Surry or Peach Bottom. There is no evidence to support this assumption; however, the other 98 reactors would have to be orders of magnitude worse than Surry and Peach Bottom for the general conclusions to be rendered invalid. As results presented in Section 7 illustrate, risks associated with space nuclear power sources tend to be less than those depicted for commercial power reactors in Figures 10.4 and 10.5. This difference is primarily due to the more limited radionuclide inventory associated with space nuclear power devices and the limited time during which this inventory is in the vicinity of people.

1.4 Consequence-Weighted Risk

One way to combine the risks associated with high, moderate, and low consequence accidents into a single overall risk measure is to use *actuarial* or *consequence-weighted risk*. The consequence-weighted risk associated with an accident is the product of the accident's probability and its consequence. The total consequence-weighted risk R is the sum of the consequence-weighted risks of the individual accidents. Mathematically,

$$R = \sum_i P_i C_i$$

(10.1)

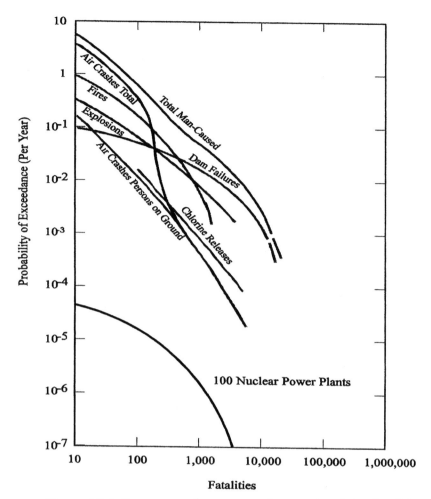

Figure 10.4 Frequency of man-caused events involving fatalities.
Courtesy of U.S. Nuclear Regulatory Commission.

Here P_i is the probability and C_i is the consequence for the i'th accident sequence and the sum is taken over all sequences. An initiating event or accident sequence probability that applies to a specified time interval (e.g., a reactor critical year in commercial reactor risk assessments) is generally called a frequency and denoted F_i rather than P_i. The consequence-weighted risk R has units of consequences (or consequences per unit time); however, because the sequence probabilities P_i (or frequencies F_i) are usually very small, the value of R is often significantly less than the value of any C_i. Consequence-weighted risk is so widely used in terrestrial nuclear power plant risk assessments that the modifier consequence-weighted (or actuarial) is often dropped, and the total consequence-weighted risk is simply called the plant risk. As explained in Section 4.4, the consequence-weighted risk can also be interpreted as the mean of the consequence probability distribution.

Example 10.1

Calculate the consequence-weighted risk for the event sequences depicted in Figure 10.1.

Solution:

The consequence-weighted risk is calculated by adding the probability-consequence products for the event sequences:

Sequence of Events i	Estimated Probability P_i	Estimated Consequence[a] C_i	Consequence-Weighted Risk $R_i = P_i C_i$	% of Consequence-Weighted Risk $100\,R_i/R$
1	1.0	0	0	0%
2	9.8×10^{-4}	0.0001	9.8×10^{-8}	4.7%
3	9.9×10^{-6}	0.1	9.9×10^{-7}	47.2%
4	9.0×10^{-6}	0.001	9.0×10^{-9}	0.4%
5	1.0×10^{-6}	0.99	9.9×10^{-7}	47.7%
	1.0×10^{-3}		$R = 2.1 \times 10^{-6}$	100%[a]

[a]Fraction of volatile core fission products released to atmosphere

In this example, the total consequence-weighted risk R is 2.1×10^{-6} of the volatile fission products.

Consequence-weighted risk is sometimes used to compare accident sequences. For example, consider Sequence 3 and Sequence 5 from Figure 10.1 and Example 10.1. In Sequence 3, power/flow control fails, water spray succeeds, and containment fails. In Sequence 5, all three safety functions fail. Sequence 3 is nearly an order of magnitude more likely than Sequence 5, but the consequence of Sequence 3 is an

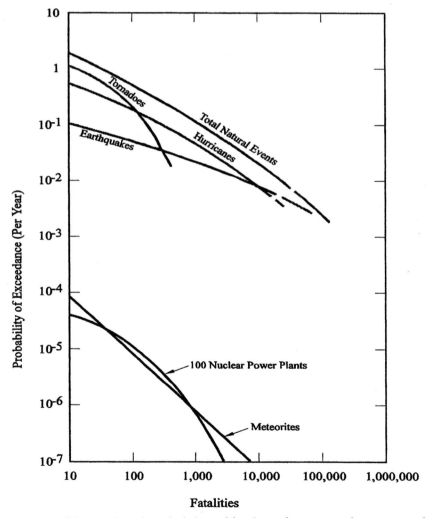

Figure 10.5 Comparison of the predicted probability of fatalities for 100 nuclear power plants with the frequency of natural events involving fatalities. *Courtesy of U.S. Nuclear Regulatory Commission.*

order of magnitude less than that for Sequence 5. As a result, Sequences 3 and 5 have very comparable consequence-weighted risks.

Data required to estimate the probabilities of accident scenarios are often very sparse, and the estimates are, therefore, uncertain. Similarly, the consequence estimates for accident scenarios are often uncertain. It is not uncommon for uncertainties to span one to three orders of magnitude. To illustrate, Figure 10.6 presents results from a comprehensive study of five U.S. nuclear power plants [3]. The five bars indicate uncertainties in estimates of the consequence-weighted risk of individual early and latent-cancer fatalities from core damage accidents initiated within electrical and mechanical systems. The analysts assign a probability of 0.9 to finding the true value for a given plant within the indicated range. Uncertainty ranges and probabilities of this type are termed subjective, state-of-knowledge, or epis-

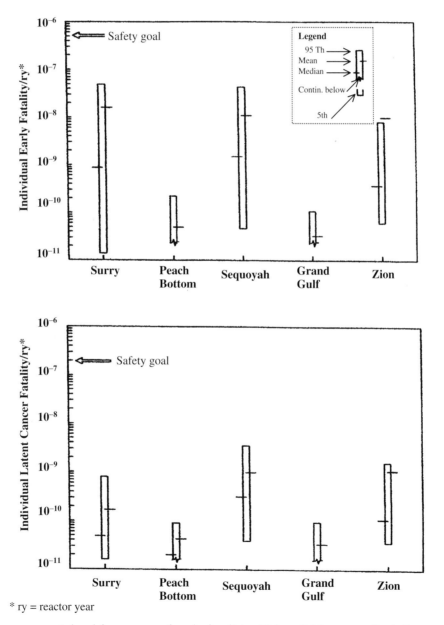

* ry = reactor year

Figure 10.6 Consequence-weighted frequency of early fatalities (5th to 95th percentiles).*Courtesy of U.S. Nuclear Regulatory Commission.*

temic. As illustrated in Sections 5 and 6, they are quantitative expressions of imperfect knowledge regarding how to predict or model infrequent and unobserved events. In spite of state-of-knowledge uncertainties, issues related to completeness, and other limitations, the information provided by risk assessments is usually very beneficial in understanding and improving the safety of man-made facilities and devices. This is illustrated by risk assessment results presented in Section 7 of this chapter.

1.5 Public Perceptions of Risk

Methods employed in risk assessments seek to systematically and objectively estimate and compare the risks of potential accidents; however, most people are unfamiliar with such methods and rely on their own, much more subjective perceptions of risks. The public receives virtually all risk-related information through the news media. News, by its very nature, emphasizes the unusual. For example roughly 40,000 people die each year in U.S. traffic accidents involving only a few deaths per accident. Such accidents are not very newsworthy, and most individuals tend to significantly underestimate their risk of dying in an automobile accident.

The latency of some hazards can also cause individuals to underestimate actual risks. For example, smoking is estimated to cause 150,000 U.S. deaths per year, but death due to smoking is typically delayed for decades. As a result, smokers tend to underestimate the risk of death due to smoking. People tend to judge an event as likely or frequent if its occurrence is easy to imagine or recall. Risks tend to be overestimated for events that are dramatic and sensational, and underestimated for events that are unspectacular. Accidents are commonly estimated by the public to cause as many deaths as diseases, even though diseases cause roughly 15 times more deaths. Rare causes of death are frequently overestimated and common causes underestimated. The magnitude of risk that individuals are willing to tolerate is influenced by the source of the risk. Individuals seem to accept a much higher risk (~1000 times higher) if it is voluntary (e.g., mountain climbing) than if it is imposed to achieve a societal benefit (e.g., electric power). The extreme of risk reduction is risk elimination, and some advocate outlawing a technology to eliminate any possibility that its use might result in deaths, injuries, or environmental damage. The difficulty with this approach is that alternative technologies may not exist or may result in more deaths, injuries, or environmental damage.

Unrealistic risk perceptions can also dominate the thinking of those familiar with a technology. Overconfidence clearly preceded the *Challenger* accident. As late as January 1986, NASA reported a very small (1/10,000 to 1/10,000,000) chance of an explosion during a Space Shuttle launch. This assessment disregarded an early, more realistic, data-based estimate of a 1/100 failure probability [4]. It is clear the more optimistic estimates were simply judgmental; they were not based on any objective analysis of actual failure data. Similarly, overconfidence in the safety of nuclear power was a preexisting mindset that contributed to both the Three Mile Island Unit 2 (TMI-2) and Chernobyl accidents. Instances in which risks have been underestimated delay a convergence of opinions between experts and the public regarding objectively estimated risks of endeavors utilizing nuclear power devices. The Three Mile Island accident caused no fatalities or injuries, and U.S. nuclear power plants are clearly much safer as a result of changes in training, procedures, hardware, regulations, and attitudes implemented after the TMI-2 accident. Yet, the negative impact of the TMI-2 accident on the public's perception of the safety of nuclear technology was immense and remains to be overcome.

Unrealistic perceived risk can create such poor public attitudes toward technologies that grossly misguided efforts in the name of public and environmental protection may result. Once incorrect perceptions of risk are formed, they can be very difficult to change by factual information. Whether one's objective is to make nuclear technologies safer or to change public perceptions of their safety, in the long run, the attitude that was recommended by the President's Commission on TMI-2 seems essential:

"Nuclear power is by its very nature potentially dangerous, and . . . one must continually question whether the safeguards already in place are sufficient to prevent major accidents." [5]

If one follows this guideline, open, honest communications with the public should, in long term, correct inaccurate risk perceptions.

2.0 Probability and Related Concepts

In Section 1, the functional failure probabilities were presumed to be known. In actual practice such probabilities must be estimated as functions of component and human failure probabilities, which, in turn, must be estimated based on available data. For such purposes, an understanding of probability models and the underlying interpretations and rules of probability are essential. A comprehensive treatment of probability theory is beyond the scope of this text. Reference [6] provides a basic mathematical treatment of probability theory.

2.1 Interpretations of Probability

Suppose an event E can occur as a result of a repeatable experiment. Let n denote the number of observations, that is, the number of times the experiment is performed. Let n_E denote the number of times event E is observed. In the *frequentist* (or von Mises) interpretation of probability, the probability $P(E)$ of event E is the frequency of occurrence of event E as the number of observations becomes very large,

$$P(E) = \lim_{n \to \infty} \frac{n_E}{n}. \tag{10.2}$$

Here the limit is not the usual pointwise limit implied in mathematics. For example, if a tack is tossed 100 times and the event E = "point up" occurs in 25 of the 100 tosses, one could estimate $P(E)$ as 25/100; however, if the point lands up in the next toss, there is no assurance that 26/101 is closer to the limiting or "true" frequency defined by Eq. 10.2 than 25/100. The result after $n = 10,000$ tosses might, for example, be $n_E/n = 0.2356$, which is closer to 25/100 than 26/101. It may be difficult to devise a physical experiment that can be repeated indefinitely without changing something that could alter the relative likelihoods of the possible outcomes. One can, nevertheless, postulate the existence of an underlying true frequency and attempt to characterize it based on available data.

Confidence intervals (see Section 5.1) or Bayesian probability intervals (see Section 5.2) can be used to characterize the uncertainty in the true frequency of an event given finite data, such as a finite number of tack tosses. Bayesian techniques are widely used in risk assessments because they can be used to answer questions like the following: Given $n_E = 25$ after $n = 100$ tosses, what is the probability that the true frequency of event E = "point up" is less than or equal to 0.25? What is the probability that the true frequency is in the interval from 0.2 to 0.3? Although the results of Bayesian analyses may be dominated by the actual data that is analyzed, the starting point is a purely subjective estimate of such probabilities that deliberately ignores the data. This requires another interpretation of probability.

A *subjective probability* is a person's degree of belief that an event E will occur or that some assertion is true or false. Subjective probabilities are often associated with situations in which data are sparse, nonexistent, or not directly applicable to the situation at hand. Subjective probabilities can be stated whether or not a repeatable experiment related to the determination of a true frequency is conceivable. If someone says the probability of life on one of Jupiter's moon's is 1/1000, this is clearly a subjective probability. There either is or is not life on one of Jupiter's moons. Odds on sporting events can be viewed as subjective probabilities for the outcomes of nonrepeatable experiments, and different people will give different odds. Some analysts reserve the term *probability* for subjective probabilities and use the term *frequency* to describe quantities based on the frequency definition. More often, however, the term *frequency* is used to describe the probability of observing an event in a unit time interval.

Regardless of whether a frequentist or subjective interpretation of probability is being applied, certain mathematical rules of probability must be obeyed. In fact mathematicians need not concern themselves with frequentist and subjective interpretations, because rules of probability can be derived using an axiomatic definition of probability. In the context of an experiment, the three axioms require that (1) the union of all events that could occur as a result of an experiment has a probability of one, (2) the probability associated with any single event must be greater than or equal to zero, and (3) two events are said to be mutually exclusive if the occurrence of one precludes the occurrence of the other.

2.2 Rules of Probability

Figure 10.7 summarizes the probability rules for the complement of an event, the intersection of two events, the union of two events, and the union of two mutually exclusive events. Venn diagram, set theory, Boolean algebra, and fault tree conventions are used to depict these event relationships. Based on the axiomatic definition, it is easy to show that the probability of the impossible event (null set ϕ) must be zero; that is, $P(\phi) = 0$. It follows that the probability of the complement E' of an event is $P(E') = 1 - P(E)$. As indicated in Figure 10.7, the probabilities of the intersection (AND, Boolean $*$) and union (OR, Boolean $+$) of two events are

$$P(A \cap B) = P(A) P(B|A) = P(B) P(A|B) \tag{10.3}$$

and

$$P(A \cup B) = P(A) + P(B) - P(A \cap B). \tag{10.4}$$

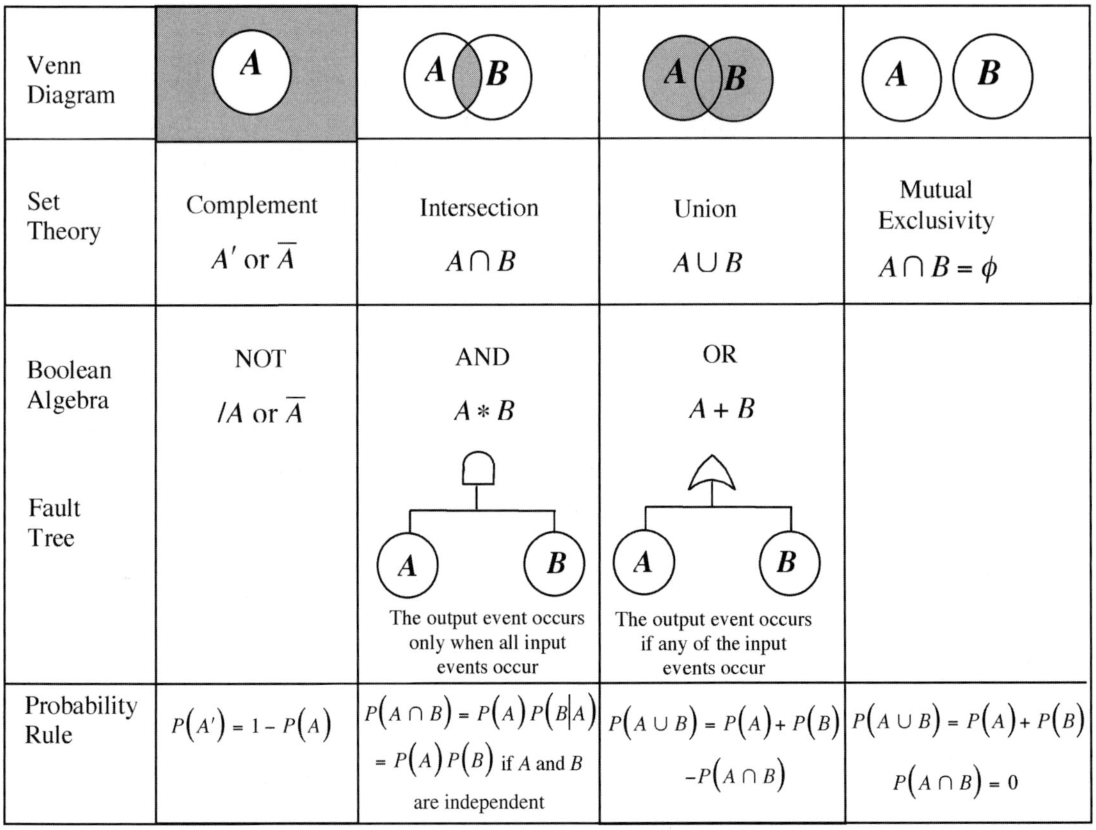

Venn Diagram	A	A B	A B	A B
Set Theory	Complement A' or $\overline{A}$	Intersection $A \cap B$	Union $A \cup B$	Mutual Exclusivity $A \cap B = \phi$
Boolean Algebra	NOT $/A$ or $\overline{A}$	AND $A * B$	OR $A + B$	
Fault Tree		A B The output event occurs only when all input events occur	A B The output event occurs if any of the input events occur	
Probability Rule	$P(A') = 1 - P(A)$	$P(A \cap B) = P(A)P(B\|A)$ $= P(A)P(B)$ if A and B are independent	$P(A \cup B) = P(A) + P(B)$ $-P(A \cap B)$	$P(A \cup B) = P(A) + P(B)$ $P(A \cap B) = 0$

Figure 10.7 Event relationships and associated probability rules.

In Eq. 10.3 $P(B|A)$ is defined as the *conditional probability* of event B given event A, that is, the probability of event B conditional on the occurrence of event A. Two events are said to be *independent* if the occurrence or nonoccurrence of one event has no effect on the probability of occurrence (or nonoccurrence) of the other event; that is, if $P(A|B) = P(A)$. In this case, the right-hand side of Eq. 10.3 reduces to $P(A)\,P(B)$. The intersection and union of n independent events $I_1, I_2, \ldots, I_n$ can be shown to have the following respective probabilities:

$$P(I_1 \cap I_2 \cap \ldots \cap I_n) = \prod_{i=1}^{n} P(I_i) \tag{10.5}$$

and
$$P(I_1 \cup I_2 \cup \ldots \cup I_n) = 1 - \prod_{i=1}^{n} [1 - P(I_i)]. \tag{10.6}$$

The probability of a sequence of events delineated in an event tree is the product of the initiating event probability and the conditional probabilities of the other events in the sequence. Section 3 discusses the use of fault tree analysis to express system failure probabilities as functions of component-failure and human-error probabilities. The probability rules summarized above suffice for these applications. A related but broader spectrum of applications is enabled by the use of probability distributions. Probability distributions and relevant applications are discussed in Sections 4 and 5.

Example 10.2

The probability of event A is 0.1, the probability of event B is 0.2, the probability of event C is 0.3, and the probability of event D is 0.4. First assume the events are mutually exclusive, what is the probability of (a) the intersection of $A * B * C * D$ and (b) the union of $A + B + C + D$? Next, assume the events are independent. What is the probability of (c) $A * B * C * D$ and (d) $A + B + C + D$?

Solution:

(a) The intersection of mutually exclusive events is the null set, which has a probability of zero.

(b) From Figure 10.7 the probability of the union of mutually exclusive events is the sum of the event probabilities, so $P(A + B + C + D) = 0.1 + 0.2 + 0.3 + 0.4 = 1.0$.

(c) By Eq. 10.5, $P(A * B * C * D) = (0.1)(0.2)(0.3)(0.4) = 0.0024$.

(d) By Eq. 10.6, $P(A + B + C + D) = 1 - (0.9)(0.8)(0.7)(0.6) = 0.6976$.

3.0 Fault Tree Analysis

Systems designed to accomplish safety functions in responding to accident initiating events are often assembled using familiar components. Extensive failure data from both testing and operating experience may be available for components such as pumps, valves, switches, relays, and so on. The failure probability of the system (or conversely the system reliability) can then be synthesized from the failure rate of the components by a number of methods. The primary method used in risk assessments is fault tree analysis [7].

3.1 Fault Tree Construction

To construct a fault tree, an analyst deductively works back from some undesired event to identify its possible causes. Fault tree logic is nearly the reverse of event tree logic. A fault tree starts with an undesired event (usually failure of a system to perform some intended function) and attempts to find its cause; whereas, an event tree starts with an undesired event and attempts to delineate the sequences of events it could cause. In constructing a fault tree, the failure event that is to be studied is called the *top*

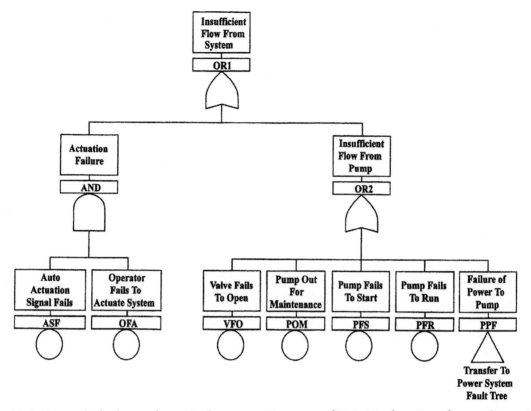

Figure 10.8 Example fault tree for a single-pump. *Courtesy of U.S. Nuclear Regulatory Commission.*

event because it is placed at the top of the fault tree. Below the top event, subordinate events that may cause the occurrence of the top event are identified and linked to the top event by simple logical relationships (AND, OR, etc.). The subordinate events themselves are then broken down, and tree construction continues in this manner until events, such as component failures or human errors, that cannot or need not be broken down further are reached. Events that can not be broken down further are called *primary events*. Events that need not be broken down further are called *undeveloped events*. Collectively the basic and undeveloped events of a fault tree are sometimes called the *base events* because they exist at the base of the tree.

A fault tree enables the analyst to visualize the system in a way that examines its functional capability. Critical components requiring high reliability can sometimes be located merely by drawing a fault tree. Figure 10.8 is an example of a fault tree for a hypothetical single-pump. The symbols used in fault trees originate from the logical operations OR (Boolean +) and AND (Boolean *). For the example, insufficient flow could result from a failure to actuate the backup system OR from insufficient flow from the pump. The actuation failure requires both that the automatic actuation signal fail AND that the operator fail to actuate the system manually. Insufficient flow from the pump can be caused by any of the failure events listed under the corresponding OR gate. Note that one of these events, failure of power to the pump, is based on another fault tree for the power system.

3.2 Minimal Cut Sets

Figure 10.8 can be used to illustrate the next product sought in fault tree analysis, the minimal cut sets. A *cut set* is any combination of base events that can cause the top event. A *minimal cut set* is a cut set that cannot be reduced in number. Events in a cut set are sufficient to cause the top event, while events in a minimal cut set are necessary to cause the top event. Minimal cut sets are sought because they do not contain events that are unnecessary for realizing the top event. For the example depicted in Figure 10.8, any of the failure events under the bottom OR gate would result in insufficient flow from the pump and hence system failure. System failure due to auto actuation signal failure (ASF) requires both events under the AND gate on the left-hand side. Hence, in Boolean notation (AND | *, OR | +), the system failure (ISF) is given by a sum over six cut sets:

$$ISF = ASF*OFA + VFO + POM + PFS + PFR + PPF$$

The first five terms on the right-hand side are minimal cut sets. The last term PPF (failure of power to the pump) is not a primary event and would have to be expressed in terms of minimal cut sets for the power system. In addition, event ASF (auto actuation system failure) is an undeveloped event, which could be modeled in more detail.

3.3 Boolean Algebra

Figure 10.8 is a very simple example of a fault tree. Fault trees for actual launch vehicle and spacecraft systems can involve hundreds of logic gates and hundreds of base events. The minimal cut sets for such complex fault trees are most often determined using Boolean algebra. Boolean algebra is applied to equations of logical variables. A logical variable can only have one of two values: zero (for FALSE) and one (for TRUE). In applying Boolean algebra to solve for the minimal cut sets of a fault tree, each event is replaced by a corresponding Boolean variable, and each logical operator is replaced by its Boolean counterpart. Each OR gate is replaced by the Boolean sum of the variables corresponding to the gate input events. Each AND gate is replaced by the Boolean product of the variables corresponding to the gate input events.

Boolean algebra is similar to ordinary algebra with respect to order, substitution, and the distributive law of multiplication over addition:

Commutative Law: $A*B = B*A$ and $A + B = B + A$

Substitution: Given $C = A*B$ if $A = G + H$ and $B = I*J$ then $C = (G + H) * (I*J)$

Distributive Law: $(A + B) * (C + D) = A*C + A*D + B*C + B*D$

In Boolean algebra, however, two important identities can be used to simplify expressions. These two Boolean identities are:

$$A*A = A \tag{10.7}$$

and
$$A + A*B = A \tag{10.8}$$

The validity of these identities can be verified by constructing a table of Boolean values. In Table 10.2, all possible combinations of values for Boolean variables A and B are listed in the first two columns. The value of $A*A$ is calculated in the third column. $A*A$ is clearly equal to A. The value of $A*B$ is cal-

Table 10.2   Truth Table for Boolean Variables

A	B	$A*A$	$A*B$	$A + A*B$
0	0	0	0	0
0	1	0	0	0
1	0	1	0	1
1	1	1	1	1

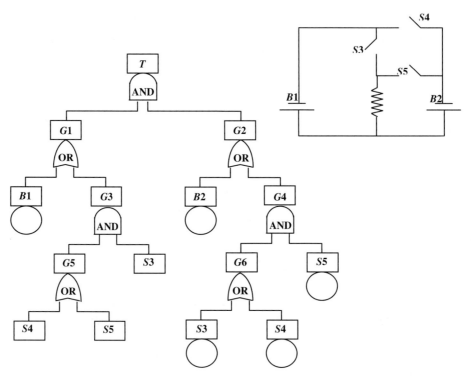

Figure 10.9 Fault tree for electrical switch system.

culated in the fourth column, and $A + A*B$ is calculated in the last column. $A + A*B$ is clearly equal to A.

3.4 Determining Minimal Cut Sets Using Boolean Algebra

Consider the example in Figure 10.9. The top event is failure to pass current through the resistor. The top event can only occur if there is no current through the resistor from battery 1 and no current through the resistor from battery 2. Current through the resistor implies current through the loop containing the battery 1, switch 3 and the resistor; the loop containing battery 1 switch 4, switch 5 and the resistor; the loop containing battery 2, switch 5, and the resistor; or the loop containing battery 2, switch 4, switch 3 and the resistor. Let gate G1 denote no current through the resistor from battery 1. Let gate G2 denote no current through the resistor from battery 2. Let $B1$ denote failure of battery 1, $B2$ denote failure of battery 2, $S3$ denote open switch 3, $S4$ denote open switch 4, and $S5$ denote open switch 5.

Step 1 in finding the minimal cut sets is to generate the intermediate event equations for the fault tree. To do this simply write each intermediate gate event as a function of its input events:

$$T = G1*G2,$$
$$G1 = B1 + G3,$$
$$G2 = B2 + G4,$$
$$G3 = S3*G5,$$
$$G4 = G6*S5,$$
$$G5 = S4 + S5, \text{ and}$$
$$G6 = S3 + S4.$$

Step 2 is to generate an equation for the top event T that is a function of only primary events. To do this, sequentially eliminate each intermediate event on the right side of the equation for T by repeated substitution. That is, replace each intermediate event by the right side of its equation from Step 1 until the top event is expressed entirely in terms of primary events:

$$T = G1*G2,$$

$$T = (B1 + G3) * (B2 + G4),$$

$$T = (B1 + S3*G5) * (B2 + G4*S5),$$

$$T = (B1 + S3 * (S4 + S5)) * (B2 + (S3 + S4) * S5).$$

Step 3 is to expand the result from Step 2 and apply the Boolean identities $P*P = P$ and $P + P*Q = P$. Expanding,

$$T = (B1 + S3*S4 + S3*S5) * (B2 + S3*S5 + S4*S5, \text{ or}$$

$$T = B1*B2 + B1*S3*S5 + B1*S4*S5 + B2*S3*S4 + S3*S4*S3*S5$$
$$+ S3*S4*S4*S5 + B2*S3*S5 + S3*S5*S3*S5 + S3*S5*S4*S5.$$

The identity $P*P = P$ reduces the last terms without a battery failure to $S3*S4*S5 + S3*S4*S5 + S3*S5 + S3*S4*S5$. The identity $P + P*Q = P$ (taking $P = S3*S5$) reduces all of these terms and the term $B1*S3*S4$ and $B2*S3*S5$ to $S3*S5$. The result is

$$T = B1*B2 + B1*S4*S5 + B2*S3*S4 + S3*S5. \qquad (10.9)$$

The minimal cut sets on the right-hand side correspond to failure of both batteries, failure of battery 1 with switches 4 and 5 open, failure of battery 2 with switches 3 and 4 open, or switches 3 and 5 open.

3.5 Determining Accident Sequence Minimal Cut Sets

When the failure logic associated with safety functions is complex, analysts build separate fault trees for these failures and use event trees to define the accident sequences that can lead to consequences of interest. Accident sequences in an event tree can be analyzed in the same way that fault trees are analyzed to determine their minimal cut sets. Each accident sequence represents a logical "ANDing" of the initiating event and subsequent functional failure events. That is, each accident sequence can be thought of as a separate fault tree with the accident sequence description as the top event being an AND gate whose inputs are the initiating event and all of the contributing safety function failures that comprise the sequence. By solving event trees in this manner, Boolean logic can be used to account for explicit dependencies.

3.6 Top Event Probability Quantification

The probability of a minimal cut set is the product of the probabilities of its base events provided that the base events are independent. If not, conditional probabilities must be used. In most applications the cut set probabilities are quite small and the top event probability can be evaluated by taking the sum of the minimal cut set probabilities. This is called the *rare event approximation*. No approximation is involved if the minimal cut sets are mutually exclusive events, but this is not the case if any base event appears in more than one minimal cut set. When minimal cut set probabilities are large, the summation rare event approximation breaks down. Probabilities in excess of one can be predicted. To illustrate, suppose each base event in Figure 10.9 has a probability of 0.6. Using the rare event approximation, each event in Eq. 10.9 is replaced by its probability. This gives a top event probability of

$$P_T = P_{B1} \, P_{B2} + P_{B1} \, P_{S4} \, P_{S5} + P_{B2} \, P_{S3} + P_{S3} \, P_{S5}$$
$$= (0.6)(0.6) + (0.6)(0.6)(0.6) + (0.6)(0.6) + (0.6)(0.6)$$
$$= 1.296 \, .$$

Another quantification approximation is the *minimal cut set upper bound approximation* which applies Eq. 10.6 to aggregate the minimal cut set probabilities thereby avoiding probabilities in excess of unity. For the current example, this approximation gives

$$P_T = 1 - (1 - 0.36)(1 - 0.216)(1 - 0.36)(1 - 0.36)$$
$$= 1 - (0.64)(0.784)(0.64)(0.64)$$
$$= 0.6788736$$

The exact top event probability for this example is $P_T = 0.648$ (see Student Exercise 5). Methods have been developed to compute exact top event probabilities [8], but such methods are usually not required because base event probabilities are sufficiently small to render either the rare event approximation or the minimal cut set upper bound approximation accurate.

3.7 Caveats Regarding Quantification

Extreme care must be taken to assure that failure probabilities from fault tree analysis are not underestimated. Several potential causes of underestimates can be identified. First, component failure rates based on laboratory data rather than data taken under actual operating conditions may result in top event failure rates that are an order of magnitude or more too low. Second, human error probabilities, which typically range from 0.1 to 0.01, must be properly taken into account in the operation, testing, and maintenance of equipment. Finally, great care must be taken in the application of fault trees to ensure that dependencies between failure rates are taken into account. Assuming that failure events are independent when in fact they are even slightly related may cause *common-mode* failures to be ignored whereas they may actually dominate the overall failure rate of a system. To illustrate, suppose that a system has four components in parallel and that any one of the four can perform the required function. The system fault tree includes an AND gate with the four component failure events as inputs. If the failure probability of one component is 0.1, the failure rate of the set of four components in parallel is 10^{-4}. But, this value assumes that the component failure events are independent. In reality, there may be a failure produced by causes that affect all four components simultaneously. Such failures may be caused, for example, by the same adverse environment, a manufacturing problem, or improper maintenance on all four components. Although, it may be possible to make such common causes small contributors to the failure rate of an individual component, they can dominate the failure rates of a highly redundant system. In the case of four components in parallel, a common mode failure probability of 10^{-3} would be insignificant for a single component, but would dominate the behavior of the system. Methods for treating common cause failures in fault tree analyses are discussed elsewhere [9].

4.0 Probability Distributions

Probability distributions are used to describe, analyze, or simulate random processes. A random (also called stochastic) process differs from a deterministic process. The outcome of a deterministic process is the same each time the process is repeated. The output of a random process varies. The adjective *random* is also applied to the variables that quantitatively describe such outcomes. Probability distributions characterize the spread and relative frequencies of occurrence of values of random variables. This section follows the common practice in which a capital letter (e.g., X) is used to denote a random variable, and the corresponding lower case letter (e.g., x) is then used to denote a value of the random variable.

4.1 Distribution Properties

A random variable X with a set $\{x_1, x_2, \ldots, x_n\}$ of discrete (mutually exclusive) values is governed by a discrete probability distribution. The number n may be finite or infinite. A discrete probability distribution assigns a probability $P(x_i) > 0$ to each of the n values such that the total probability over all values is one:

$$P(x_i) > 0, \quad i = 1, 2, \ldots, n \quad \text{and} \quad \sum_{i=1}^{n} P(x_i) = 1. \tag{10.10}$$

Other names that are used for the *discrete probability distribution* $P(x_i)$ include *probability function*, *probability mass function*, and *discrete probability density function*. Arranging the values in ascending order such that $x_1 < x_2 < \ldots < x_n$. The *cumulative distribution function (cdf)* $F(x_j)$ of a discrete random variable is the probability of observing a value less than or equal to a selected value x_j,

$$P(X \leq x_j) = F(x_j) = \sum_{i=1}^{j} P(x_i). \tag{10.11}$$

The *complementary cumulative distribution function (ccdf)* of a discrete random variable is the probability of observing a value greater than a selected value x_j,

$$P(X > x_j) = 1 - P(X \leq x_j) = 1 - F(x_j) = \sum_{i=j+1}^{n} P(x_i) \tag{10.12}$$

Continuous probability distributions apply to random variables that can assume a continuum of values. The probability distribution for a continuous random variable can be expressed in the form of a *probability density function* $f(x)$, which has the properties

$$f(x) \geq 0, \quad -\infty < x < \infty \quad \text{and} \quad \int_{-\infty}^{\infty} f(\xi)d\xi = 1. \tag{10.13}$$

The probability that a random variable X occurs in the interval dx about x is $f(x)dx$. That is, $f(x)$ distributes the unit probability of observing some outcome to values on the real axis. In stating the form of a continuous probability density function it is generally acceptable to consider only those values of x for which $f(x)$ is positive. Other names that are used for $f(x)$ include *density function, continuous density function, frequency function, integrating density function,* and *probability distribution*. The *cumulative distribution function* of a continuous random variable is the probability of observing a value less than or equal to a selected value x,

$$P(X \leq x) = F(x) = \int_{-\infty}^{x} f(\xi)d\xi. \tag{10.14}$$

The *complementary cumulative distribution function* is the probability of observing a value greater than a selected value x,

$$P(X > x) = 1 - P(X \leq x) = 1 - F(x) = \int_{x}^{\infty} f(\xi)d\xi. \tag{10.15}$$

The expectation value of a function y of a random variable X is defined as

$$E[y(X)] = \sum_{i=1}^{n} y(x_i)P(x_i) \quad \text{or} \tag{10.16-a}$$

$$E[y(X)] = \int_{-\infty}^{\infty} y(\xi)f(\xi)\,d\xi. \tag{10.16-b}$$

Equation 10.16-a applies to a discrete random variable, and Eq. 10.16-b applies to a continuous random variable. If the sum or integral on the right-hand side of the applicable equation does not converge, the expectation value does not exist. The preceding definitions permit the mean, variance, median, and mode of a random variable to be defined. The *mean μ* of a random variable is simply its expectation value. Mathematically,

$$\mu = E[X] = \sum_{i=1}^{n} x_i P(x_i) \qquad \text{or} \qquad (10.17\text{-a})$$

$$\mu = E[X] = \int_{-\infty}^{\infty} x f(x) dx. \qquad (10.17\text{-b})$$

The *variance* is a measure of the dispersion of a random variable about its mean. The variance is defined as $E[(X - \mu)^2]$,

$$\text{Var}[X] = \sum_{i=1}^{n} (x_i - \mu)^2 P(x_i) \qquad \text{or} \qquad (10.18\text{-a})$$

$$\text{Var}[X] = \int_{-\infty}^{\infty} (x_i - \mu)^2 f(x) dx \qquad (10.18\text{-b})$$

The *standard deviation σ* is the square root of the variance; that is,

$$\sigma = \sqrt{\text{Var}[X]} = \sqrt{E[(X - \mu)^2]}. \qquad (10.19)$$

The *median* or 50th percentile of a continuous random variable is the value x_{50} that satisfies

$$P(X \leq x_{50}) = F(x_{50}) = 0.5. \qquad (10.20)$$

The *mode* of a discrete random variable is the value x_j with the maximum probability. The mode of a continuous random variable is the value that maximizes the probability density function $f(x)$.

4.2 Discrete Probability Distributions

The *binomial distribution* applies when a single trial of a given process can result in only one of two mutually exclusive outcomes. The outcome of interest occurs with probability p. The other outcome occurs with probability $1 - p$. Such trials are called Bernoulli trials. For example, in a qualification test, a component can either succeed or fail to perform its intended function under the imposed set of test conditions. If a set of n trials are carried out, the number of observed failures can range from zero to n. Assuming a constant failure probability p for each trial, the probability of observing exactly x failures is governed by the binomial distribution,

$$P(x) = \frac{n!}{x!(n-x)!} p^x (1 - p)^{n-x}. \qquad (10.21)$$

The mean of the binomial distribution is $\mu = np$, and the variance is $\sigma^2 = np(1 - p)$. The cumulative distribution function for the binomial distribution is

$$P(X \leq x) = F(x) = \sum_{i=1}^{x} \frac{n!}{i!(n-i)!} p^i (1 - p)^{n-i}. \qquad (10.22)$$

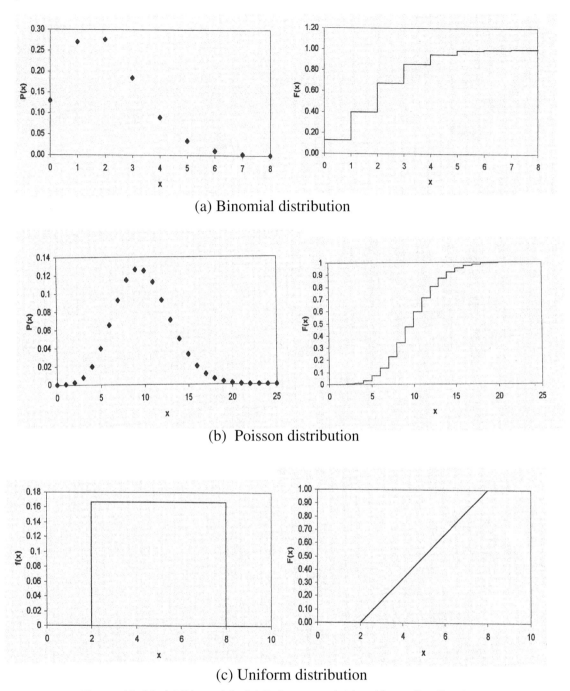

(a) Binomial distribution

(b) Poisson distribution

(c) Uniform distribution

Figure 10.10 (a) Binomial, (b) Poisson, and (c) uniform distributions.

Figure 10.10(a) shows the discrete probability density function and the cumulative distribution function for a binomial distribution with $p = 0.04$ and $n = 50$.

Example 10.3

A launch vehicle has failed twice in 50 launches. (a) What is the probability of this outcome if the true failure probability per launch is 0.02? (b) How many standard deviations from the mean is the observed result?

Solution:

(a) Using the binomial probability density function with $p = 0.02$, $x = 2$, and $n = 50$ gives

$$P(2) = \frac{50!}{2!\,(50 - 2)!} (0.02)^2 (1 - 0.02)^{48} = (1225)(0.02)^2 (0.98)^{48} = 0.1858.$$

(b) The mean is $\mu = np = 50(0.02) = 1.0$. The variance is $\sigma^2 = np(1 - p) = (1.0)(0.98) = 0.98$. The standard deviation is the square root of the variance; that is, $\sigma = 0.98995$. The observed result of two failures out of 50 observations is, therefore, $(2-1)/0.98995 = 1.01$ standard deviations greater than the mean.

The *Poisson distribution* (like the Gaussian described in Section 4.3) approximates the binomial distribution for a large number of trials. Such an approximation is useful because the factorials required to compute the binomial probabilities can easily exceed the largest integer a computer can handle. The probability function for a Poisson random variable is

$$P(x) = \frac{\mu^x e^{-\mu}}{x!}, \quad x = 0, 1, 2, \ldots; \quad \mu > 0$$

$$= 0 \qquad \text{otherwise.}$$

(10.23)

The number of events (radionuclides decaying, components failing, telephone calls received, etc.) that occur during the time interval $(0,t)$ can often be approximated as a Poisson distributed random variable. Both the mean and the variance of the Poisson distribution are equal to μ. The sum of a set of Poisson random variables is also a Poisson random variable. That is, if $\mu_1, \mu_2, \ldots, \mu_n$ are the means of n Poisson random variables, $\mu = \mu_1 + \mu_2 + \ldots + \mu_n$ is the mean of their sum, which is also a Poisson random variable. Figure 10.10(b) shows the discrete probability density function and the cumulative distribution function for a Poisson distribution with $\mu = 9.9$. This distribution arises in Example 10.4.

Example 10.4

A Pu-238 source with an activity of 10 becquerel (Bq = dis/s) is placed in a counter having an efficiency of 99%, and the number of counts in 1-second intervals are recorded. (a) What is the mean count rate? (b) What is the standard deviation of the count rate? (c) What is the probability that exactly 10 counts will be observed in any second? (d) What is the probability of observing exactly 40 counts in any 4-second interval?

Solution:

(a) The half-life of Pu-238 is 87.75 years so the change in activity over the course of the count can be neglected. Because the counter in 99% efficient, the mean number of counts recorded in 1-second intervals is 0.99 times the activity in Bq or 9.9 counts.

(b) The variance is equal to the mean, and the standard deviation is the square root of the variance; therefore, the standard deviation is $(9.9)^{1/2} = 3.146$ counts.

(c) The probability of exactly 10 counts in any second is $(9.9)^{10} \exp(-9.9)/10! = 0.1250$.

(d) The counts observed in 4 seconds is Poisson distributed with a mean of $4(9.9) = 39.6$ counts. The probability of observing exactly 40 counts in 4 seconds is $(39.6)^{40} \exp(-39.6)/40! = 0.06282$.

4.3 Continuous Probability Distributions

The *uniform distribution* is used to describe a continuous random variable X on the interval $[x_{min}, x_{max}]$. The uniform probability density function is

$$f(x) = \frac{1}{x_{max} - x_{min}} \quad \text{for} \quad x_{min} \leq x \leq x_{max} \tag{10.24}$$
$$= 0 \qquad \text{otherwise.}$$

The uniform cumulative distribution function is given by

$$P(X \leq x) = F(x) = \int_{x_{min}}^{x} f(\xi)\, d\xi = \frac{x - x_{min}}{x_{max} - x_{min}}. \tag{10.25}$$

The probability density function and cumulative distribution function for the uniform distribution are depicted in Figure 10.10(c). Because $f(x)$ is constant, X is equally likely to occur in any subinterval of a given width. Pseudo random number generators are devised so that computers can simulate values of a random variable that is uniformly distributed on the interval [0,1]. Such simulations are used in Monte Carlo analyses such as those discussed in Section 5.3.

The *exponential distribution* is a simple one-parameter distribution. The probability distribution for an exponentially distributed random variable X is

$$f(x) = \lambda e^{-\lambda x}, \quad x \geq 0, \quad \lambda > 0. \tag{10.26}$$

As indicated, λ must be a positive real number. The cumulative distribution function for the exponential distribution is

$$P(X \leq x) = F(x) = \int_{0}^{x} \lambda e^{-\lambda \xi} d\xi = 1 - e^{-\lambda x}. \tag{10.27}$$

Both the mean and the standard deviation of the exponential distribution are equal to $1/\lambda$. The median is $x_{50} = \ln(2)/\lambda$, which is obtained by setting $F(x_{50}) = 0.5$. The exponential distribution applies to the time required for a radionuclide to decay. In this case, λ is the decay constant, $1/\lambda$ is the mean decay time, and $\ln(2)/\lambda$ is the half-life. The probability density function and cumulative distribution function for the exponential distribution are depicted in Figure 10.11(a) for the case $\lambda = 1$.

The probability density function for the *Gaussian* or *normal distribution* is the familiar bell-shaped curve, which is symmetric about the mean μ:

$$f(x) = \frac{1}{\sqrt{2\pi}\,\sigma} \exp\left(\frac{(x - \mu)^2}{2\sigma^2}\right). \tag{10.28}$$

The cumulative distribution function for the normal distribution is

$$F(x) = \frac{1}{2} + \frac{1}{2} \operatorname{erf}\left[\frac{(x - \mu)}{\sigma}\Big/ \sqrt{2}\right], \tag{10.29}$$

where the error function erf (x) is

$$\operatorname{erf}(x) = \frac{2}{\sqrt{\pi}} \int_{0}^{a} \exp(-\xi^2)\, d\xi. \tag{10.30}$$

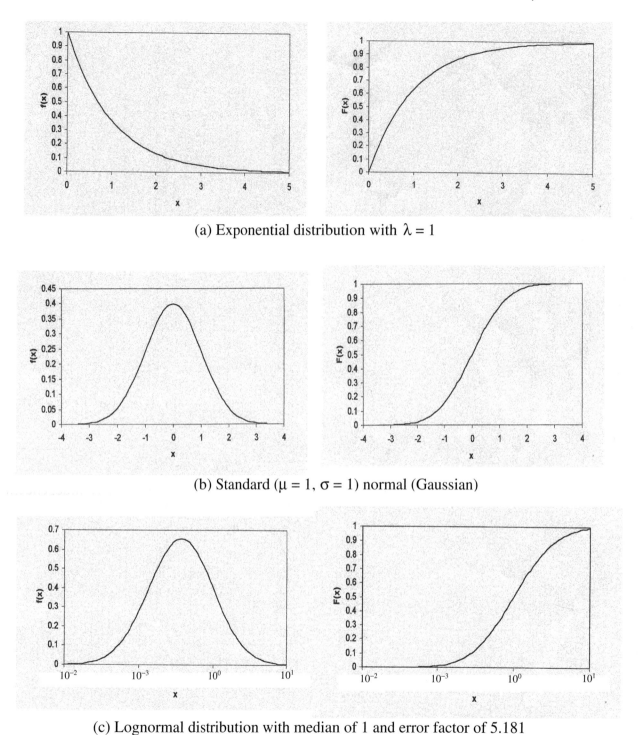

(a) Exponential distribution with $\lambda = 1$

(b) Standard ($\mu = 1$, $\sigma = 1$) normal (Gaussian)

(c) Lognormal distribution with median of 1 and error factor of 5.181

Figure 10.11 (a) Exponential, (b) standard normal, and (c) lognormal distributions.

Values of the cumulative distribution function $F(x)$ at selected values of $z = (x - \mu)/\sigma$ are listed in Table 10.3. The probability of observing a normally distributed random variable within the corresponding number of standard deviations from the mean is also indicated. The shapes of the probability density function and cumulative distribution function for the normal distribution with $\mu = 1$ and $\sigma = 1$ are depicted in Figure 10.11(b). The same shapes apply to any normal distribution under the transformation $z = (x - \mu)/\sigma$ because $f_z(z)\,dz = f_x(x)\,dx$ implies $f_z = \exp(-z^2/2)/\sqrt{2}$. As discussed in Sec-

Table 10.3 Cumulative Distribution and Related Integral of Normal Distribution

$x = \mu + z\sigma$			
z	$cdf(x) =$ $P(X \le \mu + z\sigma)$	$ccdf(x) =$ $P(X \ge \mu + z\sigma)$	$cdf(x) - ccdf(x) =$ $P(\mu - z\sigma X \le \mu + z\sigma)$
0	0.5	0.5	0
0.5	0.6915	0.3085	0.3830
1.0	0.8413	0.1587	0.6826
1.5	0.9332	0.0668	0.8664
1.645	0.0950	0.0500	0.9000
2.0	0.9772	0.0228	0.9544
2.5	0.9938	0.0062	0.9876
3.0	0.9987	0.0013	0.9974

tion 4.4, a linear combination of a set of normally distributed random variables is itself normally distributed. In addition, as illustrated in Section 5.3, for a large enough sample of values of a random variable X, the sample mean is normally distributed irrespective of the how X is distributed.

In a *lognormal distribution*, the logarithm $X = \ln(Y)$ of a positive random variable Y is normally distributed. Y is then lognormally distributed. Let the mean and variance of $X = \ln(Y)$ be $\mu_x = E[X]$ and $\sigma_x^2 = \mathrm{Var}[X]$. Setting $f_y(y)dy = f_x(x)dx$, it follows that the lognormal probability density function for the random variable Y is

$$f(y) = \frac{1}{\sqrt{2\pi}\,\sigma_x y} \exp\left(\frac{[\ln(y) - \mu_x]^2}{2\sigma_x^2}\right) \qquad y > 0. \tag{10.31}$$

Like the normal distribution, the lognormal distribution can be expressed in terms of two independent parameters. The mean μ_x and standard deviation σ_x of $X = \ln(Y)$ can be selected as the independent parameters. More commonly, however, the mean μ_y and the error factor EF_y of the lognormally distributed variable are specified. The error factor EF_y is the 95th percentile divided by the median; i.e., $EF_y = y_{95}/y_{50}$. Using the relationships

$$\sigma_x = \frac{\ln(EF_y)}{1.645} \qquad \text{and} \qquad \mu_x = \ln(\mu_y) - \frac{\sigma_x^2}{2}, \tag{10.32}$$

the mode y_m, median y_{50}, and variance σ_y^2 of the lognormally distributed random variable can be calculated as

$$y_m = e^{(\mu_x - \sigma_x^2)}, \quad y_{50} = e^{(\mu_x)} \quad \text{and} \quad \sigma_y^2 = \mu_y^2[e^{(\sigma_x^2)} - 1]. \tag{10.33}$$

The 5th and 95th percentiles are related to the median by $y_{05} = y_{50}/EF_y$ and $y_{95} = y_{50}EF_y$.

When the probability density function and cumulative distribution function are plotted for a lognormal distribution, a normal shape is obtained if a logarithmic scale is used for the y-axis. But the distribution is actually quite skewed because of the significant weights assigned to larger values. Figure 10.11(c) plots the lognormal counterpart of the normal distribution depicted in Figure 10.11(b). That is, in Figure 10.11 (b), $\ln(y)$ has a mean $\mu_x = 0$ and a standard deviation $\sigma_x = 1$. The resulting lognormal distribution for y has $y_m = 1/e = 0.36788$, $y_{50} = \exp(0) = 1$, $\mu_y = e^{0.5} = 1.6487$. For this lognormal distribution, the probability of observing a value less than or equal to the mode is $F(y_m) = 0.1587$, and the probability of observing a value less than or equal to the mean is $F(\mu_y) = 0.6915$. The product of $Y = Y_1 Y_2 \ldots Y_N$ of N lognormally distributed random variables is itself lognormally distributed, and any power of a lognormally distributed random variable is also lognormally distributed.

4.4 Risk Curve Revisited

Section 1.3 provided examples of risk curves. A risk curve is simply a complementary cumulative distribution function for some accident consequence measure. That is, a risk curve is a representation of the probability distribution of a consequence measure. In the simple event tree of Figure 10.1, the consequence measure was taken to be the fraction of volatile fission products released. To simplify the discussion, a single consequence estimate was assigned to each of the delineated accident scenarios. As a result, the event tree in Figure 10.1 implies a discrete probability distribution for the selected consequence measure.

In reality, the consequences associated with an actual accident would depend on a number of factors that are heuristically random in nature. For example, the mission elapsed time at which an explosion occurs during ascent, the temperature profile and wind patterns existing at the time of launch, the locations of individuals relative to the launch site, and human responses to radiation doses all exhibit natural variabilities. As a result, most consequence measures are appropriately treated as continuous random variables, and the resulting risk curves, unlike Figure 10.3, are smooth.

Example 10.5

For the event tree depicted in Figure 10.1, assume that the nonzero consequence values in the last column are means of normally distributed random variables, C_2, C_3, C_4, and C_5. Let the corresponding standard deviations be $\sigma_2 = 2 \times 10^{-5}$, $\sigma_3 = 0.02$, $\sigma_4 = 2 \times 10^{-4}$, and $\sigma_5 = 0.002$. Construct the probability density function and the complementary cumulative distribution function for the consequence C of a ground test.

Solution:

The probabilities of the five event tree outcomes are unchanged from Figure 10.1, $P_1 = 0.999$, $P_2 = 9.8 \times 10^{-4}$, $P_3 = 9.9 \times 10^{-6}$, $P_4 = 9.0 \times 10^{-6}$, and $P_5 = 1.0 \times 10^{-6}$. Let the probability density function for C given the ith accident sequence be denoted $f_i(c)$. The parameter $f_1(c)$ is a delta function at $c = 0$, and $f_2(c)$ is normal with mean 0.0001 and standard deviation 0.00002. The parameter $f_3(c)$ is normal with mean 0.1 and standard deviation 0.02, $f_4(c)$ is normal with mean 0.001 and standard deviation 0.0002, and $f_5(c)$ is normal with mean 0.99 and standard deviation 0.002. The probability density function for C consists of a linear combination of the probability density functions for the five event tree outcomes with each outcome weighted by its probability:

$$f(c) = P_1 \delta(c - 0) + P_2 f_2(c) + P_3 f_3(c) + P_4 f_4(c) + P_5 f_5(c). \tag{10.34}$$

The cumulative distribution function is

$$F(c) = \int_0^c f(\xi)d\xi = P_1 + P_2 F_2(c) + P_3 F_3(c) + P_4 F_4(c) + P_5 F_5(c), \tag{10.35}$$

where the P_1 term on the left reflects the unit integral over the delta function at $c = 0$. The complementary cumulative distribution function or risk curve is

$$1 - F(c) = (1 - P_1) - P_2 F_2(c) - P_3 F_3(c) - P_4 F_4(c) - P_5 F_5(c).$$

The risk curve obtained using the values listed above is shown in Figure 10.12. It is very similar to the risk curve depicted in Figure 10.3.

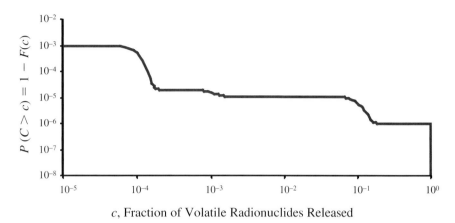

Figure 10.12 Risk curve for Example 10.5 consequence distributions.

It is also possible to treat the consequence-weighted risk R defined by Eq. 10.1 as a random variable. The probability density function $f(r)$ would be interpreted as follows: $f(r)dr$ gives the probability of an accident with a probability-consequence product in the interval dr about r. Assuming the consequence C_i of the i-th sequence is independent of the probability P_i, the mean consequence-weighted risk is

$$E[R] = \sum_{i=1}^{n} E[P_i]\, E[C_i]. \qquad (10.36)$$

If, as in the Example 10.5, one treats the accident sequence probabilities, P_i, as constants, the variance of R is

$$\text{Var}[R] = \sum_{i=1}^{n} P_i^2\, \text{Var}[C_i].$$

For the assumptions postulated in Example 10.5, R is a linear combination of normal random variables, the resulting probability distribution for R is normal. The mean consequence weighted risk $E[R]$ is 2.1 $\times$ 10^{-6}, the variance $\text{Var}[R]$ is $(9.8 \times 10^{-4})^2\,(0.00002)^2 + (9.9 \times 10^{-7})^2\,(0.02)^2 + (9.0 \times 10^{-9})^2\,(0.0002)^2 + (9.9 \times 10^{-7})^2\,(0.002)^2 = 2.2 \times 10^{-11}$, and the standard deviation $\sigma = \text{Var}[R]^{1/2}$ is 4.6×10^{-6}. While it is mathematically possible to define a random variable R in this manner, information regarding the probability distribution of accident consequences is not conveyed in the resulting probability distribution.

5.0 Uncertainty Analysis

A good heuristic introduction to the basic concepts of experimental error analysis is provided in Taylor's book entitled *An Introduction to Error Analysis, The Study of Uncertainties in Physical Measurements* [10]. In an experimental context some property of an object or process is measured and the measurements are invariably subject to experimental error. If the errors are systematic the measurements will not be accurate—they will not cluster about the true value. If systematic errors can be eliminated, the measurements will cluster about the true value. The tighter the cluster, the more precise the measurements. Following Taylor's lead, the terms *error* and *uncertainty* are used interchangeably in the discussion that follows.

From a quantitative perspective, what is sought from an uncertainty analysis is a reasonable range within which a true value of interest is likely to occur. The true value in question is often a parameter of a probability distribution. In an experimental context, the parameter might be the mean of measured values of a property of an object or a process. In a PRA, however, one can generally not make measurements of the quantities of interest. The probabilities of postulated accidents cannot be measured because the accidents generally have such low probabilities that one cannot simply wait for them to happen. Similarly, one cannot conduct experiments to directly determine the distribution of consequences of accidents. Indeed, the goal of safety engineers is to prevent accidents and their associated health and economic consequences. Finally, even if accidents do occur, their consequences may not be evident. For example, cancers induced by radiation exposure have significant latency periods and the number of cancers induced by even a major accident can be small compared to the number of cancers occurring naturally from all causes. As a result, PRA uncertainties are those associated with the model predictions. The concepts of experimental error analysis can be extended to modeling. Uncertainties in the values of model inputs give rise to corresponding uncertainties in the values of model outputs. Similarly, just as systematic errors can bias experimental results, systematic modeling errors can bias model predictions.

Uncertainty analysis as it is conducted in PRA is best illustrated by example. Sections 5.1 and 5.2 illustrate two statistical approaches to the characterization of uncertainties in failure probabilities based on available data. Classical confidence intervals are discussed in Section 5.1, and Bayesian probability intervals are discussed in Section 5.2. Section 5.3 extends the Bayesian approach to the analysis of uncertainties in model predictions using the Monte Carlo method. Section 5.4 provides a categorization of the types uncertainty encountered in PRAs, some of which are not amenable to quantitative analysis result for the Cassini mission.

5.1 Confidence Interval on a Bernoulli Probability

A confidence interval is a statistical statement about a parameter. It provides an interval of values that is apt to include the true value of the parameter. A level of confidence is associated with the interval. To demonstrate these concepts, consider the case in which n Bernoulli trials are to be conducted to develop a two-sided confidence interval for parameter p, the failure probability of a particular component. The number of observed failures x will be governed by the Bernoulli distribution. (A random variable which can take on only two values is said to have a Bernoulli distribution.) The idea is to find two functions of the data, a lower confidence limit $p_L(x,n,\gamma)$ and an upper confidence limit $p_U(x,n,\gamma)$ such that, prior to observing the sample, it can be shown that the probability of obtaining an interval that contains the true value p is greater than or equal to the confidence level $(1 - \gamma)$,

$$P[p_L(x,n,\gamma) \leq p \leq p_U(x,n,\gamma)] \geq 1 - \gamma. \quad (10.37)$$

The interval (p_L, p_U) is a two-sided confidence interval of level $(1 - \gamma)$ or a two-sided $100(1 - \gamma)\%$ confidence interval. The upper and lower confidence limits for the binomial distribution are computed using the cumulative distribution $F(x|n,p)$, which is the probability $P(X \leq x|n, p)$ of observing x or less failures out of n trials given failure probability p. Specifically, using Eq. 10.22, $p_L(x,n,\gamma)$ and $p_U(x,n,\gamma)$ are the values p_L and p_U that satisfy

$$P(X \geq x|n, p_L) = 1 - F(x - 1|n, p_L) = 1 - \sum_{i=0}^{x-1} \frac{n!}{i!(n - i)!} (p_L)^i (1 - p_L)^{n-i} = \frac{\gamma}{2} \quad (10.38)$$

and
$$P(X \leq x|n, p_U) = F(x - 1|n, p_U) = \sum_{i=0}^{x} \frac{n!}{i!(n - i)!} (p_U)^i (1 - p_U)^{n-i} = \frac{\gamma}{2}. \quad (10.39)$$

For example, for a two-sided 90% confidence interval, $\gamma = 0.1$ and the upper confidence limit when $x = 0$ failures are observed in $n = 50$ trials is the solution to $F(0|50, p_U) = (1 - p_U)^{50} = 0.05$, which is $p_U = 1 - (0.05)^{1/50} = 0.058155$. In most other cases, an iterative numerical solution is required. The lower and upper confidence limits for 0 to 10 failures in 50 trials are listed in Table 10.4.

Table 10.4 Two-sided 90% Confidence Interval (p_L, p_U) and Bayes'
Probability Interval (p_1, p_2) for x out of 50 Failures

x	p_L	p_1	$x/50$	p_2	p_U
0	0	0.00100525	0	0.0570480	0.0581551
1	0.00102534	0.00701255	0.02	0.0896715	0.0913982
2	0.00715372	0.0162234	0.04	0.118349	0.120614
3	0.0165519	0.0272337	0.06	0.145075	0.147837
4	0.0277877	0.0394309	0.08	0.170559	0.173791
5	0.0402366	0.0524945	0.10	0.195151	0.198833
6	0.0535714	0.0662330	0.12	0.219054	0.223170
7	0.0675967	0.0805208	0.14	0.242399	0.246936
8	0.0821850	0.0952719	0.16	0.265275	0.270220
9	0.0857619	0.110423	0.18	0.287748	0.293091
10	0.100302	0.125928	0.20	0.309864	0.315596

Example 10.6

A launch vehicle has failed twice in 50 launches. Develop a two-sided 90% confidence interval for the failure probability.

Solution:

Setting $n = 50$ and $x = 2$, the lower confidence limit is determined from Eq. 10.38:

$$1 - \frac{50!}{0!50!} p_L^0 (1 - p_L)^{50} - \frac{50!}{1!49!} p_L^1 (1 - p_L)^{49} = 0.05$$

$$(1 - p_L)^{49} (1 + 49\, p_L) = 0.95$$

$$p_L = 0.00715372$$

and the upper confidence limit is determined from Eq. 10.39,

$$\frac{50!}{0!50!} p_U^0 (1 - p_U)^{50} + \frac{50!}{1!49!} p_U^1 (1 - p_U)^{49} + \frac{50!}{2!48!} p_U^2 (1 - p_U)^{48} = 0.05$$

$$(1 - p_U)^{48} [(1 - p_{UL})^2 + 50(p_U)(1 - p_U) + 1225(p_U^2)] = 0.95$$

$$p_U = 0.120614$$

Once x is observed, the resulting confidence interval either contains the true value of p or it does not contain the true value of p. For example if the true value is $p = 0.1$ and one failure is observed in 50 trials, the 90% confidence interval [0.00102534, 0.0913982] will not contain p; however, if two failures are observed, the 90% confidence interval [0.00715372, 0.120614] will contain p. That is, once x failures are observed in n trials, the probability that the confidence interval is correct is either 0 or 1. This is why the term *level of confidence* is used rather than probability.

The probability of obtaining a confidence interval that contains the true value of p is called the coverage. The coverage is the sum over all possible values of x of the probability of observing the x times a delta function $\delta_p(x)$ indicating whether p is included in the xth interval or not,

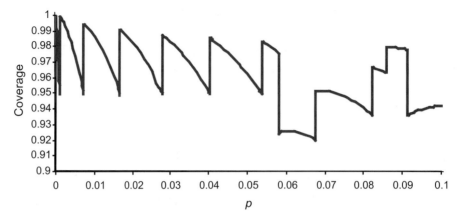

Figure 10.13 Coverage for 90% binomial confidence interval, 50 trials.

$$\text{Coverage} = P[\text{Correct Interval}] = \sum_{x=0}^{n} \frac{n!}{x!(n-x)!} p^x (1-p)^{n-x} \delta_p(x) \geq (1-\gamma). \qquad (10.40)$$

The coverage for the $n = 50$ trials is plotted as a function of p in Figure 10.13. The coverage varies with p, but is never less than $(1 - \gamma) = 0.9$, the level of confidence.

Confidence interval estimation for the failure probability of a system using failure data for the system components is more difficult than the preceding example. The problem can be solved for very simple systems, such as a parallel system in which the system failure rate is the product of two component failure probabilities [11]. But for systems complex enough to warrant the use of fault tree analysis to develop an expression for the system failure probability as a function of numerous base event probabilities, confidence intervals can usually only be estimated using approximate methods. The Bayesian approach described in the following sections is more widely applied. The transition to a Bayesian approach requires an important shift in the interpretation of the parameters of probability distributions. Instead of treating parameters like p in the binomial distribution as unknown constants, the Bayesian approach treats them as random variables whose values can be characterized by state-of-knowledge probability distributions.

5.2 Bayesian Analysis of a Bernoulli Probability

Bayes' theorem [12] follows naturally from the notion of conditional probability, which was introduced in Eq. 10.3. Rearranging Eq. 10.3 to solve for the conditional probability of event A given event B yields

$$P(A|B) = \frac{P(B|A)P(A)}{P(B)}. \qquad (10.41)$$

More generally, Bayes' theorem says: If $A_1, A_2, \ldots, A_n$ are mutually exclusive events and if B is any other event (subset of the union of A_i) such that $P(B) > 0$, then

$$P(A_i|B) = \frac{P(B|A_i)P(A_i)}{P(B)}, \qquad (10.41)$$

where

$$P(B) = \sum_{j=1}^{n} P(B|A_j)P(A_j). \qquad (10.43)$$

Example 10.7

A bin contains computer chips produced by three manufacturers, M_1, M_2, and M_3. M_1 produced 20%, M_2 produced 30%, and M_3 produced 50% of the chips. The defect rates are 0.01, 0.02, and 0.04 respectively for M_1, M_2, and M_3. If a randomly selected chip is found to be defective, what is the probability that it was supplied by M_1?

Solution:

The problem statement provides the following probabilities: $P(M_1) = 0.2$, $P(M_2) = 0.3$, $P(M_3) = 0.5$, $P(D|M_1) = 0.01$, $P(D|M_2) = 0.02$, and $P(D|M_{23}) = 0.04$. First we find

$$P(D) = P(D|M_1)P(M_1) + P(D|M_2)P(M_2) + P(D|M_3)P(M_3)$$

$$= (0.01)(0.2) + (0.02)(0.3) + (0.04)(0.5)$$

$$= 0.028.$$

That is, 2.8% of the computer chips are defective. Then, by Bayes' theorem,

$$P(M_i|D) = P(D|M_i)\,\frac{P(M_i)}{P(D)}, \quad i = 1, 2, 3.$$

Therefore, $P(M_1|D) = (0.01)(0.2)/0.028 = 0.071,$

$P(M_2|D) = (0.02)(0.3)/0.028 = 0.214,$

$P(M_3|D) = (0.04)(0.5)/0.028 = 0.714.$

The probability that the randomly selected defective chip was supplied by M_1 is 0.071.

In Bayesian statistics, the $P(A_i)$ are called *prior* probabilities for the events A_i and the $P(A_i|B)$ terms are called *posterior* probabilities for the events A_i. For the preceding example, $P(M_1)$, $P(M_2)$, and $P(M_3)$ are the prior probabilities that a randomly selected computer chip was produced by manufacturers M_1, M_2, and M_3 respectively. $P(M_1|D)$, $P(M_2|D)$, and $P(M_3|D)$ are the posterior probabilities that a randomly selected defective chip was produced by manufacturers M_1, M_2, and M_3 respectively. For example, manufacturer M_3 produces 50% of all computer chips but 71.4% of all defective computer chips.

The preceding statement of Bayes' theorem is for mutually exclusive discrete events; however, Bayes' theorem can also be stated for probability density functions. If X is a continuous random variable whose probability density function depends on a variable θ so that the conditional probability density function of X, is given by $f(x|\theta)$, and if the prior probability density function of θ is $g(\theta)$, then for every x such that $f(x) > 0$ exists, the posterior probability density function of θ given $X = x$ is

$$g(\theta|x) = \frac{f(x|\theta)g(\theta)}{f(x)}, \tag{10.44}$$

where

$$f(x) = \int_{-\infty}^{\infty} f(x|\theta)g(\theta)d\theta \tag{10.45}$$

is the marginal probability density function of X. In the continuous form of Bayes' theorem, the probability density functions $g(\theta)$ and $g(\theta|x)$ denote the beliefs concerning the likelihood of various values of the random variable Θ prior to and posterior to observing a value of another random variable X.

The continuous form of Bayes' theorem can be applied to develop a probability density function for the parameter p of the Bernoulli distribution that was considered in Section 5.1. Treat the component failure probability as a random variable P, and take the prior probability density function to be uniform on the interval $[0,1]$. In the nomenclature of the continuous form of Bayes' theorem, the prior distribution is

$$g(p) = 1 \quad \text{for} \quad 0 \le p \le 1$$
$$= 0 \quad \text{otherwise.}$$

The underlying probability model is the binomial distribution,

$$f(x|p) = \frac{n!}{x!(n-x)!} p^x (1-p)^{n-x}.$$

The marginal probability density function for X is

$$f(x) = \int_0^1 f(x|p)g(p)dp$$

$$= \int_0^1 \frac{n!}{x!(n-x)!} p^x (1-p)^{n-x}(1)dp$$

$$= \frac{1}{n+1}.$$

The posterior probability density function for P is

$$g(p|x) = \frac{f(x|p)g(p)}{f(x)} = \frac{(n+1)!}{x!(n-x)!} p^x (1-p)^{n-x}. \tag{10.46}$$

In contrast to the classical confidence interval discussed in Section 5.1, a Bayesian analyst views the parameter p of the binomial distribution as a random variable P. Any two values of the variable, say p_1 and p_2, that satisfy

$$P(p_1 \le P \le p_2) = \int_{p_1}^{p_2} g(p|x)dp = 1 - \gamma \tag{10.47}$$

define a $(1 - \gamma)$ level or $100(1 - \gamma)\%$ two-sided Bayes' probability interval. Typically, p_1 and p_2 are selected to be the values for which the cumulative distribution is equal to $\gamma/2$ and $1 - \gamma/2$ respectively. Table 10.4 lists the values of p_1 and p_2 for x ranging from 0 to 10 failures in $n = 50$ trials.

Example 10.8

Determine the 5th and 95th percentiles of the posterior distribution for the parameter p of a Bernoulli distribution given zero failures in 50 trials. Assume a uniform prior. Repeat for two failures in 50 trials.

Solution

Using Eq. 10.46 with $x = 0$ failures in $n = 50$ trials, the posterior distribution is

$$g(p|0,50) = \frac{51!}{0!(50-0)!} p^0 (1-p)^{(50-0)} = 51(1-p)^{50}.$$

The posterior cumulative distribution function is

$$F(p|0,50) = \int_0^p 51(1-\xi)^{50}d\xi = 1-(1-p)^{51}.$$

$F(p|0,50) = 0.05$ is satisfied when $p = 0.001005$. $F(p|0,50) = 0.95$ is satisfied when $p = 0.05705$. Therefore, [0.00105, 0.05705] is a 90% two-sided Bayesian probability interval. Given zero failures in 50 trials, the probability that P falls in this interval is 0.9.

For two failures in 50 trials, the posterior probability density function from Eq. 10.46 is

$$g(p|2,50) = \frac{51!}{2!(50-2)!}p^2(1-p)^{(50-2)} = 62475\, p^2(1-p)^{48}.$$

The posterior cumulative distribution function is the integral of the preceding density function from 0 to p. The analytic result is rather messy, but the limits of the 90% Bayesian probability interval can be determined numerically [13]. From Table 10.4, the resulting interval is (0.0162234, 0.118349).

In Examples 10.3, 10.6, and 10.8, for the case in which a certain type of launch vehicle fails twice in 50 launches, a nominal estimate of the probability of failure is obviously 2/50 or 0.04. It is clear, however, that the probability of failure is uncertain. With one more launch the nominal estimate will either be 3/51 or 2/51 depending on whether or not another failure occurs. Figure 10.14 show the posterior probability density function and cumulative distribution function that would be used to express the uncertainty in the probability of failure given two failures in 50 events. In this uncertainty representation 0.04 is the most probable value of the failure probability, but there is a 5% chance that the true value is less than 0.0162234, a 5% chance that the true value is greater than 0.118349, and a corresponding 90% chance that the true value is between 0.0162234 and 0.118349. The uncertainty analysis performed for the Cassini risk assessment, which is discussed in Section 6.5, refers to such levels and intervals as *confidence levels* and *confidence intervals*. *Confidence* in this sense is simply state-of-knowledge probability. This use of the term *confidence* avoids awkward discussions of state-of-knowledge probability distributions of frequentist failure probabilities. It should be emphasized that *confidence,* in this context, depends on the distributions assigned to reflect state-of-knowledge uncertainties in model parameters and is not quantified from classical statistical analysis of experimental data as described in Section 5.1.

5.3 Parametric Uncertainty Analysis Using Monte Carlo

Monte Carlo is a powerful method that can be applied to many problems that would otherwise be intractable. In Monte Carlo, a game of chance is devised so that the outcomes from a large number of plays yield the answers being sought. Pseudorandom number generators permit such games of chance to be played on computers. A game of chance can be a direct analog of a process being studied or relatively artificial. In some cases, different games can be devised to solve the same problem. The art of Monte Carlo is in devising a suitably efficient game.

Although Monte Carlo is powerful, one would generally prefer an analytic solution or a deterministic numerical solution. The difficulty is that, because Monte Carlo simulates random processes, the answer always involves a statistical error. Yet, often Monte Carlo is the only practical method for getting a realistic estimate of the answer. This is often the case for problems involving complex geometries or more than six to eight independent variables. A brief discussion of Monte Carlo methods for neutronics calculations was provided in Chapter 8. A full treatment of Monte Carlo is well beyond the scope

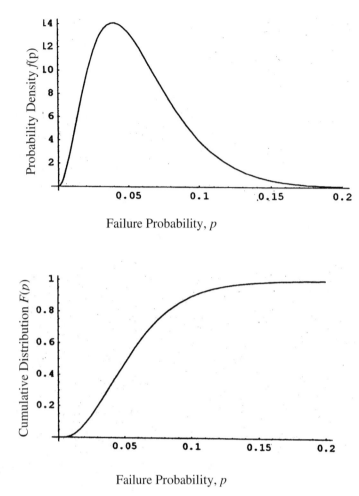

Figure 10.14 Posterior probability density and cumulative distribution function for failure rate p given two failures in 50 events.

of this book. In this section, some basic principles of Monte Carlo are explained, and the method is used to propagate parametric uncertainties through a simple system failure probability model.

   At the heart of the Monte Carlo method, values of random variables are sampled from their known probability distributions. A straightforward sampling process is illustrated in Figure 10.15, which plots the cumulative distribution function $F(x)$ of a random variable X as a function of observed value x. To sample a value of x, one first uses a pseudorandom number generator. A pseudorandom number generator simulates the random selection of a value r_i of a random variable R that is uniformly distributed on the interval [0,1]. The value r_i is used as $F(x_i)$, a value of the cumulative distribution function. The inverse of the cumulative distribution function yields x_i, the sampled value of the random variable X. The process is repeated for $i = 1, 2, 3, \ldots, n$, until the desired sample size n is obtained.

Example 10.9

Sample the posterior distribution for p given zero failures in 50 trials (see Example 10.8) given the successive values $u_1 = 0.356$, $u_2 = 0.768$, and $u_3 = 0.595$ from a pseudorandom number generator.

Solution:

The cumulative distribution function from Exercise 10.8 is $F(p) = 1 - (1 - p)^{51}$. The inverse function is $p_i = F^{-1}(u_i) = 1 - (1 - u)^{1/51}$, so

$$p_1 = 1 - (1 - 0.356)^{1/51} = 0.00859,$$

$$p_2 = 1 - (1 - 0.768)^{1/51} = 0.02824, \text{ and}$$

$$p_3 = 1 - (1 - 0.595)^{1/51} = 0.01756.$$

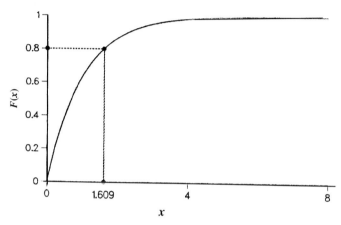

Figure 10.15 Exponential cumulative distribution function showing inverse selection process used in Monte Carlo analysis.

Let the Y denote a random variable that is generated in a Monte Carlo simulation. Random variable Y will generally be a function $y(X_i)$ of a vector X_i of sampled random variables. Let N denote the sample size. The fundamental theorem of Monte Carlo says that the sample mean

$$\overline{Y}_N = \frac{1}{N} \sum_i y(x_i) \tag{10.48}$$

is an estimator of the true mean (the expectation value $E[Y]$); that is, as N becomes large, the sample mean approximates the true mean,

$$\lim_{N \to \infty} \overline{Y}_N = E[Y]. \tag{10.49}$$

The larger the sample size, the better the approximation. In particular, the standard deviation $\sigma_{\overline{Y}_N}$ of the sample mean can be shown to be equal to the standard deviation of the underlying random variable Y divided by the square root of the sample size,

$$\text{Var}(\overline{Y}_N) = \frac{\text{Var}(Y_N)}{N} \quad \text{or} \quad \sigma_{\overline{Y}_N} = \frac{\sigma_y}{\sqrt{N}} . \tag{10.50}$$

Finally, the central limit theorem shows that repeated means from samples of the same size follow a Gaussian distribution provided N is sufficiently large. Properties of the Gaussian distribution can then be applied. In particular, 68.27% of the time the sample mean would be expected to be within one standard deviation of $E[Y]$, 95.45% of the time the sample mean would be expected to be within two standard deviations of $E[Y]$, 99.73% of the time the sample mean would be expected to be within three standard deviations of $E[Y]$.

Example 10.10

Apply the Monte Carlo method to generate 10 values of $Y = U_1 + U_2$, where U_1 and U_2 are independent random variables, which are uniformly distributed on the interval [0,1].

Solution:

Twenty consecutive values were selected from a table of random numbers. In Table 10.5, the first 10 values are listed in the column labeled $(u_1)_i$, and the second 10 are listed in the column labeled $(u_2)_i$. The 10 corresponding values of $Y = U_1 + U_2$ are listed in the last column of Table 10.5. The sample means are compared to the corresponding expectation values in the last two rows of Table 10.5. The sample means are not particularly close to the expectation values because the sample size was only 10. A larger sample size would produce sample means closer to the expectation value.

Figure 10.16 shows the result of the preceding simulation for 200 random samples of size $n = 10$. The simulation results produced a minimum value of $\bar{Y} = 0.636$ and a maximum value of $\bar{Y} = 1.298$. As predicted by the central limit theorem, the distribution for $\bar{Y}$ shown in Figure 10.16 is closely approximated by a normal distribution. The overall mean was 1.0034 and the variance over the 200 samples of size 10 was 0.01748, which compare to the true values of 1 and 0.01667 respectively.

Various methods have been devised to greatly improve the efficiency (reduce the variance) of Monte Carlo methods. One method that is commonly employed in uncertainty analyses is Latin hypercube sampling, which is described by Iman and Shortencarier [14].

5.4 Types of Uncertainty

Aleatory uncertainty refers to events or phenomenon being modeled that are characterized as occurring in a "random" or "stochastic" manner, and probabilistic models are adopted to describe their occurrences. Aleatory uncertainty is built into the structure of the PRA model. This aspect of uncertainty gives PRA the probabilistic part of its name.

Epistemic or state-of-knowledge uncertainty is associated with the analyst's confidence in the predictions of the PRA model. It reflects the analyst's assessment of how well the PRA model represents

Table 10.5 Monte Carlo Sample for Simple Linear Model
$Y = U_1 + U_2$, Sample Size $n = 10$ From Example 10.10

i	$(u_1)_i$	$(u_2)_i$	$y_i = (u_1)_i + (u_2)_i$
1	0.10480	0.15011	0.25491
2	0.22368	0.46573	0.68941
3	0.24130	0.48360	0.72490
4	0.42167	0.93093	1.35260
5	0.37570	0.39975	0.77545
6	0.77921	0.06907	0.84828
7	0.99562	0.72905	1.72467
8	0.96301	0.91977	1.88278
9	0.89579	0.14342	1.03921
10	0.85475	0.36857	1.22332
Sample Mean	0.58555	0.46600	1.05155
Expectation Value	0.5	0.5	1.0

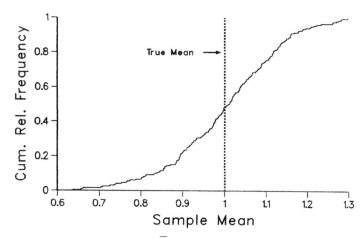

Figure 10.16 Sampling distribution of $\overline{Y}_1$ based on 200 random samples of size $n = 10$.

the actual system being modeled. As such, it generally varies from analyst to analyst. Uncertainty in the results obtained from the PRA model is epistemic. The confidence intervals and Bayesian probability intervals discussed in the preceding sections are quantitative estimates of epistemic uncertainties.

Epistemic (state-of-knowledge) uncertainties are commonly divided into three classes: parameter uncertainty, model uncertainty, and completeness uncertainty. Parameter uncertainties are those associated with the values of parameters of the PRA models. They are typically characterized by establishing probability distributions on the parameter values. These distributions express the analyst's degree of belief in the values these parameters could take, based on the current state of knowledge. It is reasonably straightforward to propagate the distributions representing uncertainties in base event probabilities to obtain probability distributions on accident sequence probabilities. Quantitative analysis of uncertainties in radionuclide releases and health and economic consequences is generally difficult but can be done as illustrated for the Cassini mission in Section 6 of this chapter.

Model uncertainties are those associated with incomplete knowledge regarding how models used in PRAs should be formulated. Such uncertainties arise, for example, in modeling common cause failures, human errors, and mechanistic failures of structures, systems, and components. Model uncertainties grow in number and magnitude as one proceeds from accident probability estimation to health and economic consequence evaluation. In some cases, where well-formulated alternative models exist, PRAs have addressed model uncertainty by using discrete distributions over the alternative models, with the probability (or weight) associated with a specific model representing the analyst's degree of belief that the model is the most appropriate. Another approach to addressing model uncertainty is to adjust the results of a single model through the use of an adjustment factor. Using such approaches, model uncertainty can be propagated through the analysis in the same way as parameter uncertainty. More typically, although model uncertainties are recognized, they are not quantified. Assumptions are made and specific models are adopted. Unquantified model uncertainty also arises because PRAs bin the continuum of possible system states in a discrete way. Such approximations may introduce biases into the results.

In interpreting the results of a PRA, it is important to develop an understanding of the impact of specific assumptions or model choices on the predictions of the PRA. This is true even when the model uncertainty is treated probabilistically, since the probabilities, or weights, given to different models are subjective. The impact of using alternative assumptions or models may be addressed by performing appropriate sensitivity studies, or they may be addressed using qualitative arguments, based on an understanding of the contributors to the results and how they are impacted by the change in assumptions or models. The impact of making specific modeling approximations may be explored in a similar manner.

Completeness uncertainty refers to things that are not modeled in a PRA. This includes risk contributors that can be modeled but are often excluded such as risks associated with transportation to the launch site. It also includes considerations for which methods of analysis have not been developed, for example, acts of sabotage, heroic acts, and influences of organizational performance. Finally, completeness uncertainty includes initiators and accident scenarios that have not been conceived (in spite of some analyst's delusions of omniscience). Incompleteness in a PRA can be addressed for those scope items for which methods are in principle available, and therefore some understanding of the contribution to risk exists. This may be accomplished by supplementing the analysis to enlarge the scope, by using more restrictive acceptance criteria, or by providing arguments that, for the application of concern, the out-of-scope contributors are not significant. Additional design conservatism can also be used to compensate for completeness uncertainty.

6.0 The Cassini Mission Risk Assessment

To illustrate the application of the risk assessment process in the context of space nuclear power, the results obtained from an assessment of radiological risks to the public posed by the Cassini mission are discussed in this section. Public controversy regarding the safety of launching RTG power sources intensified following the *Challenger* accident. This led to significant safety testing in support of the Ulysses and Cassini missions, and the risk analyses performed for these missions were quite comprehensive. In general, the scope and level of a risk assessment should be that required to support a correct decision. A risk assessment as comprehensive as that performed for Cassini may not be required in other cases. For example, other space-reactor risk insights described in Section 7 are based on far less comprehensive risk assessments. Nevertheless, Cassini results provide a practical illustration of the risk assessment concepts introduced in the preceding sections.

6.1 The Cassini Mission

Cassini, along with the European Space Agency's Huygens probe, is a $3.4 billion joint mission of NASA, the European Space Agency (ESA), and the Italian Space Agency. Cassini successfully lifted off aboard a Titan IV launch vehicle from Cape Canaveral Air Station at 4:43 a.m. on October 15, 1997. Figure 10.17 depicts the key features of the launch vehicle including first and second stage liquid rocket engines, solid rocket motor upgrades (SRMUs), the Centaur upper stage, and the payload faring. The two-story tall Cassini spacecraft arrived at Saturn in 2004, after making gravity-assist swingbys past Venus twice, then past Earth, and finally, past Jupiter. Cassini is scheduled to spend 4 years orbiting Saturn gathering data about Saturn's atmosphere, rings, and magnetosphere. On January 14, 2006, the Huygens probe, with another six instruments, separated from Cassini and parachuted into the thick atmosphere of Saturn's moon Titan. The package dropped through the mostly methane gas atmosphere, taking measurements of chemistry, temperature, pressure, and density during its decent. Titan is the only celestial body besides Earth to have an atmosphere rich in nitrogen. Scientists hope to learn more about the origins of Earth by sampling the chemistry and photographing the surface of Titan.

Electrical power for the Cassini spacecraft is provided by three general purpose heat source (GPHS) radioisotope thermoelectric generators (RTGs) (see Figures 2.14 and 2.15 in Chapter 2). As discussed in Chapter 2, these RTGs convert heat released from the radioactive decay plutonium dioxide fuel into electricity. The total mass of plutonium dioxide contained within the three RTGs is 32.7 kilograms. Pu-238 dioxide is also used as the heat source in 82 small radioisotope heater units (RHUs) on the Cassini orbiter. Another 35 RHUs are onboard the Huygens probe (totaling about 0.3 kg of Pu-238 for RHUs on the orbiter and probe). Each RHU produces about 1 watt of heat to keep nearby electronics at their proper operating temperatures.

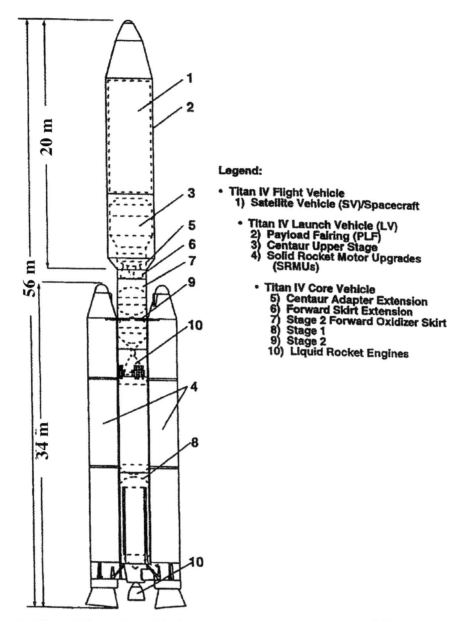

Figure 10.17 Cassini Titan IV launch vehicle features (not to scale). *Courtesy of the U.S. Department of Energy.*

Both RTGs and RHUs have a long and safe heritage of use and high reliability in NASA's planetary exploration program. More than three decades have been invested in the engineering, safety analysis, and testing of RTGs. Safety features are incorporated into the RTG design, and extensive testing has demonstrated that they can withstand physical conditions more severe than those expected from most accidents. The fuel is heat-resistant, ceramic plutonium dioxide, which reduces its chance of vaporizing in fire or reentry environments. This ceramic fuel is also highly insoluble, has a low chemical reactivity, and primarily fractures into chunks and particles too large to be inhaled. These characteristics help to reduce the number of potential health effects from accidents involving fuel releases.

Multiple layers of protective materials, including iridium capsules and high-strength graphite blocks, are used to protect the fuel and prevent its accidental release. The modular design reduces the chance of a large fuel release in an accident because all modules would not experience equal impact damage (see Chapter 7). Each PuO_2 pellet is sealed in an iridium clad which is then locked inside carbon composite canisters within bricks of the same material (see Figure 2.14 in Chapter 2). Iridium metal has a melting point of 2727 K, which is well above the temperature it would reach on atmospheric reentry. Iridium is strong, corrosion-resistant, and chemically compatible with plutonium dioxide. These characteristics make iridium useful for protecting and containing each fuel pellet. Graphite is used because it is lightweight and highly heat-resistant. The RTGs are located at the top of the rocket, away from the propellant.

The Cassini mission was controversial because of the quantities of Pu-238 used in its power and heat sources. Some worried that an accident would scatter radioactive plutonium through Earth's atmosphere. In particular, concerns were expressed about the possibility of Cassini making an error during its swing past Earth, reentering Earth's atmosphere at high speed, and dispersing plutonium through the atmosphere as it burned up during reentry. Robotic planetary spacecraft have performed numerous similar gravity-assist maneuvers with extraordinary precision, and NASA took specific actions to design the spacecraft and mission to ensure a very low probability of Earth impact. Until 7 days before the Earth swingby, the spacecraft trajectory, without any alterations, would miss Earth by thousands of kilometers. This trajectory strictly limits the possibility that random external events, such as a micrometeoroid puncture of a spacecraft propellant tank might lead to Earth impact. Originally, NASA planned for Cassini's trajectory to pass by Earth at an altitude of 500 km and calculated the odds of reentry at 1 in 1 million. NASA later adjusted the trajectory to take Cassini past Earth at 850 km, reducing the odds of reentry to 1 in 1.25 million. The redundant design of Cassini's systems and navigational capability allows control of the swingby altitude to within an accuracy of 3–5 km at an altitude of 800 km or higher.

In seeking approval, NASA undertook a comprehensive risk assessment to identify potential accident scenarios and to estimate their likelihoods and potential consequences. John Gibbons, the director of the U.S. Office of Science and Technology Policy (OSTP) said, "NASA and its interagency partners have done an extremely thorough job of evaluating and documenting the safety of the Cassini mission. I have carefully reviewed these assessments and have concluded that the important benefits of this scientific mission outweigh the potential risks" [15].

6.2 Cassini Accident Scenarios

Table 10.6 indicates the timeline for the Cassini mission. The risk analysis considers accidents that can occur from the time of RTG attachment to the spacecraft 2 days before launch until the final interplanetary trajectory is established following final Earth pass-by. It is convenient to group the accident scenarios into four main categories by timeline:

1. On-pad (prelaunch) accidents
2. Early launch phase accidents (mission phases 1 and 2)
3. Late launch phase (mission phases 3 through 8)
4. Reentry during earth gravitational assist

Prelaunch accidents involve on-pad explosions caused by inadvertent mixing of propellants in the Centaur or core vehicle. Some prelaunch accident scenarios also involve SRMU propellant fires. Early launch phase accidents are those initiated prior to jettison of the payload faring (PLF) at 16.5 km. In such accidents the RTG or its components may be damaged by an explosion, by fragments generated in an explosion, and/or by striking the ground. After PLF jettison, the ambient air pressure is so low that the RTGs cannot be damaged by the postulated in-air explosion environments. Late launch phase accidents are those initiated after PLF is jettisoned. These accidents subject the RTGs to high aerodynamic loads

Table 10.6 Cassini Mission Timeline
Courtesy of U.S. Department of Energy.

Mission Phase	Event Description Start	Finish	Elapsed Time (s) Start	Finish
0	Complete RTG installation	SRMU ignition	−48 hours	0
1	SRMU ignition	SRMU jettison	0	143
2	SRMU jettison	PLF jettison	143	208
3	PLF jettison	Stage 1 jettison	208	320
4	Stage 1 jettison	Stage 2 jettison	320	554
5	Stage 2 jettison	Centaur main engine cutoff 1	554	707
6	Centaur main engine cutoff 1	Centaur main engine start 2	707	1889
7	Centaur main engine start 2	Earth escape	1889	2277
8	Earth escape	Centaur main engine cutoff 2	2277	2349

and thermal stresses associated with reentry. Accidents initiated prior to the attainment of parking orbit lead to suborbital reentries and result in potential surface impacts of RTG components along the nominal flight trajectory over the Atlantic Ocean, southern Africa, and Madagascar, as shown in Figure 10.18. Accidents initiated after the PLF is jettisoned and before the start of the first Centaur burn would result in impact in the Atlantic Ocean, which would not cause fuel to be released from an RTG or its components. Late launch phase accidents initiated after attainment of parking orbit lead to orbital decay reentries. RTG components subject to orbital decay reentry can impact anywhere on the globe between latitude boundaries determined by the parking orbit inclination and the type of failure. For the most probable orbital decay reentry scenarios, the impact latitude boundaries are 38°N and 38°S. Collectively, the suborbital reentry and orbital decay reentry accident scenarios are termed "out-of-orbit" reentries.

The baseline trajectory incorporates a Venus-Venus-Earth-Jupiter gravity assist (VVEJGA) trajectory. During the Earth gravity assist swingby, malfunctions could cause the spacecraft to reenter the Earth's atmosphere, imposing aerodynamic loads and thermal stresses too great for the RTGs to withstand. A reentry occurring during Earth swingby is considered a *short-term Earth impact*. Loss of spacecraft control during the interplanetary cruise could potentially result in a *long-term Earth impact* decades to millennia later as the spacecraft orbits around the Sun.

Analyses were conducted to identify accident cases of potential significance in each of the four main timeline categories. Table 10.7 lists the cases that were identified based on the existing failure database as documented in the Cassini Titan IV/Centaur RTG Safety Databook and the Cassini Earth Swingby Plan [16, 17]. The probabilities of occurrence of the cases listed in Table 10.7 cannot be estimated with certainty. A distribution is used to characterize the uncertainty in the probability of occurrence for each case. The mean probability of occurrence, which is provided for each case in Table 10.7, is a distribution parameter that is used as a "best" or "nominal" estimate because it is neither optimistically nor pessimistically biased. The Cassini uncertainty analysis is discussed in Section 6.5.

6.3 Cassini Source Term Analysis

The Cassini Titan IV/Centaur RTG Safety Databook provides information about the severity of the environment to which the RTGs would be exposed during prelaunch and early launch accidents [16]. Any such accident would involve an explosion that could damage the RTGs either directly or indirectly. Mathematical models are used to determine the response of the GPHS-RTGs and the characteristics of potential PuO_2 releases. These models are based upon physical principles, known mechanical properties of the components of the GPHS-RTGs, and results of tests on the GPHS-RTGs and their components. Such models are integrated into computer codes to predict the effect of explosions, fragment impacts, and ground impacts on GPHS-RTGs and their components.

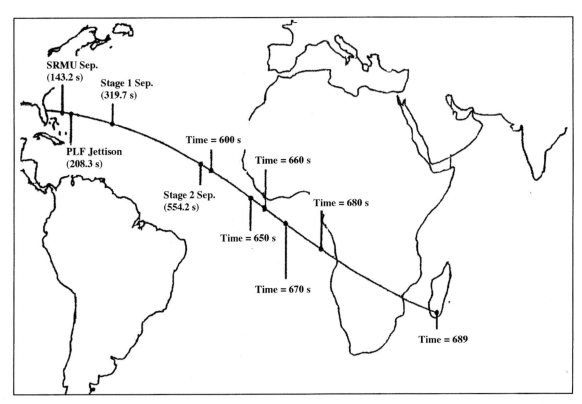

Figure 10.18 Cassini instantaneous impact points. (Times shown are loss of thrust times which lead to impact at the indicated point along the flight path without accounting for atmospheric drag.) *Courtesy of the U.S. National Aeronautics and Space Administration.*

Table 10.7 Cassini Accident Case Descriptions
Courtesy of U.S. Department of Energy.

Mission Stage	Case Number	Case Description	Mean Probability
Pre-Launch	0.1	On-Pad Explosion	1.4×10^{-4}
	0.2	On-Pad Explosion with SRMU Aft Segment Impact	4.3×10^{-6}
Early Launch	1.1	Total Boost Vehicle Destruct (TBVD)	4.2×10^{-3}
	1.2	Command Shutdown and Destruct	6.6×10^{-4}
	1.3	TBVD with SRMU Aft Segment Impact	8.0×10^{-4}
	1.4	SRMU Explosion	1.2×10^{-4}
	1.5	Space Vehicle (SV) Explosion	7.6×10^{-4}
	1.6	TBVD without Payload Faring (PLF)	9.0×10^{-6}
	1.7	CSDS* without PLF	9.0×10^{-6}
	1.8	SV Explosion without PLF	1.4×10^{-6}
	1.9	Centaur Explosion	1.4×10^{-4}
	1.10	Space Vehicle/RTG Impact	2.4×10^{-4}
	1.11	Payload Faring/RTG Impact	1.9×10^{-6}
	1.12	Payload Faring/RTG Impact, RTG Falls Free	1.9×10^{-6}
Late Launch	3.1	Suborbital Reentry	1.4×10^{-3}
	5.1	Suborbital Reentry from CSDS Configuration 5	1.2×10^{-2}
	5.2	Orbital Reentry, Nominal	7.1×10^{-3}
	5.3	Orbital Reentry, Off-Nominal Elliptic Decayed	8.9×10^{-3}
EGA Swingby		Reentry During Earth Gravitational Assist Swingby	8.0×10^{-7}

* Command Shutdown and Destruct System

The variability in the time and strength of the explosion and the variability in the response of the GPHS-RTGs to the explosion require a probabilistic approach to response modeling. Whenever a value is required of a quantity that has a probability distribution, a random number distributed according to that probability distribution is generated and used. The simulation is repeated thousands of times for each accident case. The set of results from all the repetitions (trials) provide probability distributions of the possible outcomes for each accident case. This simulation process is embodied in a computer code entitled the Launch Accident Safety Evaluation Program-Titan IV (LASEP-T). This code calculates the effects of various damaging environments (e.g., explosion blast, fragment collision, and ground collision) on the fueled clads. The location of each fueled clad and each released module is tracked individually.

Analyses performed by NASA-Jet Propulsion Laboratory (JPL) indicate that the space vehicle will break up on reentry. The break-up of the space vehicle frees the GPHS modules, which then fall to Earth. The GPHS module was designed to survive out-of-orbit reentry. During reentry, the composite carbon shell of the module ablates, carrying away the heat generated by friction with the atmosphere. Calculations performed specifically for the Cassini mission show that 71% of the module shell would ablate as a result of the most severe possible out-of-orbit reentry. This level of ablation would not result in structural failure of the module during its descent. Therefore, ground impact is the only potential cause of plutonium release for out-of-orbit reentry.

GPHS modules impacting the Earth from out-of-orbit reentry would strike the ground at their terminal velocity, approximately 49 meters per second. Tests in which GPHS modules impacted various targets at terminal velocity have shown that PuO_2 would only be released from the fueled clads if the GPHS modules strike hard materials such as steel, rock, or concrete. Even in those situations, only a fraction of the fueled clads would be breached, and only small amounts of PuO_2 would be released. A worldwide surface type database is used to calculate the probability of a module striking rock (see Chapter 7). The amount and particle size distribution of PuO_2 releases were calculated from a model based on GPHS component test data. As in the case of launch accidents, a probabilistic sampling approach is used to model the fuel release characteristics that could result from out-of-orbit reentry scenarios.

Inadvertent reentry during the Earth gravity assist swingby would occur at a velocity of approximately 19.4 kilometers per second, which is considerably greater than the maximum velocity for out-of-orbit reentry. Calculations have shown that a number of failure modes are possible in this situation: in-air failure of the GPHS module and/or the graphite impact shells (GISs), failure upon impact of modules or GISs containing melted fueled clads, and failure upon impact of modules or GISs containing intact clads. The actual sequence of events and condition of the iridium clad would depend upon the angle of reentry and the dynamic condition, including orientation, of the modules during descent. In-air failure of individual modules or clads would result in a full release of fuel in those modules or clads at high altitudes (23 to 32 km). Impact of modules or GISs containing melted clads would result in a full release of fuel if a hard surface (rock) is hit and may result in a full release if impact soil is hit. Rock impact of modules or GISs containing intact clads may result in a small fuel release. The possible sequences of events are modeled with an event tree. The source term release predicted from an EGA reentry accident is determined from the distribution of module failure types, based in part on a worldwide surface database. The mean fuel release predicted in the event of an EGA reentry accident which produces a fuel release is 2600 grams.

Figure 10.19 provides source term complementary cumulative distribution functions (*ccdfs*) for the combined phase 0/1 launch accident cases, the combined out-of-orbit accident cases, and EGA short-term reentry. These *ccdfs* give the probability of the source term exceeding any given magnitude. For example, there is about a 10^{-5} (1 in 100,000) probability that a phase 0/1 launch accident would result in a PuO_2 release exceeding 1 gram. The probability of an EGA source term exceeding 10,000 grams is about 10^{-7}. Table 10.8 summarizes source term results for individual launch accident cases

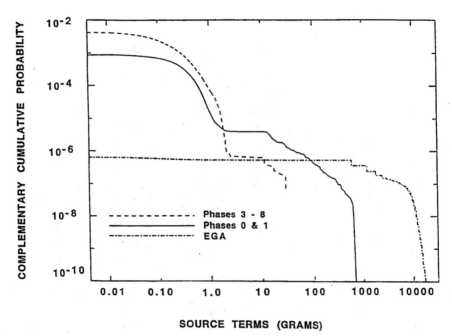

Figure 10.19 Source terms by mission segment. *Courtesy of the U.S. Department of Energy.*

Table 10.8   Cassini Source Terms
Courtesy of U.S. Department of Energy.

Mission Segment	Case	Description	Mean POF	Release Prob.	Mean PuO$_2$ Release and Percentiles* (grams)				
					Mean	5th	50th	95th	99th
Pre-Launch	0.1	On-pad Explosion	1.4×10^{-4}	1.4×10^{-4}	0.299	0.075	0.255	0.524	0.694
	0.2	On-pad Explosion with SRMU Aft Segment Impact	4.3×10^{-6}	4.3×10^{-6}					
Early Launch	1.1	Total Boost Vehicle Destruct	4.2×10^{-3}	5.3×10^{-4}	0.567	0.025	0.219	0.790	1.305
	1.2	Command Shutdown & Destruct	6.6×10^{-4}	1.4×10^{-5}	0.811	0.022	0.174	0.934	17.79
	1.3	Total Boost Vehicle Destruct with SRMU Aft Segment Impact	8.0×10^{-4}	1.4×10^{-4}	0.532	0.045	0.335	0.848	1.192
	1.4	SRMU Explosion	1.2×10^{-4}	1.9×10^{-5}	0.843	0.046	0.337	0.829	1.097
	1.6	Total Boost Vehicle Destruct without Payload Faring	9.0×10^{-6}	1.0×10^{-6}	0.445	0.027	0.218	0.827	1.447
	1.9	Centaur Explosion	1.4×10^{-4}	2.4×10^{-5}	0.654	0.040	0.316	0.858	1.512
Late Launch	3.1	Suborbital Reentry	1.4×10^{-3}	6.0×10^{-5}	0.097	0.014	0.063	0.212	0.298
	5.1	CSDS Suborbital Reentry	1.2×10^{-2}	3.6×10^{-5}					
	5.2	Orbital Reentry (Nominal)	7.1×10^{-3}	1.8×10^{-3}	0.218	0.019	0.132	0.696	1.299
	5.3	Orbital Reentry (Off-Nominal, Elliptic Decayed)	8.9×10^{-3}	2.3×10^{-3}					

* For trials that produced fuel releases

with probabilities of release greater than 10^{-6} (one in a million) and expected contributions to risk greater than 1%. The first data column in Table 10.8 gives the mean probability of failure (POF) for each accident case. These mean POF values are used to calculate the nominal results. The remaining columns in Table 10.8 provide information about the mean mass of PuO_2 released and the 5th, 50th, 95th and 99th percentiles of this released mass for each of the accident cases.

6.4 Cassini Consequence and Risk Analysis

Probability distributions for various consequence measures are calculated from the source term probability distributions. For Cassini, distributions were developed for the following consequence measures: (1) latent cancer fatalities induced by exposure to released PuO_2 (health effects); (2) collective radiation dose to the Earth's population; (3) maximum dose delivered to a single individual in each accident; (4) the maximum potential dose irrespective of the predicted proximity of released PuO_2 to individuals or population centers; and (5) land area contaminated above 0.2 $\mu Ci/m^2$ (the threshold above which the EPA recommends cleanup measures be considered).

Collective dose and health effects were calculated with and without de-minimis. Predicted consequences with de-minimis assume that exposure to radiation dose levels below 1 mrem per year have no discernible effect to the health of an individual. In contrast, consequence estimates without de-minimis assume that any radiation dose, no matter how small, incrementally increases the probability of latent cancers. The proportionality constant used in the Cassini risk assessment postulated 1 additional cancer for every 5000 person rem. Radiological consequence modeling is the topic of Chapter 11. The *ccdf*s for 50-year health effects by mission segment are presented in Figures 10.20 and 10.21 without and with de-minimis respectively. The curves show the overall probability that health effects would exceed a given value. These probabilities include the probability of failures initiating an accident and the probability of PuO_2 release given an accident. Clearly the low-consequence risk is dominated by Phase 1 launch accidents, whereas the high-consequence risk is dominated by EGA reentry. From the combined results of all mission phases, the probability of having more than one health effect is 1 in 100,000 (Figure 10.20 or 10.21).

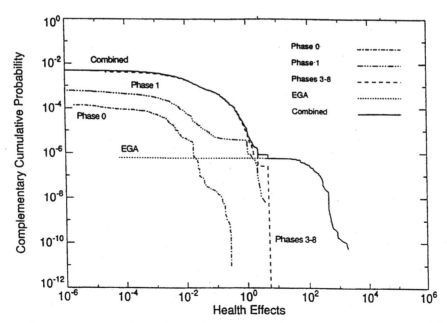

Figure 10.20 50-year health effects without de-minimis (by mission segment). *Courtesy of the U.S. Department of Energy.*

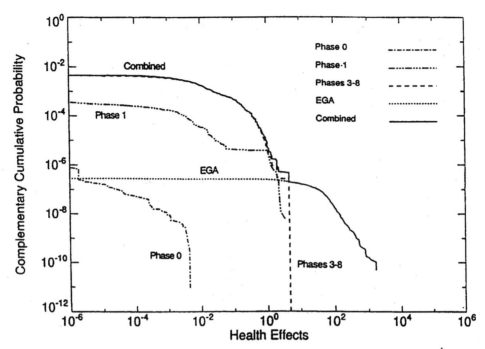

Figure 10.21 50-year health effects with de-minimis (by mission segment). *Courtesy of the U.S. Department of Energy.*

Table 10.9 gives the mean probabilities of release, the mean number of health effects, and the mean actuarial risks with and without de-minimis by mission segment. The tabulated risk is the product of the mean probability of release and the mean number of latent cancers. As indicated in Table 10.9, the average number of incremental latent cancer fatalities for all accidents involving release of PuO_2 is 0.055. If a de-minimis dose level of 1 mrem per year is postulated, this decreases to 0.037 incremental latent cancer fatalities. Accidents occurring near the launch site produce the lowest expected health consequences of any mission segment. Accidents postulated during late launch (Phases 3 through 8) would lead to out-of-orbit reentry. The nominal probability of release associated with these accidents is 4.1×10^{-3}, which is greater than the probability of release for accidents postulated during all other phases. The mean number of incremental cancer fatalities associated with late launch accidents is 0.045. The greatest expected consequence is from inadvertent reentry during Earth swingby, with 130 incremental cancer fatalities predicted without de-minimis and 15 predicted with de-minimis. The effect of de-minimis in these scenarios reflects the preponderance of low exposure doses expected for high altitude releases.

Table 10.9   Mean 50-Year Health Effects by Mission Segment for the Cassini Mission
Courtesy of U.S. Department of Energy.

Mission Segment	Mean Probability	Mean Latent Cancers		Latent Cancer Actuarial Risk	
		Without De-minimis	With De-minimis	Without De-minimis	With De-minimis
Prelaunch	1.4×10^{-4}	1.8×10^{-3}	2.8×10^{-7}	2.5×10^{-7}	3.9×10^{-11}
Early Launch	6.9×10^{-4}	0.011	8.410^{-3}	7.3×10^{-6}	5.8×10^{-6}
Late Launch	4.1×10^{-3}	0.045	0.040	1.9×10^{-4}	1.7×10^{-4}
EGA Reentry	6.3×10^{-7}	130	15	8.0×10^{-5}	9.8×10^{-6}
Total	5.0×10^{-3}	0.055	0.037	2.7×10^{-4}	1.8×10^{-4}

The mission actuarial risk without de-minimis is 2.7×10^{-4} incremental latent cancers. Since the consequence results are dominated by the late launch and Earth swingby reentry scenarios, this risk is applicable to the global population. For accidents near the launch site (phases 0 and 1), the probability of having more than one health effect is about 1 in 300,000 and the contribution to mission actuarial (consequence-weighted) risk is only 7.6×10^{-6} latent cancers without de-minimis. The largest contribution to the mission actuarial risk without de-minimis is 1.9×10^{-4} latent cancers from the late launch phases (3 through 8). Late launch accidents have about the same range of health effects as on-pad and early launch phase accidents, but are more likely to occur and therefore contribute more significantly to total mission risk. The second largest contribution to mission actuarial risk without de-minimis is 8.0×10^{-4} latent cancers from earth gravitational assist (EGA) reentry.

The risks posed to individuals from the Cassini mission are low when compared to those posed by other radiation sources. A typical background radiation dose to an individual is 0.3 rem/year. Assuming 0.0005 incremental cancer fatalities per person rem, the risk to an individual of developing a fatal cancer from a 50-year exposure to background radiation is 7.5×10^{-3} or 1 in 133. This estimated lifetime risk from background radiation is over 8 orders of magnitude greater than the 3.2×10^{-12} or 1 in 300 billion value for the Cassini mission segment with the highest average individual risk (early launch).

6.5 Cassini Uncertainty Analysis

The analysis results presented in the preceding sections are based on mathematical models that predict source terms and consequences associated with a wide spectrum of potential accidents. These models are based on experiments, observations, and known physical principals. The model inputs include variables related to timing, orientation, weather, and other stochastic processes. The model outputs are probability distributions and associated expectation values and percentiles of source term characteristics and consequence estimates.

Because of the limitations in the amount of data available and the level of understanding of the processes that occur, the models are not perfect. Output probability distributions significantly different from those presented in the preceding sections are, therefore, possible. Uncertainty analysis is undertaken to establish ranges within which actual distributions are expected. Uncertainty analysis also indicates those model parameters that have the strongest influence on predicted results. This permits potential model improvements to be prioritized in a rational manner.

The mathematical models used in the Cassini risk assessment were analyzed in order to distinguish sources of aleatory or stochastic variability (variables) from sources of epistemic uncertainty (parameters). Distributions were developed to describe state-of-knowledge uncertainties in the parameters. The parameters were held at their best estimate values to obtain the results discussed in preceding sections. The results obtained in this manner are also called "best estimates" meaning they are neither optimistically nor conservatively biased. For the uncertainty analysis the model parameters as well as input variables were selected by sampling from their distributions. Comparison of variability plus uncertainty results with the variability-only results yields uncertainty multiplier (sample estimate divided by best estimate) distributions for the model outputs.

Uncertainty analyses were performed on two consequences: 50-year health effects (without de-minimis) and 50-year collective dose (without de-minimis). The results of the analyses are: (1) probability distributions for these two consequence measures at various confidence levels, and (2) expectation values (averaged over variability) of the consequences at various confidence levels. Figures 10.22 and 10.23 show total-mission complementary cumulative distribution functions (*ccdf*s) for latent cancers and collective dose, respectively, at the 5%, 50%, and 95% confidence levels. The near coincidence of the variability-only *ccdf* and the 50% confidence level *ccdf* indicates that the variability-only result is indeed

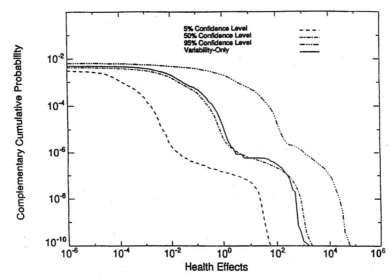

Figure 10.22 Total mission 50-year health effects (without de-minimis) from uncertainty analysis. *Courtesy of the U.S. Department of Energy.*

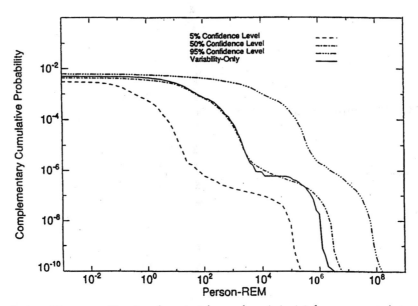

Figure 10.23 Total mission 50-year collective dose (without de-minimis) from uncertainty analysis. *Courtesy of the U.S. Department of Energy.*

unbiased. The width of the 5% to 95% interval depends on the consequence level of interest. For example, for one latent cancer, the exceedance probability goes from 5% level of about 10^{-7} to a 95% level approaching 3×10^{-3}. The uncertainty in the maximum number of latent cancers goes from a 5% level approaching 100 to a 95% level approaching 100,000.

Figure 10.22 indicates the uncertainty associated with latent cancer risk (without de-minimis). Both best estimate (variability-only) and 5%, 50% and 95% confidence-level expectation values are listed. Figure 10.23 gives similar results for 50-year collective dose (without de-minimis). Total mission latent cancer risk is 2.7×10^{-4} from the variability-only analysis and 2.1×10^{-4} at the 50% confidence level. The 95% confidence level is about 2 orders of magnitude higher than the best estimate, and the 5% confidence level is about 2 orders of magnitude less than the best estimate. The mission segment with the largest contribution to risk is the late launch for the best estimate, 50% confidence level, and 95% confidence level. Short-term reentry during the EGA swingby is the largest contributor to risk at the 5% confidence level.

Regression analyses of the variability plus uncertainty results were undertaken to determine the dominant sources of uncertainty. The majority of the uncertainty arises from the consequence models, which are based more on observational data than experimental data. The major parameters contributing to uncertainty were health-effects related (dose conversion factor and health effects estimator) and fuel-transport related (particle deposition velocity and particle resuspension factors).

7.0 Additional Space Nuclear Risk Perspectives

To provide additional perspective, results from the Cassini risk assessment are compared to results from other risk assessments in Table 10.10. Nominal (mean or point estimate) person rems per mission are shown for the Galileo, Ulysses, and Cassini RTG missions, four proposed SP-100 reactor missions, and a proposed Topaz II (Enisey) reactor mission. In addition, results from five commercial reactor risk assessments are included.

The nature of the mission often determines the particular accident scenarios that dominate the risk. In particular, accidents that disperse radioactive material during launch and assent or early reentry from a low orbit are often risk dominant. For sufficiently high orbital missions, the times required for orbital decay preclude any significant risk contribution from reentry accidents. Consider the Galileo and Ulysses RTG missions. The Galileo mission involves a Shuttle launch into low Earth orbit followed by a chemical boost to a Venus-Earth-Earth Gravity Assist (VEEGA) trajectory and final orbit around Jupiter [18]. The dominant accident scenario (98% of nominal risk) involves loss of solid rocket booster thrust in the first 30 seconds after launch, followed by a fireball and ground impact. The Ulysses mission involves a Shuttle launch into a low Earth orbit followed by a chemical boost into a Jupiter fly-by and solar pole overflight trajectory. The dominant accident scenario (64% of nominal risk) involves a solid rocket booster case rupture 105 to 120 seconds after ignition with large fragments impacting the RTGs causing a release of plutonium dioxide fuel at altitude. Two different estimates of the Ulysses risks were made, one by the NUS Corporation and one by the Interagency Nuclear Safety Review Panel [19, 20]. Results from both of these risk assessments are included in Table 10.10.

Risk assessments were performed for four proposed SP-100 reactor missions. Risks associated with the first three missions were analyzed by the NUS Corporation [21]. First, for a Titan launch into a >300-year orbit, the dominant accident scenario (80% of nominal risk) involved a 292.7-year orbital decay after 7.3 years of reactor operation at 2400 kWth with a reactivity excursion upon land impact. Second, for a Shuttle launch into a low (300 km) orbit in preparation for a nuclear electric propulsion (NEP) mission to Neptune, the dominant scenario (91% of nominal risk) involved failure of electric propulsion and subsequent 80-day orbital decay after 2 days of reactor operation at 2400 kWth. Third, for a Shuttle launch into a 700 km (50-year) orbit, the dominant accident scenario was a 42.7-year orbital decay after 7.3 years of reactor operation at 2400 kWth. General Electric performed the risk analysis for the fourth SP-100 mission, which involved a Shuttle launch into a >300-year orbit. For this mission, the dominant accident scenario (100% of nominal risk) involved a 100-day orbital decay after 7 years of reactor operation at 2400 kWth due to an errant boost to a high orbit [22].

A preliminary risk assessment was performed for a Nuclear Electric Propulsion Space Test Program (NEPSTP) mission, which was proposed to evaluate nuclear electric propulsion for orbital transfer applications using the Russian TOPAZ II nuclear reactor as a power source [23]. The Topaz II/NEPSTP spacecraft would be launched by a medium class launch vehicle into a high (>5200 km) orbit followed by 2 years nuclear electric propulsion to a 50,000 km disposal orbit. The very high initial orbit effec-

Table 10.10 Space versus Commercial Nuclear Power Consequence-Weighted Risks

Mission or Plant	Person-rem/ Mission	Notes	Ref.
		RTG-Powered Space Missions	
Galileo	0.13	Shuttle launch to low Earth orbit followed by chemical boost to VEEGA trajectory and final orbit around Jupiter, possible loss of solid rocket booster thrust in first 30 s followed by fireball and ground impact.	GE/NUS
Ulysses	3.1×10^{-4}	Shuttle launch to low Earth orbit followed by chemical boost into a Jupiter fly-by and solar pole overflight trajectory, possible solid rocket booster case rupture 105 to 120 s after ignition causing large fragments to impact RTGs.	NUS
Ulysses	0.062		INSRP
Cassini	0.57	Titan IV launch to parking orbit, Centaur boost to VVEJGA trajectory to Saturn orbit, possible late launch reentry or Earth gravity assist reentry.	LMMS
		Reactor-Powered Space Missions	
SP-100 (HOM)[1]	3.0	Titan launch into >300-year orbit, possible reactivity excursion after orbital decay and land impact.	NUS
SP-100 (NEP)[2]	0.51	Shuttle launch into low (300 km) orbit, nuclear electric propulsion to Neptune, possible failure of electric propulsion and 80-day orbital decay after 2-day reactor operation.	NUS
SP-100 (LOM)[3]	0.59	Shuttle launch into 700 km (50-year) orbit, possible reactivity excursion after orbital decay and land impact.	NUS
SP-100 (GE)	18	Shuttle launch into >300-year orbit, possible 100-day orbital decay due to errant boost to higher disposal orbit.	
Topaz II NEPSTP[4]	4.4×10^{-3}	Launch into high orbit for nuclear electric propulsion tests, possible failure of anticriticality device given launch abort and ground impact.	SNL
		Commercial Plants (mean accidental person rem within 80 km for 40 years)	
Grand Gulf	21	Full power, system-internal initiators	NUREG-1150
Peach Bottom	2.1×10^4	Full power, system-internal, seismic and fire initiators	
Sequoyah	4.8×10^2	Full power, system-internal initiators	
Surry	2.0×10^3	Full power, system-internal, seismic and fire initiators	
Zion	2.2×10^3	Full power, system-internal initiators	

1. High Orbit Mission   2. Nuclear Electric Propulsion   3. Low Orbit Mission   4. Nuclear Electric Propulsion Space Test Program

tively precludes any early reentry accidents after reactor startup (the orbital decay time for the lowest orbit is millions of years). The dominant accident scenario (100% of nominal risk) involved preorbital impact and failure of the anti-criticality device to prevent a critical excursion on impact.

For comparison, consequence-weighted risks (for collective doses) are also included in Table 10.10 for the five NUREG-1150 risk assessments for five commercial nuclear power plants (Grand Gulf, Peach Bottom, Sequoyah, Surry, and Zion) [3]. All five risk assessments considered accidents initiated by failures of system components during full power operation. Risks associated with earthquakes and fires were only considered for Peach Bottom and Surry. Because space nuclear power risks are calculated for a complete mission, the commercial reactor risk results, which are stated per reactor critical year, were multiplied by the nominal 40-year life of a nuclear power plant. None of the five risk assessments considered startup, shutdown, or decommissioning risks. Even with the indicated scope limitations, the results provided in Table 10.10 clearly indicate that risks associated with space nuclear power missions are significantly less than commercial nuclear power risks (recall that Figure 10.5 shows that the risk for commercial nuclear plants are small compared to other human-caused risks). The person-rems listed for commercial nuclear power plants in Table 11.10 as well as the results presented in Figure 10.5 are consistent with the Nuclear Regulatory Commission's safety goals, which hold that nuclear power plant operation should not significantly add to individual or societal risks. For example, the risk to the population within 10 miles of a commercial nuclear power plant of cancer fatalities that might result from plant operation should be less than one tenth of one percent of the cancer fatality risk resulting from all other sources.

8.0 Chapter Summary

Probabilistic risk assessments of space nuclear missions identify accidents that could endanger the public health and safety, assess their probabilities, and quantify their public health consequences. Key outputs of PRAs are risk curves, which consist of complementary cumulative distribution functions for health and economic consequence measures. Event trees are used to delineate sequences of functional failure events that can lead to accidents. Fault trees and Boolean logic are used to delineate sets of component failures and other basic events that can cause the functional failures. Rules of probability are used to predict the functional failure probabilities from the minimal cut set probabilities. Probability distributions are often used to account for uncertainties in the inputs to PRA models. Using Bayesian techniques, such distributions reflect the state-of-knowledge regarding the various model inputs. Monte Carlo methods can then be used to assess uncertainties in PRA outputs. Treatments of model and completeness uncertainties tend to be more qualitative. Results from risk assessments of a number of space nuclear power missions indicate that space nuclear risks are limited by the inventory of radionuclides, the launch, orbital and space trajectories, safety features, and the brief time intervals during which members of the public could be exposed. Risks associated with space nuclear power missions are significantly less than commercial nuclear power risks and other human-caused risks.

Symbols

c, C	accident consequence measure	$P(E)$	probability of event E
$erf(x)$	error function	P_i	probability of i-th event or sequence of events
E	event E		
E'	complement of event E	R	consequence-weighted risk
$E[X]$	expectation value of random variable X	T	top event
EF	error factor	$Var[X]$	variance of random variable X
$f(x)$	probability density function	x, θ, u	a value of continuous random variables X
$F(x)$	cumulative distribution function, $P(X \neq x)$	x_i	i-th value of discrete random variable X
F_i	frequency	x_{50}	median of random variable X
$g(\theta, x)$	posterior probability density of parameter θ given observation x	X, Θ, U	random variables
		$Y(X)$	function of random vector X
n, N	number of observations, number of trials, or sample size	Y	sample mean
		z	$(x - \mu)/\sigma$
n_E	number of occurrences of event E		
P	probability	γ	1 − confidence level
p	true failure probability	$\delta()$	delta function
p_L	lower confidence limit	λ	parameter of exponential distribution
p_U	upper confidence limit	μ	mean of a random variable
p_1	lower Bayesian probability limit	σ	standard deviation
p_2	upper Bayesian probability limit	ξ	dummy variable

References

1. Center for Chemical Process Safety, *Hazard Evaluation Procedures: Second Edition With Worked Examples.* American Institute of Chemical Engineers, New York, ISBN 0-8169-0491-X, 1992.
2. U.S. Nuclear Regulatory Commission, *Reactor Safety Study: An Assessment of Accident Risks in U.S. Commercial Nuclear Power Plants.* WASH-1400, Oct. 1975.
3. U.S. Nuclear Regulatory Commission, *Severe Accident Risks: An Assessment for Five U.S. Nuclear Power Plants.* NUREG-1150, Washington DC, 1990.
4. Carlson, D. C., S. W. Hatch, R. L. Iman, and S. C. Hora, *Review and Evaluation of Wiggins' and SERA's Space Shuttle Range Safety Hazards Reports for the Air Force Weapons Laboratory.* SAND84-1579, Sandia National Laboratories, 1984.
5. Rogovin, M. and G. T. Frampton, Jr., Nuclear Regulatory Commission Special Inquiry Group, *Three Mile Island, A Report to the Commissioners and to the Public.* Rogovin, Stern & Huge law firm report (January 1980); summarized in *Nucl. Safety* **21**, 389, 1980.
6. Scheaffer, R. L., *Introduction to Probability and Its Applications.* The Duxbury Advanced Series in Statistics and Decision Science, ISBN 0-534-91970-7, PWSKENT Publishing Company, 1990.
7. McCormick, N. J., *Reliability and Risk Analysis: Methods and Nuclear Power Applications.* Boston: Academic Press, ISBN 0-12-482360-2, 1981.
8. Corynen, G. C., *Evaluating the Response of Complex Systems to Environmental Threats: The $\Sigma\Pi$ Method.* UCRL-53399, Lawrence Livermore National Laboratory, University of California, Livermore, CA, May 1983.
9. *Procedure for Testing Common Cause Failures in Safety and Reliability Studies.* U.S. Nuclear Regulatory Commission, NUREG/CR-4780, Jan. 1988.
10. Taylor, J. R., *An Introduction to Error Analysis, The Study of Uncertainties in Physical Measurements.* University Science Books, ISBN 0-935702-07-5, 1982.
11. Mann, N. R., R. E. Schafer, and N. D. Singpurwalla, *Methods for Statistical Analysis of Reliability and Life Data.* New York: John Wiley & Sons, 1974.
12. Bayes, T. "Essay Towards Solving a Problem in the Doctrine of Chances." *Philosophical Transactions*, Essay LII, p. 370–418, 1763.
13. Press, W. H., S. A. Teukolsky, W. T. Vetterling, and B. P. Flannery, *Numerical Recipes in FORTRAN, The Art of Scientific Computing.* 2nd ed., Cambridge University Press, ISBN 0 521 43064, 1992.
14. Iman, R. L. and M. J. Shortencarier, "A FORTRAN 77 Program and User's Guide for the Generation of Latin Hypercube and Random Samples for Use with Computer Models." NUREG/CR-3624, SAND83-2365, Sandia National Laboratories, Albuquerque, NM.
15. *Nuclear News*, Nov. 1997, p. 18.
16. Martin Marietta Technologies, Inc., a Lockheed Martin Company, *Cassini Titan IV/Centaur RTG Safety Databook.* Revision A, Report NAS3-00031, June 1996.
17. Jet Propulsion Laboratory (JPL), *Cassini Program Environmental Impact Statement Supporting Study, Volume 3: Cassini Earth Swingby Plan.* JPL Publication Number D-10178-3, Pasadena, CA, Nov. 18, 1993, with addendum dated Aug. 24, 1994.
18. General Electric Astro-Space Division Spacecraft Operations and NUS Corporation, *Final Safety Analysis Report for the Galileo Mission.* DOE/ET/32043-T26, Department of Energy, Washington, DC, 1988.
19. NUS Corporation, *Final Safety Analysis Report for the Ulysses Mission.* ULS-FSAR-006, Gaithersburg, MD, 1990.
20. Interagency Nuclear Safety Review Panel (INSRP), *Safety Evaluation Report for Ulysses.* INSRP 99-01, Washington, DC, 1990.
21. Bartram, B. W. and A. Weitzberg, *Radiological Risk Analysis of Potential SP-100 Space Mission Scenarios.* NUS-5125, NUS Corporation, Gaithersburg, MD, 1988.
22. General Electric, *SP-100 Mission Risk Analysis.* GESR-00849, GE Aerospace, San Jose, CA, 1989.
23. Payne, A. C. and F. E. Haskin, "Risk Perspectives for Potential Topaz II Space Applications." International Nuclear Safety Conference, CONF-940101, *American Institute of Physics*, 1994.

Student Exercises

Section 1

1. The financial cost of the accident sequences in Figure 10.1 are listed below.

	Probability	Financial Cost ($)
2	9.8×10^{-4}	1×10^{6}
3	9.9×10^{-6}	3×10^{7}
4	9×10^{-6}	5×10^{6}
5	1×10^{-6}	1×10^{9}

(a) Construct a risk curve. (b) Compute the consequence-weighted risk.

2. In the example depicted in Figure 10.1, assume that noble gas fission products are unaffected by the fission removal system. Construct a risk curve for the fraction of noble gas fission products released from containment. Compute the corresponding consequence weighted risk.

Section 2

3. Events A_1 and A_2 are mutually exclusive. The probability of event A_1 is 0.1. The probability of event A_2 is 0.9. The conditional probability of event B given event A_1 is 0.2. Events A_2 and B are independent. Compute (a) $P(A_1*A_2)$; (b) $P(A_1+A_2)$; (c) $P(A_1*B)$; (d) $P(A_1+B)$; (e) $P(A_2*B)$; (f) $P(A_2+B)$; (g) $P(A_1|B)$; (h) $P(A_2|B)$.

4. A certain component is supplied by two manufacturers. Manufacturer A supplies 30% of the components. The failure rate of components supplied by manufacturer A is 1 in 100. Manufacturer B supplies 70% of the components. The failure rated of components supplied by Manufacturer B is 1 in 50. A component selected at random from the warehouse fails. What is the probability that the component was provided by Manufacturer B?

Section 3

5. Assume that the five events in the example depicted in Figure 10.9 are independent. (a) Delineate the $2^5 = 32$ mutually exclusive combinations of success and failure events. (b) Compute the probability of each combination. (c) Indicate which combinations lead to the top event. (d) Calculate the exact top event probability assuming each of the basic events has a probability of 0.6.

6. Construct a fault tree in which the top event is success of current to the resistor in Figure 10.9. Use Boolean algebra to solve for the minimal cut sets.

Section 4

7. The failure rate of memory bits due to radiation damage in a particular orbit can be approximated by a Poisson distribution with a mean of 1.5 failures per day. (a) What is the expected number of failures in a 2 mission? (b) What is the probability of observing exactly 512 failures? (c) What is the probability of less than $2^8 = 512$ failures?

8. The time to failure for Stage 1 of a certain launch vehicle can be treated as a random variable that follows an exponential distribution with parameter $\lambda = 0.002$ s^{-1}. Stage 1 is designed to burn for 300 s. (a) Estimate the probability of launch vehicle failure during Stage 1. (b) Estimate the probability of failure within 100 s of ignition.

9. The minimum launch site temperature during January can be approximated by a normal distribution with a mean of 6°C and a standard deviation of 3°C. (a) What is the 95th percentile temperature? (b) What is the probability of freezing?

10. The failure probability of a certain type of valve can be approximated by a lognormal distribution with a mean of 0.01 and an error factor of 3. A hydraulic system contains two of these valves. Evaluate the mean and error factor of the distribution for the probability that both valves fail.

Section 5

11. Test data on a certain battery indicate 1 failure in 70 tests. (a) Find a 90% confidence interval for the failure probability. (b) Find a 90% Bayes probability interval for the failure probability.

12. Assume that the time-to-failure (T) of a pump has an exponential distribution with parameter λ. Further assume that Λ is a random variable with prior probability density function

$$g(\lambda) = 2\,e^{-2\lambda}.$$

Find the posterior probability density function $g(\lambda|t)$.

13. The cumulative distribution function plotted in Figure 10.15 is $F(x)=1 - e^{-x}$. What values of x would be obtained given the following pseudo-random numbers: 0.1045, 0.7563, and 0.9652?

14. The cumulative distribution function for the failure probability of event A is $F(p_A) = 5(1 - p_A)^4$. The cumulative distribution function for the failure probability of event B is $F(p_B) = 4(1 - p_B)^3$. Assume the two events are independent. (a) What is the expectation value of event $C = A + B$? (b) Evaluate P_C by Monte Carlo using a sample of size 10.

Chapter 11

Accident Consequence Modeling

F. Eric Haskin and Albert C. Marshall

The objective of this chapter is to provide the reader with an understanding of processes and phenomena that influence the radiological consequences of postulated accidents involving the release of radionuclides from space nuclear power sources. Simple consequence models are provided as well as a discussion of detailed models used in computer code consequence analysis.

1.0 Factors Influencing Accident Consequences

Radiological consequences quantified in analyses of space nuclear power accident scenarios typically include latent cancers, population dose, number of people whose radiation exposure exceeds selected levels, and land areas contaminated above selected levels (e.g., 0.2 Ci/m^2 of Pu-238). Although acute fatalities may be possible for some types of postulated space reactor accidents, for accident scenarios involving RTG power sources estimated radiation doses are generally insufficient to cause acute fatalities.

The analysis of radiological consequences involves modeling potential accident scenarios and physical processes that may affect radionuclide release, characterizing associated releases of radionuclides, and estimating transport and deposition of radionuclides in the environment. Ultimately, consequence analyses predict the resulting radionuclide contamination levels and associated radiation doses and health effects. Random variabilities as well as parameter and modeling uncertainties influence accident consequence estimates.

The actual doses received by an individual as a result of an accidental release of radioactive material depend on the following:

1. the source term; i.e., the quantities of specific radionuclides released, and their physical and chemical forms,
2. the physical and chemical environment,
3. the location of the release, in particular, the altitude of the release,
4. chemical and physical transformations that occur after the initial release,
5. the weather during and after the release, which determines the concentrations of airborne radionuclides and ground contamination, and
6. protective actions such as evacuation, sheltering, and decontamination.

Three basic altitude regimes are important to consequence analysis; i.e., the troposphere, the stratosphere, and the mesosphere. The troposphere is identified as the 10 km zone directly above Earth's surface, characterized by a relatively steady decrease in temperature with increasing altitude. The stratosphere extends from an altitude of 10 km to about 45 km. The temperature in the stratosphere increases with increasing altitude. The mesosphere, located in the altitude region between 45 km and about 80 km, is characterized by a relatively steady decrease in temperature with increasing altitude [1].

Gases and particles released within the troposphere are transported and dispersed in a manner that is strongly influenced by the local weather (temperature, wind field, atmospheric stability, and rainfall patterns), topography, and ground surface cover. Radioactive clouds released in the stratosphere (or radioactive clouds that rise to these altitudes by momentum or buoyancy) travel several thousand kilometers before ground deposition is complete. If released at very high altitudes within the mesosphere, small fuel particles will spread throughout the globe and may take several years to completely deposit on Earth's surface. Larger particles fall to Earth's surface in a relatively short time, depositing over a well-defined footprint. Worldwide weather and surface characteristics influence deposition patterns associated with high and very high level releases.

Section 2 of this chapter provides a brief review of exposure pathways, dose, and health effect models; these models are applicable to all exposure scenarios. Sections 3 through 5 present radioactive release and dispersion models. Section 3 discusses models applicable to accident scenarios in which radionuclide release and transport occur at elevations less than 10 km (troposphere). Sections 4 and 5 discuss models applicable to, respectively, accident scenarios in which radionuclide dispersion and transport occur at intermediate elevations (stratosphere) and accident scenarios in which releases occur at very high altitudes (mesosphere). The consequence models presented in this chapter are simple; however, they serve to illustrate concepts that are central to consequence analysis. Section 6 provides a brief description of computer codes that are commonly used to perform more sophisticated consequence calculations for safety analysis reports.

2.0 Pathways, Doses, and Health Effects Models

In this section, exposure pathways and simple models are discussed for cloud shine, skin contamination, ground shine, inhalation, and ingestion.

2.1 Exposure Pathways

As indicated in Figure 11.1, an individual can receive a radiation dose from a radioactive cloud or plume in several ways, which are commonly called pathways. Doses can be received externally from the radiation given off by the passing cloud or by ground contamination. Such doses are called cloud shine and ground shine, respectively. The dose due to radioactive particles that settle directly onto the skin or clothing of persons immersed in the cloud is called the skin dose. A radiation dose can also be received by inhaling the radioactive material in the plume; this is called inhalation dose. Some of the inhaled material may concentrate in particular organs such as the lungs or thyroid and thus become a special threat to those organs. Cloud shine, ground shine, and inhalation are collectively considered parts of the plume exposure pathway. Doses received from a particular radionuclide via the plume exposure pathways are generally proportional or related to the time-integrated air concentration. The time-integrated air concentration experienced at any fixed location or by any individual is, in turn, generally proportional to the quantity of the radionuclide released in the accident. If an exposed individual does not remain stationary, the appropriate air concentration integral is that experienced by the moving individual.

Longer-term pathways include inhalation of resuspended radionuclides; the ingestion pathway; and ground shine, which continues after passage of the radioactive cloud. Resuspension is the process by which particles deposited on the ground may become airborne due to the action of surface winds and from human or animal activities. Dose received from the ingestion pathway typically results from eating or drinking contaminated food or water. For some scenarios, hand-to-mouth contact can result in ingestion of contaminated soil. As in the case of inhaled material, ingested material can concentrate in various organs. Ingestion of milk is particularly important because radioiodine from a plume can contaminate grass eaten by dairy herds and can be greatly concentrated in the milk. Radioiodine can then concentrate in the thyroid gland of an individual drinking the contaminated milk. To model the inges-

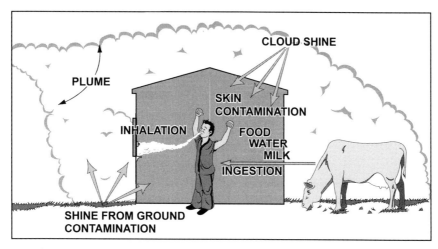

Figure 11.1 Radiation dose pathways. *Courtesy of U.S. Nuclear Regulatory Commission* [2].

tion pathway, concentrations of radionuclides deposited on crops, ingested by animals, and transported to bodies of water must be tracked. Such models are not discussed in detail in this text; however, they are described in the cited references.

Doses and resulting health effects associated with inhalation of radioactive material depend on whether the material is in the form of a gas or particulate. For particulate, the particle size is an important consideration and the chemical form of the radioactive material is important for both gases and particulate. The exposure pathway is also very important. For example, the health effects associated with ingested plutonia are generally much smaller than than those attributable to inhaled plutonia because most of the ingested plutonia passes quickly through the body and is excreted. Inhaled plutonia may reside in the lungs for some time before it is removed by blood absorption and by other mechanisms. For the ingestion pathway, only a small fraction of the ingested plutonia is absorbed by the blood, and only a small fraction of the radioactive material carried by the blood stream is subsequently deposited in body organs. Another example is the dose received from released radioactive iodine in nuclear reactor accidents. The dose typically results from inhalation of radioiodine, which is easily transported to the thyroid where it tends to stay. Doses received by people and associated health effects may be strongly influenced by protective actions such as evacuation, sheltering, and decontamination. To simplify the discussion, such protective actions are ignored in the models presented here.

2.2 Cloud Shine

In considering the dose received by an individual as a result of exposure to radiation emitted by an airborne radionuclide, the dose rate is proportional to the concentration χ (Bq/m^3) of the radionuclide in the ambient air. The corresponding time-integrated dose is proportional to the time-integrated air concentration χ_t (Bq·s/m^3). Consider an infinite cloud containing a uniform concentration χ of a particular radionuclide. Let f_j denote the fraction of radionuclide decays that result in the emission of the j-th radiation, which is emitted with average energy $\overline{E}_j$. Because the cloud is infinite in extent, every emitted radiation must be fully absorbed in the cloud. The dose rate in the cloud from the j-th radiation is the energy emission rate per unit volume $\chi\overline{E}_jf_j$ divided by the cloud density ς_{air}; i.e.,

$$\dot{D}_{\infty,j} = \frac{\chi\overline{E}_jf_j}{\varsigma_{air}}.$$

(11.1)

If a person is immersed in the cloud, the dose equivalent rate to a small volume of tissue will be the product of the dose rate in air, the ratio k of the energy absorption rates per unit mass in tissue and air, the radiation weighting factor W_R and a geometric attenuation factor g_a that depends on the energy of the emitted radiation and the organ for which the dose is sought,

$$\dot{H}_j = \frac{\chi \bar{E}_j W_R k f_j}{\varsigma_{air}} g_a(\bar{E}_j, d). \tag{11.2}$$

The attenuation factor g can be expressed as an empirical function of the emitted energy $\bar{E}_j$ and the distance into the tissue d. For alpha particles and low energy beta particles $g_a = 0$ because these particles cannot penetrate to the basal layer of the epidermis (at a depth of 70 μm), nor can they penetrate to the lens of the eye (at a depth of 3 mm). For more energetic beta particles and for low-energy gamma rays, g_a is approximately 0.5 near the body surface and decreases roughly exponentially with increasing depth. To be conservative, the skin dose attributable to betas and low-energy gammas emitted by airborne radionuclides is often computed with $g_a = 0.5$. For high-energy gamma rays, which are very penetrating, g_a is approximately equal to one throughout the body.

If the cloud contains a radionuclide that emits a single beta or gamma per disintegration $f_j = 1$. For these conditions, the radiation factor W_R and the ratio of energy absorption rates in tissue and air are also near unity. Taking the density of air to be 1.293×10^{-3} g/cm^3, the dose equivalent rate (Sv/s) can be approximated as

$$\dot{H} = \chi\left(\frac{\text{dis}}{\text{s·m}^3}\right)\bar{E}\,\frac{(\text{MeV})}{\text{dis}}\left(\frac{1.603 \times 10^{-10}\,\text{Sv/s}}{\text{MeV/g·s}}\right)(0.5)\left(\frac{\text{cm}^3}{1.293 \times 10^{-3}\,\text{g}}\right)\left(\frac{10^{-6}\,\text{m}^3}{\text{cm}^3}\right),$$

or

$$\dot{H} \approx 6.20 \times 10^{-14}\,\chi\bar{E}. \tag{11.3}$$

The factor of 0.5 in the preceding equation reflects $g_a = 0.5$ for betas and low-energy gammas and accounts for the semi-infinite extent of the cloud seen by a person at ground level for high-energy gammas.

The preceding approximation applies to the dose equivalent rate from external radiation only. Furthermore, for higher-energy gamma rays it is generally important to consider the finite extent of a cloud above ground level, any spatial variations of radionuclide concentrations within the cloud, and the contribution of scattered radiation to the dose a person at ground level would receive. The attenuation of penetrating radiations generally depends on the specific organ for which the dose equivalent is sought. The height of various organs with respect to the ground must also be considered. In general, however, the cloud shine dose rate attributable to an airborne radionuclide is still proportional to a representative concentration of the radionuclide and the total dose received is proportional to the time integrated air concentration seen by the exposed individual. Cloud shine dose conversion factors, provided by the Nuclear Regulatory Commission, are presented in Table 11.1 for the lung, bone marrow, and thyroid for several important radionuclides.

Table 11.1   Representative Cloud Shine Dose Rate Conversion Factor (Sv/s)/(Bq/m^3) [3]

	Kr-88	Sr-90	I-131	Xe-133	Cs-137	Pu-238
Lung	1.14×10^{-13}	0	1.41×10^{-14}	1.11×10^{-15}	2.17×10^{-14}	1.01×10^{-18}
Red Bone Marrow	1.16×10^{-13}	0	1.45×10^{-14}	7.27×10^{-16}	2.22×10^{-17}	4.55×10^{-19}
Thyroid	1.37×10^{-13}	0	1.77×10^{-14}	1.72×10^{-15}	2.73×10^{-14}	1.47×10^{-18}
ICRP-60 Effective Dose Equivalent	1.16×10^{-13}	9.51×10^{-17}	1.46×10^{-14}	1.19×10^{-15}	2.10×10^{-14}	2.01×10^{-18}

2.3 Skin Contamination

The preceding cloud shine model is appropriate for skin dose from an airborne noble gas like xenon, but it ignores the potential for deposition of radionuclides onto the skin. The surface concentration S_{skin} (Bq/cm^2) of radionuclides deposited on exposed skin is generally approximated as being proportional to the time-integrated air concentration to which the skin is exposed. The proportionality factor is called a deposition velocity v_d because it has units m/s from the ratio of Bq/cm^2 to (Bq/cm^3)·s. A typical deposition velocity is about 0.01 m/s. Surface deposition brings beta emitting materials next to the skin, typically resulting in higher skin doses than calculated above.

In practice, dose conversion factors that account for deposition on the skin are computed for each radionuclide. The resulting dose equivalent rate is proportional to the surface contamination level in Bq/cm^2. Dose from skin contamination by radionuclides is generally attributable to beta emitters. Alpha particles cannot penetrate the basal layer of the epidermis, and gamma rays are usually too penetrating to deposit much energy in the thin layer of skin. If the skin is contaminated with a beta-emitting radionuclide, the dose equivalent rate to the contaminated skin can be estimated by assuming that half of the betas will be emitted in directions away from the skin while the other half will be emitted in directions that intercept the skin; i.e., the flux of beta particles ϕ_β (β/cm^2·s) at the skin surface is equal to half the skin contamination $S_{skin} f_\beta$ (Bq/cm^2). The dose rate to the skin can then be computed using the basic approach used in Section 1.4 of Chapter 4 for gamma ray exposures. This methodology gives

$$\dot{H}_j = \phi_\beta \bar{E}_j \tilde{\mu}_{\beta,j}(1.603 \times 10^{-10}) \exp(-\mu_{\beta,j}d),$$

$$= 0.801 \times 10^{-10} S_{skin} \bar{E}_j f_{\beta,j} \tilde{\mu}_{\beta,j} \exp(-\mu_{\beta,j}d). \tag{11.4}$$

Here, $\mu_{\beta,j}$ and $\tilde{\mu}_{\beta,j} = \tilde{\mu}_{\beta,j}/\varsigma_T$ are the linear and mass attenuation coefficients for beta particles and ς_T is the density of the overlying tissue, respectively. The mass attenuation coefficient can be approximated as $\tilde{\mu}_{\beta,j} = 18.6(E_{\beta max,j} - 0.036)^{-1.37}$ cm^2/g.

An approximation to the preceding approach is often adequate. Over the range of beta energies from 0.2 to 2 Mev the dose rate from material deposited on the skin surface does not show significant variability at the critical depth of human skin. For betas emitted in this range of energies, the dose rate at the critical depth of skin is roughly 5.4×10^{-14} Sv per emitted beta per square meter. For a particular isotope i, let $f_{\beta,i}$ denote the fraction of decays that result in emission of a beta particle. The skin dose conversion factor is $(5.4 \times 10^{-14}) f_\beta$ Sv/s per Bq/m^2. If the time over which deposition occurs is short relative to the radionuclide half-life, the skin dose equivalent in Sv attributable to the isotope can then be expressed as

$$H_i = \chi_{t,i} v_d(5.4 \times 10^{-14}) f_{\beta,i} [1 - \exp(-\lambda_i \cdot t_r)]. \tag{11.5}$$

Here, $\chi_{t,i}$ is the time integrated air concentration of the isotope (Bq·s), v_d is the deposition velocity (m/s), λ_i is the decay constant of the isotope (s^{-1}), and t_r is the residence time of radioactive material on the skin.

2.4 Ground Shine

In estimating doses from ground-deposited radionuclides, it is generally adequate to consider only the gamma dose. The dose from β particles is ignored because β particles generally contribute only to skin dose and the contribution to skin dose from β-rays emitted at ground level is generally small because of β attenuation in air and clothing.

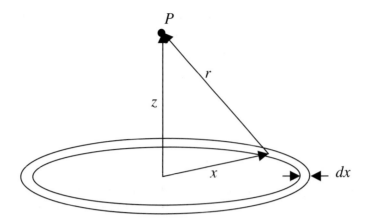

Figure 11.2 Ground shine model variables.

Consider an infinite plane surface containing S Bq/m^2 of a particular radionuclide. Let f_j denote the fraction of disintegrations that result in the emission of the j-th gamma ray with energy E_j. For an annular ring of radius x on the plane surface, as illustrated in Figure 11.2, the contribution to the gamma ray flux (gamma/m^2/s) at point P is

$$d\phi_{\gamma,j} = \frac{S f_j 2\pi x}{4\pi r^2} \, dx \, B(\mu_j r) \exp(-\mu_j r). \tag{11.6}$$

Here, $B(\mu_j r)$ is an empirical buildup factor that accounts for the contribution of scattered gammas to the flux at point P, and μ_j is the linear attenuation coefficient of the j-th gamma. Because $r^2 = x^2 + z^2$, where z is the elevation of point P with respect to ground, $x \, dx = r \, dr$, and the preceding equation becomes

$$d\phi_{\gamma,j} = \frac{S f_j}{2r} \, dr \, B(\mu_j r) \exp(-\mu_j r) \tag{11.7}$$

The contribution of the j-th type of gamma radiation to the total gamma flux at point P is then

$$\phi_{\gamma,j} = \frac{S f_j}{2} \int_x^\infty \frac{B(\mu_j r)}{r} \exp(-\mu_j r) dr. \tag{11.8}$$

For many empirical representations of the buildup factor, the integral can be evaluated analytically. The contribution of the j-th gamma to the dose rate at point P is proportional to the flux derived above, and the total gamma dose rate at point P is obtained by summing over all gammas emitted by radionuclides on the ground. Table 11.2 provides dose rate conversion factors derived in the manner described above for various radionuclides and organs of interest for an individual standing on the ground.

2.5 Inhalation

Unlike cloud shine dose, inhalation dose is attributable to radionuclides that decay within body organs. Computation of the dose from inhalation of radioactive substances is a two-step process. The first step is to compute the quantity of the radionuclide inhaled and its subsequent time-dependent distribution to the organs of the body using a biokinetic model. For the second step, the time-dependent quantity of the radioisotope in the various organs is used to compute the dose to the various organs.

Table 11.2 Representative Ground Shine Dose Rate Conversion Factor (Sv/s)/(Bq/m²) [3]

	Kr-88	Sr-90	I-131	Xe-133	Cs-137	Pu-238
Lung	—	0	2.97×10^{-16}	—	4.34×10^{-16}	1.05×10^{-19}
Red Bone Marrow	—	0	3.06×10^{-16}	—	4.42×10^{-16}	3.87×10^{-20}
Thyroid	—	0	3.75×10^{-16}	—	5.44×10^{-16}	1.29×10^{-19}
ICRP-60 Effective Dose Equivalent	—	1.48×10^{-18}	3.08×10^{-16}	—	4.29×10^{-16}	3.46×10^{-19}

Biokinetic Model

The quantity of radioactive material inhaled (in units of mass or activity) depends on the airborne concentration χ and the breathing rate R_b. The average breathing rate during an 8-hour working day, as defined for a standard man by the International Commission on Radiological Protection (ICRP) is 3.47×10^{-4} m³/s, while the average for a full 24-hour day is 2.32×10^{-4} m³/s. For short periods of time a person may breathe more rapidly, and breathing rates as high as 5×10^{-4} m³/s are sometimes used in accident consequence calculations.

The duration of inhaled radioactive substances in body organs depends upon the physical and chemical form of the radioisotope. Although inhaled radioactive gases are subsequently exhaled, a significant fraction of chemically reactive gases may be absorbed by the body. Inhaled radioactive particulate can deposit in the respiratory system, and is subsequently transported by ciliary action or absorbed by the blood and transported through the body. Ciliary action is the clearance of particulate by the movement of hair-like cells along the respiratory system walls. Radioactive materials absorbed by the blood can be deposited in other body organs. The ICRP has established a number of models for particle deposition, transport, blood absorption, blood-deposition to organs, and excretion for many possible radioactive materials. Here, we will use a more simplified model for predicting organ doses.

We define an organ-specific organ retention factor $R_T(t - \tau)$, where t is the time of interest and τ is the time of inhalation. The retention function can typically be expressed as

$$R_T(t - \tau) = f_T \exp[-\lambda_T(t - \tau)]. \tag{11.9}$$

Here, f_T is the fraction of the radionuclide that goes to the particular organ T and λ_T is the decay constant for biological clearance from that organ. The incremental activity $d\mathcal{A}_T$ in organ T of a radionuclide inhaled in time interval $d\tau$ about τ is the breathing rate $R_b(\tau)$ times the airborne concentration $\chi(\tau)$ times the time interval $d\tau$; i.e.,

$$d\mathcal{A}_T(t,\tau) = R_b(\tau)\chi(\tau)R_T(t - \tau) \exp[-\lambda(t - \tau)]d\tau, \tag{11.10}$$

where λ is the radioactive decay constant. For the case of exposure to a cloud of radioactive material in which the period of inhalation is long compared to the half-life or duration in the body, the activity in the organ at time t as a result of inhalation over the time interval $(0, t)$ is

$$\mathcal{A}_T(t) = \int_0^t R_b(\tau)\chi(\tau) \exp[-\lambda(t - \tau)]f_T \exp[-\lambda_T(t - \tau)]d\tau. \tag{11.11}$$

If the breathing rate and the airborne concentration are constant, integration of the preceding equation gives

$$\mathcal{A}_T(t) = \frac{R_b f_T \chi}{\lambda_{ef}}[1 - \exp(-\lambda_{ef}t)], \tag{11.12}$$

where λ_{ef} is the effective decay constant of the radionuclide in the organ, $\lambda_{ef} = \lambda + \lambda_T$.

For many scenarios, radioactive aerosols or gases are released as a single *puff* that is transported from the release site by wind. If the half-life and duration in the body is long compared to the puff duration at a particular location P, then the initial activity in organ T is

$$A_{T_o} = \int_0^{\tau_{\max}} R_b(\tau)\chi(\tau)f_T\,d\tau. \tag{11.13}$$

Here, $\tau_{\max}$ is the time required for the puff to pass the location P of the exposed individual. The concentration $\chi(\tau)$ at P increases from zero to a maximum concentration then decreases to effectively zero at time $\tau_{\max}$. The subsequent time dependence of the activity in organ T for puff inhalation is

$$A_T(t) = A_{T_o}\exp(-\lambda_{ef}t). \tag{11.14}$$

For this scenario, t is the time following the inhalation of the radionuclide. Equation 11.14 is a good approximation for PuO_2 particulate deposited in a compartment (region) of the lung. Note that Eqs. 11.9 through 11.14 assume that the inhaled radioisotope is instantly deposited in organ T. This is usually a good assumption for the lung or for radioactive substances rapidly absorbed by the blood and deposited in an organ. For the general case, however, accurate predictions require the inclusion of the time-dependent transport of the radioisotope from each lung compartment to organ T. The methodology is straightforward but tedious, and is usually analyzed using established computer codes.

Organ Dose

Most radionuclides decay by the emission of a number of alpha or beta particles, each of which may be accompanied by one or more gamma rays as well as internal conversion or Auger electrons. The method used to compute radiation dose from internalized radioactive materials will depend on the nature of the emitted radiation. Because the betas and alphas are charged particles, their penetration through tissue is very shallow, and essentially all their energy is typically absorbed within the organ containing the radioactive material. Gamma rays, however, do not have a finite range in tissue, and much of their energy is deposited in regions external to the organ. Computations of the fraction of the gamma ray energy retained within an organ have been performed for a variety of organs. These computations are used in the following way. Let $F_{j,TT'}$ denote the fraction of the energy E_j that is deposited in organ T' from organ T. Because dose is energy absorbed per gram, the dose rate attributable to a radionuclide within an organ is the activity of the radionuclide in Becquerel times the average energy absorbed in the organ per decay divided by the organ mass $m_{T'}$,

$$\dot{D}_{T'}(t) = A_T(t)\sum_j \frac{f_j F_{j,TT'}E_j}{m_{T'}}. \tag{11.15}$$

The dose equivalent rate, however, must consider the radiation-weighting factor W_R of each emitted radiation. It follows that the dose equivalent rate in organ T' associated with activity A_T of a radionuclide in organ T is

$$\dot{H}_{TT'}(t) = A_T(t)\sum_j \frac{f_j F_{j,TT'}E_j W_R}{m_{T'}} = A_T(t)\frac{E_{TT'}}{m_{T'}} = A_T(t)\tilde{E}_{TT'}. \tag{11.16}$$

The quantity $E_{TT'}$ is called the effective energy equivalent and $\tilde{E}_{TT'}$ is called the specific effective energy. Values of $E_{TT'}$ for some important nuclides are given in Table 11.3. For children, the time dependence of $F_{j,TT'}$ and m_T should be considered in the dose prediction. For a continuous plume, if the breathing rate and the airborne concentration are constant, Eq. 11.12 can be used with Eq. 11.16 to give

Table 11.3 Physical and Biological Data for Selected Radionuclides [3]

i Nuclide	Half-life (days) Radioactive	Biological	Organ	$E_{TT'}$ (MeV)	W_T Inhalation
Sr-90	1.1×10^4	1.3×10^4	Total body	1.1	0.4
		1.8×10^4	Bone	5.5	0.12
I-131	8.04	138	Total body	0.44	0.75
			Thyroid	0.23	0.23
Cs-137	1.1×10^4	70	Total body	0.59	0.75
		140	Bone	1.4	0.03
Pu-238	3.2×10^4	7.3×10^4	Bone	270	0.2
Pu-239	8.9×10^6	7.3×10^4	Bone	270	0.2

$$\dot{H}_{TT'}(t) = \frac{\tilde{E}_{TT'} R_b f_T \chi}{\lambda_{ef}} [1 - \exp(-\lambda_{ef} t)]. \tag{11.17}$$

When the duration of inhalation is much greater than the effective half-life $[\ln(2)/\lambda_{ef}]$, the exponential term in Eq. 11.17 becomes very small; consequently, the steady state, equilibrium dose rate for persons continually inhaling a radionuclide is

$$\dot{H}_{TT'}(t) = \frac{\tilde{E}_{TT'} R_b f_T \chi}{\lambda_{ef}}. \tag{11.18}$$

Equation 11.18 can also be expressed as

$$\dot{H}_{TT'}(t) = (DFI)R_b \chi, \tag{11.19}$$

where $DFI \equiv (\tilde{E}_{TT'} f_T)/\lambda_{ef}$ is called the *dose inhalation factor*.

The committed equivalent dose is obtained by integrating the dose rate expressed by Eq. 11.16 over time; i.e.,

$$H_{TT'}(t) = \int_0^t A_T(t)\tilde{E}_{TT'} dt. \tag{11.20}$$

For a puff,

$$H_{TT'}(t) = A_{To}(DFI)[1 - \exp(-\lambda_{ef} t)]; \tag{11.21}$$

and at $t = \infty$, $H_{TT'}(t) = A_{To}(DFI)$. The U.S Nuclear Regulatory Commission computes dose commitment factors over a period of 50 years, rather than for an infinite time. Consequently, the NRC inhalation dose factors are slightly smaller than the dose from the preceding equation for nuclides with long radiological and biological half-lives. For the case of a long inhalation duration and short effective half-life,

$$H_{TT'}(t) = [(DFI)R_b \chi]t. \tag{11.22}$$

The total dose rate to organ T' from all other organs is given by

$$\dot{H}_{T'}(t) = \sum_T \dot{H}_{TT'}(t). \tag{11.23}$$

Table 11.4   Representative Inhalation Dose Conversion Factor (Sv/Bq) for Chronic Exposures [3]

	Kr-88	Sr-90	I-131	Xe-133	Cs-137	Pu-238
Lung	4.22×10^{-12}	3.42×10^{-9}	6.56×10^{-10}	3.72×10^{-13}	8.80×10^{-9}	3.19×10^{-4}
Red Bone Marrow	3.67×10^{-13}	3.05×10^{-7}	6.26×10^{-11}	1.69×10^{-13}	8.30×10^{-9}	5.78×10^{-5}
Thyroid	3.54×10^{-13}	2.33×10^{-9}	2.91×10^{-7}	1.38×10^{-13}	7.92×10^{-9}	3.85×10^{-10}
ICRP-60 Effective Dose Equivalent	8.34×10^{-13}	4.57×10^{-8}	1.47×10^{-8}	1.75×10^{-13}	8.49×10^{-9}	6.13×10^{-5}

From Chapter 4 we recall that the effective dose rate accounts for the organ doses and organ sensitivities of all organs. Thus the effective dose rate is given by

$$\dot{H}_E(t) = \sum_{T'} \dot{H}_{T'}(t)\, W_{T'}, \tag{11.24}$$

where $W_{T'}$ is the organ-weighting factor. The appropriate weighting factors for affected organs are also provided in Table 11.3.

Table 11.4 provides factors, for chronic inhalation, to convert from inhaled activity (Bq) to effective dose equivalent and to dose equivalent for various organs. The doses obtained from these conversion factors are suitable for estimating chronic health effects such as latent cancers.

2.6 Ingestion

Radioactive substances may be ingested by eating contaminated food, drinking contaminated water, or by hand-to-mouth transfer from contaminated hands. In addition, inhaled radioactive particulate may be transported by ciliary action from the lungs to the throat where it may be swallowed. The standard ICRP model assigns rate constants for the passage of ingested radioactive material from the stomach and sequentially to the small intestines, upper large intestines, and the lower large intestines. In the small intestines, a fraction of the radioactive material f_1 is absorbed by the blood, where

$$f_1 = \frac{\lambda_B}{\lambda_B + \lambda_{SI}}. \tag{11.25}$$

Here, λ_B is the rate constant for blood absorption and λ_{SI} is the rate constant for transport from the small intestine to the upper large intestine. The rate constants for the GI tract are presented in Table 11.5.

For chronic ingestion at constant rate R_i(g/s), the time-dependent mass in the GI organs will quickly reach equilibrium concentrations. The time-dependent masses of a radioactive substance in all GI tract organs $m_g(t)$ are predicted by equations of the form

Table 11.5   GI System Transport Rates

Organ	Symbol	To	λ_t (d^{-1})
Stomach	ST	Small intestines	24
Small Intestines	B	Blood	0.088
	SI	Upper large intestines	6
Upper Large Intestines	ULI	Lower large intestines	1.8
Lower Large Intestines	LLI	Feces excretion	1

Table 11.6 Representative Ingestion Dose Conversion Factor (Sv/Bq) [3]

	Kr-88	Sr-90	I-131	Xe-133	Cs-137	Pu-238
Lung	0	1.33×10^{-9}	1.02×10^{-10}	0	1.27×10^{-8}	8.64×10^{-14}
Red Bone Marrow	0	1.75×10^{-7}	9.44×10^{-11}	0	1.32×10^{-8}	1.27×10^{-8}
Thyroid	0	1.33×10^{-9}	4.75×10^{-7}	0	1.26×10^{-8}	7.99×10^{-14}
ICRP-60 Effective Dose Equivalent	0	2.73×10^{-8}	2.39×10^{-8}	0	1.33×10^{-8}	9.45×10^{-9}

$$m_g(t) = \frac{R_i}{\lambda_g}[1 - \exp(-\lambda_g t)], \tag{11.26}$$

where the subscript g refers to a particular organ in the GI tract. For acute ingestion, the time-dependent mass in the stomach is

$$m_{ST}(t) = m_0 \, e^{-\lambda_{ST} t}. \tag{11.27}$$

Here, m_0 is the quantity of the ingested radioactive substance. Writing the differential equation for the rate of change of the mass in the small intestines and solving for the time dependence of the mass in the small intestines give

$$m_{SI}(t) = \frac{\lambda_{ST}}{(\lambda_{SI} + \lambda_B - \lambda_{ST})}[e^{-\lambda_{ST} t} - e^{-(\lambda_{SI} + \lambda_B) t}]. \tag{11.28}$$

The time-dependent radioisotope masses in the upper and lower large intestines are obtained in the same manner. Table 11.6 provides factors used to convert from ingested activity (Bq) to dose equivalent for various organs and to effective dose equivalent.

2.7 Health Effects

In considering releases from reactor accidents, both doses to the bone marrow and thyroid are important. Dose to the bone marrow (mostly from cloud shine and ground shine) is a dominant cause of potential early health effects resulting from accidents for reactors that have been operated at high power levels; however, this scenario is improbable for most missions involving space reactors. Thyroid dose is important because inhalation or ingestion of small amounts of radioiodine can result in damage to the thyroid. Unlike the bone marrow, radiation exposure of the thyroid will not be fatal in the short term in most cases. Nonetheless, radiation exposure of the thyroid presents an increased risk of death due to thyroid cancer. In considering RTG accidents, the potential for early health effects is generally negligible, and doses resulting from inhalation of PuO_2 tend to dominate the potential for latent health effects. Chapter 4 provides a discussion of the possibility of radiation-induced cancer induction and radiation-induced birth defects.

3.0 Low Altitude Releases and Transformation Processes

Radioisotope power sources are designed and tested to assure integrity of containment barriers during accidental reentry. The rugged RTG design also provides containment of the radioactive fuel during many postulated launch pad explosions and fires accidents. However, for some postulated fire and explosion accidents, release of radioactive materials may be possible and an analysis must be performed to determine the potential consequences.

The specific activity of unirradiated uranium fuel used in space reactors is much smaller than that for a typical RTG. Nonetheless, an analysis is performed to determine the effects of accidental low-altitude release of unirradiated fuel. Space missions typically preclude reactor operation until a safe, stable orbit or trajectory has been established; thus, post-reactor-operation fire, explosion, and impact accidents are generally not credible. However, some accident scenarios can be postulated that could result in the release of fission product gases and radioactive particulate (generated during reactor operation). An analysis is often performed to determine the radiological consequence of accidental fission product release for these postulated space reactor accidents.

Release characteristics depend on the type of space nuclear system (RTG or reactor) and the type of postulated accident. Low-altitude release and dispersion of fission products for space reactors are similar to the release and dispersion of fission products from postulated terrestrial reactor accidents. Terrestrial reactor accident analysis methods are well established and discussed in many texts and reports. As a consequence, the focus of low-altitude release in this chapter will be on the release and dispersion of RTG fuel particulate.

3.1 Fire Release Considerations

In this section, some of the principal considerations relating to RTG fuel release characteristics are discussed. The following discussion is intended to provide some idea of the complexity of the accident environment and the type of analysis required. We assume here that the radioisotope fuel is typical U.S. RTG fuel; i.e., ^{238}PuO$_2$. This discussion addresses aerosol and particulate physics; plutonia vaporization, condensation, and thermodynamics; PuO$_2$ vaporization kinetics; structure response; soot generation; and dirt entrainment.

Aerosol and Particulate Physics

For the purpose of this discussion, particles in a cloud or fireball are divided into two types depending on their size; i.e., aerosols and "rocks." Aerosols are fine liquid or solid particles suspended in a gas. Fog, smoke, and mist are examples of aerosol-bearing gases. Aerosols have volume-equivalent sphere diameters of less than about 0.01 cm (100 µm), roughly the diameter of a human hair. Aerosols can agglomerate and aerosol settling is affected by air currents. Aerosols < 1 µm in diameter are also subject to Brownian motion. "Rocks" are larger particles that settle according to their terminal velocities in the surrounding gas and generally do not agglomerate.

The discipline of aerosol physics deals primarily with the agglomeration of aerosol particles. The agglomeration of two or more aerosols produces a larger aerosol, which settles more rapidly. Agglomeration, like vaporization and condensation, modifies the size distribution of radionuclide-bearing aerosol particles, which in turn influences their potential to cause health effects if inhaled. Particles of dirt, soot, and aluminum oxide may also be present and can serve as condensation sites for radioactive vapors or as agglomeration sites for radioactive aerosols. Rock particles consist only of plutonia and are assumed to not agglomerate. The heat transfer response of particles in both size classes must be modeled to allow calculation of vaporization and condensation of radioactive species.

Particle heat transfer for both aerosols and rock particles can be modeled using a lumped-capacitance approach in which spatial temperature gradients within the particles are ignored. This approach is justified because the resistance to heat transfer at the surface of the particles is much greater than the internal conductive resistance. Because aerosols have effective thermal time constants on the order of about 10 milliseconds or less, the temperatures of aerosols can be modeled using a quasi-static approach; that is, aerosol particles are assumed to instantaneously reach the temperature of the surrounding gas. This approximation dispenses with the need to track the time history of a particle and its components as the particle changes in size due to vaporization, condensation, and agglomeration. Such a problem would be almost intractable. For rock particles, because of their larger size (and hence greater heat capacity) compared to aerosol particles, a quasi-static solution is not appropriate and the full transient governing heat transfer equation must be solved.

Plutonia Vaporization, Condensation, and Thermodynamics

The hypothetical launch accident envisages fragmentation of nuclear fuel into particles (including aerosols) and the dispersal of these particles into the fireball. The nuclear fuel particle will be suddenly exposed to a very high-temperature, chemically reactive environment. In this environment it is possible for the nuclear fuel to vaporize. The vapors may subsequently nucleate to form very fine aerosols or condense and contaminate other particles in the fireball. Both the fine nuclear-fuel-rich particles nucleated from the vapor and the coarser particles contaminated with nuclear fuel by condensation must be considered in the analyses of the consequences of a hypothetical launch accident.

A brief opportunity for the vaporization of plutonium dioxide arises when the particle temperature is high enough to create a thermodynamic driving force for substantial vaporization. The first step in the analysis of plutonium dioxide vaporization is to estimate the thermodynamic driving force. The thermodynamic driving force is a function of both the temperature and the chemical composition of the fireball gases. The second analysis step requires an evaluation of the vaporization kinetics in response to the thermodynamic driving force. The rate of plutonium dioxide vaporization depends on the plutonium dioxide fragment size, heat transport to the fragment, and mass transport away from the fragment.

Plutonium dioxide is not a stoichiometric compound. It is most accurately designated as $PuO_{2-x'}$, where x' assumes temperature-dependent values as large as about 0.35. The stoichiometry of plutonium dioxide, in recent U.S. RTGs, is adjusted to minimize oxygen loss at the normal operating temperature. The fragments of plutonium dioxide dispersed into the fireball during an accident can be expected to have initial compositions ranging from $PuO_{1.96}$ to $PuO_{1.98}$. In the fireball, plutonium dioxide particles encounter an environment that is much hotter than the normal operating environment within the GPHS. The fireball atmosphere is chemically reactive and plutonium dioxide loses oxygen to the fireball atmosphere, thereby reducing the chemical potential of oxygen in the plutonia particulate. The changes in the stoichiometry of the plutonium dioxide can be expected to affect the vapor pressure, which is the thermodynamic driving force for vaporization. The chemical potential of oxygen in the fireball atmosphere is determined by the details of the combustion process and the extent to which air is entrained as the fireball rises.

Vaporization of plutonium dioxide is complicated by the existence of multiple plutonium-bearing species. Widely recognized are gaseous PuO_2, PuO, and Pu. A realistic estimate of the vapor pressure, which is the thermodynamic driving force for vaporization, must incorporate the effects of both temperature and the ambient chemical potential of oxygen; i.e.,

$$PuO_{2-x}(solid) + \frac{x}{2}O_2 \Leftrightarrow PuO_2(gas), \tag{11.29}$$

$$PuO_{2-x}(solid) \Leftrightarrow PuO(gas) + \frac{(1-x)}{2}O_2, \tag{11.30}$$

$$PuO_{2-x}(solid) \Leftrightarrow Pu(gas) + \frac{(2-x)}{2}O_2, \tag{11.31}$$

and
$$2PuO_{2-x}(solid) \Leftrightarrow Pu_2(gas) + (2-x)O_2. \tag{11.32}$$

Kinetics of PuO$_2$ Vaporization

Vaporization of plutonium dioxide from fragments dispersed in the fireball requires mass transport of vapor from the fragment surface. This mass transport comes about because of vapor diffusion enhanced by the relative motions of fragments and the fireball gases. Fragment sizes produced by the impact of a

GPHS in a hypothetical accident are expected to vary over a very broad range [4]. The convective enhancement of diffusion can then be brought on by the ballistic motions of very large fragments falling through the fireball or the natural convection of gas around small fragments suspended in the fireball gas.

Vaporization is an endothermic process; e.g., vaporization of PuO_2 requires 2234 J/g at 2000 K. Consequently, sustained vaporization requires heat transport to the fragment and heat loss due to vaporization results in a sustained temperature difference between the fragment and the surrounding gas. This temperature difference can affect the diffusive mass transport from the fragment surface. Either mass transport of vapor or heat transport to the fragment can limit the extent of vaporization from plutonium-dioxide fragments dispersed in the fireball. For the purposes of estimating vaporization rate, fragments are approximated as spheres and a quasi-steady vaporization rate is assumed to initiate at the instant the fragments are dispersed in the fireball.

The formation of vapor at the surface of the fragment is assumed to be controlled by surface processes approximated by Hertz-Knudsen vaporization (the vaporization in a vacuum) to an interfacial boundary at a vanishingly small distance from the geometrical surface of the fragment. Vapor diffuses from this interface across a boundary layer (of thickness δ) and into the bulk fireball atmosphere. With the quasi-static assumption, the vapor flux from the surface due to Hertz-Knudsen vaporization is equal to the vapor flux away from the fragment by convection-enhanced diffusion. A serial resistance to vaporization is thus hypothesized to exist. Inclusion of the Hertz-Knudsen resistance assures that at the very high temperatures possible in fireballs, physically unrealistic vaporization rates will not be calculated.

Aluminum Structure Response

Launch vehicles consist mainly of aluminum alloy structures, which have the potential of being vaporized by a fireball. The combustion of the vaporized aluminum would then lead to the formation of aluminum-oxide (Al_2O_3) particles. These particles would provide additional agglomeration and condensation sites in the fireball. It is therefore desirable to incorporate models of aluminum heat transfer, vaporization, and combustion in a fireball analysis.

As with the particle heat transfer models, a lumped-capacitance model is adequate to estimate the transient temperature response of the aluminum alloy structures. If the temperature of an aluminum alloy structure is sufficiently high (greater than about 1300 K), aluminum vaporization from the surface into the fireball will occur with subsequent combustion of the vapor to form aluminum oxide particles.

Soot Generation and Dirt Entrainment

The presence of soot in the fireball provides additional sites for condensation and agglomeration and thus can alter the final plutonium size distribution. The actual amount of soot generated from large-scale combustion almost always exceeds the theoretical amount of soot. Thermodynamically, soot should only form when the carbon-to-oxygen (C/O) ratio is greater than one. Experimentally, the limits of soot formation are usually equated with the onset of luminosity and this usually occurs when the C/O ratio is about 0.5 [5]. This limit is called the critical C/O ratio. Many factors affect the generation of soot such as localized nonstoichiometric regions due to hydrodynamic effects, localized pressure and temperature regions, the presence of diluents, the presence of nitrogenous species, and the presence of metals. Factors affecting soot generation are discussed in References [6 and 7].

The entrainment of dirt and sand into the fireball directly affects the fireball emissivity, particle agglomeration, particle heat transfer, and plutonium condensation processes. Thus dirt entrainment can appreciably alter the size distribution of plutonium-bearing particles. A significant body of literature deals with dirt entrainment models. Such models consider adhesive and cohesive forces between the particles and the ground. These forces are functions of many variables such as humidity, sunshine, particle size and roughness, chemical composition, surface tension, intermolecular and electrostatic attractions, and local flow velocity vectors. Experimental measurements show adhesive forces that vary over eight orders of magnitude for particles in the 10 to 100-μm-diameter size range for various surface conditions, materials, and humidity levels [6]. In general, the models have many adjustable parameters and do not have reliable predictive capability. Complicating the problem is the potential for crater formation for several launch-abort scenarios. Therefore, the issue of dirt entrainment is generally modeled parametrically.

3.2 Meteorology

In the absence of significant heat transfer with the ground or between adjacent layers of air, the temperature in a well-mixed atmosphere decreases linearly with altitude at a rate of about 1 K/100 m. This is called the adiabatic lapse rate (or adiabatic temperature distribution) because it is derived by treating the expansion of air with altitude as an adiabatic expansion [8]. As indicated in Figure 11.3, other temperature distributions such as isothermal, superadiabatic, and inversions may exist over particular ranges of altitudes. The actual temperature profile at any time is determined by a number of factors including heating and cooling of Earth's surface, the movements of large air masses, the existence of cloud cover, and the presence of large topographical obstacles. For example, on clear days with light winds, superadiabatic conditions may exist in the first few hundred meters of the atmosphere due to the heat transferred to the air from the hot surface of the Earth. Conversely, on a cloudless night, when the Earth radiates energy most easily, Earth's surface may cool down faster than the air immediately above it, resulting in a *thermal inversion*.

The degree to which pollutants are dispersed in the atmosphere depends to a large extent on the atmospheric temperature profile. Consider the case of dispersion in a superadiabatic atmosphere. If a small parcel of contaminated air is released at some altitude *h* and at the same temperature as the atmosphere, as indicated in Figure 11.4 (a), the parcel will remain in equilibrium at that point if not disturbed. Suppose, however, that a fluctuation in the atmosphere moves the parcel upward. The parcel will cool adiabatically as it rises; that is, the temperature of the parcel will follow the adiabatic curve shown by the dashed lines in Figure 11.4 (a). Because the surrounding superadiabatic atmosphere is cooler, the parcel is less dense than the surrounding atmosphere. This means the parcel becomes increasingly buoyant, causing it to move more rapidly upward. On the other hand, if the par-

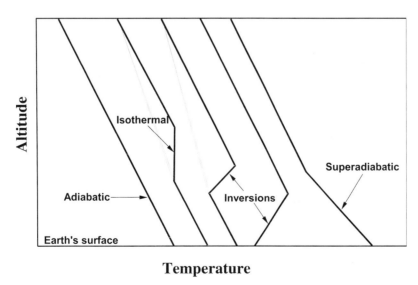

Figure 11.3 Examples of low-level temperature distribution in the atmosphere. *Courtesy of U.S. Nuclear Regulatory Commission* [2].

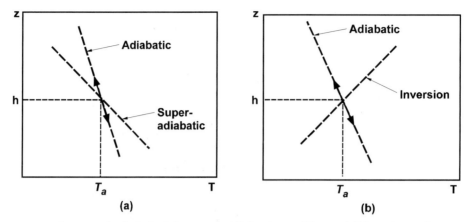

Figure 11.4 Movement of a parcel of air in (a) a superadiabatic profile, and (b) an inversion profile. *Courtesy of U.S. Nuclear Regulatory Commission* [2].

cel is pushed downward, its temperature will fall more rapidly than the surrounding atmosphere and it will become increasingly denser than the surrounding superadiabatic air. Thus, the parcel will accelerate downward. Clearly, the superadiabatic atmospheric conditions are inherently unstable and are favorable for dispersing pollutants. In contrast, if the parcel is released into an isothermal or inversion profile, as indicated in Figure 11.4 (b), a fluctuation upward will make it cooler and hence more dense than the surrounding atmosphere, tending to return the parcel to its original position. Similarly, a downward fluctuation will make the parcel hotter and more buoyant than the surrounding air. These conditions will also tend to return the parcel to its equilibrium point. Atmospheres characterized by isothermal or inversion profiles are therefore said to be stable. This is unfavorable for pollutant dispersal.

Frequently, the parcel is hotter than its surroundings when released, and it will initially rise due to its greater buoyancy. Various types of dispersal patterns can be observed depending on the conditions in the surrounding atmosphere, as illustrated in Figure 11.5. Plumes emitted into an inversion layer (stable atmosphere) disperse horizontally much more rapidly than they disperse vertically (vertical dispersion is inhibited in an inversion layer). Therefore, the plume spreads out horizontally but not vertically, which produces a fan shape when viewed from below (fanning). If a hot plume is emitted into an unstable atmosphere that is capped by an inversion layer, the plume rises to the inversion layer and then spreads rapidly downward, fumigating the ground below (fumigation). Plumes emitted into an uncapped unstable atmosphere tend to break up because vertical displacements of plume parcels are enhanced (looping). Plumes emitted into a neutral atmosphere (lapse rate equal to the adiabatic lapse rate) are dispersed smoothly both vertically and horizontally, and therefore have a conical profile in the crosswind direction (coning). Plumes emitted into a neutral layer that overlies an inversion layer can spread upward but not downward (lofting).

It is possible to estimate the stability conditions in the lower atmosphere by simply measuring the temperature at two or more heights on a meteorological tower. The slope of the temperature profile can then be compared by dividing the temperature difference ΔT by the difference in height Δz of the measurements. Based on experimental data on atmospheric dispersion, stability regions are often divided into the seven stability classes listed in Table 11.7, depending on the indicated ranges of $\Delta T/\Delta z$. Other meteorological conditions that can have a strong impact on atmospheric dispersion or ground contamination include wind speed, precipitation and humidity. Data on these factors are also measured on the meteorological tower. The significance of such factors is discussed in the following sections.

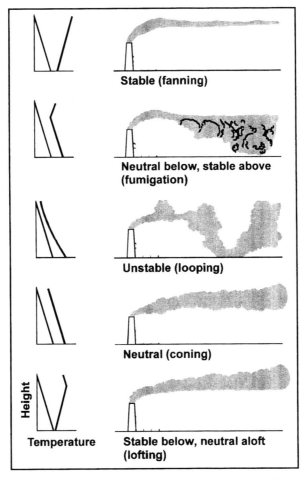

Figure 11.5 Various types of smoke plume patterns. *Courtesy of U.S. Nuclear Regulatory Commission* [2].

3.3 Dispersion and Deposition at Low Altitudes

Materials released into the atmosphere disperse as they are transported downwind. Dispersion causes released materials (droplets, light particles, aerosols, and gas molecules) to move away from the cloud centerline in a random series of steps. As a result, material concentrations tend to assume bell-shaped normal (or Gaussian) distribution in both the vertical and horizontal directions. The characteristic width of these distributions is the standard deviation, which is also called the dispersion coefficient. The rate of spreading (increase in the dispersion coefficients) depends on wind speed and atmospheric stability. Spreading is more rapid for greater wind speeds and unstable meteorological conditions. The rates of spreading in the vertical and horizontal directions generally differ.

Table 11.7   Relationship between Pasquill Class and $\Delta T/\Delta z$ [9]

Pasquill Stability Class	$\Delta T/\Delta z$ (K/100 m)
A—Extremely Unstable	$\Delta T/\Delta z \leq -1.9$
B—Moderately Unstable	$-1.9 < \Delta T/\Delta z \leq -1.7$
C—Slightly Unstable	$-1.7 < \Delta T/\Delta z \leq -1.5$
D—Neutral	$-1.5 < \Delta T/\Delta z \leq -0.5$
E—Slightly Stable	$-0.5 < \Delta T/\Delta z \leq 1.5$
F—Moderately Stable	$1.5 < \Delta T/\Delta z \leq 4.0$
G—Extremely Stable	$4.0 < \Delta T/\Delta z$

Models of atmospheric dispersion range in complexity from simple to sophisticated. Perhaps the simplest model is the straight-line Gaussian plume model. This model assumes materials are released continuously at a constant rate and transported from their release point by wind blowing in a fixed horizontal direction at a constant speed. The model also assumes a flat terrain and invokes a Gaussian-shaped spreading of the resulting plume, due to turbulent diffusion, with downwind distance. This spreading is characterized by empirically determined dispersion coefficients in the vertical and cross wind directions. The straight-line Gaussian plume model is derived in this section based on a superposition of a series of continuously emitted puffs all transported by a wind of fixed speed and direction.

For postulated space nuclear power accidents, radioactive material is most frequently emitted in puffs rather than continuously. The straight-line Gaussian plume model can be used to represent puffs as plumes of finite length; however, the Gaussian puff-trajectory model provides a better representation by treating the puff as an elliptical cloud transported by a wind that can periodically change directions. Dispersion and deposition calculations begin when the puff achieves thermal equilibrium with the surrounding atmosphere. By this time, with the exception of washout by rain, transformations to the physical and chemical form of radioactive materials in the puff are generally neglected. In the puff-trajectory model, successive displacements of the puff in the wind field and the cumulative horizontal travel distance are tracked. The wind field is modeled based on meteorological data.

The concentration of radioactive material within the puff is assumed to be distributed with Gaussian shape profiles in the horizontal and vertical directions. Consider a puff release of Q Bq of a radionuclide. Let x_c, y_c, and z_c denote the Cartesian coordinates of the puff center at time t with z_c denoting the altitude. The puff radionuclide concentration χ at location x, y, z can be expressed as

$$\chi(x,y,z) = Q[G_x(x - x_c)\, G_y(y - y_c)\, G_z(z - z_c)]. \tag{11.33}$$

Here, the Gaussian shape functions have the form

$$G_x(x - x_c) = \frac{1}{(2\pi)^{1/2}\sigma_x} \exp\left(\frac{-(x - x_c)^2}{2\sigma_x^2}\right), \tag{11.34}$$

$$G_y(y - y_c) = \frac{1}{(2\pi)^{1/2}\sigma_y} \exp\left(\frac{-(y - y_c)^2}{2\sigma_y^2}\right), \tag{11.35}$$

and
$$G_z(z) = \frac{1}{(2\pi)^{1/2}\sigma_z}\left[\exp\left(\frac{-(z_c - z)^2}{2\sigma_z^2}\right) + f \exp\left(\frac{-(z_c + z)^2}{2\sigma_z^2}\right) + R\right], \tag{11.36}$$

where R is the mixing term and f is the ground reflection parameter.

To simplify the equations, the x-axis is typically selected to coincide with the puff horizontal trajectory so that y_c can be set to zero. The parameters σ_x, σ_y, and σ_z are the along-wind, crosswind and vertical dispersion coefficients, respectively. These dispersion coefficients are empirically determined standard deviations of the respective Gaussian shape functions. The values of σ_x, σ_y, and σ_z increase monotonically with the total horizontal distance x traveled by the puff and decrease with atmospheric stability. The values depend on the type of release (continuous plume or instantaneous puff), the stability class, and the ground cover. Formulas developed by Briggs to describe stability classes A through F dispersion coefficients for continuous, open-country releases are summarized in Table 11.8. A seventh stability class, G, extremely stable, can be approximated as indicated in Table 11.8. In applying the Gaussian puff trajectory model, it is common to assume horizontal spreading is the same in the x (downwind) and y (crosswind) directions; i.e., $\sigma_y = \sigma_x$.

Table 11.8 Briggs Formulas for Continuous Release, Open-Country Conditions [10, 11]

Pasquill Stability Class	σ_y (m)	σ_z (m)
A—Extremely Unstable	$0.22\,x\,/\,(1 + 0.0001\,x)^{1/2}$	$0.20\,x$
B—Moderately Unstable	$0.16\,x\,/\,(1 + 0.0001\,x)^{1/2}$	$0.12\,x$
C—Slightly Unstable	$0.11\,x\,/\,(1 + 0.0001\,x)^{1/2}$	$0.08\,x\,/\,(1 + 0.0002\,x)^{1/2}$
D—Neutral	$0.08\,x\,/\,(1 + 0.0001\,x)^{1/2}$	$0.06\,x\,/\,(1 + 0.0015\,x)^{1/2}$
E—Slightly Stable	$0.06\,x\,/\,(1 + 0.0001\,x)^{1/2}$	$0.03\,x\,/\,(1 + 0.0003\,x)^{1/2}$
F—Moderately Stable	$0.04\,x\,/\,(1 + 0.0001\,x)^{1/2}$	$0.016\,x\,/\,(1 + 0.0003\,x)^{1/2}$
G—Extremely Stable	$(2/3)\,\sigma_y(F)$	$(3/5)\,\sigma_y(F)$

Contaminants released at low elevations tend to be restricted and mix vertically between ground elevation and the so-called mixing layer height $\mathcal{H}$. During daytime, the mixing layer height corresponds roughly to the inversion height of the planetary boundary layer. At nighttime, the mixing layer height is about a factor of five smaller than the daytime value. This difference is due in part to the presence of a stable temperature gradient above the mixing height. The terms f and R in the puff concentration equation account for reflection off of the ground and the top of the mixing layer. The factor f in the axial Gaussian profile is set to zero when the equilibrium puff altitude at the beginning of dispersion is above the mixing layer height, otherwise, f is set to unity. When f is set to unity, mixing between ground level and the mixing layer height $\mathcal{H}$ transforms the Gaussian axial shape to a uniform $(1/\mathcal{H})$ axial shape as dispersion proceeds. The mixing term R can be approximated as

$$R = f \sum_{k=1}^{4} \exp\left[\frac{-Z_k^2}{2\sigma_z^2}\right], \tag{11.37}$$

where

$$Z_1 = (2\mathcal{H} - z_c - z),$$
$$Z_2 = (2\mathcal{H} - z_c + z),$$
$$Z_3 = (2\mathcal{H} + z_c - z),$$
and
$$Z_4 = (2\mathcal{H} + z_c + z). \tag{11.38}$$

Example 11.1

At time $t = 0$, a released puff reaches equilibrium with the surrounding atmosphere at altitude $z_c = 100$ m. The wind is blowing in the x direction at speed $u = 2$ m/s under slightly stable meteorological conditions (Pasquill stability class E), and the mixing layer height $\mathcal{H}$ is 1000 m. Plot the normalized ground level concentration χ/Q at downwind distance $x = 4000$ m as a function of time.

Solution:

For the conditions stated the location of the puff center as a function of time is $x_c = ut = 2\,t$, $y_c = 0$, and $z_c = 100$ m. The coordinates of the exposure location are $x = 4000$ m, $y = 0$, and $z = 0$. Assume $\sigma_x = \sigma_y$. Assuming that the puff diameter at $t = 0$ is negligible, Table 11.8 gives $\sigma_y = 0.06\,x_c/(1 + 0.0001\,x_c)^{1/2}$, and $\sigma_z = 0.03\,x_c/(1 + 0.0003\,x_c)^{1/2}$.

The factor f is set to 1 to account for reflections, and the distances used in determining the reflection coefficient R are $Z_1 = Z_2 = 1900$ m and $Z_3 = Z_4 = 2100$ m. Evaluating the puff concentration

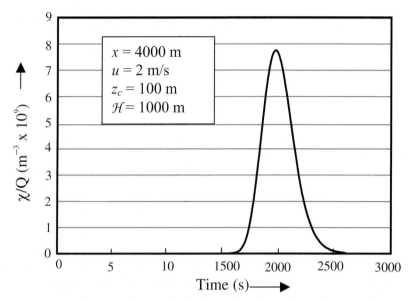

Figure 11.6 Normalized ground-level concentration as function of time (from Example 11.1).

equation at the exposure location as a function of time gives the values plotted in Figure 11.6. Note that the maximum concentration occurs at the time of closest approach; i.e.,

$$t = x/u = \frac{(4000 \text{ m})}{2 \text{ m/s}} = 2000 \text{ s}.$$

The inhalation and immersion doses that would be received by an individual standing in the path of the puff at coordinates x, y, z and the amount of material deposited on skin or ground at this location are proportional to the time-integrated airborne concentration χ_t (Ci/m³). Integration of the time-dependent air concentration at each exposure location can always be performed numerically; however, if the puff is not stationary and moves at constant altitude, the integral can be approximated analytically. The approximation is most accurate when (as in Example 11.1) the puff moves in a straight line and at constant speed u relative to the exposure location in question. The approximation is obtained by setting $x_c = ut$ and integrating $\chi(x,y,z)$ from $t = 0$ to infinity while holding σ_x, σ_y, and σ_z at their values at the point of closest approach; that is, where x_c equals the downwind coordinate x, which occurs when $t = x/u$. This gives the following equation for the relative concentration χ_t/Q, with units of s/m³,

$$\frac{\chi_t(x,y,z)}{Q} = \frac{G_y(y)G_z(z - z_c)}{2u}\left[1 + \text{erf}\left(\frac{x}{\sqrt{2}\sigma_x}\right)\right]. \tag{11.39}$$

Here the error function is defined as

$$\text{erf}(\xi) \equiv \frac{2}{\sqrt{\pi}}\int_0^\xi \exp(-\tau^2)d\tau.$$

Note that $\text{erf}(0) = 0$ and $\text{erf}(\infty) = 1$. In the preceding approximation, the integrated air concentration $\chi_t(x,y,z)$ is proportional to the released quantity Q and the instantaneous air concentration at the point of closest approach, $\chi(0,y,z-z_c)$. The time-integrated concentration $\chi_t(x,y,z)$ is inversely proportional to the wind speed u.

Setting the along-wind dispersion coefficient σ_x to zero in the preceding equation for χ_T/Q gives the straight line Gaussian plume model; that is,

$$\frac{\chi_T(x,y,z)}{Q} = \frac{\chi(x,y,z)\Delta t}{\dot{Q}\Delta t} = \frac{G_y(y)G_z(z - z_c)}{u}. \qquad (11.40)$$

Here, $\dot{Q}$ denotes the fixed continuous release rate that results in the plume and Δt denotes the duration of the release. Because along-wind dispersion is neglected, Δt is also the time duration spent by the plume over each downwind location. During the time interval in which the plume passes (x,y,z), the concentration $\chi(x,y,z)$ is constant. Before and after plume passage, $\chi(x,y,z)$ is zero.

Figure 11.7 provides a plot of χ_t/Q at ground level along the centerline of a straight-line plume trajectory. The puff height z_c is taken to be 30 m and the mixing layer height $\mathcal{H}$ is taken to be 100 m. Results are shown for Pasquill stability classes B, C, and D for a 3 m/s wind. χ_t/Q is also shown for a 1 m/s wind speed and stability class D. In all cases, at sufficient distances, χ_t/Q decreases rapidly. Ground level concentrations are low initially because of the time required for materials to disperse downward. The ground level concentration reaches a maximum and then begins to fall off very rapidly with distance. With the more unstable conditions (stability class B), the maximum of χ_t/Q occurs closer to the release point (within a few hundred meters), then drops rapidly to very low values. Under neutral conditions (stability class D), the peak χ_t/Q is located much further from the source. Given an explosion during ascent, ground level concentrations in populated offsite locations can, therefore, be greater under neutral or stable conditions than under unstable conditions. The impact of wind speed is illustrated by the results for neutral stability conditions. The relative concentration at a wind speed of 1 m/s is three times that at a wind speed of 3 m/s at all downwind distances.

The preceding discussion ignores several effects. Changes in wind speed and atmospheric stability cause the rate at which plume concentrations decrease with distance to change but do not cause the preceding generalizations to be seriously violated. Wind stagnation, however, can cause high local air or ground concentrations. Wind stagnation causes cloud shine, inhalation, and skin doses at the stagnation distance to increase at exposure locations near the stationary puff. In addition, prolonged stagnation can produce a hot spot on the ground because of the greatly increased time period during which deposition of radionuclides to the ground occurs near the stagnation point. Radioactive decay and dep-

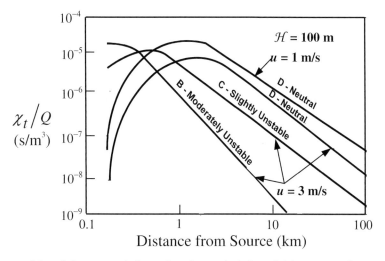

Figure 11.7 χ_t/Q at ground level for materials emitted at a height of 30 m, as a function of distance from the release point. *Courtesy of U.S. Nuclear Regulatory Commission* [2].

osition, both wet and dry, are each first order processes; i.e., their rates are proportional to the local concentration. For example the deposition rate $R_d(x,y)$ of material deposited to the ground may, in the absence of rain, be modeled as

$$R_d(x,y) = v_d \, \chi(x,y,0). \tag{11.41}$$

The deposition rate can have units of g/cm^2·s or Bq/cm^2·s, depending on how χ is defined. Here, v_d is called the dry deposition velocity (m/s), which models the combined effects of gravitational settling, impaction, and diffusion. Radioactive decay and dry deposition cause atmospheric concentrations to decrease more rapidly with distance. In effect, radioactive decay and dry deposition introduce additional exponential decay terms in the equations for airborne concentrations. Rain (i.e., wet deposition) also decreases plume concentrations and associated cloud shine, inhalation, and skin doses, but rain can result in rapid removal of material from a puff or plume and very high local ground concentrations (hot spots) distributed in very complex patterns.

In a 1981 study conducted at the Idaho National Engineering Laboratory, a nonradioactive tracer (SF$_6$) was released and the resulting air concentrations were compared with predictions made by various models to evaluate their potential use in emergency response situations. Figure 11.8 compares the predicted surface deposition using (a) a simple, straight-line Gaussian plume model; (b) a Gaussian-puff trajectory model; and (c) a more sophisticated wind field and topographic model with (d) the actual integrated air concentration profile observed for one of the tests. The Gaussian-puff trajectory model included the effects of wind shifts. The wind field and topographic model is used in the DOE's Atmospheric Release Advisory Capability (ARAC) program. Even the ARAC model could not reproduce what actually occurred. The initial transport of radioactive material from a site after it is released to the atmosphere may be dominated by local conditions (e.g., hills, valleys, lakes, and precipitation). A single source of weather information (e.g., a single meteorological tower) cannot give a definitive in-

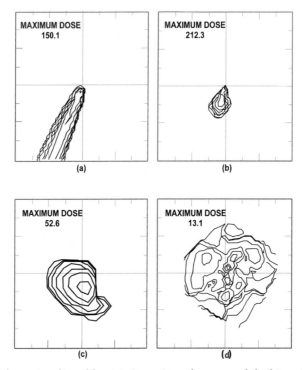

Figure 11.8 One-hour surface doses predicted by (a) Gaussian plume model, (b) puff-trajectory model, (c) complex numerical model, and (d) doses actually observed in test. *Courtesy of U.S. Nuclear Regulatory Commission* [2].

dication of winds away from the release point. Launch sites are typically located in very complex areas (e.g., on the coast), where wind direction and flows can vary considerably within a short distance. As an example, a 180 difference in wind direction could result from sea breeze effects at some coastal sites. Given these considerations, protective actions are taken in all directions near (within a few km) potential or actual accident sites.

It is clear that one should not expect close agreement when comparing dose projections with early field monitoring data. Dose projections should be viewed only as rough estimates, and knowledge of local meteorological conditions and trends (e.g., the winds shift every morning at about 9:00 a.m.) may be more important than more sophisticated modeling. The analyst needs to understand the problem, the models, and the results. Indiscriminate use of dose projection models without access to individuals who understand the unpredictability of local conditions can provide misleading input to protective action decision-making.

Deposited radioactive particulate may be resuspended by wind or human and animal activity. Civilians or animals in the vicinity may then inhale resuspended particulate long after the radioactive cloud has passed. Resuspended particulate may also settle on vegetation that may be eaten by humans or animals. Resuspension is discussed in Section 5.3.

4.0 Intermediate Altitude Releases

Some accident scenarios are predicted to result in the dispersion of radionuclides at intermediate altitudes between ~10 and ~45 km (stratosphere). Examples include an explosion during ascent of a spacecraft, an accident at the launch pad with a plume reaching mid-troposphere altitude, or disintegration of a spacecraft during reentry in the upper atmosphere. When released at intermediate altitudes, aerosols with 10 μm to 1000 μm diameters take from several days to several minutes to reach the ground. For such particles, the locations, sizes, and shapes of surface areas in which deposition occurs are largely determined by gravitation settling. Models to predict such settling and associated ground concentrations are discussed in this section based primarily on the GEOTRAP computer code [12]. Once the ground concentrations and integrated ground-level air concentrations are computed, global demographic data can be used to estimate radiological consequences. Short-term exposures are computed for the cloud shine, ground shine, and inhalation pathways. Long-term exposure includes inhalation of resuspended material, ground shine due to irradiation from ground deposition, and ingestion of contaminated foodstuffs. For aerosols less than approximately 10 μm in diameter, gravitation-settling times are relatively slow and rainout or cloud scavenging at low altitudes can be the dominant effect on the ground-level concentrations and depletion of particulate from the puff. For such small particles, the dispersion model described in Section 5 is more appropriate. Ground concentrations resulting from radioactive gases (rather than particulate) released or dispersed at intermediate altitudes are generally insignificant because gases are transported by diffusion only, that is, they have infinite settling times.

4.1 Terminal Velocity

When a particle is allowed to fall in a viscous fluid, the force of gravity acts in the downward direction. The buoyancy force and the drag force due to fluid motion around the particle act in the upward direction. A constant, terminal velocity is reached when the upward forces equal the downward force. The velocity of an aerosol particle relative to a surrounding gas is well approximated by the terminal velocity because aerosols have very small mobility (inertial) relaxation times. A 100-μm-diameter particle in air at standard conditions reaches its terminal velocity in less than 0.1 s while a 10-μm-diameter particle requires less than 1 ms.

If the particle is falling at its terminal velocity relative to a surrounding fluid, a force balance on the particle requires that

Gravitational Force = Buoyancy Force + Drag Force.

Let V_p denote the volume of the particle in question, ς_p denote its density, and v_t denote its terminal velocity. The gravitational force on the particle is simply $V_p\varsigma_p g$, which is the particle mass $V_p\varsigma_p$ times the gravitational acceleration g. The buoyancy force is $V_p\varsigma_f g$ where ς_f denotes the fluid density. The drag force may be expressed as the product of the intercepted area A_p, a characteristic kinetic energy per unit volume of fluid $(\varsigma_f v_t^2/2)$, and a dimensionless drag coefficient C_D. The force balance is then

$$V_p\varsigma_p g = V_p\varsigma_f g + C_D A_p \varsigma_f \frac{v_t^2}{2}. \tag{11.42}$$

Let d_p denote the volume-equivalent sphere diameter such that $V_p = \pi d_p^3/6$. Then, the intercepted area is the product of the cross-sectional area $\pi d_p^2/4$ of the sphere midplane and a dynamic shape factor ξ_p. If the particle is spherical $\xi_p = 1$, otherwise, $\xi_p > 1$. The force balance then becomes

$$\frac{\pi}{6} d_p^3 \varsigma_p g = \frac{\pi}{6} d_p^3 \varsigma_f g + C_D \xi_p \frac{\pi}{4} d_p^2 \frac{1}{2} \varsigma_f v_t^2. \tag{11.43}$$

Solving for the terminal velocity yields

$$v_t = S \left[\frac{4(\varsigma_p - \varsigma_f) d_p g}{3 C_D \varsigma_f \xi_p} \right]^{1/2}. \tag{11.44}$$

A slip correction factor S has been included in the preceding equation to account for the fact that for very small particles or at high altitudes where the air is not very dense, the mean free path of gas molecules is not negligible. Slip flow rather than continuum flow then occurs, and a faster settling velocity applies. The slip correction factor can be estimated as

$$S = \frac{\lambda_f}{d_p} \left[1.764 + (0.562) \exp\left(\frac{-0.785 \, d_p}{\lambda_f} \right) \right]. \tag{11.45}$$

where λ_f is the mean free path of molecules in the fluid. The ratio λ_f/d_p is Kn, the Knudsen number.

In order to determine the terminal velocity using the preceding expression, both the material properties of the gas and the drag coefficient C_D must be determined. The density, temperature, and pressure of air are given as a function of altitude in Table 5.9 in Chapter 5. Formulas for the density and viscosity of air, as well as the speed of sound and mean free path in air are given as a function of altitude in Table 11.9. The drag coefficient depends on the flow regime as delineated by the Reynolds number

$$\text{Re} = \frac{d_p v_t \varsigma_f}{\mu_f}. \tag{11.46}$$

Here μ_f is the fluid viscosity, which can be expressed in poise (1 poise = 1 g cm^{-1} s^{-1}). In the Stokes flow regime (Re $\leq$ 1), the drag coefficient can be set to 24/Re, which is the result derived for slow flow of an incompressible fluid about a solid sphere. Using C_D = 24/Re and S = 1, the expression for the terminal velocity in Stokes flow reduces to

$$v_t = \frac{(\varsigma_p - \varsigma_f) d_p^2 g}{18 \mu_f \chi_p}. \tag{11.47}$$

Table 11.9 Properties of Air as a Function of Altitude [13]

Density ς_f (g/cm^3) Formula:	Altitude (km)
$\varsigma_f = (0.001) 10^x$	All
where $x = [0.087 - 0.035836\, z - 0.0015724\, z^2$	
$\quad + (3.0103 \times 10^{-5})\, z^3 - (1.78 \times 10^{-7})\, z^4]$	

Viscosity μ_f (g cm^{-1} s^{-1}) Formula:	Altitude (km)
$1.7931 \times 10^{-4} - 3.3368 \times 10^{-6}\, z$	$z \leq 10.0163$
$1.5959 \times 10^{-4} - 2.2402 \times 10^{-6}\, z$	$10.0163 < z < 32.7269$
$\quad + 8.1044 \times 10^{-8}\, z^2 + 6.024 \times 10^{-10}\, z^3$	
$2.375 \times 10^{-4} - 1.3233 \times 10^{-6}\, z$	$z > 32.7269$

Speed of Sound c_s (cm/s) Formula:	Altitude (km)
$34{,}077 - 409.87\, z$	$z \leq 11.3595$
$27{,}068 + 451.87\, z - 30.097\, z^2 + 0.8364\, z^3 - 0.0073423\, z^4$	$11.3595 < z < 50.143$
$41{,}286 - 163.54\, z$	$z > 50.143$

Mean Free Path λ_f (μm) Equation:	Altitude (km)
$30{,}000\, \mu_f / (1.34867\, \varsigma_f c_s)$	All

At sufficiently high values of Re the drag coefficient can be approximated by a constant value, typically $C_D = 0.44$. The flow regime for high values of Re is called the Newton's drag regime. Various empirical correlations have been proposed to express C_D as a function of Re between the Stokes flow regime and the Newton's drag regime; however, because the Reynolds number depends on the terminal velocity, which in turn depends on the drag coefficient, an iterative scheme must generally be applied with such correlations. An alternative, non-iterative approach relies on the fact that the product $C_D \text{Re}^2$ is independent of the particle velocity; i.e.,

$$C_D \text{Re}^2 = \frac{4 d_p^3 \varsigma_p \varsigma_f g S^2}{3 \mu_f^2}. \tag{11.48}$$

In the Stokes flow regime this simplifies to Re $= C_D \text{Re}^2/24$ with $S=1$. Also, the following empirical expressions can be used to determine Reynolds number as a function of $C_D \text{Re}^2$:

for $C_D \text{Re}^2 < 138$,

$$\text{Re} = \frac{C_D \text{Re}^2}{24} - (2.34 \times 10^{-4})(C_D \text{Re}^2)^2$$

$$+ (2.015 \times 10^{-6})(C_D \text{Re}^2)^3 + (6.91 \times 10^{-9})(C_D \text{Re}^2)^4; \tag{11.49-a}$$

and for $C_D \text{Re}^2 > 138$,

$$\log_{10}(\text{Re}) = -1.29536 + 0.986 \log_{10}(C_D \text{Re}^2)$$

$$- 0.046677 \left[\log_{10}(C_D \text{Re}^2)\right]^2 + 0.0011235 \left[\log_{10}(C_D \text{Re}^2)\right]^3. \tag{11.49-b}$$

The Reynolds number is determined from Eq. 11.49 and the terminal velocity is determined by solving Eq. 11.46 for v_t and applying the slip correction factor.

Example 11.2:

Compute the terminal velocity for a 10 μm spherical particle with a density of 10 g/cm^3, released at an altitude of 30 km.

Solution:

Substituting $z = 30$ m into the equations for air properties from Table 11.9 gives

$$\varsigma_f = 1.842 \times 10^{-5} \text{ g/cm}^3,$$

$$\mu_f = 1.816 \times 10^{-4} \text{ g/cm/s},$$

$$c_s = 3.017 \times 10^4 \text{ cm/s, and}$$

$$\lambda_f = 7.267 \text{ micron.}$$

Substituting $d_p = 0.001$ cm, $\varsigma_p = 10$ g/cm³, $g = 980.665$ cm²/s, and the preceding values for ς_f and λ_f into Eq. 11.48 gives

$$C_D Re^2 = 7.305 \times 10^{-3}.$$

Since $C_D Re^2$ is less than 138, Re is obtained from Eq. 11.49-a as

$$Re = 3.044 \times 10^{-4}.$$

From Eq. 11.45, the slip correction factor is $S = 2.421$.

Solving Eq. 11.46 for v_t and applying the slip correction factor yields

$$v_t = 7.262 \text{ cm/s.}$$

For increasingly high altitude air becomes less dense and the continuum flow regime gradually transitions into the free molecular flow regime. For altitudes above 50 km, the speed of sound is less than 310 m/s, and heavy or large particles can enter the supersonic regime. In terms of dispersion, however, the particles in question respond poorly to air turbulence and tend to cluster in small areas of deposition. As a result, the radiological consequences associated with such particles tend to be much less important than the consequences associated with smaller particles. For completeness, the fall velocity can be linearly interpolated with Opik's equation, as recommended by Hage, for regimes above the speed of sound:

$$v_t = \sqrt{[2d_p \varsigma_p g/(3\varsigma_a)] - 1.07\, c_s^2}\,. \tag{11.50}$$

Here, c_s denotes the speed of sound at the altitude in question.

The preceding formulations for the terminal fall velocity neglect the effects of turbulence during the transport through different atmospheric layers. Because the applied drag force is a function of the relative velocity of the particle with respect to the surrounding air, the particle experiences a retarding effect on fall speed when crossing from one turbulent eddy structure to another. However, in terms of parameter ranges applicable to aerosol settling, the retarding effect of turbulence is significant only under certain conditions. When considering the settling of aerosols in the atmosphere, such conditions only apply to shallow turbulent layers near Earth's surface where strong turbulent eddies could be associated with strong winds. Over the range of applicability of the model, aerosols spend a relatively short time falling through the surface layer and, therefore, the retarding effect of turbulence on fall velocity is neglected.

Figure 11.9 shows the estimated terminal velocities for spherical aerosols with 1, 10, 100, and 1000 μm diameters assuming a particle density of 10 g/cm³ at altitudes up to 80 km (solid lines). The selected density is between the nominal densities of PuO_2 (9.6 g/cm³) and UO_2 (10.2 g/cm³). Terminal velocities decrease rapidly with decreasing altitude, especially for the 1-μm and 10-μm particles. Even for the 100-μm and 1000-μm particles, the terminal velocity decreases by about a factor of 10 between 50 km and 5 km. The dependence of settling velocity on particle size is clear. The 1000-μm particle typically

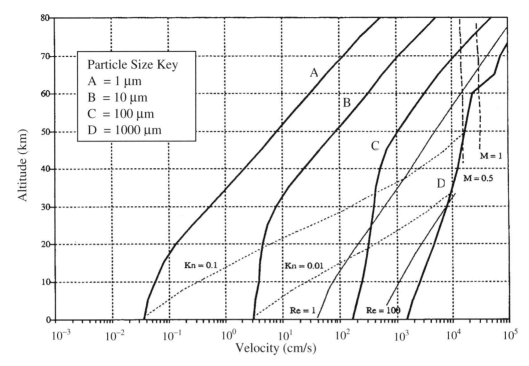

Figure 11.9 Terminal velocities vs. altitude for various-size spherical particles with density 10 g/cm³. *Courtesy of U.S. Department of Energy* [14].

settles a factor of 1000 or more times as fast as the 1-μm particle. Figure 11.9 also provides plots of altitude vs. particle terminal velocity for constant values of Knudsen number, Reynolds number, and Mach number (allowing the particle size to vary). The constant-Knudsen number curves show that slip flow (Kn > 0.01) increases rapidly with increasing altitude. The Stokes regime, which is applicable for Re < 1 and Kn < 0.01, is confined to a narrow area delimited by particle sizes from 10 to 50 μm under 20 km. The 1000-μm particle reaches the supersonic flow regime above about 60 km, and the associated terminal velocity curve then has some abrupt slope changes that result from a combination of large variations in atmospheric properties and crude interpolations between flow regimes. The model is not very accurate in this large heavy particle, high-altitude regime. Nonetheless, this model is sufficient from the point of view of dispersion analysis because large heavy particles fall rapidly to lower altitudes and the fall regime is fairly well described once such particles fall below 50 km.

Figure 11.10 illustrates the cumulative travel time for four spherical particle sizes (volumes of 1 μm, 10 μm, 100 μm, and 1,000 μm) released at 60 km. The solid lines and dotted lines correspond to particle densities of 10 g/cm³ and 1 g/cm³, respectively. While the 1000-μm particle takes less than 2 hours to reach the ground level, the 1-μm particle needs more than a year to descend. In the latter case, the dispersion would be expected to cover the entire globe and the high altitude release model discussed in Section 5 would be more appropriate. Particles with intermediate sizes have travel times ranging from half a day to 2 weeks. For a density of 10 g/cm³, the 10-μm particle spends about 5 days (out of a total 9 days of travel time) under the 10 km level. In this case, for under 10 km altitude, wet deposition (rainout and cloud scavenging) would be more effective than gravitational setting and the high altitude release model would again be more appropriate.

4.2 Transport and Dispersion

The horizontal velocity of the center of a cloud containing aerosols of a particular size is taken, at all times, to be equal to the local horizontal wind velocity. The wind field can be approximated by inter-

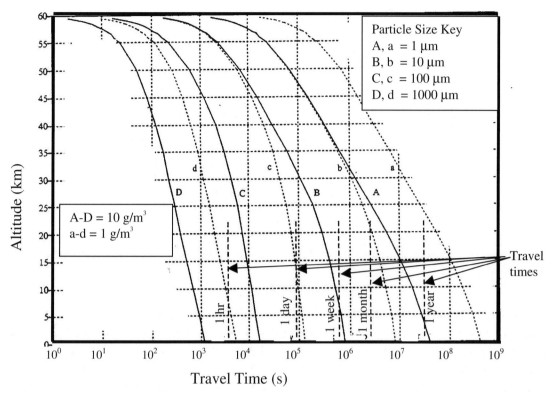

Figure 11.10 Altitude vs. velocity and cumulative travel time for various particle sizes and densities released at 60 km. *Courtesy of U.S. Department of Energy* [14].

polating results from an appropriate global wind field model. The interpolated data typically represent monthly averages. Global wind field models are beyond the scope of this chapter; however, Figure 11.11 illustrates the nature of averaged wind field data on a 10° latitude by 10° longitude grid at one altitude. Such data are typically specified at 5 km altitude increments.

Figure 11.12 shows trajectory results for a tropospheric release, at 10 km altitude and at 40°N latitude, − 45°E longitude. Trajectories for 10 g/cm³ spherical particles with 6, 8, 10, 13, and 20 μm diameters are shown. The trajectory for each particle size is presented as a function of longitude and lat-

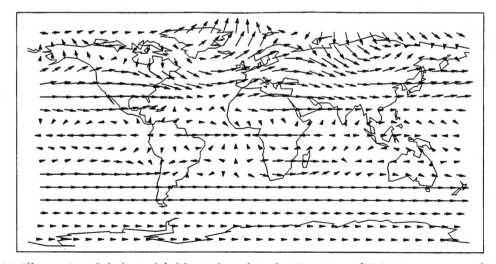

Figure 11.11 Illustrative global wind field at 5 km altitude. *Courtesy of U.S. Department of Energy* [14].

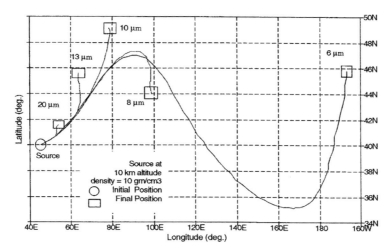

Figure 11.12 Trajectories for aerosols released at 10 km, 40°N latitude, −45°E longitude. *Courtesy of U.S. Department of Energy* [14].

itude in a top-down view to show the effects of the meridional and longitudinal wind components. The final positions of the particles indicate that deposition not only could occur several thousand kilometers away from the source, but also could reach regions far north or far south of the release latitude. Figure 11.12 illustrates why, for a polydisperse source term, the tracking of several selected particle sizes is required to obtain reasonable final footprints at ground level. Separations of deposition locations can be very large even for small increments of particle size. For high release altitudes it is clear that a three-dimensional wind field is needed to predict final deposition locations. A uniform, unidirectional wind field cannot be assumed.

It is generally advisable to evaluate the travel times before using a trajectory model for global dispersion of small particles. While the 10-μm, 13-μm, and 20-μm particles are deposited on the ground within 3 days, the 6-μm particle requires nearly 9 days. For such a long travel time, considering the importance of wet deposition at low altitudes, the 6-μm particles might not reach the final destination as predicted. For such particles, the problem should be treated with a high altitude dispersion model described in Section 5.

Dispersion Coefficients

In parallel with the transport process, atmospheric turbulent eddies act to diffuse and grow the size of the initial clouds. Eddies smaller in size than the source cloud are primarily responsible for the rate of expansion, while eddies larger than the source cloud move the cloud through the atmosphere as a whole and those of comparable size effectively distort the source cloud shape. Let D_e and D_c denote the diameters of the eddies and the cloud respectively. As the size of the cloud increases, larger and larger eddies from the transport flow will enter the deformation scale (D_e nearly equal to D_c) and the diffusion scale ($D_e << D_c$). Later, when vertical motions are inhibited by the shallow depth of the atmosphere, the diffusion process proceeds more rapidly in the horizontal direction than it does in the vertical direction.

The GEOTRAP computer code models the growth of a cloud as it is transported in the following manner. Gaussian shape profiles are used to characterize the distribution of mass concentration relative to the puff center with the assumption that the standard deviation in the direction of cloud motion is equal to that in the lateral direction. The following approximation of an empirical expression for growth in the horizontal or lateral dispersion coefficient is used:

$$\log_{10}(\sigma_y) = -2.82 + 0.12\,(\log_{10}t) + 0.27\,(\log_{10}t)^2 - 0.024\,(\log_{10}t)^3, \qquad (11.51)$$

where t denotes the cloud travel time in seconds. In spite of vertical wind shear, vertical dispersion is modeled as Fickian diffusion; i.e., the flux of material is assumed proportional to the vertical concentration gradient so that

$$\sigma_z = \min\,[(2K_z t)^{1/2},\ 30\ \text{km}\,]. \tag{11.52}$$

The maximum value of $\sigma_z = 30$ km is imposed to avoid unrealistic upward dispersion. Different values of K_z are used in different altitude ranges; therefore, in applying the preceding equation t is interpreted as current time minus the time at which the puff center enters the layer in question. As shown in Table 11.10, the uncertainty regarding the vertical dispersion coefficient within the mesosphere is high. The mesosphere is the zone located between 45–100 km above Earth's surface; it is a highly anisotropic eddy-mixing layer with characteristics that change seasonally, and dispersion data in this layer are sparce.

Example 11.3

Estimate the horizontal and vertical dispersion coefficients at 15 km for a puff of 100-μm spherical aerosols with density 10 g/cm^3 released at 60 km. Use GEOTRAP values for K_z given in Table 11.10.

Solution:

From Figure 11.10, the time spent in the mesosphere is approximately 1000 s and the total time to reach 15 km is approximately 10,000 s. The time spent in the stratosphere, between 10 and 45 km, is $10,000\ \text{s} - 1000\ \text{s} = 9000$ s.

Neglecting the initial size of the puff, and setting $\log_{10}(t) = \log_{10}(10^4) = 4$, hence,

$$\log_{10}\sigma_y = -2.82 + 0.12\,(4) + 0.27\,(4)^2 - 0.024\,(4)^3 = 0.444.$$

Thus, $\sigma_y = 2.78$ km.

Also,

$$\sigma_z^2 = \min\,[2\,(80\times10^{-6})\,1000 + 2\,(0.1\times10^{-6})\,(9000),\ 900\ \text{km}^2] = 0.16\ \text{km}^2,$$

or $\sigma_z = 0.40$ km.

4.3 Concentrations at Ground Level

When a cloud released at intermediate altitudes descends to lower levels, more and more mass is transferred from above into the mixing layer and then deposited to the ground along the advection path. Figure 11.13 illustrates this gradual mass exchange. The cloud cross section is depicted as circular consistent with the assumption that dispersion occurs equally in all lateral directions. To simplify the computations, complete vertical mixing is often assumed to occur within the mixed layer because tur-

Table 11.10   Cited Range and GEOTRAP Values of Vertical Fickian Diffusivity K_z [14]

Altitude Range	Low Cited Value of K_z [m^2/s]	GEOTRAP Value of K_z [m^2/s]	High Cited Value of K_z [m^2/s]
Mesosphere: 45 km − 80 km	1	80	1000
Stratosphere: 10 km − 45 km	0.1	0.1	1
Troposphere: 0 km − 10 km	5	5	10

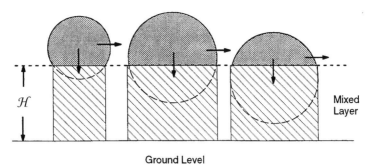

Figure 11.13 Mass transfer from a descending cloud to the mixing layer. *Courtesy of U.S. Department of Energy* [14].

bulence levels are higher and most of the clouds have long travel times. At time t, the mass within the mixing layer that is available for deposition is then

$$Q_{mix} = \frac{Q}{2}\left[1 - \text{erf}\left(\frac{\mathcal{H} - z_c}{\sqrt{2}\sigma_z}\right)\right] - m_D \qquad \text{for } z_c > \mathcal{H}, \quad \text{and} \qquad (11.53)$$

$$Q_{mix} = \frac{Q}{2}\left[1 + \text{erf}\left(\frac{\mathcal{H} - z_c}{\sqrt{2}\sigma_z}\right)\right] - m_D \qquad \text{for } z_c < \mathcal{H}, \quad \text{and} \qquad (11.53)$$

where Q_{mix} is the particle mass in the portion of the cloud within the mixing layer, Q is the initial particle mass in the cloud, z_c is the instantaneous height of the cloud center, σ_z is the vertical standard deviation of particles in the cloud, and m_D is the total mass of particles deposited up to time t. The height $\mathcal{H}$ of the mixing layer is typically approximated as 1 km.

To model ground-level air concentrations and deposited ground concentration profiles, the longitude, latitude, horizontal dispersion coefficient, and mass available for deposition must be tracked for each cloud reaching the mixing layer. Since clouds with various particle sizes may be involved, cloud footprints may overlap. The overlap of footprints is illustrated in Figure 11.14. In this figure, the deposition area is defined by five overlapping cloud footprints, each represented by a circle of radius $3\sigma_y$.

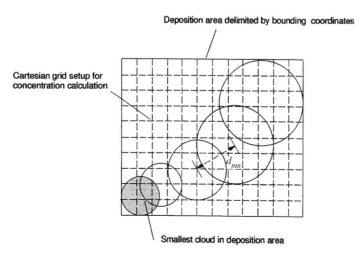

Figure 11.14 Deposition area resulting from overlapping cloud footprints. *Courtesy of U.S. Department of Energy* [14].

This configuration could result from a release of five closely spaced clouds containing different particle sizes. The criteria for two footprints to be connected can be expressed as

$$d_{mn} < 3(\sigma_{ym} + \sigma_{yn}),$$ (11.55)

where σ_{ym} and σ_{yn} are the lateral dispersion coefficients of clouds m and n and d_{mn} is the horizontal distance between the cloud centers. The factor 3 accounts for 99% of total mass within the footprints.

To estimate the consequences associated with intermediate level releases, regions of the globe affected by cloud deposition must be identified. Unlike short-range dispersion problems where the domain of deposition can be set a-priori, the dispersion of multi-particle sizes released at intermediate altitudes can produce complex deposition patterns of various sizes such that pre-determined receptor grids are excluded. Given a list of clouds with overlapping footprints, the boundaries of a deposition area can be defined by the bounding coordinates and a Cartesian reference grid can be selected with a scale based on the smallest standard deviation σ_{min}. This approach allows predictions of concentration profiles resulting from overlapping footprints.

Once the reference grid for each deposition area is set up, the time-dependent ground-level concentration [g/km³] of particles from a cloud at a grid point with Cartesian coordinates (x,y) is

$$\chi(x,y) = \frac{Q_{mix}}{2\pi\sigma_y^2 \mathcal{H}} \exp\left(\frac{(x - x_c)^2 + (y - y_c)^2}{2\sigma_y^2}\right).$$ (11.56)

The ground deposition rate (g km⁻² s⁻¹) at the grid point is

$$R_d(x,y) = \chi(x,y)\, v_D,$$ (11.57)

where v_D is the deposition velocity (km/s). The total ground-level air concentration and the total deposition rate at each grid point are the sum over all clouds within $3\sigma_y$ of the grid point. The deposition velocity v_D may be modeled as the terminal velocity at ground elevation or by an empirical relationship that accounts for mixing layer and ground surface characteristics. If particles with different sizes affect the same location, the calculated concentrations are tracked separately, because inhalation doses depend on the inhaled particle size. To adequately estimate integrated ground-level air concentrations and ground contamination levels, time steps are selected such that the advection distance remains less than a preset value, typically one horizontal standard deviation σ_y. Given appropriate time-integrated ground-level concentrations, doses can be estimated for the cloud shine, inhalation, ground shine, and food chain pathways.

A simple case analyzed by the GEOTRAP code illustrates the nature of the results predicted for intermediate level releases. In this illustration, source clouds are released at an altitude of 45 km, 15°N latitude, and 60°W longitude. Ten different clouds, corresponding to 10 different particle sizes (10, 12, 13, 14, 15, 20, 30, 50, 100, and 150 μm), are tracked. Each cloud initially contains 100 g of Pu-238. In Figure 11.15, the positions of various cloud centers in the 1 km sub-layer near ground level are depicted with dark spots. The cross lines represent the initial coordinates of release. As shown, the gravitational sorting is quite significant and the separation between deposition areas can be very large. Carried by the predominant zonal wind eastward, the two smallest particle sizes, 10 and 12 μm, have the longest trajectories. Travel times range from 2.6 hours, for the 150-μm particle, to 9.3 days, for the 10-μm particle.

Given the large separation between different zones of deposition and various cloud dimensions, an adequate evaluation of ground concentrations requires a variable-gridding setup, and GEOTRAP created seven zones of deposition. In Figure 11.16, the surface concentrations and their corresponding contour plots are generated in relative scale for four different particle sizes. Note that the contour plots

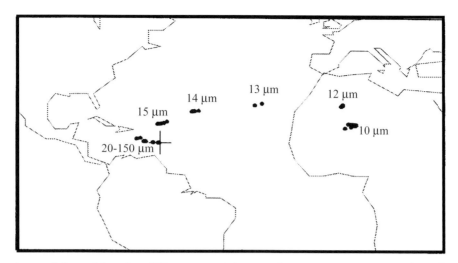

Figure 11.15 Locations of deposition for different particle sizes from a GEOTRAP analysis for a specific release case with 10 different particle sizes. *Courtesy of U.S. Department of Energy* [14].

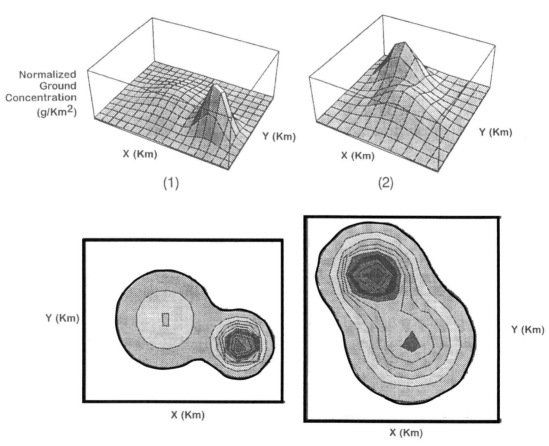

Figure 11.16 Ground level concentration profiles (1) for 20 and 30 μm particles, (2) for 10 and 12 μm particles. *Courtesy of U.S. Department of Energy* [14].

are normalized to the maximum concentration value and only a relative scale is displayed. The grid spacings in Figure 11.16 are 36 km and 152 km for zone 1 and zone 2, respectively. The large gap between 20 μm and 30 μm introduced two distinct footprints with a factor of 4 difference in the ratio of peak concentrations. In contrast, the more diffusive 10 and 12 μm particles created overlapping footprints with a larger deposition area and a ratio of peak concentrations of about 1.7. For heavy particles above 50 μm, the footprints are nearly circular due to very short horizontal transport and all deposition areas are separated.

5.0 High Altitude Releases

Accident scenarios that involve the release of very small aerosol particles at high altitude include (1) debris injection into a plume that is lofted by momentum and buoyancy effects, (2) rupture of a heat source container due to an accident during ascent, and (3) reentry burn-up. Very small aerosols (e.g., PuO_2 aerosols with diameters less than 10 μm) have negligible settling velocities. If released at very high altitudes, such particles will spread around the globe and take several years to completely deposit on the ground or bodies of water.

Transport of small particles released at high altitude involves two processes: (1) latitudinal dispersion and (2) downward transport to lower altitudes. Latitudinal dispersion is governed by worldwide atmospheric circulation patterns and results in a final latitudinal distribution, which is highly dependent on the latitude at which the release occurs. Except for very high releases into the mesosphere (> 45 km) and releases into the stratosphere (10 to 45 km) near the equator, very small particles tend to ultimately deposit in the hemisphere (northern or southern) in which they are released. Within each hemisphere, latitudinal distributions are skewed toward the mid-latitudes. Downward transport of small particles is influenced by circulation patterns in the mesosphere, stratosphere, and troposphere. Different residence times can be used to characterize downward transport through these successive layers [15,16,17].

5.1 Latitudinal Dispersion

Studies of debris from aboveground nuclear weapons tests have contributed significantly to the understanding of upper atmosphere transport processes. Clouds of radioactive particles generated by above ground nuclear weapons tests spread over a full range of altitudes. These particles spread throughout the atmosphere on a global scale. Very small particles and gases released into the mesosphere between 45 km and 80 km are rapidly redistributed by large-scale circulation or eddy diffusion. Complete mixing in the mesosphere occurs within a few months; whereas, the mean residence time of particles released in this layer ranges from 4 to 20 years depending on altitude. At lower altitudes of the stratosphere, radiation heating of the increasingly denser atmosphere produces large-scale eddies and mixing patterns that exhibit large latitudinal, seasonal, and even yearly variations. The influence of these complex circulation patterns on very small particles depends on several competing factors including gravitational sedimentation, atmospheric mean motion, turbulent eddy mixing, and molecular diffusion. Far from being a well-mixed reservoir, the stratosphere tends to move very small particles poleward, inhibiting mixing between the northern and southern hemispheres. Unless the particles enter the stratosphere from the well-mixed mesosphere above or are injected into the stratosphere near the equator, they are predominately confined to the hemisphere (northern or southern) where they are released.

The ultimate longitudinal (east–to-west) distribution of very small particles released at high altitude can be assumed to be uniform. In contrast, the ultimate latitudinal (north-to-south) distribution tends to be Gaussian in shape with a peak at mid-latitudes. Table 11.11 gives the fraction of very small particles deposited in each of 20 equal area latitude bands due to releases to the stratosphere in consecutive 10° latitude ranges. The latitudinal deposition distributions given in the table are plotted in Figure 11.17 for releases into the stratosphere near the equator (latitude range 10) and within six different

Table 11.11   Fraction of Stratospheric Release Deposited in Equal
Area Latitude Bands for Successive Release Latitudes [18]

		Release Latitude Range									
		1	2	3	4	5	6	7	8	9	10
		85°N–	75°N–	65°N–	55°N–	45°N–	35°N–	25°N–	15°N–	5°N–	5°S–
Equal Area Latitude Band		90°N	85°N	75°N	65°N	55°N	45°N	35°N	25°N	15°N	5°N
1	64°9′30″N–90°0′0″N	0.0727	0.0727	0.0725	0.0718	0.0700	0.0672	0.0636	0.0588	0.0524	0.0437
2	53°7′30″N–64°9′30″N	0.1477	0.1477	0.1473	0.1450	0.1398	0.1316	0.1208	0.1064	0.0876	0.0622
3	44°26′30″N–53°7′30″N	0.1787	0.1787	0.1783	0.1752	0.1686	0.1582	0.1444	0.1262	0.1023	0.0702
4	36°52′30″N–44°26′30″N	0.1745	0.1745	0.1741	0.1712	0.1647	0.1547	0.1412	0.1236	0.1003	0.0691
5	30°30′0″N–36°52′30″N	0.1606	0.1606	0.1602	0.1576	0.1518	0.1427	0.1304	0.1145	0.0935	0.0652
6	10 23°35′30″N–30°30′0″N	0.1109	0.1109	0.1107	0.1091	0.1055	0.1000	0.0927	0.0830	0.0703	0.0532
7	17°27′30″N–23°35′30″N	0.0823	0.0823	0.0822	0.0812	0.0789	0.0753	0.0707	0.0644	0.0563	0.0453
8	11°32′30″N–17°27′30″N	0.0474	0.0474	0.0474	0.0471	0.0464	0.0454	0.0439	0.0421	0.0397	0.0363
9	5°44′30″N–11°32′30″N	0.0178	0.0178	0.0179	0.0182	0.0188	0.0199	0.0213	0.0231	0.0255	0.0288
10	Equator–5°44′30″N	0.0074	0.0074	0.0075	0.0080	0.0091	0.0108	0.0133	0.0164	0.0205	0.0260
11	Equator–5°44′30″S	0	0	0.0001	0.0008	0.0024	0.0049	0.0082	0.0126	0.0183	0.0260
12	5°44′30″S–11°32′30″S	0	0	0	0.009	0.0027	0.0060	0.0091	0.0139	0.0202	0.0288
13	11°32′30″S–17°27′30″S	0	0	0.0002	0.0011	0.0034	0.0068	0.0115	0.0176	0.0256	0.0363
14	17°27′30″S–23°35′30″S	0	0	0.0002	0.0014	0.0043	0.0085	0.0143	0.0219	0.0319	0.0453
15	23°35′30″S–30°30′0″S0	0	0	0.0002	0.0017	0.0049	0.0100	0.0168	0.0257	0.0374	0.0532
16	20 30°30′0″S–36°52′30″S	0	0	0.0003	0.0021	0.0060	0.0122	0.0205	0.0315	0.0458	0.0652
17	36°52′30″S–44°26′30″S	0	0	0.0003	0.0022	0.0064	0.0129	0.0218	0.0333	0.0486	0.0691
18	44°26′30″S–53°7′30″S	0	0	0.0003	0.0022	0.0065	0.0131	0.0221	0.0339	0.0494	0.0702
19	53°7′30″S–64°9′30″S	0	0	0.0003	0.0019	0.0058	0.0116	0.0196	0.0300	0.0437	0.0622
20	64°9′30″S–90°0′0″S	0	0	0	0.0013	0.0041	0.0082	0.0138	0.0211	0.0307	0.0437

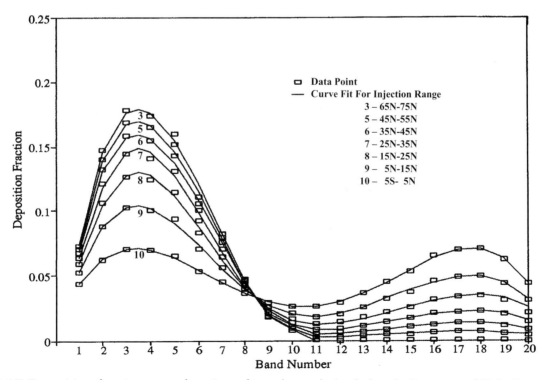

Figure 11.17 Deposition fractions as a function of equal-area latitude band. *Courtesy of U.S. Department of Energy* [14].

northern hemisphere latitude ranges (3 and 5 through 9). The distributions depicted in Figure 11.17 are for ground level deposition; they should not be confused with the atmospheric distribution of particulate during the dispersion phase. The horizontal axis indicates the centers of the latitude bands, and the rectangles plotted with each distribution indicate average fractions ultimately deposited in each band.

Distributions for releases into the stratosphere occurring in southern latitude ranges are assumed to be symmetric to those shown for releases into corresponding northern latitude ranges (most nuclear fallout data are for releases into the northern hemisphere). These distributions are bounded by two cases. The plotted distribution with the greatest peak occurs when the release to the stratosphere is in the 65°N–75°N latitude range. In this case, very small particles are virtually all deposited to the ground in the northern hemisphere. As the latitude of the release to the stratosphere shifts toward the equator, the relative deposition between both hemispheres tends to equalize. The maximum spread occurs for a release at the equator (5°S to 5°N). In this case, deposition to the ground is symmetric in the northern and southern hemispheres. As noted earlier, very small particles released into the mesosphere (> 45 km) are well mixed in this layer. Inter-hemisphere mixing and long residence within the mesosphere lead to the assumption that the deposition distribution for very small particles released into the mesosphere is equivalent to that for a release into the stratosphere near the equator.

Example 11.4

What is the fraction of very small particles deposited in equal-area latitude band 3 for: (a) a release to the stratosphere between 25°N to 35°N, (b) a release to the stratosphere between 25°S and 35°S, and (c) a release to the mesosphere?

Solution:

(a) From Table 11.11, for a release into the stratosphere between 25°N and 35°N, the fraction deposited in equal area band 3 is 0.1444.

(b) The deposition distribution for a release to the stratosphere between 25°S and 35°S is symmetric to that for a release to the stratosphere between 25°N and 35°N. The fraction for equal area band 18 (symmetric to equal area band 3) in column 7 of Table 11.11 is 0.0221.

(c) The latitudinal deposition distribution for release into the mesosphere is equivalent to that for a release into the stratosphere near the equator. For this distribution, the last column of Table 11.11 indicates the fraction deposited in equal area band 3 is 0.0702.

5.2 Downward Transport

As illustrated in Figure 11.18, downward transport of very small particles can be modeled using four compartments. In order of decreasing altitude these compartments are the mesosphere (> 45 km), the stratosphere (10 to 45 km), the troposphere (0 to 10 km), and the ground surface. The rate of downward transport from each compartment is taken to be proportional to the inventory in the compartment. The proportionality constants are called compartment transfer constants. They are denoted and defined as follows:

λ_3— mesospheric transfer constant, fractional loss of mesospheric inventory to stratosphere per unit time;

λ_2— stratospheric transfer constant, fractional loss of stratospheric inventory to troposphere per unit time;

λ_1— tropospheric transfer constant, fractional loss of tropospheric inventory to the Earth's surface per unit time; and

λ_0— weathering constant, fractional loss of inventory on ground surface per unit time due to weathering.

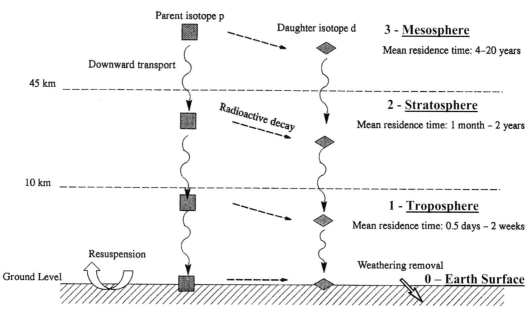

Figure 11. 18 Compartments and removal processes for high-altitude releases. *Courtesy of U.S. Department of Energy* [14].

Note that the compartment indices increase with altitude. Just as the decay constant of a radionuclide is the inverse of its mean lifetime, the transfer constant of a compartment is the inverse of the mean residence time of stable species in the compartment. That is, the mean residence time of stable species in the mesosphere is $\tau_3 = 1/\lambda_3$, the mean residence time of stable species in the stratosphere is $\tau_2 = 1/\lambda_2$, and the mean residence time of stable species in the troposphere is $\tau_1 = 1/\lambda_1$. Empirical values for the weathering rate constant λ_0 are discussed in conjunction with Example 11.3.

Appropriate numeric values of the residence times can be selected based on nuclear fallout data. These data are applicable to suspended particulate debris in its entirety, rather than to a particular particle size or specific release altitude. Individual particles may, in the process of transport, lose their identity due to attachment to larger particles. Ranges of mean residence times estimated for atmospheric layers at various altitudes are presented in Table 11.12. The widest range is for the mesosphere because for this layer the data are sparse and processes are poorly understood. One comparison of a three-compartment model against cumulative ground level fallout data concluded that a mean residence time of 14 years should be used for releases of aerosols above 100 km, while a value of 4 years is appropriate for releases at lower mesospheric altitudes [16]. The mean residence time of tropospheric particles is limited mainly by wet deposition. The presence of clouds below 5 km causes particles to deposit within days as a result of either condensation or washout effects. Significant variations in residence times are possible with season, climate, and longitude; however, accounting for such variations would require a significantly more sophisticated model. The default mean residence times utilized in the HIAD computer code are also summarized in Table 11.12.

The rates of change in the quantities of radionuclides present in each compartment may be expressed as ordinary differential equations. To illustrate, consider the rates of change for a single parent nuclide and a single daughter nuclide. Let Q_{pc} and Q_{dc} denote the quantities of parent and daughter present in the c-th compartment at any instant in time. Then arranging terms in parent and daughter differential equations by transport constant, decay constant, and source term gives

Table 11.12 Mean Residence Time of Aerosols in the Atmosphere [19]

	Atmospheric Layer		Mean Residence Time τ Range	HIAD Default
1	Troposphere	Below 1.5 km	0.5 day–2 days	1 week
		Lower	2 days–1 week	
		Mid & Upper	1 week–2 weeks	
		Tropopause	3 weeks–1 month	
2	Stratosphere	Lower	1 month–2 months	14 months
		Upper	1 year–2 years	
3	Mesosphere	Lower	4 years–14 years	4 years
		Upper	14 years–20 years	

$$
\begin{aligned}
dQ_{p3}/dt &= -\lambda_3 Q_{p3} - \lambda_p Q_{p3} + S_{p3}(t),\\
dQ_{d3}/dt &= -\lambda_3 Q_{d3} + \lambda_p Q_{p3} - \lambda_d Q_{d3} + S_{d3}(t),\\
dQ_{p2}/dt &= \lambda_3 Q_{p3} - \lambda_2 Q_{p2} - \lambda_p Q_{p2} + S_{p2}(t),\\
dQ_{d2}/dt &= \lambda_3 Q_{d3} - \lambda_2 Q_{d2} + \lambda_p Q_{p2} - \lambda_d Q_{d2} + S_{d2}(t),\\
dQ_{p1}/dt &= \lambda_2 Q_{p2} - \lambda_1 Q_{p1} - \lambda_p Q_{p1} + S_{p1}(t),\\
dQ_{d1}/dt &= \lambda_2 Q_{d2} - \lambda_1 Q_{d1} + \lambda_p Q_{p1} - \lambda_d Q_{d1} + S_{d1}(t),\\
dQ_{p0}/dt &= \lambda_1 Q_{p1} - \lambda_0 Q_{p0} - \lambda_p Q_{p0} \quad \text{and}\\
dQ_{d0}/dt &= \lambda_1 Q_{d1} - \lambda_0 Q_{d} + \lambda_p Q_{p0} - \lambda_d Q_{d0}(t).
\end{aligned} \tag{11.58}
$$

Here λ_p and λ_d denote the radioactive decay constants of parent and daughter nuclides, respectively. $S_{pc}(t)$ and $S_{dc}(t)$ denote accident-related sources of parent and daughter nuclides in compartment c. The solution to these ordinary differential equations depends on the nature of the source term. For an instantaneous release to the mesosphere, the source term is a delta function at time $t = 0$, $S_{p3}(t) = Q_{p3}(0)\delta(t - 0)$ and the solutions for the parent isotope are

$$
Q_{p0}(t) = \lambda_1\lambda_2\lambda_3 Q_{p3}(0)\left\{\begin{array}{l}
\dfrac{\exp[-(\lambda_p + \lambda_3)t]}{(\lambda_2 - \lambda_3)(\lambda_1 - \lambda_3)(\lambda_0 - \lambda_3)} + \dfrac{\exp[-(\lambda_p + \lambda_2)t]}{(\lambda_3 - \lambda_2)(\lambda_1 - \lambda_2)(\lambda_0 - \lambda_2)}\\[2ex]
+ \dfrac{\exp[-(\lambda_p + \lambda_1)t]}{(\lambda_3 - \lambda_1)(\lambda_2 - \lambda_1)(\lambda_0 - \lambda_1)} + \dfrac{\exp[-(\lambda_p + \lambda_0)t]}{(\lambda_3 - \lambda_0)(\lambda_2 - \lambda_0)(\lambda_1 - \lambda_0)}
\end{array}\right\},
$$

$$
Q_{p1}(t) = \lambda_2\lambda_3 Q_{p3}(0)\left[\frac{\exp[-(\lambda_p + \lambda_3)t]}{(\lambda_2 - \lambda_3)(-\lambda_1 - \lambda_3)} + \frac{\exp[-(\lambda_p + \lambda_2)t]}{(\lambda_3 - \lambda_2)(\lambda_1 - \lambda_2)} + \frac{\exp[-(\lambda_p + \lambda_1)t]}{(\lambda_3 - \lambda_1)(\lambda_2 - \lambda_1)}\right],
$$

$$
Q_{p2}(t) = \lambda_3 Q_{p3}(0)\left[\frac{\exp[-(\lambda_p + \lambda_3)t]}{\lambda_2 - \lambda_3} + \frac{\exp[-(\lambda_p + \lambda_2)t]}{\lambda_3 - \lambda_2}\right], \quad \text{and}
$$

$$
Q_{p3}(t) = Q_{p3}(0)\exp[-(\lambda_p + \lambda_3)t]. \tag{11.59}
$$

Example 11.5

Write the solution to the ordinary differential equations for $Q_{pc}(t)$ given an instantaneous release of quantity $Q_{p2}(0)$ to the stratosphere at time $t = 0$.

Solution:

For a release to the stratosphere, the quantity in the mesosphere remains zero at all times because upward transport is neglected. Eliminating terms involving the mesosphere, the three remaining ordinary differential equations for the stratosphere, troposphere, and ground surface are identical in form to the original equations for the mesosphere, stratosphere, and troposphere, respectively. By analogy, the solutions for an instantaneous release to the stratosphere are, therefore,

$$Q_{p1}(t) = \lambda_2 Q_{p2}(0)\left[\frac{\exp[-(\lambda_p + \lambda_2)t]}{\lambda_1 - \lambda_2} + \frac{\exp[-(\lambda_p + \lambda_1)t]}{\lambda_2 - \lambda_1}\right] \quad \text{and}$$

$$Q_{p2}(t) = Q_{p2}(0)\exp[-(\lambda_2 + \lambda_p)].$$

Example 11.6

One curie of ^{238}Pu is released to the mesosphere. Using the HIAD default residence times from Table 11.12 and a weathering constant of $\lambda_0 = 1.132$ y^{-1}, plot the fraction of the original activity that resides in the mesosphere, stratosphere, troposphere, and deposited to the Earth's surface as a function of time after the release.

Solution:

From Table 11.12, the mean residence times for the mesosphere, stratosphere, and troposphere are 4 years, 14 months, and 1 week, respectively. The half-life of ^{238}Pu is 87.7 years. Hence, the various rate constants are:

$$\lambda_p = \ln(2)/(87.7 \text{ years}) = 7.9036 \times 10^{-3} \text{ y}^{-1},$$
$$\lambda_3 = 1/(4 \text{ years}) = 0.25 \text{ y}^{-1},$$
$$\lambda_2 = 1/(14 \text{ months}) = 0.871 \text{ y}^{-1}, \text{ and}$$
$$\lambda_1 = 1/(1 \text{ week}) = 1.916 \times 10^{-2} \text{ y}^{-1}.$$

Substituting these values into the compartment inventory solutions for an instantaneous release to the mesosphere yields the values plotted in Figure 11.19.

Empirical data indicate that the weathering removal effect for material deposited to the ground can be modeled as a sum of exponential decay terms. For example, the Reactor Safety Study recommended the following equation for the fraction of deposited material residing on the ground t years after deposition [20]:

$$f_w(t) = 0.63\exp(-0.693\,t/0.612) + 0.37\exp(-0.693\,t/92.6). \tag{11.60}$$

The two exponential terms reflect two different modes of removal with participating fractions 0.63 and 0.37, weathering half-lives of 0.612 year and 92.6 years, and weathering rate constants of $\ln(2)/0.612 = 1.1326$ y^{-1} and $\ln(2)/92.6 = 7.4854 \times 10^{-3}$ y^{-1}, respectively. The effect of such a two-mode deple-

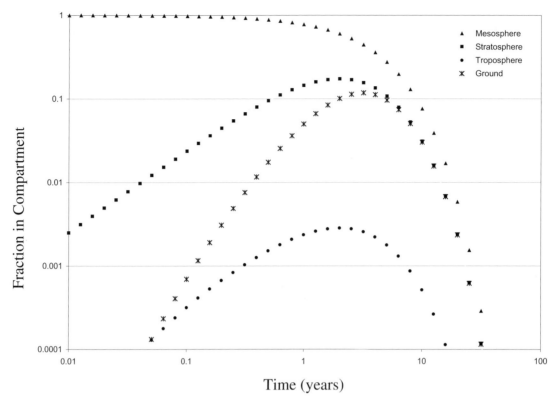

Figure 11.19 Compartment inventories for Pu-238 release to mesosphere. From Example 11.6.

tion mechanism can be simulated by linearly combining two solutions for the deposited inventory $Q_{p0}(t)$. The product of 0.63 times the solution using $\lambda_0 = 1.1326$ y^{-1} is added to the product of 0.37 times the solution obtained using $\lambda_0 = 7.854 \times 10^{-3}$ y^{-1}.

Example 11.7

Applying the two-term weathering model from the Reactor Safety Study for the problem posed in Example 11.6, compute and plot the fraction of ^{238}Pu on the ground as a function of time.

Solution:

Figure 11.20 presents a plot of the fraction deposited on the ground for $\lambda_0 = 1.1326$ y^{-1}, for $\lambda_0 = 7.854 \times 10^{-3}$ y^{-1}, and for the linear combination of these two solutions weighting the first by 0.63 and the second by 0.37. Note that the choice of the weathering model does not impact the solutions for the airborne inventories.

5.3 Ground Level Concentrations

Estimates of doses and potential health effects require time integrated air and ground concentrations at ground level. Methods used to estimate these concentrations are described below.

Ground Level Air Concentration

The concentration of radioactive material in the troposphere near ground level is needed to compute cloud shine and inhalation doses. There is little information regarding the concentration profile of nuclear fallout as a function of altitude within the troposphere. However, as noted earlier, the concentra-

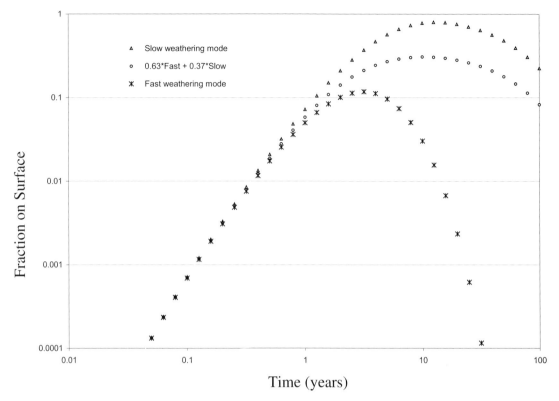

Figure 11.20 Surface inventory from release of Pu-238 to mesosphere for slow, average, and fast weathering. From Example 11.7.

tion tends to decrease rapidly due to wet deposition processes as clouds are encountered below 5 km. It is, therefore, generally considered conservative for the purpose of estimating ground level concentrations to assume that the concentration profile does not vary with altitude within the troposphere. The mean air concentration within a given latitude band is then the troposphere inventory $Q_{p1}(t)$ times the fraction d_b ultimately deposited in the band divided by the volume V_{1b} of the troposphere within the band. For the isotope with index p in equal-area latitude band b,

$$\chi_p(t) = \frac{Q_{p1}(t)d_b}{V_{1b}}. \tag{11.61}$$

Taking the troposphere thickness h_1 to be 10 km, the troposphere volume is

$$V_1 = \frac{4\pi}{3}\left[(R_E + h_1)^3 - R_E^3\right] = 5.109 \times 10^{18} \text{ m}^3, \tag{11.62}$$

where R_E is Earth's radius. The volume of the troposphere above each of the 20 equal-area latitude bands is $V_{1b} = V_1/20 = 2.5545 \times 10^{17}$ m^3. The time-integrated air concentration χ_{tp} (Bq s/m^3) between times t_1 and t_2 is

$$\chi_{tp} = \int_{t_1}^{t_2} \frac{Q_{p1}(t)d_b}{V_{1b}} \, dt. \tag{11.63}$$

For an instantaneous release to the mesosphere, this gives

$$\chi_{tp} = \frac{\lambda_2 \lambda_3 Q_p(0) d_b}{V_{1b}} \left[\frac{e^{-\lambda_{p3}t_1} - e^{-\lambda_{p3}t_2}}{(\lambda_2 - \lambda_3)(\lambda_1 - \lambda_3)\lambda_{p3}} + \frac{e^{-\lambda_{p2}t_1} - e^{-\lambda_{p2}t_2}}{(\lambda_3 - \lambda_2)(\lambda_1 - \lambda_2)\lambda_{p2}} + \frac{e^{-\lambda_{p1}t_1} - e^{-\lambda_{p1}t_2}}{(\lambda_3 - \lambda_1)(\lambda_2 - \lambda_1)\lambda_{p1}} \right]. \quad (11.64)$$

where $\lambda_{p3} = \lambda_p + \lambda_3$, $\lambda_{p2} = \lambda_p + \lambda_2$, and $\lambda_{p1} = \lambda_p + \lambda_1$.

Example 11.8

For the instantaneous release of one curie of ^{238}Pu to the mesosphere as described in Example 11.4, plot the time-integrated ground-level air concentration in equal-area latitude band 3 as a function of time after the release.

Solution:

Using the fraction deposited to the ground from Example 11.4, the fraction $d_3 = 0.0702$ from Example 11.4, and the preceding value of $V_{1b} = 2.5505 \times 10^{17}$ m³, and setting $t_1 = 0$, values of the time-integrated ground-level air concentrations as functions of time are plotted in Figure 11.21.

Ground Concentration

The ground concentration of an isotope is estimated as the total amount deposited times the fraction deposited on the area in question divided by the area. For nuclide p the ground concentration at time t in equal-area latitude band b $S_{Gp}(t)$ is

$$S_{Gp}(t) = \frac{Q_{p0}(t) d_b}{A_b}. \quad (11.65)$$

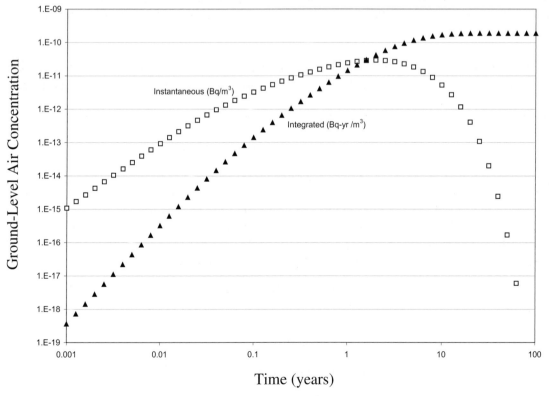

Figure 11.21 Instantaneous and integrated ground-level air concentrations from release of Pu-238 to mesosphere. From Example 11.8.

Here A_b denotes the area of one of the 20 equal area latitude bands. Using an Earth radius of 6731.229 km, the surface area of the Earth is

$$A_E = 4\pi(6371.229 \text{ km})^2 = 5.101 \times 10^8 \text{ km}^2, \text{ then} \tag{11.66}$$

$A_b = (1/20) A_E = 2.5505 \times 10^{13} \text{ m}^2.$

The time-integrated ground concentration S_{tGp} (Bq·s/m^2) over the interval t_1 to t_2 is

$$S_{tGp} = \int_{t_1}^{t_2} \frac{Q_{p0}(t)d_b}{A_b} \, dt. \tag{11.67}$$

This integral can be evaluated analytically. For a release to the mesosphere,

$$
\begin{aligned}
S_{tGp} &= \frac{\lambda_2 \lambda_3 Q_{p3}(0) d_b}{S_{1b}} \\
&= \left[\frac{e^{-\lambda_{p3}t_1} - e^{-\lambda_{p3}t_2}}{(\lambda_2 - \lambda_3)(\lambda_1 - \lambda_3)(\lambda_0 - \lambda_3)\lambda_{p3}} + \frac{e^{-\lambda_{p2}t_1} - e^{-\lambda_{p2}t_2}}{(\lambda_3 - \lambda_2)(\lambda_1 - \lambda_2)(\lambda_0 - \lambda_2)\lambda_{p2}} \right. \\
&\left. + \frac{e^{-\lambda_{p1}t_1} - e^{-\lambda_{p1}t_2}}{(\lambda_3 - \lambda_1)(\lambda_2 - \lambda_1)(\lambda_0 - \lambda_1)\lambda_{p1}} + \frac{e^{-\lambda_{p0}t_1} - e^{-\lambda_{p0}t_2}}{(\lambda_3 - \lambda_0)(\lambda_2 - \lambda_0)(\lambda_1 - \lambda_0)\lambda_{p0}} \right].
\end{aligned}
\tag{11.68}
$$

Example 11.9

For the instantaneous mesospheric release described in Example 11.8, express the time-integrated ground concentration in equal-area latitude band 3 as a function of time after the release.

Solution:

Using the fraction $d_3 = 0.0707$ from Example 11.6, and the preceding value of $A_b = 2.55 \times 10^{13}$ m^2, the instantaneous and time-integrated ground concentrations as functions of time are plotted in Figure 11.22.

Resuspension at Ground Level

Ground level resuspension can be estimated by considering the sublayer adjacent to the ground. It is assumed that resuspended material is confined to this sublayer and has no influence on concentrations in the troposphere. The preceding equations describe a continuous deposition process, which occurs over several years rather than a short dispersion-deposition process that occurs for low-altitude releases. Deposition from the troposphere is dominated by rain scavenging; therefore, the amount of material available for resuspension usually appears in the same time in certain areas.

Resuspension at ground level is a complex process; however, the resuspended air concentration of an isotope p at time t following deposition at time t_o can be approximated as

$$\chi_{rp}(t) = \Delta S_{Gp}(t_o) \, \mathcal{K}(t - t_o) \exp(-\lambda_p t) \tag{11.69}$$

where $\mathcal{K}(t-t_o)$ is an empirically determined resuspension factor with units of m^{-1}. In the rural INSRP model [21],

$$\mathcal{K}(t - t_o) = k_r/(t - t_o) \tag{11.70}$$

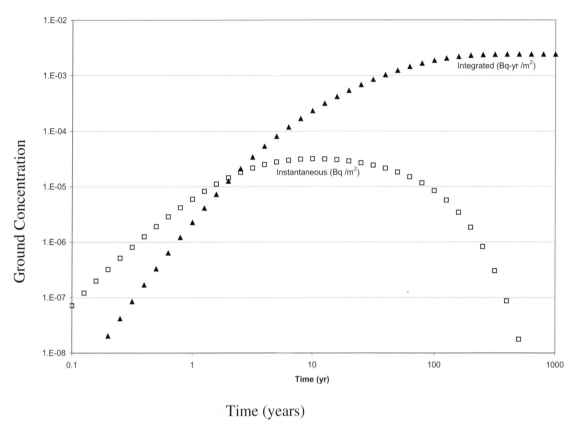

Time (years)

Figure 11.22 Instantaneous and integrated ground contamination from release of Pu-238 to mesosphere assuming average weathering. From Example 11.9.

where k_r is 10^{-6} day/m. Significant differences in resuspension would certainly occur in different deposition zones; however, such considerations are beyond the scope of this chapter. The ground concentration due to deposition in interval Δt about t_o is

$$\Delta S_{Gp} = \frac{\lambda_1 Q_{p0}(t_0) d_b \Delta t}{A_b}. \tag{11.71}$$

Using the INSRP form of the resuspension factor, the total resuspended air concentration at time t is therefore

$$\chi_{rp}(t) = \int_0^1 \frac{\lambda_1 Q_{p0}(t_o) d_b k_r}{A_b(t - t_o)} \exp[-\lambda_p(t - t_o)] dt_o. \tag{11.72}$$

Typically, the integral in Eq. 11.72 must be evaluated numerically. The time-integrated resuspended air concentration is simply the integral, which must also be evaluated numerically, of $\chi_{rp}(t)$ over the exposure interval.

Example 11.10

For an instantaneous release to the mesosphere of 1 curie of ^{238}Pu described in Example 11.6, plot the resuspended air concentration and the tropospheric air concentration as functions of time after the release.

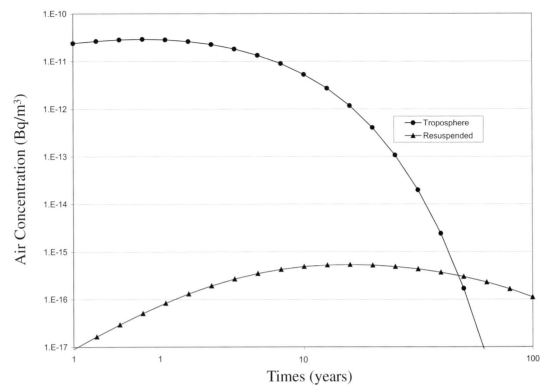

Figure 11.23 Troposphere and resuspended air concentrations from release of Pu-238 to mesosphere assuming rapid weathering. From Example 11.10.

Solution

Using the rate constants specified in the solution to Example 11.10 and the resuspension model discussed above, the requested air concentrations are plotted in Figure 11.23. Note that the resuspended air concentration generally begins to dominate only after the ground level tropospheric air concentration peaks.

6.0 Computer Methods

Table 11.13 summarizes the computer codes that have been developed and used to predict the consequences of accidents for U.S. launches of space nuclear devices. As indicated in the table, different codes are used to model different types of accidents. The nature and altitude of the accidental release determine which computer models are most appropriate.

For low (< 5 km) altitude releases, both transformation in the physical form of the released radionuclides and rise of the hot cloud that contains released radionuclides must be modeled as discussed in Section 3.1. For releases induced by low altitude explosions or ground impact upon reentry, SFM (Sandia Fireball Model) and PUFF can be applied to model the physical transformations and cloud rise, respectively [22,23]. For accidents involving solid rocket propellant ground fires and coincident impact of solid rocket propellants and PuO$_2$ fueled devices, preliminary models of physical transformation processes have been implemented in PEVACI (Plutonium Entrainment and Vaporization After Coincident Impact) [14], and PUFF has been used to model the rise of the cloud of combustion products resulting from ground fires.

Table 11.13   Computer Codes Used for U.S. Consequence Analysis

Release Elevation	Transformation of: Physical Cloud	Transformation of: Form Elevation	Dispersion & Deposition	Doses & Health Effects
Very High Altitudes (>50 km)	-	-	HIAD	PARDOS
Intermediate Altitudes (5–50 km)	-	-	GEOTRAP or HIAD	PARDOS
Low Altitudes (< 5 km)			GEOTRAP	PARDOS
Propellant Explosion	SFM	SFM, PUFF	or SATRAP	
SRP Ground Fire	PEVACI	SFM, PUFF	SATRAP	PARDOS
Reentry Ground Impact	SFM	SFM	SATRAP	PARDOS

The U.S. computer codes capable of modeling the dispersion and deposition of radioactive materials released in space nuclear accidents include SATRAP, GEORAD, and HIAD. SATRAP (Site-Specific Analysis of Transport and Dispersion of Radioactive Particles) models dispersion and deposition when the cloud containing released radionuclides is confined to low altitudes (< 5 km) [25]. The empirical puff-trajectory and resuspension models discussed in Sections 3.3 and 3.4 are implemented in SATRAP. For accidents initiated near the launch pad SATRAP utilizes a database that contains information about the population distribution, surface characteristics, and weather patterns specific to the launch and surrounding areas. A different version of SATRAP is used to model dispersion and deposition of radionuclides released from ground impacts associated with reentry accidents. Such impacts may occur in bands of locations over part of the Earth's surface. The database for this version of SATRAP contains worldwide information about weather, population density, and surface types.

GEOTRAP (Global Transport and Dispersion of Radionuclide Particles) models dispersion and deposition when the cloud containing released radionuclides is released into or rises to intermediate altitudes (5 to 50 km) [12]. GEOTRAP implements the type of model discussed in Section 4; that is, a Lagrangian-trajectory model with Gaussian puffs, to predict the final location and deposition patterns of particles with diameters of several microns and above. For smaller particles released at intermediate levels diffusion times can be several orders of magnitude longer and HIAD (High Altitude Aerosol Dispersion) is more appropriate. HIAD models the dispersion and deposition associated with high altitude (> 50 km) releases and small particles at intermediate altitudes [26]. HIAD implements the type of models discussed in Section 5; that is, HIAD uses three exponential transfer compartments: a mesospheric layer bounded at 45 km and higher altitudes, a stratospheric layer between 45 km and 10 km, and a tropospheric layer for lower altitudes. Both GEOTRAP and HIAD use a database of worldwide information about weather, population density, and surface types.

The PARDOS (Particulate Dose) code models the pathways by which radionuclides reach humans and the associated doses and health effects [27]. The dominant pathways are those discussed in Section 2. For releases of plutonium dioxide fuel, the inhalation from the passing cloud is the principal contributor to health effects. Longer-term pathways include inhalation due to resuspension, ground shine, and ingestion of contaminated water and food. Consequence measures typically computed by PARDOS include (1) collective radiation dose (person rems), (2) latent cancer fatalities, (3) number of people whose radiation exposure exceeds selected levels, and (4) land area contaminated above selected levels.

7.0 Summary

Consequences to the public of an accident involving a space nuclear power source depend on the amount of each radionuclide released, the chemical and physical characteristics of the released gases and aerosols, the cause of the release (explosion, ground fire, impact), the coordinates (longitude, latitude, and especially altitude) of the release, weather conditions that influence dispersion, the types of ground-level surfaces, the population distribution, and any actions (evacuation, sheltering, etc.) taken to protect the public.

The pathways whereby members of the public could receive dose from an accidental release of radionuclides include cloud shine, skin contamination, ground shine, inhalation, and ingestion. For each of these pathways, the dose received can be related to the to the time-integrated air concentration of each radionuclide at the exposure location. Relatively simple representations of such concentration to dose conversion factors are provided in Section 2.

Dispersion of radionuclides released in an accident varies strongly with altitude and with the composition and size distribution of the aerosols that are dispersed. For low-altitude releases buoyancy or momentum driven rise of the aerosol-laden cloud as well as chemical and physical processes impacting aerosols during cloud rise must be considered. After the cloud rises to an equilibrium altitude, dispersion is strongly influences by local weather conditions. Low-altitude dispersion can be represented by a relatively simple puff trajectory model in which the cloud containing radioactive particles follows the local wind trajectory and grows with distance. The concentration profiles within the cloud are assumed to be Gaussian in shape. The empirically determined widths of the Gaussian shape profiles increase with distance traversed. As a result, ground level concentrations decrease rapidly with distance. The rate of puff growth is strongly influenced by meteorological conditions. The more unstable the meteorological conditions, the more rapid the growth of the cloud and the associated dispersion of radionuclides.

At intermediate (5 to 45 km) altitudes, aerosols with 10 to 1000 micron diameters take from several days to several minutes to reach the ground, and the locations, sizes, and shapes of surface areas in which deposition occurs are largely determined by gravitational settling. Models to predict deposition patterns for such aerosols are discussed in Section 4. As discussed in Section 5, very small aerosols released at intermediate or high altitudes have negligible settling velocities, spread around the globe, and take several years to completely deposit on the ground or bodies of water. Downward transport of small particles is influenced by circulation patterns in the mesosphere, stratosphere, and troposphere. Such particles tend to deposit in the hemisphere in which they are released. Dispersion patterns depend on latitude but not longitude. The latitudinal distributions are skewed toward the mid-latitudes.

As discussed in Section 6, to perform the safety analyses required for launching space nuclear power sources in the United States, computer codes have been developed to model chemical and physical transformations of released materials; cloud rise; aerosol dispersion in the mesosphere, stratosphere, and troposphere; ultimate deposition to ground-level surfaces; and doses health effects associated with various pathways. To cover the spectrum of postulated accidents, these codes require information regarding population distributions, and land types both near the launch site and worldwide. The computed consequence measures generally include collective population dose, latent cancers, number of people whose radiation exposure exceeds selected levels, and land areas contaminated above selected levels. As discussed in Chapter 10, the predicted public consequences of accidents involving space nuclear power sources are generally much less than those predicted for severe accidents at terrestrial nuclear power plants. This is because of the more limited radionuclide inventory associated with space nuclear power sources and the dispersive nature of explosions, impacts, and altitudes associated with accidents involving space nuclear power sources.

Symbols

$\mathcal{A}$	surface area	k_r	resuspension parameter
$\mathcal{A}_T$	activity in organ T	Kn	Knudsen number
B	gamma ray buildup factor	m_D	deposited mass
c_s	speed of sound	$m_{T'}$	mass of organ T'
C_D	drag coefficient	Q	quantity of radioactive material
D_∞	dose rate for infinite cloud	$\dot{Q}$	material release rate
d	distance into tissue	Q_{mix}	particle mass in mixing layer
d_b	deposited fraction in band b	R	mixing term
d_p	particle diameter	Re	Reynolds number
DFI	dose inhalation factor	R_E	Earth's radius
$\overline{E}$	average radiation energy	R_b	breathing rate
$\widetilde{E}_j$	jth radiation specific energy	R_d	deposition rate
$E_{TT'}$	effective energy equivalent	R_i	chronic ingestion rate
$\widetilde{E}_{TT'}$	effective specific energy	r	distance from source
f	ground reflection parameter	S	slip coefficient
f_1	fraction of material in blood from SI	S	source concentration
f_j	fraction of decays producing radiation j	S_{skin}	skin surface concentration (radioactive)
$f_w(t)$	weathering factor	S_{tGp}	time-integrated ground concentration
$F_{j,TT'}$	fraction of radiation energy in organ T' from T	t	time
G	Gaussian shape function	u	wind speed
G_i	defined by Eqs. 11.34–11.36	V_p	particle volume
g	acceleration due to gravity	v	velocity
g_a	geometric attenuation factor	v_d	deposition velocity
h	height or thickness in atmosphere	v_t	terminal velocity
H	dose equivalent	W_R	weighting factor for radiation type R
$\dot{H}_{TT'}$	dose equivalent rate from organ T to T'	$W_{T'}$	weighting factor for organ T' for radiation
$\mathcal{H}$	mixing height	x	along wind coordinate in horizontal plane
K	Fickian diffusivity	y	cross wind coordinate in horizontal plane
$\mathcal{K}$	resuspension factor	z	vertical coordinate
k	absorbed radiation energy ratio (tissue/air)		
χ	airborne radioactive concentration	λ_f	mean free path in fluid
χ_{rp}	resuspended concentration	μ_f	fluid viscosity
χ_t	time-integrated air concentration	μ_j	linear attenuation coefficient
Δt	time increment	$\tilde{\mu}_j$	mass attenuation coefficient
ΔT	temperature increment	σ_x	along-wind dispersion coefficient (m)
Δz	altitude increment	σ_y	crosswind dispersion coefficient (m)
ϕ	particle flux	σ_z	vertical dispersion coefficient (m)
λ	decay or rate constant	τ	time of inhalation
λ_{ef}	$\lambda + \lambda_T$	ξ	dynamic shape factor
λ_T	biological rate constant	ζ	mass density

Special Subscripts/Superscripts

B	blood	LLI	lower large intestines
b	latitude band	m	cloud index number
c	puff center	n	cloud index number
d	daughter nuclide	o	initial value
E	Earth	p	parent nuclide
ef	effective	SI	small intestines
f	fluid	ST	stomach
G	ground	T	tissue or organ type index
g	gastrointestinal track	ULI	upper large intestines
I	time-integrated	TT'	from T to T'
i	nuclide index	β	beta particle
j	radiation type index	γ	gamma ray
k	mixing term index		

References

1. Tascione, T. F., *Introduction to the Space Environment*. Orbit Book Co., Malabar, FL, 1988.
2. Haskin, F. E. and A. L. Camp, "Perspectives on Reactor Safety." NUREG/CR-6042, SAND93-0971, prepared for the U.S. Nuclear Regulatory Commission by Sandia National Laboratories, Albuquerque, NM, Mar. 1994.
3. Young, M. L. and D. Chanin, "DOSFAC2 User's Guide." NUREG/CR-6547, SAND97-2776, Sandia National Laboratories, Appendix A - Sample Output Files Distributed with DOSFAC, Dec. 1997.
4. Pavone, D., "GPHS Safety Tests Particle Size Data Package." LACP-86-62, Los Alamos National Laboratory, Los Alamos, NM, May 1986.
5. Haynes, B. S. and H. G. Wagner, "Soot Formation." *Prog. Energy Combust. Sci.*, Vol. 7. p. 229–273, Pergamon Press, 1981.
6. Powers, D. A., K. K. Murata, D. C. Williams, J. B. Rivard, C. D. Leigh, D. R. Bradley, R. J. Lipinski, J. M. Griesmeyer, and J. E. Brockmann, "Uncertainty in Radionuclide Release Under Specific LWR Accident Conditions." Volume II, *TMLB' Analyses*, Sandia National Laboratories, SAND84-0410/2, Feb. 1985.
7. Davies, C. N., "Definitive Equations for the Fluid Resistance of Spheres." *Proc. Phys. Soc.*, 57, p. 259–270, 1945.
8. Hage, K. D., G. Arnason, N. E. Browne, P. S. Brown, H. D. Entrekin, M. Levitz, and J. A. Serkorski, Particle Fallout and Dispersion in the Atmosphere Final Report. Sandia Corporation, SC-CR-66-2301, 1966.
9. Regulatory Guide 1.23. U.S. Nuclear Regulatory Commission, 1980.
10. Briggs, G. A., "Diffusion Estimation for Small Emissions." ATDL Contribution File No. 79, Atmospheric Turbulence and Diffusion Laboratory, 1973.
11. Hanna, S. R., G. A. Briggs, and R. P. Hosker, Jr., *Handbook of Atmospheric Dispersion*. DOE/TIC-11223, Department of Energy, Washington, DC, 1982.
12. Lockheed Martin, GEOTRAP Model Description. Appendix G to Volume 3 of Cassini GPHS-RTG Final Safety Analysis Report, June 1997.
13. NASA, U.S. Standard Atmosphere 1976. NASA-TM-X-74336, 1976.
14. Haskin, F. E., "Plutonium Entrainment and Vaporization After Coincident Impact (PEVACI) Version 1.0. Letter Report ERIC-0004 to V. J. Dandini, Sandia National Laboratories, Oct. 10, 2001.
15. Volchock, H. L., "The Anticipated Distribution of Cd-109 and Pu-238 (from SNAP-9A) Based upon the Rh-102 Tracer Experiment." HASL-165, p. 312–331, U.S. Atomic Energy Commission, 1966.
16. Leipunskii, O. I., J. E. Konstantinov, G. A. Fedorov, and O. G. Scotnikova, "Mean Residence Time of Radioactive Aerosols in the Upper Layers of the Atmosphere Based on Fallout of High-Altitude Tracers." *J. of Geophysical Research*, 75, p. 3569–3574, 1970.
17. Bartram, B. W. and D. K. Dougherty, "A Long Term Radiological Risk Model for Plutonium-fueled and Fission Reactor Space Nuclear Systems." NUS Corporation, NUS-3845, 1981.
18. Pruppacher, H. R. and J. D. Klett, Microphysics of Clouds and Precipitation. Reidel, Dortrecht, 1978.
19. Lockheed Martin Missiles & Space, Cassini GPHS-RTG Final Safety Analysis Report (FSAR). Appendix H, HIAD Model Description, June 1997.
20. U.S. Nuclear Regulatory Commission, Reactor Safety Study. Appendix VI, WASH-1400, 1975.
21. Interagency Nuclear Safety Review Panel, "The Role of Resuspension of Radioactive Particles in Nuclear Assessments." Proceedings of Technical Interchange Meeting, Cocoa Beach, FL, 1993.
22. Dobranich, D., D. A. Powers, and F. T. Harper, "The Fireball Integrated Code Package." SAND97-1585, Sandia National Laboratories, Albuquerque NM, July 1997.
23. Boughton, B. A. and J. M. DeLaurentis, "An Integral Model for Plume Rise from High Explosive Detonations." ASME/AIChE National Heat Transfer Conference, Pittsburgh, PA, 1987.
24. Haskin, F. E., Plutonium Entrainment and Vaporization after Coincident Impact. Letter Report to Vincent J. Dandini, Sandia National Laboratories, Oct. 2001.
25. Lockheed Martin, SATRAP Model Description. Appendix F to Volume 3 of Cassini GPHS-RTG Final Safety Analysis Report, June 1997.
26. Lockheed Martin, HIAD Model Description. Appendix H to Volume 3 of Cassini GPHS-RTG Final Safety Analysis Report, June 1997.
27. Lockheed Martin, PARDOS Model Description. Appendix I to Volume 3 of Cassini GPHS-RTG Final Safety Analysis Report, June 1997.

Student Exercises

1. Using Eq. 11.3, estimate the cloud shine dose rate conversion factors for Kr-88, Cs-137, and I-131. Compare to the values in Table 11.1.

2. An individual is exposed to a cloud containing 1 Bq/m^3 of Sr-90. (a) What is the skin contamination in Bq/m^2 after 4 hours of exposure assuming a deposition velocity of 0.01 m/s? (b) What is the beta dose rate to the skin at 4 hours? (c) What skin dose would an individual exposed for 4 hours receive?

3. Repeat Exercise 2 assuming I-131 instead of Sr-90. Assume deposition to the skin ceases at 4 hours but the individual is not decontaminated. What is the additional skin dose?

4. The ground shine skin dose rate conversion factor for Cs-137 is 2.75×10^{-16} (Sv/s)/(Bq/m^2). Assume an individual is a cloud contaminated with Cs-137. Assume a deposition velocity to both skin and ground. What is the ratio of the skin dose rate due to ground shine to that from skin contamination?

5. What is the fraction of very small particles deposited in equal-area latitude band 5 for (a) a release to the stratosphere between 35°N to 45°N, (b) a release to the stratosphere between 35°S and 45°S, (c) a release to the mesosphere?

6. For an instantaneous release of a parent nuclide to the mesosphere, solve the ordinary differential equation for the activity of the daughter nuclide in the mesosphere.

7. One curie of ^{238}Pu is instantaneously released to the stratosphere at 40° latitude. Using the HIAD default residence times from Table 11.12 and a weathering constant $\lambda_0 = 7.4854 \times 10^{-3}$ y^{-1}, plot the fraction of the original activity that resides in the mesosphere, stratosphere, troposphere, and on that deposited to the Earth's surface as a function of time after the release.

8. For an instantaneous release of a parent nuclide to the stratosphere develop a solution for (a) the time-integrated air concentration at ground level as a function of time and (b) the time-integrated ground surface contamination as a function of time.

9. For the instantaneous 1 Ci release of ^{238}Pu to the stratosphere described in Exercise 7, plot the time-integrated ground-level air concentration in equal-area latitude band 5 as a function of time after the release.

10. For the instantaneous 1 Ci release of ^{238}Pu to the stratosphere described in Exercise 7, plot the time-integrated ground concentration in equal-area latitude band 5 as a function of time after the release.

11. Develop an expression for the time-integrated ground concentration when the integration limits range from zero to infinity. Find the value of the weathering rate constant λ_0 that gives the same integral as the two-mode Reactor Safety Study weathering model for a mesospheric ^{238}Pu release and the HIAD default values of λ_3, λ_2, and λ_1.

12. For the instantaneous 1 Ci release of ^{238}Pu to the stratosphere described in Exercise 7, plot the resuspended ground level air concentration as a function of time after the release.

Chapter 12

Safety Implementation

Joseph A. Sholtis, Albert C. Marshall, Gary L. Bennett,
Neil W. Brown, Veniamin A. Usov, and Sandra M. Dawson

The objective of this chapter is to provide the reader with an understanding of the safety implementation process; i.e., this chapter will describe how safety issues discussed in previous chapters are addressed in space nuclear safety programs. The structure and approach of safety programs and the U.S. and Russian safety and environmental review processes are discussed. International space nuclear safety guidelines, lessons learned, future directions, and concluding observations are presented.

1.0 Overview of Review Process

Safety approaches for terrestrial nuclear systems are not, in general, appropriate for nuclear systems launched into space. Both the United States and Russia have established safety review and approval processes for space nuclear missions. As mentioned in Chapter 3, the U.S. Interagency Nuclear Safety Review Panel (INSRP) was established in the 1960s as a standard process for reviewing the safety of U.S. space nuclear missions. Representatives from the DOE, DoD, NASA, the Environmental Protection Agency (EPA), and the Nuclear Regulatory Commission (NRC) form the INSRP. The INSRP representatives are supported by experts with extensive backgrounds and experience in launch vehicle design and operations, reentry response, nuclear systems design and operations, nuclear systems accident response, meteorology, radiological and environmental effects, probabilistic risk assessment, and uncertainty analysis. The mission sponsoring agency or organization is responsible for preparing Preliminary, Draft Final (previously called "Updated"), and Final Safety Analysis Reports (PSAR, Draft FSAR, and FSAR, respectively) for review by the INSRP. The INSRP uses the SARs and other sources of information (e.g., launch vehicle reliability data and nuclear safety test data) to prepare a Safety Evaluation Report (SER) for the Executive Office of the President. Approval by the Executive Office of the President is required for launch of any nuclear system or significant quantity of nuclear material.

Although the focus of this text is on topics addressed by the INSRP process, other review processes address radiological issues. In 1971, a requirement for an Environmental Assessment (EA) or an Environmental Impact Statement (EIS) was established under U.S. Public Law 91-190 (known as the National Environmental Policy Act, or NEPA). An EIS is required for those activities that could have a significant adverse effect on the environment. For space nuclear missions, the EIS must address potential radiological and non-radiological environmental impacts. Other safety review and approval processes include approvals for transportation of nuclear materials to the launch site and launch site safety approval. A timeline of the INSRP review and approval process is presented in Figure 12.1. The figure also presents the approximate timeline for the NEPA process, range safety, range operations, and orbital safety review and approval processes [1]. Figure 12.1 does not include all documentation and approval requirements.

The current Russian space nuclear safety review and approval process is very similar to the U.S. process. In particular, the Russian process requires the development of Preliminary, Intermediate, and

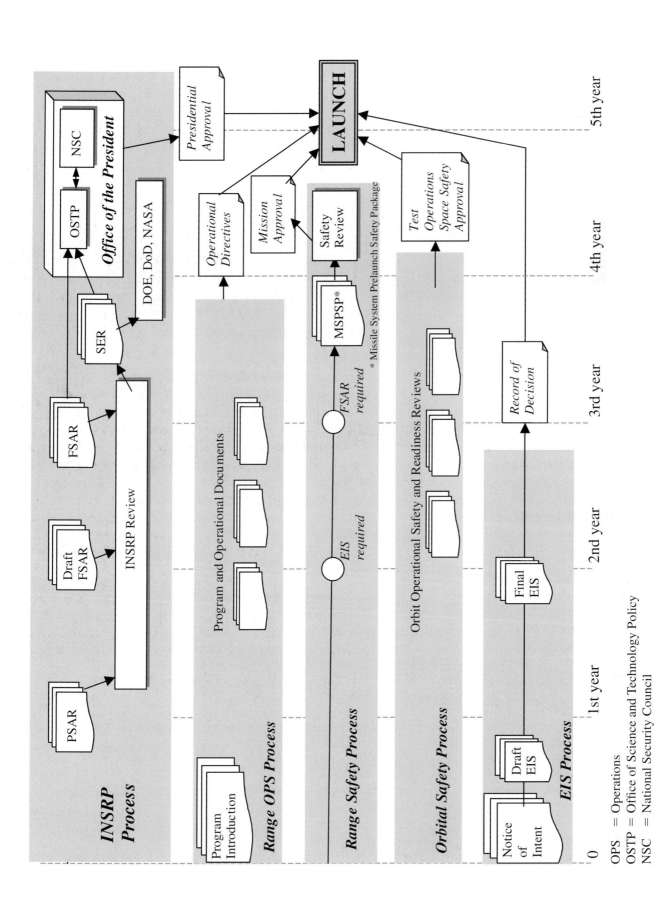

Figure 12.1 Approximate timeline for launch approval processes for U.S. space nuclear missions.

OPS   = Operations
OSTP = Office of Science and Technology Policy
NSC   = National Security Council

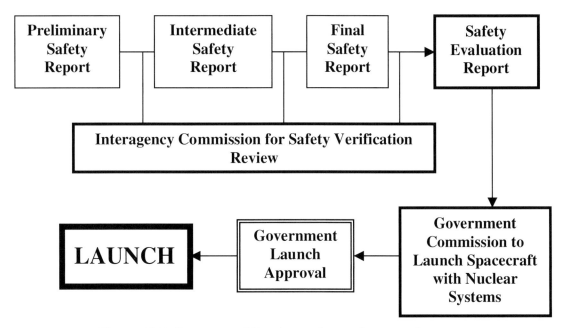

Figure 12.2 Summary of Russian review and approval process.

Final Safety Reports. Review is carried out by the Interagency Commission for Safety Verification. The government agency for oversight of nuclear and radiation safety presides over interagency safety committees for space nuclear safety missions. Safety issues are addressed by various agencies of the Ministry of Health, the agency for oversight of nuclear and radiation safety, the government agency for environmental protection, and other organizations. These organizations provide oversight during design and construction of the space nuclear system. A summary of the basic Russian review and approval process is presented in Figure 12.2.

2.0 Safety Programs

Safety programs are established to assure that space nuclear missions will be conducted safely and to assure that launch approval will be obtained. Space nuclear safety programs must be initiated at the concept design phase and continue through system development, system deployment, and operation. Addressing safety issues during system development permits incorporation of safety functions into the system design, rather than adding safety features after the system design and development have been completed. The design-in safety approach generally provides a more functional safety approach while reducing costs, relative to a post-design safety approach.

A well-planned safety program should foster a safety culture among all program participants. This philosophy encourages all program participants to consider safety as a normal matter of course while performing their jobs, rather than assuming that all safety issues are only the safety team's concern. Making safety an integral aspect of the program also enhances safety communication and teamwork. Safety program activities must be planned to interface with design, quality assurance, and other mission activities. Internal safety reviews and surveys are also an essential feature of safety programs.

The organizational structure of space nuclear programs can assume a number of possible forms. One possible organizational structure, adopted by the U.S. SP-100 space reactor program, is illustrated by the solid lines and boxes in Fig. 12.3. The space nuclear system program manager has overall responsibility for guiding the development of the space nuclear system. In the United States, development of the space nuclear system is primarily the responsibility of the Department of Energy. For SP-100, the

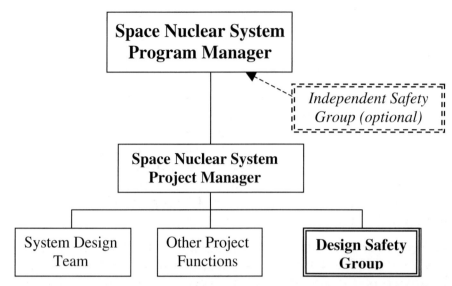

Figure 12.3 Possible basic nuclear safety organizational structure (designer is typically the contractor).

detailed direction and implementation of the development program were carried out at the Project Office at Jet Propulsion Laboratory with Los Alamos National Laboratory support for the reactor subsystem. Other organizations may take on the Project Office role, such as NASA headquarters or the integrating contractor. A private contractor generally carries out system development and manufacture; however, the project office and other national laboratories often play active roles in analysis and testing of systems and system components. For space nuclear missions, the design safety group for the system design contractor performs the primary safety functions. Flight nuclear safety functions include:

1. Interfacing with the Program and Project Offices on safety matters,
2. Development of a Safety Program Plan,
3. Identifying applicable safety guidance and requirements,
4. Identifying safety documentation and approval requirements,
5. Interpretation of safety guidance,
6. Development of detailed safety specifications,
7. Contributing to safety component design,
8. Performance of in-depth safety analysis,
9. Performance of safety testing,
10. Development of safety procedures,
11. Preparation and Safety Analysis Reports for the Program Office,
12. Presentation of safety findings to INSRP and defense of findings,
13. Interface with design team and quality assurance team on safety matters,
14. Conduct internal safety audits and reviews,
15. Preparation of much of the EIS,
16. Preparation of other safety documentation (e.g., for range safety), and
17. Serving as public interface on safety matters.

In addition, the design safety team develops safety procedures and performs safety analysis for ground nuclear safety.

Although the contractor design safety team typically carries out much of the detailed safety work, the U.S. national laboratories have traditionally had major roles in safety programs, safety analysis, and safety testing. The organizational structure of the safety program for a particular mission will depend on many factors; e.g., interagency agreements, contractual arrangements, and precedent.

For space nuclear missions, INSRP provides the required independent safety review. In order to provide an additional measure of independence, the design safety team is funded by and reports to a line organization different from the design team. Although not required, the program office may choose to establish its own independent safety team to provide top-level safety guidance, to monitor and assess safety activities, to carry out independent analysis, and to suggest modifications relating to safety. The independent safety team can also serve as the public interface for safety matters. The system development contractor may also choose to establish its own independent safety group (this option was taken for the SP-100 program by General Electric, the system developer).

Like the United States, Russian space nuclear programs establish safety programs that emphasize a culture of safety and follow similar safety review and approval approaches. Russian nuclear systems for use in space undergo rigorous safety verification procedures to minimize risk to the public and the environment.

3.0 Safety Documentation

Safety documentation is an essential element of the safety process. Safety documentation includes both guiding documentation provided by the government and documentation developed during the safety process for a particular mission.

3.1 Safety Guidance Documentation

A number of government issued documents are relevant to space nuclear safety programs. A partial list of documents potentially relevant for U.S. missions includes:

- U.S. Department of Energy. *DOE Order 5481.B: Safety Analysis and Review System,* September 23, 1986, Revised May19, 1997.

- U.S. Department of Energy. *DOE Order 5480.6: Safety of Department of Energy-Owned Reactors,* September 23, 1986.

- HNUS Corporation. *Overall Safety Manual,* Prepared by the U.S. Department of Energy (Updated 1982).

- NASA Std 3000. *Man-Systems Integration Standards, Rev. B Vol. I,* July 1, 1995.

- ESMC 127, Eastern Space and Missile Center. *Range Safety,* July 30, 1984.

- ESMC S-Plan 28-74, Eastern Space and Missile Center. *Major Radiological Support (HOT SHOT) Revision # 1,* August 7, 1990.

- ESMC 160-1, Eastern Space and Missile Center. *Radiation Protection Program, change* 1, October 22, 1985.

- *Nuclear Safety Review and Launch Approval for Space or Missile Use of Radioactive Material and Nuclear Systems,* Air Force Regulation 122-16, August 21, 1992.

- United Nations. *Report of the Legal Subcommittee on the Work of its Thirty-First Session, Chairman's Text, Principles Relevant to the Use of Nuclear Power Sources in Outer Space,* U.N. Document A/AC.105/L,198. June 23, 1992.

The principal guidance for Russian space nuclear safety includes national standards of radiation safety and basic medical/clinical regulations, as well as mission-specific safety requirements developed during the design and fabrication of the space nuclear systems. More recently, safety criteria have been incorporated based on recommendations by the U.N. committee for utilization of space for peaceful objectives. Safety requirements are specified in the national documents:

1. *Standards of Radiation Safety,*
2. *Basic Health Rules of Assuring Radiation Safety,* and
3. *Health Rules of Assuring Radiation Safety of Space Nuclear (Reactor and Radioisotope) Power Sources.*

3.2 Program-Developed Safety Documentation

A number of other documents are required before launch of space nuclear systems. A detailed discussion of much of this documentation is found in Ref. [1]. For the United States, a list of important safety documentation includes:

1. *Launch Vehicle Data Book*
2. Space Nuclear Mission Safety Documentation
   - *Safety Program Plan*
   - *Flight Safety Requirements Documents*
   - *Safety Analysis Reports* (PSAR, Draft FSAR, and FSAR)
   - *Safety Evaluation Report*
   - *Radiological Protection Plan*
   - *Emergency Preparedness and Response Plan*
   - *Safety Analysis Report for Packaging* (SARP)
   - *Ground Safety Analysis Report* (GSAR)
3. Range Safety Documentation
   - *Program Introduction Plan* (PI)
   - *Statement of Capabilities Document* (SOC)
   - *Program Requirements Document* (PRD)
   - *Operational Requirements* (OR)
   - *Operational Directive* (OD)
   - *Missile System Prelaunch Safety Package* (MSPSP)
4. Orbital Safety Documentation
   - *Test Operational Risk Assessment* (TORA)
   - *Preliminary Hazards List* (PHL)
   - *Test Operations Space Safety Approval* (TOSSA)
   - *Orbital Readiness Space Safety Review* (ORSSR)
5. Environmental Protection Documentation
   - *Notice of Intent*
   - *Environmental Impact Statement (Draft and Final)*
   - *Notice of Availability*
   - *Record of Decision*

A partial list of Russian program-developed safety documentation includes:

1. *Preliminary, Intermediate,* and *Final Safety Reports*
2. *Safety Evaluation Report*
3. *Health Rules of Assuring Radiation Safety of Space Nuclear Power Sources* (issued by Ministry of Health)
4. *Rules for Nuclear Safety of Space Nuclear Power Sources* (agreed upon by the Science and Technology Council of the Ministry of Nuclear Energy in view of inspections of results and reviews by experts, and approved by the state oversight committee for nuclear and radiation safety)

5. *Rules for Notification of Executive Authorities on the Launch of a Spacecraft with a Nuclear Power Source, Notification of Local Authorities and Rendering Help to Population as Necessary at Accidental Reentry of the Spacecraft (approved by government decree)*

3.3 Safety Analysis Reports

The safety analysis considerations discussed in this book are relevant, in particular, to the development of Safety Analysis Reports; consequently, a brief discussion of SARs is presented in this section.

The first formal safety analysis document delivered to INSRP is the Preliminary Safety Analysis Report. The PSAR should be issued after the design concept has been selected for a mission, and the Draft FSAR is issued after the design freeze (prior to 2002, the Draft FSAR was referred to as the Updated Safety Analysis Report). The FSAR should be issued about two years before the planned launch date. Safety Analysis Reports are organized into three volumes. The following summary of the SAR content was adapted from Reference [2]:

Volume 1. *Reference Design Document*
- Mission/Flight System Summary
- Nuclear Power Source Description
- Ground Support Equipment
- Mission Profile
- Launch Vehicle (including flight safety and tracking plans)
- Launch and Trajectory Characteristics
- Launch Site

  (including demographic, topographic, and meteorological characteristics)
- Range and Radiological Safety Operations
- Safety-Related Systems and Components

Volume 2. *Accident Model Document*
- Accident and Radiological Models and Data (including test data and validated computer codes that support the analysis)
- Vehicle and Nuclear System Failure Mode Analysis (from prelaunch through final disposition, with a description of potential accident environments and flight contingency options)
- Nuclear System Response to Accident Environments
- Mission Failure Evaluation (includes accident probabilities and quantity of radioactive material potentially released to provide risk profile)

Volume 3. *Nuclear Risk Analysis Document*
- Presentation of Mission Radiological Risk

4.0 Safety Review and Launch Approval

Safety review for launch approval is the principal focus of space nuclear safety activities. Very similar processes for safety review have evolved in the United States and Russia.

4.1 U.S. Safety Review and Launch Approval

For any U.S. space mission involving the use of nuclear energy systems with significant quantities of radioactive or fissile material, launch approval must be obtained from the Executive Office of the President. The launch decision is based on a consideration of the projected benefits and risks of the mission. It is also based on an established and proven in-depth technical nuclear safety analysis and risk evaluation that includes both the program's analysis and an independent characterization of mission radiological risk by an Interagency Nuclear Safety Review Panel.

The current safety review and launch approval process used in the United States is an outgrowth of the original separate NASA, DoD, and the Atomic Energy Commission (AEC) safety reviews. Because of statutory requirements, DOE is responsible for the safety of the nuclear systems that it designs and produces; DoD, through its principal launching agency—the Air Force—has range safety responsibility. Both DoD and NASA are responsible for the overall safety of their respective missions. As a consequence, DoD, NASA, and the AEC each conducted their own separate safety reviews for nuclear-powered space missions flown through 1965.

To avoid duplication of effort, in the mid-1960s a joint NASA/DoD/AEC safety review process replaced the three separate reviews. The process ensures independence and objectivity of all its participants by requiring that each individual involved have no responsibilities or interests relating to the mission or the nuclear system. Overall, the current process involves three basic activities: (1) preparation of several increasingly detailed Safety Analysis Reports by the mission sponsor and participating contractors; (2) review and evaluation of those SARs by INSRP (specifically empanelled for the purpose); and (3) characterization of the mission radiological risks by the INSRP in a Safety Evaluation Report for the decision-maker within the Executive Office of the President. This process permits an informed decision about whether to proceed with the mission. The decision is made at the highest level of the U.S. Government based on risk, benefit, and uncertainty considerations buttressed by the best available information and carefully scrutinized analysis and test results.

During the late 1960s up through the early 1990s, empanelled INSRPs were chaired by three INSRP coordinators appointed by the NASA Administrator, the Assistant to the Secretary of Defense for Atomic Energy, and the Secretary of Energy from within their respective agency's Inspector General or Safety Offices. (When the functions previously carried out by the former AEC were split between the Department of Energy and the NRC, the AEC INSRP coordinator was replaced a representative from the DOE.) Presidential Directive/National Security Council Memorandum #25 (PD/NSC-25), which establishes the process and serves as the authorization for empanelling an INSRP, was revised in 1995—adding a coordinator from the Environmental Protection Agency (EPA) and a technical advisor from the Nuclear Regulatory Commission (NRC). Thus, the current coordinators include representatives from NASA, DoD, DOE, and the EPA, with the NRC serving as a technical advisor. Collectively, the INSRP coordinators and technical advisor are responsible for conducting a comprehensive, independent nuclear safety evaluation for each space mission employing a nuclear power system or significant quantities of radioactive materials. The INSRP documents their evaluation in a Safety Evaluation Report, and the SER is provided to the mission-sponsoring agency as an input to their formally requesting launch approval from the White House. More importantly, the SER is provided to the White House Office of Science and Technology Policy (OSTP) to ensure that an independent characterization of the mission radiological risks is available to the decision-maker to permit an informed launch decision on the basis of risk-benefit considerations.

Technical experts from government, national laboratories, private industry, and academia assist the INSRP coordinators and technical advisor, as needed. These technical experts have no direct ties with or any conflicting interest in the program, mission, or nuclear systems to be used. Up through 1997, the technical experts were grouped into five working groups. The five INSRP working groupss were the Launch Abort working group (LASP), the Reentry working group (ReSP), the Power Systems working group (PSSP), the Meteorology working group (Met), and the Biomedical & Environmental Effects working group (BEES). Experts that made up the working groups are now referred to as consultants; however, the basic function and working areas of the expert consultants remain unchanged. For convenience, we will group the discussion of the experts' activities into the same functional work areas:

- **Launch Abort:** Experts in this work area conduct a review and evaluation to ensure that a complete matrix of credible prelaunch, launch, and ascent accidents has been identified and characterized. The matrix of postulated accidents that can pose a threat to the nuclear systems includes accident type, probabilities, and environments. In addition, these experts ensure that all postulated accidents are properly characterized and all relevant information is conveyed to other experts supporting the INSRP review.

- **Reentry:** These technical consultants review and evaluate all accidents that can potentially result in the reentry of a nuclear system into the Earth's atmosphere. Their evaluation ensures that the likelihood of occurrence, environmental conditions, and response of the nuclear systems during atmospheric reentry are properly characterized. If radioactive material release is predicted as a result of a postulated atmospheric reentry, this group ensures that the release probability and release description are properly characterized and passed to meteorological and biomedical experts for further evaluation. If the nuclear system at least partially survives the reentry and impacts Earth's surface, the group ensures the conditions of the impact are properly characterized and conveyed to the power systems experts.

- **Power Systems:** These consultants ensure that the response of the nuclear systems to all postulated prelaunch, launch, ascent, and post-reentry Earth impact accidents and accident environments is properly characterized. This includes proper characterization of any postulated nuclear fuel releases, their associated conditional probability of occurrence, and any modification to those initial fuel releases due to post-release environments.

- **Meteorology:** These experts ensure proper treatment of dispersal and transport of any postulated nuclear fuel releases within the biosphere. In addition, they ensure that the predicted ground surface deposition of fuel materials from a postulated release has been properly analyzed (to permit projections of land contamination), and assess resuspension from areas of ground contamination. The results of this evaluation are passed to experts responsible for evaluating biomedical and environmental effects.

- **Biomedical and Environmental Effects:** These technical experts ensure that all significant dose pathways are analyzed and that individual and collective doses are properly estimated. They also ensure proper characterization of mission radiological risk and ground contamination that could potentially involve cleanup.

These expert reviews and evaluations are conducted in three main stages; i.e., after the issuance of the PSAR, Draft FSAR (USAR), and FSAR. Information is shared among the technical experts and with the INSRP coordinators and technical advisor. In addition, the experts typically document the results of their review and evaluation in reports to the INSRP coordinators and technical advisor.

The INSRP coordinators, and technical advisor, sometimes with the help of a probabilistic risk assessment and uncertainty expert, use the experts' findings when developing the SER. The SER characterizes INSRP's independent evaluation of the mission radiological risk, i.e., INSRP's independent assessment of the state of knowledge about the mission's radiological risk, including uncertainty. The results are typically presented in the SER in the form of complementary cumulative distribution functions (*ccdfs*) illustrating the probability of exceeding specified consequences at various levels of confidence (e.g., at the 5th, 50th, mean, and 95th percentiles). These results are presented along with descriptions of several scenarios representing the central tendency and the high end. The risks associated with other types of activities are also included to provide decision-makers a perspective of the mission radiological risk relative to the other types of risks.

Upon completion, the SER is provided to the OSTP within the Executive Office of the President. The SER is also provided to the INSRP parent agencies for their review. Management within the mission-sponsoring agency uses the SER to assist in its decision whether to formally request launch approval from the White House; or the sponsoring agency may choose to modify the program, mission, or systems to further reduce risk. The OSTP uses the SER to characterize and weigh mission risks against mission benefits in arriving at a launch determination decision. The SER, therefore, serves as an independent assessment of the mission radiological risks and as a basis for an informed launch decision, based on risk-benefit considerations, at the highest level of the U.S. government.

At present (2003), INSRP is empanelled for systems containing more than 20 curies of radioactive materials. The fuel activity for some reactor designs falls below this limit; however, it should be assumed that all U.S. space reactor missions will undergo INSRP review regardless of the activity of their nuclear fuel.

4.2 Russian Safety Oversight, Review, and Launch Approval

A number of organizations provide oversight and control functions and resolve safety issues during design and construction of space nuclear systems. The organizations include various government agencies and arms of the Ministry of Health as well as the government oversight agency for nuclear and radiation safety and the agency for environmental protection. The Science and Technology councils of various ministries, agencies, companies, and institutes, as well as the Council of Chief Designers of participating organizations, also provide control and oversight during design and construction. Interagency safety committees are created at the request of the customer. The interagency committee is presided over by the representative of the government oversight agency for nuclear and radiation safety. The interagency committee assures the independent review of safety documents developed for the nuclear systems. This committee evaluates the documents and, if necessary, includes the findings from its own analysis. Safety criteria used in the design and safety evaluation of space nuclear systems are based on national standards and rules, as well as guidance from the International Atomic Energy Agency (IAEA) and the United Nations.

The interagency review begins with the issuance of the Preliminary Safety Report. The Preliminary Safety Report is developed from an analysis of safety systems and design features for the proposed system. Following the preliminary design, safety systems and design features are selected to form the basis for safety testing. At this stage, an Intermediate Safety Report is submitted to the interagency committee for review and comment. A Final Safety Report is submitted following the completion of the final design and supporting technical documentation. Verification of the effectiveness and reliability of safety systems and system components is also completed at this stage. After interagency committee review of the Final Safety Report, a Safety Evaluation Report is prepared for publication for the United Nations and the IAEA.

A special decree of the Russian government establishes the government commission that makes the final decision (with government approval) to allow the launch of spacecraft with nuclear systems onboard. The membership of the commission includes representatives of the government customer, ministries, industry, and organizations developing the system. The structure of this decision-making process can be modified, however, to respond to future developments in applicable regulations, including guidance by appropriate U.N. committees.

5.0 U.S. Environmental Assessment Regulations

The National Environmental Policy Act (NEPA) is a law requiring all federal agencies to consider and document the consideration of any agency actions and activities that could have a significant impact on the quality of the human environment. Launch and operation of a spacecraft may meet these criteria. Implementing procedures and guidelines provide detailed guidance on how projects and programs can meet the NEPA requirements.

NEPA implementing regulations require that projects with potentially significant impacts complete an environmental analysis, either an environmental assessment (EA) or environmental impact statement (EIS). The EA is a concise public document that provides sufficient information and analyses to determine if an EIS needs to be prepared. If it is determined that there are no significant environmental impacts, a Finding of No Significant Impact (FONSI) is published in the Federal Register and NEPA compliance for the project is completed. If the determination is that there are potential significant impacts, an EIS is prepared. An EIS may be done without an EA preceding it if it is reasonably certain that an EIS will be required.

An EIS is a detailed and rigorous document, which provides full and fair discussion of significant environmental impacts, based on the best available information at that time. It presents a discussion of the purpose of the proposed action and details the need for the action. The EIS provides a general understanding of how the action will be implemented, and what potential environmental impacts, both positive and negative, could result. It is also designed to inform decision makers and the public of a reasonable range of alternatives that are compatible with the purpose and need of the action. These alternatives must be compared with the proposed action in terms of their potential environmental impacts. In addition to alternatives, the EIS must also consider the no-action alternative. The EIS undergoes two pubic comment periods. The first comment period follows after a notice of intent (NOI) to prepare an EIS is published in the Federal Register, with a request for comments on the scope of the project. The NOI is also sent to any person or group who has indicated an interest in the project.

The second public comment opportunity is a review of the draft EIS (DEIS). Copies of the DEIS are sent to anyone who responded to the NOI and to any others who have indicated an interest in the project. Additionally, a Notice of Availability (NOA) is published in the Federal Register. Also, a draft EIS is provided to any interested or involved federal and state agencies for review. Anyone can provide comments on this draft. Comments received are considered by the sponsoring agency; and agency responses, as appropriate, are published in the final EIS. At the conclusion of the EIS process a record of decision (ROD) is filed, documenting the agency's decision on whether to go forward with the action, an alternative, or no action.

An EIS for space nuclear missions must provide a detailed analysis of the impact of a normal launch on the air, water, plants, etc., of the launch area. It must also discuss a reasonable range of postulated accidents and their potential impact. Alternatives considered include alternative launch vehicles, alternative power sources, and alternative flight trajectories.

The launch radiological impact analysis for accidents at the launch site, down range, and from Earth orbit reentry is based on best available information with respect to launch vehicle accident probabilities, the resulting potential accident environments, the nuclear system responses to those environments, and an assessment of the potential health impacts. Postulated reentry accidents during an Earth swing-by are also considered. The analysis is a combined effort of the launch vehicle owner, the launching organization (NASA or DoD), the Department of Energy, and other involved government agencies and supporting contractors. Although the analysis performed for an EIS is similar to those performed for INSRP, NEPA does not require a worst-case analysis.

For missions that use radioisotope heater units with no RTGs, either an EA or an EIS is required. The decision on which will best satisfy NEPA compliance requirements is based on a number of factors, most importantly whether it is likely that the process will end with a FONSI. If the EA is completed, and it is determined that there is a potential for significant environmental impacts, a DEIS will be prepared.

The NEPA process differs significantly in several ways and is wholly separate from the INSRP review and Presidential launch approval process. Because the NEPA process is completed early in the program cycle, the agency may have to use preliminary data and analysis instead of the later, more comprehensive or refined analysis used for the INSRP review process. In addition, the NEPA process differs in that it involves the public and a large number of federal, state, and local agencies.

6.0 Other Safety Program Responsibilities

In addition to the responsibilities discussed in the preceding sections of this chapter, the mission space nuclear safety program has several other important safety responsibilities. These other responsibilities are summarized in this section for U.S. programs.

6.1 Ground Operations

Although the emphasis of this text is on flight safety, it is important to recognize that the safety program must also address nuclear safety considerations relating to ground activities. For example, the contractor or user organization must make a safety assessment for planned operations at each facility where the fueled nuclear system or radioactive materials will be located. The assessment is documented in a Ground Safety Analysis Report (GSAR). DOE approval is required prior to ground operations with nuclear systems or radioactive materials. The minimum contents of the GSAR [2] are a discussion of:

- The facility organization,
- Facilities and equipment,
- Postulated accidents involving radioactive materials or the nuclear system,
- Consequences of the postulated accidents,
- Mitigation of the postulated accidents,
- Personnel operational dose analysis, and
- Steps taken to assure ALARA radiation doses compliance with DOE 5480.1B.

In addition to the GSAR, a comprehensive, ongoing operational safety program must be established to assure safe operation for the protection of workers, the public, and the environment. The operational safety program is conducted in compliance with DOE orders relating to radiation safety, occupational safety, and environmental protection. The operational safety program includes an operational safety analysis, a quality assurance program, a radiation protection plan and radiation protection program, operational reviews and surveys, a contractor self-appraisal program, provisions for DOE reviews surveys and approvals, and readiness reviews for each specific facility.

If a full ground reactor test program is planned, a separate safety program may be established to comply with appropriate DOE orders. The safety program requirements for extensive ground nuclear testing is beyond the scope of this book.

6.2 Transportation Safety

DOE orders and DOT regulations, based on 10CFR 71 [3], establish the requirements for packaging and transport of special nuclear materials and nuclear systems. To transport significant quantities of radioisotope fuel or highly enriched fresh uranium fuel, a certified shipping container is required. Certification requires that the container must pass a series of extremely demanding tests. The development and certification of shipping containers can be a slow and very expensive process. The need to move systems to the launch site must account for the fact that existing certified shipping containers may not accommodate large space nuclear power systems containing nuclear fuel.

6.3 Emergency Preparedness

User agencies and contractors must incorporate emergency planning, preparation, and response in their safety program. An emergency response plan is required. The safety program must provide personnel training and ready necessary resources to assure effective response to emergency situations. Specific requirements are contained in DOE orders 5500.2, 500.3, 500.4 and in the Minimum Radiological Health and Criticality Safety Criteria (AL). Emergency preparedness plans are required for all sites and situations involving the utilization, handling, transporting, or storing of special nuclear materials.

Contingency plans and emergency response resources are also required for possible launch aborts and for other postulated scenarios that could result in reentry of the system. Unique features of launch abort and reentry emergency response include plans for nuclear source or nuclear system recovery and long-term post-contingency assessments and recovery.

7.0 International Considerations

By its very nature, space nuclear safety is an international consideration. The international community, through the United Nations and other international entities, has been taking an increasingly active role in establishing agreements on safety and environmental issues relating to space activities and on international nuclear power considerations. In the 1960s through the early 1980s, five international treaties were established to address space safety and environmental issues; i.e.,

- Treaty on Principles Governing the Activities of States in the Exploration and Use of Outer Space, Including the Moon and Other Celestial Bodies (known as the *Outer Space Treaty*) 10 October 1967;
- Agreement on the Rescue and Return of Astronauts and the Return of Objects Launched into Outer Space (known as the *Rescue Agreement*) 3 December 1968;
- Convention on International Liability for Damage Caused by Space Objects (known as the *Liability Convention*) 9 October 1973;
- Convention on Registration of Objects Launched into Outer Space (known as the *Registration Convention*) 15 September 1976; and
- Agreement Governing the Activities of States on the Moon and Other Celestial Bodies (known as the *Moon Treaty*, not signed by the United States) 11 July 1984.

The first four treaties govern activities in space whether nuclear or non-nuclear while the Moon Treaty does mention the emplacement of radioactive materials on the Moon.

In the late 1980s, in the wake of the Chernobyl nuclear power plant accident, the United Nations adopted two conventions relating to nuclear power:

- Convention on Early Notification of a Nuclear Accident, 27 October 1987, and
- Convention on Assistance in the Case of a Nuclear Accident or Radiological Emergency, 26 February 1987.

Development of U.N. principles relating to space nuclear power began in the early1980s. However, it was not until 1992 that the Scientific and Technical Subcommittee (STSC) of the United Nations Committee on the Peaceful Uses of Outer Space (COPUOS) adopted a preamble and 11 principles directly addressing space nuclear power [4]. A brief description of the preamble and principles follows:

Preamble—The preamble recognizes that for some missions nuclear power systems are essential and affirms that the principles only apply to nuclear power systems "... devoted to generation of electric power on board space objects for non-propulsive purposes, which have characteristics generally comparable to those of systems used and missions performed at the time of the adoption of the Principles . . . ." In other words, the principles do not apply to nuclear propulsion systems or to new types of nuclear power systems. The preamble also recognizes "... that this set of Principles will require future revision in view of emerging nuclear-power applications and of evolving international recommendations on radiological protection."

Principle 1—*Applicability of international law*—basically states that the use of nuclear power systems will be carried out in accordance with international law.

Principle 2—*Use of terms*—defines a number of terms, in particular "... the terms 'foreseeable' and 'all possible' describe a class of events or circumstances whose overall probability of occurrence is such that it is considered to encompass only credible possibilities for purposes of safety analysis." In addition the definition of the term "general concept of defense-in-depth" allows flexibility in achieving this goal by allowing consideration of "... the use of design features and mission operations in place of or in addition to active systems, to prevent or mitigate the consequences of system malfunctions. Redundant safety systems are not necessarily required for each individual component to achieve this purpose.

Given the special requirements of space use and of varied missions, no particular set of systems or features can be specified as essential to achieve this objective."

Principle 3—*Guidelines and criteria for safe use*—this principle sets forth general goals for radiation protection and nuclear safety followed by specific safety criteria for nuclear reactors and for radioisotope generators.

Principle 4—*Safety assessment*—requires a "thorough and comprehensive" safety assessment which is to be made publicly available prior to each launch.

Principle 5—*Notification of reentry*—requires a timely notification of the reentry of radioactive materials to the Earth and provides a format for such notification.

Principle 6—*Consultations*—requires States providing information under Principle 5 to respond promptly to requests for further information or consultations sought by other States.

Principle 7—*Assistance to States*—requires States with tracking capabilities to provide information to the Secretary-General of the United Nations and to the State concerned and requires the launching State to promptly offer assistance. After reentry, other States and international organizations with relevant technical capabilities should also provide assistance to the extent possible when requested by the affected State.

Principle 8—*Responsibility*—States shall bear international responsibility for their use of space nuclear power systems.

Principle 9—*Liability and compensation*—holds the launching State and the State procuring such a launch internationally liable for any damage, including restoration ". . . to the condition which would have existed if the damage had not occurred". Compensation includes ". . . reimbursement of the duly substantiated expenses for search, recovery and clean-up operations, including expenses for assistance received from third parties."

Principle 10—*Settlement of disputes*—disputes ". . . shall be resolved through negotiations or other established procedures for the peaceful settlement of disputes, in accordance with the Charter of the United Nations."

Principle 11—*Review and revision*—requires that "These Principles shall be reopened for revision by the Committee on the Peaceful Uses of Outer Space no later than two years after their adoption.".

Principle 3 is perhaps the most important and the most controversial of the principles. During the adoption of the principles, the U.S. delegation formally expressed reservations about the technical validity of these principles to the United Nations. For example, in Section 1.3 of Principle 3, dose limits were established for accidents. Although the dose limits did not apply to "low probability accidents with potentially serious consequences" the U.S. delegation pointed out that Principle 3 should address risk (probability of exposure times consequence) rather than numerical dose limits [5]. In particular, the U.S. delegation made the point that ". . . this modification, by taking into account the probabilistic concept of risk, which is a central feature of a thorough safety assessment, relates the recommendation directly to the well-proven [space nuclear power system] practices of the United States"[6]. "In November 1990 the International Commission on Radiological Protection published new recommendations in the form of ICRP-60, which supersede the approach taken in Principle 3 when it was developed earlier that year" [7]. The IAEA independently supported the U.S. position by stating that "The sole use of the individual-related dose limits, rather than the complete ICRP system of radiation protection (including source-related constraint), is, in the Agency's view, inappropriate and does not conform with the aims of the ICRP recommendations. Secondly, as the ICRP has recently issued new recommendations on dose limitation . . . It might, therefore, be problematic to issue guidelines and criteria of safe use of [nuclear power systems] in outer space that would be outdated from their inception"[8].

The United States also proposed clarifying language in 1991 with the statement that "We believe that this clarification removes any doubts as to the intent behind the application of the term 'defense-in-depth'. As was clear at the time that the Legal Subcommittee reached consensus on this principle, the Subcommittee did not intend to apply the terrestrial standards as such to space systems." In 1991 the U.S. delegation proposed changes to the wording of Section 3.2 (relating to reentry of radioisotope sources) ". . . to take into account the fact that the probability of accidental re-entry from a hyperbolic or highly elliptical orbit can be reduced to a very low value by mission design and operations" and to recognize ". . . the fact that the practical design objective for RTG containment systems is localization rather than zero release under all circumstances, and that there are practical limits from a cost-versus-risk standpoint on 'complete' clearing of radioactivity by a recovery operation"[6].

At a meeting of the Special Political Committee, on 28 October 1992, the U.S. representative stated "The United States did not block the consensus recommendation of the Committee to forward the principles to the General Assembly, nor will the United States oppose their adoption here. On some points, however, it remains our view that the principles related to safe use of nuclear power sources in outer space do not yet contain the clarity and technical validity appropriate to guide safe use of nuclear power sources in outer space. The United States has an approach on these points which it considers to be technically clearer and more valid and has a history of demonstrated safe and successful application of nuclear power sources. We will continue to apply that approach" [9].

Although the U.N. principles are nonbinding, the United States and other nations are working together to establish generally acceptable international safety and environmental guidance relating to space nuclear power. International cooperation in space enterprises and the voice of the international community is expected to increase during the 21st century.

8.0 Retrospective on the Safety Process

The process for ensuring safety and environmental protection during space nuclear missions has been highly successful. The United States and Russia launched nearly 60 major nuclear systems into space over a period of 4 decades. Not a single injury has resulted from a space nuclear mission over the entire history of the U.S. and Russian programs. This statement does not imply, of course, that the system is perfect. A number of launch or reentry accidents have occurred with nuclear systems onboard. For all but one of these incidents, safety systems performed as planned; and no environmental damage resulted. In the case of the reentry of Cosmos 954, the reactor system did not fully disperse as designed during the reentry accident; and a significant area of the Canadian wilderness was contaminated with radioactive debris. Although no one was injured, and the environment was satisfactorily restored, human safety could have been at risk had the reentry occurred over a populated area. Lessons were learned, however, and the system design and operational practices were modified to reduce the possible risk of a similar incident. This learning process is an essential element of space nuclear safety programs.

8.1 Lessons Learned

We can also ask what lessons have been learned relating to safety programs and the safety process. The authors have made the following observations.

Planning

- An effective safety program requires careful planning. Planning should include the establishment of safety objectives, agreements on the kinds of information to be provided for safety reviews, and charters that define the roles of participants.
- Planning must determine the funding required to carry out an effective safety program. Top-level management commitment is required to assure that adequate funding, resources, and support are provided.

Design

- Safety considerations must be included in the design process, rather than trying to make post-design fixes to meet safety requirements.
- The program must develop a set of system design specifications, each of which may have safety requirements. These requirements must be used in support of safety assessments and may lead to component and system testing for both development and flight system acceptance.
- Initial safety guidance should be nonprescriptive. Prescriptive safety requirements should be developed during design development and after the basic safety approach has been selected.

Analysis and Testing

- Care must be taken to select appropriate analysis tools. Validation needs for these tools must be identified. Validation will be required for both operational environments as well as postulated accident environments. Testing will be needed to support design decisions and safety approvals.
- Safety reviewers often use bounding tests and analysis to scope out the response of nuclear systems to postulated accidents. Unfortunately, these tests and analysis are not representative of the most likely environments and responses. Testing and analysis should focus on those conditions of most use to the overall safety analysis, leaving the upper limit cases to sensitivity analysis and test only as necessary.

Presentation of Findings

- Presentation of safety documentation is very important. Historically, when SARs and SERs were not widely available (they sometimes contained classified information), the results could be presented in a form appropriate to the program, INSRP, and decision-makers. Beginning with Galileo and Ulysses, much of the process became public; consequently, it is important that information is presented in a form that is both understandable to the general public and useful to safety experts. Care must be taken to assure that appropriate qualifying language is used to provide the proper context for discussion of safety evaluations.
- A general policy of openness and responsiveness should be adopted for conveying safety findings to the public. Agency requirements must be addressed early to avoid delays in the release of safety findings. This policy must acknowledge, of course, that classified information cannot be revealed; and preliminary findings must be verified before information is released. Although some parties may choose to misuse the released information as a means to prevent program execution, this possibility must be accepted as one of the unfortunate consequence of the necessity for openness.

8.2 Future Directions

If current trends continue, we can expect increased public involvement, more international discussion, and international agreements on space nuclear safety considerations. This more inclusive environment is a positive outcome, but it remains the safety community's responsibility to provide reasoned and appropriate safety guidance rather than simply responding to popular perceptions. As a minimum, nations sponsoring space nuclear missions should take a more proactive role in forging international safety guidance.

Another important step focuses on establishing an appropriate methodology for reporting safety information. The common practice of reporting overly conservative, worst-case assessments is inappropriate and highly misleading when discussing safety findings with the public. In particular, we need to reexamine the use of the linear dose response hypothesis to estimate health effects among very large populations of people receiving very small radiation doses. The de minimus approach of using a low dose cut-off has been suggested [10] as a method for estimating population doses from very low radiation exposures. The de-minimis approach should be carefully explored to determine if the method is appropriate for predicting potential consequences of space nuclear missions.

9.0 Concluding Remarks

With the arrival of the 21st century and the rapid pace of technological advancement, we can envision a future filled with space missions that once seemed to fall squarely in the realm of science fiction. Satellites will locate ships and planes with pinpoint accuracy, process and transmit billions of bits of information in a few seconds, and monitor Earth's environmental changes in astonishing detail. Space transportation will become routine, reliable, and economical; and we may soon set foot on Mars. Advances in technology will undoubtedly produce space missions not yet envisioned. Although many space projects will benefit from technological advances that reduce power requirements, our more ambitious missions will require more power and energy to carry out their functions than can be practically provided by conventional solar cells and chemical power supplies. These requirements suggest that the high power capability of nuclear energy will be essential to our future in space. The outstanding accomplishments made possible by space nuclear power over the past several decades have placed humankind on a path to realize this vision.

Nonetheless, two major issues must be addressed to advance the field of space nuclear energy. First, as mentioned in the introduction, the widespread concern over nuclear safety has intensified since the Three Mile Island and Chernobyl accidents. The second issue relates to the low level of activity in space nuclear enterprises. In the absence of a vigorous space nuclear program, the expertise needed for future space nuclear missions could be lost. Consequently, this textbook on space nuclear safety was written both to enhance understanding of safety issues and methods in existing space nuclear programs and to preserve vital knowledge for future space nuclear programs. Without the firm belief that space nuclear missions can be conducted safely, this book would not have been attempted and completed. The authors firmly believe that, as in the past, space nuclear programs and missions will continue and will enable us to safely take the next giant step into space.

References

1. Mehlman, W. F., *Nuclear Space Power Safety and Facility Guidelines Study.* Prepared for the Department of Energy, 1995.
2. U.S. Department of Energy, *Nuclear Safety Criteria and Specifications for Space Nuclear Reactors.* OSNP-1, Rev. 0, Aug. 1982.
3. Title 10 CFR Part 71, *Packaging and Transportation of Radioactive Materials.* Jan. 1, 1992.
4. United Nations, *Report of the Legal Subcommittee on the Work of Its Thirty-First Session, Chairman's Text, Principles Relevant to the Use of Nuclear Power Sources in Outer Space.* U.N. Document A/AC.105/L,198. June 23, 1992.
5. Lange, R., *Statement by Robert Lange, U.S. Adviser at the Twenty-Eighth Session of the Scientific and Technical Subcommittee of the United Nations Committee on the Peaceful Uses of Outer Space, on Agenda Item 7, Nuclear Power Sources in Outer Space.* USUN Press Release 08-(90), U.S. Mission to the United Nations, New York, Feb. 26, 1991.
6. Lange, R., *Statement by Robert Lange, U.S. Adviser at the Thirtieth Session of the Legal Subcommittee of the United Nations Committee on the Peaceful Uses of Outer Space, to the Working Group on Nuclear Power Sources in Outer Space.* USUN Press Release 18 (91), U.S. Mission to the United Nations, New York, Apr. 10, 1991.
7. Smith, P. G., *Statement by Peter G. Smith, U.S. Representative to the Twenty-Ninth Session of the Scientific and Technical Sub-Committee of the United Nations Committee on the Peaceful Uses of Outer Space.* February 25, 1992.
8. IAEA, *IAEA Statement to the Scientific and Technical Sub-Committee of the Committee on the Peaceful Uses of Outer Space.* United Nations, New York, Feb.–Mar. 1991.
9. Hodgkins, K., *Statement by Kenneth Hodgkins, U.S. Adviser to the 47th Session of the United Nations General Assembly, in the Special Political Committee, on Item #72, International Cooperation in the Peaceful Uses of Outer Space.* USUN Press Release #116-(92), U.S. Mission to the United Nations, New York, Oct. 28, 1992.
10. Rossi, H. H., "The Threshold Question and the Search for Answers." *Radiat. Res.,* **119**, 1989.

Appendix A

Acronyms Used in Text

AEC	(US) Atomic Energy Commission
ALARA	As Low As Reasonably Achievable
AMTEC	Alkali Metal Thermoelectric Converter
ARAC	Atmospheric Release Advisory Capability
ASTHMA	Axi-Symmetric Transient Heat Conduction and Material Ablation code
BCI	Bare Clad Impact Tests
BEES	Biomedical and Environmental Effects Subpanel
BLEVE	Boiling Liquid Expanding Vapor Explosion
BMI	Bare Module Impact Tests
CBGS	Confined by Ground Surface
CBM	Confined by Missile
CCB	Common Core Booster
ccdf	Complimentary Cumulative Distribution Function
CFR	Code of Federal Regulations
C-J	Chapman and Jouquet
CMA	Charring Material Ablation
COPUOS	(UN) Committee on Peaceful Uses of Outer Space
CPV	Cold-Process Verification Tests
CSDS	Command Shutdown and Destruct System
CST	Converter Segment Tests
DDT	Deflagration to Detonation Transition
DEIS	Draft Environmental Impact Study
DFI	Dose Inhalation Factor
DIT	Design Iteration Tests
DoD	(U.S.) Department of Defense
DOE	(U.S.) Department of Energy
EA	Environmental Assessment
EGA	Earth Gravitational Assist
EIS	Environmental Impact Study
EM	Electromagnetic
EPA	(U.S.) Environmental Protection Agency
ESA	European Space Agency
ESMC	Eastern Space and Missile Center
FICP	Fireball Integrated Code Package
FONSI	Finding of No Significant Impact
FSAR	Final Safety Analysis Report
FTG	Fragment Tests in Gas Gun

GEM	Graphite Epoxy Motor
GEO	Geostationary Earth Orbit
GEOTRAP	Global Transport and Dispersion of Radioactive Particulate
GI	Gastro-Intestinal
GIS	Graphite Impact Shell
GPHS	General Purpose Heat Source
GSAR	Ground Safety Analysis Report
GTO	Geosynchronous Transfer Orbit
HANDI	Heating Analysis Done Interactively
HIAD	High Altitude Aerosol Dispersal
HTPB	Hydroxyl-Terminated Polybutadiene
HVI	High Velocity Impact
HYTEC	Hydrogen Thermal-to-Electric Converters
IAEA	International Atomic Energy Agency
ICRP	International Commission on Radiological Protection
INSRP	Interagency Nuclear Safety Review Panel
LAS	Launch Abort Subpanel
LASEP-T	Launch Accident Scenario Evaluation Program-Titan
LD	Lethal Dose
LEO	Low Earth Orbit
LET	Linear Energy Transfer
LFT	Large Fragment Tests
LH$_2$	Liquid Hydrogen
LOCA	Loss-of-Coolant Accident
LOM	Low Orbit Mission
LOX	Liquid Oxygen
LWRHU	Light Weight Radioisotope Heater Unit
Met	Meteorological Subpanel
MET	Mission Elapsed Time
MHD	Magneto-Hydro-Dynamics
MHG	Multi-Hundred Watt Generator
MMH	Monomethyl Hydrazine
MPRE	Medium Power Reactor Experiment
MSPSP	Missile System Prelaunch Safety Package
NACA	(U.S.) National Advisory Committee for Aerodynamics
NASA	(U.S.) National Aeronautics and Space Administration
NASC	(U.S.) National Aeronautics and Space Council
NCRP	(U.S.) National Council on Radiological Protection and Measurement
NEP	Nuclear Electric Propulsion
NEPA	National Environmental Policy Act
NEPSTP	Nuclear Electric Propulsion Space Test Program
NERVA	Nuclear Engine for Rocket Vehicle Applications
NOA	Notice of Availability
NOI	Notice of Intent
NRC	(U.S.) Nuclear Regulatory Commission
NSPWG	Nuclear Safety Policy Working Group
OD	Operational Directive
OR	Operational Requirements
ORSRP	Orbital Readiness Space Safety Review

OSHA	(U.S.) Occupational Safety and Health Administration
OSTP	(U.S.) Office of Science and Technology Policy
PHL	Preliminary Hazards List
PI	Program Introduction
PLF	Payload Fairing
PRA	Probabilistic Risk Assessment
PRD	Program Requirements Document
PSAR	Preliminary Safety Analysis Report
PSSP	Power System Subpanel
RBE	Relative Biological Effectiveness
ROD	Record of Decision
RTG	Radioisotope Thermoelectric Generator
RTS	Random Tumble and Spin
SARP	Safety Analysis Report for Packaging
SATRAP	Site-Specific Analysis of Transport and Diffusion of Radioactive Particles
SER	Safety Evaluation Report
SHO	Sufficiently High Orbit
SNAP	Systems for Auxiliary Nuclear Power
SOC	Statement of Capabilities
SRB	Solid Rocket Booster
SRMU	Solid Rocket Motor Upgrade
STSC	Science and Technical Subcommittee
SVT	Safety Verification Tests
TBVD	Total Boost Vehicle Destruct
TFE	Thermionic Fuel Element
TMI	Three Mile Island
TORA	Test Operational Risk Assessment
TOSSA	Test Operations Space Safety Approval
TSAP	Trajectory Simulation and Analysis Program
UDMH	Unsymmetrical Dimethyl Hydrazine
UN	United Nations
USAR	Updated Safety Analysis Report
VCE	Vapor Cloud Explosion
VEEGA	Venus-Earth-Earth Gravity Assist
VVEJA	Venus-Venus-Earth-Jupiter Assist
ZND	Zeldovich, von Neumann, and Doring

Appendix B

Units Conversion

Length:	micrometer	μm	$= 10^{-6}$ m
	millimeter	mm	$= 10^{-3}$ m
	centimeter	cm	$= 10^{-2}$ m
	meter	m	$= 3.2808$ feet
	kilometer	km	$= 10^{3}$ m
	inch	in	$= 2.540$ cm
Area:	barn	b	$= 10^{-24}$ cm^2
Mass:	atomic mass units	u	$= 1.6605 \times 10^{-24}$ g
			$= 931.494$ Mev
	gram	g	$= 10^{-3}$ kg
	kilogram	kg	$= 2.205$ pounds
Energy:	electron volt	eV	$= 1.603 \times 10^{-19}$ J
	million electron volts	Mev	$= 10^{6}$ eV
	joule	J	$= $ W·s $= 10^{7}$ ergs
	calorie	cal	$= 4.186$ J
Power:	watt	W	$= 1$ Newton·m/s $= 3.413$ BTU/hr
	kilowatt	kW	$= 10^{3}$ W
Force:	dyne	dyne	$= 1$ g·cm/s^2
	newton	N	$= 1$ kg·m/s$^2 = 10^{5}$ dyne
	newton	N	$= 0.2248$ pound-force
Pressure:	pascal	Pa	$= 10^{-5}$ bar
	megapascal	MPa	$= 10^{6}$ Pa
	atmosphere	atm	$= 1.013 \times 10^{5}$ Pa
	pound-force/inch^2	psi	$= 6.894 \times 10^{3}$ Pa
	torr	torr	$= 133.3$ Pa
Radiological:	curie	Ci	$= 3.7 \times 10^{10}$ disintegrations/second
	becquerel	Bq	$= 1$ disintegrations/second
	radiation absorbed dose	rad	$= 0.01$ J/kg
	gray	Gy	$= 100$ rad $= 1$ J/kg
	sievert	Sv	$= 100$ rem

Appendix C

Basic Constants

Avogadro's number	N_a	6.022×10^{23} atoms/mole
Boltzmann constant	k	8.617×10^{-5} eV/K
Plank constant	h	6.6256×10^{-34} Js
Speed of light	c	2.99793×10^{10} cm/s
Stefan-Boltzmann constant	σ	5.669×10^{-8} W/m^{2}K^4
Gas constant	R	8.315 J/K·mole
Newton gravitational constant	G	6.670×10^{-8} cm^3/g·s
Acceleration due to gravity	g	980.665 cm/s^2

Appendix D

Nuclear Properties of Selected Isotopes

Atomic Number	Element	Isotope Symbol	Half-life	Relative Atomic Mass	Natural Abundance (%)	Absorption Cross Section (b − 2200 m/s)
1	Hydrogen	^{1}H	—	1.007825	99.985	0.332
		^{2}H (D)	—	2.01410	0.015	0.00050
		^{3}H (T)	12.6 y	3.01605	—	$< 6.7 \times 10^{-6}$
2	Helium	^{3}He	—	3.01603	0.00013	5.327×10^3
		^{4}He	—	4.00260	99.99987	0
3	Lithium	^{6}Li	—	6.01512	7.42	945 (α)
		^{7}Li	—	7.01601	92.58	0.037
4	Beryllium	^{9}Be	—	9.01218	100	0.009
5	Boron	^{10}B	—	10.0129	19.78	3.837×10^3
		^{11}B	—	11.0093	80.22	5×10^{-3}
		^{12}B	0.02 s	12.0143	—	—
6	Carbon	^{12}C	—	12.0000	98.89	0.0034
		^{13}C	—	13.00335	1.11	9×10^{-4}
		^{14}C	5730 y	14.00323	—	—
7	Nitrogen	^{14}N	—	14.00307	99.64	1.82
		^{15}N	—	15.00011	0.36	4×10^{-5}
		^{16}N	7.2 s	16.00656	—	—
8	Oxygen	^{16}O	—	15.99491	99.759	0.000178
		^{17}O	—	16.99914	0.037	0.04
		^{18}O	—	17.99915	0.204	1.6×10^{-4}
		^{19}O	29 s	19.00344	—	—
11	Sodium	^{23}Na	—	22.98977	100	0.534
		^{24}Na	15 hr	23.99102	—	—
13	Aluminum	^{27}Al	—	26.98153	100	0.231
14	Silicon	^{28}Si	—	27.97693	92.21	0.17
		^{29}Si	—	28.97649	4.70	0.30
		^{30}Si	—	29.97376	3.09	0.11
17	Chlorine	^{35}Cl	—	34.96885	75.53	44
		^{37}Cl	—	36.96590	24.47	0.43
18	Argon	^{40}Ar	—	39.96238	99.60	0.61 (α)
19	Potassium	^{39}K	—	38.96371	93.10	2.1
		^{40}K	1.28×10^9 y	39.9740	0.0118	70.3
		^{41}K	—	40.96184	6.88	1.46
20	Calcium	^{40}Ca	—	39.96259	96.97	0.22
25	Manganese	^{55}Mn	—	54.9381	100	13.3
26	Iron	^{54}Fe	—	53.9396	5.82	2.3
		^{55}Fe	2.6 y	54.9386	—	—
		^{56}Fe	—	55.9349	91.66	2.7
		^{57}Fe	—	56.9354	2.19	2.5
27	Cobalt	^{59}Co	—	58.9332	100	37.2
		^{60}Co	5.27 y	59.9334	—	—

Atomic Number	Element	Isotope Symbol	Half-life	Relative Atomic Mass	Natural Abundance (%)	Absorption Cross Section (b − 2200 m/s)
29	Copper	^{63}Cu	—	62.9296	69.17	4.5
		^{64}Cu	12.9 hr	—	—	—
		^{65}Cu	—	64.9278	30.83	2.2
36	Krypton	^{85}Kr	10.76 y	84.9126	—	—
38	Strontium	^{89}Sr	51 d	88.9057	—	—
		^{90}Sr	29 y	89.9072	—	—
40	Zirconium	Zr	—	91.22	—	0.18
42	Molybdenum	^{98}Mo	—	97.9055	24.4	0.15
		^{99}Mo	66.7 hr	98.9069	—	—
48	Cadmium	^{113}Cd	9×10^{15} y	112.9046	12.26	20×10^3
		^{114}Cd	—	113.9036	28.86	0.3
		^{115}Cd	53.5 hr	114.9076	—	—
49	Indium	^{115}In	6×10^{14} y	114.9041	95.72	—
53	Iodine	^{131}I	8.07 d	130.9060	—	24.5
		^{135}I	6.585 hr	—	—	—
54	Xenon	^{34}Xe	—	133.9054	10.44	0.228
		^{135}Xe	9.2 hr	—	—	2.6×10^6
55	Cesium	^{137}Cs	30 y	136.9073	—	0.11
56	Barium	^{137}Ba	—	136.9061	11.32	5.1
		^{138}Ba	—	137.9050	71.66	0.35
		^{139}Ba	82.9 m	138.9079	—	—
62	Samarium	^{149}Sm	—	148.9169	13.83	41×10^3
73	Tantalum	^{181}Ta	—	180.9840	99.988	21
		^{182}Ta	115 d	181.94	—	8.2×10^3
80	Mercury	^{199}Hg	—	198.9683	16.84	2.5×10^3
82	Lead	Pb	—	207.18	—	0.17
84	Polonium	^{210}Po	138.4 d	209.9829	—	—
86	Radon	^{222}Rn	3.823 d	222.0175	—	0.73
88	Radium	^{226}Ra	1.6×10^3 y	226.0245	—	<0.0001
90	Thorium	^{232}Th	1.41×10^{10} y	232.0382	100	7.56
92	Uranium	^{233}U	1.65×10^5 y	233.0396	—	576 ($\sigma_f = 530$)
		^{234}U	2.47×10^5 y	234.0409	0.005	95
		^{235}U	7.1×10^8 y	235.0439	0.720	678 ($\sigma_f = 580$)
		^{236}U	2.39×10^7 y	236.0457	—	5.1
		^{238}U	4.51×10^9 y	238.0508	99.275	2.73
94	Plutonium	^{238}Pu	87.8 y	238.0496	—	500 ($\sigma_f = 16.6$)
		^{239}Pu	2.44×10^4 y	239.0522	—	1.014×10^3 ($\sigma_f = 742$)
		^{240}Pu	6.54×10^3 y	240.0540	—	295 ($\sigma_f = 0.08$)
		^{241}Pu	15 y	241.13154	—	1.38×10^3 ($\sigma_f = 1010$)
		^{242}Pu	3.87×10^5 y	242.0587	—	19
		^{243}Pu	4.96 hr	243.0601	—	300
		^{244}Pu	8.3×10^7 y	244.0630	—	1.7

Data from El-Wakil, M. M., *Nuclear Power Engineering*, Van Nostrand Reinhold Co., New York, 1967, General Electric Co., *Chart of the Nuclides* (11th Edition), San Jose, Ca, 1972, and other sources.

Appendix E

Properties of Selected Materials

Approximate* Values at 1 Atmosphere

Material	M_r (u)	ς (g/cm³)	c_p J/g·K	k W/cm·K	T_{melt} (K)
U (metal)	238.07	18.9	0.18	0.35	1405
UO$_2$	270.03	10.0	0.34	0.035	3150
UN	252.03	13.5	0.23	0.23	2900
Zr	91.22	6.51	0.38	0.23	2100
ZrH$_x$	93.24	5.6	0.42	0.18	**
Be	9.01	1.85	1.8	2.0	1550
BeO	25.01	3.0	1.0	2.1	2800
Graphite	12.01	1.7	1.60	0.35	3866
LiH	7.95	0.77	4.3	0.10	960
Stainless Steel	—	7.86	0.60	0.25	1700
W	183.84	19.3	0.13	1.21	3680
B$_4$C	55.26	2.52	0.50	0.02	2620
Air	—	0.0012	1.0	0.00026	—
H$_2$O	18.02	1.0	4.16	0.0069	273

* Some properties are strongly temperature dependent. Air and water properties are at ambient temperature. Other properties are at elevated temperatures (typically ~1000 K). Most densities are typical rather than theoretical. The thermal conductivity for graphite depends strongly on the type of graphite.

** Decomposes at elevated temperatures.

M_r = relative molecular mass
ς = mass density
c_p = specific heat at constant pressure
k = thermal conductivity
T_{melt} = melt temperature

Appendix F

Properties of Coolants

(at ~800 K, 1 atm)

Coolant	Li	Na	NaK (22% Na)	H_2	He
Atomic or molecular mass	6.94	22.997	—	2.016	4.0
Density (g/cm^3)	0.479	0.823	0.742	3.0×10^{-5}	6.1×10^{-5}
Viscosity $\times 10^3$ (g/cm·s)	3.4	2.1	1.5	0.17	0.37
Prandtl number	0.020	0.004	0.005	0.71	0.74
Melting point (K)	454	371.2	262	13.81	0.95
Boiling point (K)	604	1154	1057	20.28	4.26
Specific heat c_p (J/g·K)	1.286	0.389	0.270	14.6	5.19
Thermal conductivity (W/cm·K)	0.30	0.65	0.27	0.0036	0.0024
Heat of fusion (J/g)	88.96	23.31	—	—	—
Heat of vaporization (J/g)	3995	822.2	—	—	—
Fire/explosion hazard	High	High	High	High	None

Index

461